PRINCIPLES

OF

FIELD

CROP

PRODUCTION

THIRD EDITION

JIM PRATLEY

Melbourne

OXFORD UNIVERSITY PRESS

Oxford   Auckland   New York

SYDNEY UNIVERSITY PRESS
in association with
OXFORD UNIVERSITY PRESS

National Library of Australia
Cataloguing-in-Publication data:

Principles of field crop production.
    Rev. ed.
    Bibliography.
    Includes index.
    ISBN 0 424 00200 0

    1. Field crops — Australia. I. Pratley, J. E.

633′.00994

Typeset by Solo Typesetting
Printed in Malaysia by SRM Production Services Sdn. Bhd.
Published by Oxford University Press,
253 Normanby Road, South Melbourne, Australia

CONTENTS

Contributors

R. J. Banyer, RDA, BSc, PhD(Adel)

Consultant—Education and Training
Formerly Senior Lecturer in Plant Protection
School of Agriculture, Charles Sturt University
Wagga Wagga. NSW

K. G. Beirne, BScAgr(Syd)

Formerly Lecturer in Plant Science
School of Agriculture
Riverina–Murray Institute of Higher Education
Wagga Wagga. NSW

Late O. G. Carter, BScAgr(Hons)(Syd), MS, PhD(Cornell)

Formerly Deputy Principal
Hawkesbury Agricultural College
Richmond. NSW

J. W. Chudleigh, BScAgr(Syd)

Associate Professor and Head
School of Rural Management
Orange Agricultural College
Orange. NSW

E. J. Corbin, BScAgr(Syd)

Director, NSW Department of Agriculture
Richmond. NSW

E. G. Cuthbertson, BScAgr(Syd)

Formerly Special Research Officer (Weeds)
NSW Department of Agriculture
Agricultural Research Institute
Wagga Wagga. NSW

D. W. Glastonbury, BScAgr(Hons)(Syd), DipFBA(Lond)

Senior Lecturer in Farm Management
School of Agriculture
Charles Sturt University
Wagga Wagga. NSW

G. M. Halloran, BSc, DAgrSci(Melb)

Reader in Plant Breeding and Genetics
School of Agriculture and Forestry
The University of Melbourne
Parkville, Victoria

F. M. Kelleher, BScAgr(Hons), PhD(Syd)

Senior Lecturer in Agronomy
University of Western Sydney
Hawkesbury. NSW

D. J. Luckett, BSc(Hons)(Sheff) MPhil, PhD(Camb)

Research Scientist (Tissue Culture)
NSW Agriculture
Agricultural Research Institute
Wagga Wagga. NSW

J. W. McGarity, BScAgr(Hons), MScAgr, PhD(Syd)

Formerly Associate Professor of Soil Science
Department of Agronomy and Soil Science
University of New England
Armidale, NSW

J. E. Pratley, BSc(Hons), PhD(NSW)	Associate Professor and Dean Faculty of Science and Agriculture Charles Sturt University Wagga Wagga. NSW
P. D. Slater, BScAgr(Hons)(Dunelm), DipTertEd(UNE), MSc(App.Ent), DIC(Lond)	Lecturer in Applied Entomology School of Agriculture Charles Sturt University Wagga Wagga. NSW
R. R. Storrier, BScAgr(Hons)(Syd), PhD(Lond)	Formerly Professor of Agriculture and Dean School of Agriculture Charles Sturt University Wagga Wagga. NSW
R. H. Wilson, BAgEc, DipTertEd, MEcon(UNE)	Senior Lecturer in Rural Accounting Orange Agricultural College Orange. NSW

INTRODUCTION

J. E. Pratley

Australia is renowned for its production of agricultural field crops, particularly because of its consistent, clean, high-quality produce with low levels of pesticide residues. In the past, export prices for a range of rural products have been sufficient, albeit variable, to enable farming families to earn a reasonable living, with subsidies from government.

High inflation rates in the 1970s and 1980s, higher input costs, large fluctuations in market demand, removal of support subsidies by government and the floating of the Australian dollar have all contributed to a more financially demanding climate for Australian farmers. This is further compounded by the support subsidies of international competitors such as the USA and the European Community.

Farmers must therefore raise productivity to be financially viable, ensuring of course that the product satisfies the market's requirements. Increasingly, farmers also have to address the issue of environmental sustainability because of community demands and because of the increasing incidence of soil salinisation and soil acidification.

As a consequence, field crop production in Australia requires much higher management skills than before to ensure compatability between productivity and environmental sustainability. Farm managers must therefore have a strong understanding of all the factors which contribute to or interact with that production.

The book addresses many of the factors which contribute to this process. These factors can be classified into the natural resource factors, climatic factors, crop factors and the socio-economic and political components. For a successful outcome of the process, the agronomist and farm manager must appreciate the importance of each of the factors, particularly their interrelationships.

NATURAL RESOURCES

In the past, farmers utilised the natural resources, specifically soil and water, with no particular regard for their long-term maintenance and improvement.

Australian soils, because of their structural fragility, reacted poorly to the intensity of cultivation imposed on them by agricultural operations. Major soil erosion by both water and wind has been a regular feature of the Australian landscape, removing valuable topsoil from fields and silting up waterways. Degradation of soils was also expressed in the crusting of the surface, which impeded water infiltration and seedling emergence, and in compacted layers at the base of the plough layer, thus preventing root elongation and encouraging waterlogging.

A survey of land degradation in Australia (Table 1) in the mid 1970s (Anon., 1978) demonstrated that farm practices had to change if the future for agriculture was to be protected.

TABLE 1   The extent and severity of degradation of Australian soils used for crop production (Anon, 1978)

	Extensive Cropping	Intensive Cropping	Total
Area used (thousand ha)	443	24	467
% area needing treatment (management practices only)	32	30	32
% area needing treatment (with works)	34	34	34
% total area needing treatment	66	64	66

Further, farmers were experiencing problems with soils which were acidifying, whilst salinisation was a problem in the irrigation areas and, increasingly, under dryland agriculture, as a result of rising watertables.

Watercourses and storage facilities also deteriorated because, in addition to siltation problems, salinity levels increased, particularly in the Murray River, as did nitrate and phosphate levels. This resulted in the proliferation of blue-green algae, with consequent toxicity problems for livestock and human consumption. This process, known as eutrophication, focused attention on the contribution of these elements from farming pursuits and from urban sewage effluent.

Because farming is no longer just concerned with the growing of crops, but also with environmental management, the sustainability of farming soils and the associated water resources have become major components of field crop production and must be addressed.

CLIMATIC FACTORS

Agronomists must be aware of the relationships between the physiological requirements of the plant and the likelihood of the climatic zone being able to satisfy those requirements. A knowledge of seasonality and reliability of the climate inputs are paramount to achieving success. In particular, an understanding of the water available to plants and the efficiency of its utilisation holds the key to high yields and also the management of water-induced land degradation.

In the future, climatic change brought about by ozone depletion and the greenhouse effect will need to be monitored in order that appropriate management strategies are put in place.

CROP FACTORS

Crop production will always be limited by the most limiting input, particularly moisture and nutrients.

In the case of nutrients, fertiliser additions are normally necessary for high yields to be attained. Phosphorus, in particular, is an essential input, and soil nitrogen levels need to be augmented in many cases. There needs to be sufficient fertiliser input to allow the realisation of high yields, but their utilisation depends on the removal of other impediments to yield, such as disease or weeds. Unused fertiliser, particularly nitrates, becomes a source of eutrophication and hence an environmental problem.

Pest and disease management becomes a critical component in achieving high quality and quantity of produce. Community demands for low pesticide produce has placed pressure on farmers to minimise use of these chemicals which still remain an integral part of the farming program. However, pesticide resistance is now a factor to be considered and provides a constraint on the extent of chemical usage.

SOCIO-ECONOMIC AND POLITICAL FACTORS

Crop production processes are not only determined by physical factors but also influenced by those not related to agronomy. The anticipated profitability of a particular crop for example, often overrides decisions that ought to be made on good agronomic bases. Similarly, concern about likely returns inhibits expenditure on inputs thus restricting the yield potential of the crop and worsening the income component of the budget. Such decisions also need to be taken in the context of the longer term rather than a season-by-season consideration, since no account is taken of the economic benefits of sound rotation practices.

Profitability also depends on off-farm activities, including foreign exchange rates and government subsidies in competitive countries. Farm products therefore need to be available and of quality good enough to satisfy the markets in order to maintain the custom. Farmers therefore need to focus on crop quality as an essential output and put in place mechanisms for quality to be realised and markets to be satisfied.

THE BEST FARMING SYSTEM

In ecological systems there are no simple answers, since simplicity is usually associated with instability. There can therefore be no one best farming system.

Farming systems evolve as the knowledge base increases, and as technology develops. Attitudes also change and farmers are now more environmentally conscious than ever before, in line with community expectations. They have to operate within the constraints imposed at the time—be they economic, regulatory or knowledge constraints. However, evolving farming systems must be intensively monitored to measure performance so that adjustments can be made where undesirable trends are identified.

The final outcome will be determined by the number of factors which have been applied appropriately. The top-performing farmers are those who get all the factors right. This is evidenced by the relationship established with a crop monitoring group at Finley, NSW (Figure 1).

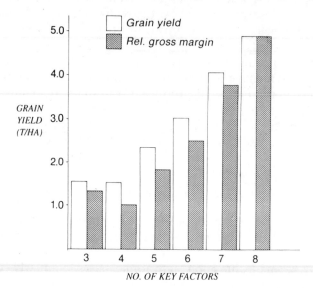

NO. OF KEY FACTORS

The key factors in any field, crop or season will vary in their importance. The message, however, remains—crop productivity depends on an holistic and systems approach where all the factors are correctly applied.

This book aims to provide the necessary background for practitioners and advisers to achieve economically and environmentally sustainable production.

Further Reading

Cornish, P. S. and Pratley, J. E. (eds) (1987), *Tillage—New Directions in Australian Agriculture*, Inkata Press, Melbourne.

French, R. J. (1987), 'Future productivity on our farmlands', in *Proceedings of the 4th Australian Agronomy Conference*, Melbourne, 140–149.

Loomis and Connor, D. J. (1992), *Crop Ecology—Productivity and Management in Agricultural Systems*, Cambridge University Press, UK.

Squires, V. and Tow, P. (eds) (1991), *Dryland Farming—A Systems Approach*, Sydney University Press, Sydney.

Reference

Anon. (1978), *A Basis for Soil Conservation Policy in Australia*. Commonwealth and State Governments Collaborative Soil Conservation Study 1975–1977, Report No. 1, AGPS, Canberra.

CHAPTER 1

THE SOCIO-POLITICAL ASPECTS OF CROP PRODUCTION

J. W. Chudleigh

Two main groups of people exert influences on the crop production and distribution pattern in the Australian agricultural system.

First, there are the farmers themselves, usually fiercely independent, who must use their technical skills to produce their crops as efficiently and as profitably as possible. The agronomic principles which farmers must employ in the successful production of their crops are detailed in following chapters.

Second, there is the remainder of the population who, through government controls and influences, demands for food, recreation demands, environmental concerns, housing demands and the desire for an ordered and democratic society, influence the alternatives available to the farmers as to what they can and will produce. These direct and indirect influences are illustrated in this chapter.

The development of agricultural production in Australia has been based principally on the free enterprise system. In such a system, time ensures that commodities are produced if there is a genuine demand for them at an economic price. Time also ensures that the centres of production will generally develop in areas most suited agronomically and economically to that particular crop and that the crops will be produced by those farmers who are the most efficient.

Whilst this broad principle explains the location and production of most field crops in Australia, as outlined in later chapters, there are many exceptions which cannot be simply explained by agronomic or economic considerations. These socio-economic influences are generally complex in their effects, but usually result either from political implementation of certain economic or marketing policies or from some particular historical or social background.

Political influences may come from federal, state or local government or from industry representative groups. These influences probably have the greatest impact on the pattern of agricultural production in Australia. Many of these effects involve legal restraint or control by statute, whereas others act as an incentive or disincentive to production.

1

Economic influences virtually dictate the whole production pattern of crops. However, many extraordinary economic influences may have a greater bearing on production patterns or developments than cost/price or supply and demand relationships. Such influences are usually in the form of statutory production controls or marketing schemes.

Marketing has always received considerable attention and much has been done, and no doubt still can be done, to improve marketing methods for agricultural products. Government influence and assistance have been significant in this area in Australia, but the worldwide move to deregulation has resulted in a considerable reduction in direct marketing support by government and government agencies.

Historical developments of many kinds have influenced the spread of various field crops. Although some of these developments may have been politically motivated initially, their ramifications are still felt although the political influence has long since been removed.

Social pressures caused by population distribution and demands also affect crop production areas. Proximity to markets, amongst other factors, is important in determining where certain crops are grown.

Environmental concerns, especially in highly populated areas, are increasing in importance as a key social issue.

There is undoubtedly a great deal of interaction between influences categorised above. Each is discussed separately, and examples are given of how various aspects affect crop production. Many of these influences may traverse the full range, being an historical political decision aimed at improving marketing and thus the economics of the crop's production. (An example of such an influence is the Australian and Queensland Governments' Sugar Agreement, which is discussed later.)

Where possible, examples given are seen to fit best into the category under which they are discussed. However, readers should be aware that socio-political pressures on crop production involve an interaction of many factors which may belong to several of the categories mentioned above.

POLITICAL INFLUENCES

These influences can be categorised into two main areas.

First, there are those decisions that directly restrict, prevent, control or allocate the production or marketing of certain crops in a certain manner. These decisions are affected by statute which generally relates to a specific industry and often to a specific aspect of that industry. The decisions may be taken by state or federal governments or by an industry body. Examples of such influences are found in marketing schemes where price support and stabilisation are common and, to a lesser extent, in laws affecting land use and product disposal.

Second, there are those decisions that indirectly encourage or discourage the production of a certain crop or crops. These effects are generally industry-wide or are associated with a specific project development which encourages certain crops to be grown in a specific area.

Examples of such decisions are found in the application of tariffs, bounties and concessions in various industries, capital investment in irrigation, land development projects, transport systems, support for research, trade agreements, and financial assistance to help producers remain viable.

Before discussing examples of these influences in more detail it is important to recognise how government policies are formulated and against what background most decisions are generally made.

Agricultural production is a slow and tedious process in comparison to many manufacturing industries. Most field crops take a full year to produce from the start of ground preparation to the final harvest. As a result of this, most agricultural industries cannot adapt quickly to marked changes in demand and price without widespread dislocation of the industry in question. Farmers will quickly cease to produce crops admirably suited agronomically to an area if returns are more viable from alternative land uses.

Governments, on the other hand, are elected generally for only three years in Australia and as such have tended to make many decisions based on short-term expediency rather than providing long-term solutions or assistance to an industry. Governments are generally reluctant to change the *status quo* unless an industry is apparently fully behind government action. Despite this, both state and federal governments have historically taken long-term initiatives which have had a large bearing on the agricultural industries.

Governments have a responsibility to the whole of the community and not just one sector, so decisions must be seen to be, where possible, in the national interests and not as favoured treatment to sectional interests. Because of this, many stabilisation schemes have been based on ensuring a stable home consumption price generally subsidised by farmers in times of high export prices and providing additional income to farmers in times of low export prices.

As mentioned before, Australian agriculture is based on a free enterprise system where choice of crop or livestock is, in the main, a producer's decision.

Government influences are generally not intended to change this freedom of choice but to complement or supplement it. Some exceptions do occur, but even then the right to produce can usually be negotiated. In some of these cases government policy can ossify an industry to the extent that it is insulated from the supply and demand mechanisms that the original policy was designed to promote.

The policies of governments in the above context cannot create markets or demand, but can assist in finding and developing new markets. Similarly, governments cannot create the supply of a commodity, but they can stimulate it by assistance and encouragement to producers within an industry.

The broadest aim of most major decisions of government relating to their agricultural policy has been to ensure income stability to producers engaged in a particular industry. In pursuit of this aim, government policy ranges from that of complete control of the industry from production to processing through quotas, licences and guarantees (e.g. the sugar industry), to virtually no control (e.g. the edible bean industry, the lupin industry).

The wheat industry falls between these two extremes of control and this stability and orderly marketing over a number of years has enabled wheat production to become the cornerstone of the Australian grain industry.

Some of the earlier political decisions were based on a desire for development and were not always based on market forces. This does not mean that their influence has not had a profound effect on the location and production of the cropping industries. Examples of this include the railways, irrigation dams, irrigation schemes and road transport systems.

In 1974, the Australian government set up a Working Group[1] to produce a paper on all aspects of rural policy in Australia. The document was for use and debate by all sectors of the community. Some of the concluding comments of this document explained why governments have become involved in agricultural policy making.

> A national rural policy is an integral part of a national economic policy and it must be judged in terms of its contribution to national economic and social objectives with equity not only between those in rural areas and those in the rest of the community, but also among those within rural areas.
>
> One objective of government intervention in the economy is to remove barriers to the efficient working of the market and to compensate for its sometimes harsh consequences. Government involvement in agriculture will remain substantial for both reasons. In some respects it is likely to be less than in the past, since it is important that the rural sector should become more market-oriented. Intervention may be greater in other respects however, such as in the provision of welfare support measures.
>
> The second major reason for the Government's intervention in agriculture is the large number of relatively small competitive producing units dispersed over a wide geographic area. This has two broad implications. It is often not possible for farm unincorporated enterprises to organise themselves or to provide individually many of the services they require to maintain and increase their efficiency. A further application arises from the fact that in most of his business dealings, particularly in marketing his product, the farmer is dealing with firms or organisations substantially larger than his own.
>
> The third reason for government involvement in agriculture concerns the

---

[1] *Rural Policy in Australia*, commonly known as the 'Green Paper', was prepared by a Working Group comprising: Dr S. Harris, Department of Overseas Trade, Canberra; Sir John Crawford, formerly Secretary of the Department of Trade and Director of the Bureau of Agricultural Economics; Professor F. H. Gruen, Professor of Economics at the Australian National University; and Mr N. Honan, Director of the Bureau of Agricultural Economics.

slow and painful adjustment to economic change which normally takes place in agriculture. The Government's involvement with assistance for farm adjustment is therefore designed to reduce the income problems which arise when help is not provided, and to lessen the inefficiency in the use of the nation's resources that the slow rate of adjustment implies.

In 1986 the Australian government, through the Department of Primary Industry, produced a new Economic and Rural Policy statement clarifying the then current policy and initiatives towards rural industries. This policy statement occurred during a serious world market oversupply for many products and contained several initiatives of a welfare nature whilst still preserving the essence of the 1974 Working Group's report. This basically was to allow rural industries to be market-oriented and minimally influenced by government intervention or support. Australia's economic problems in the 1980s and 1990s forced an acceleration of the withdrawal of government support from rural industries and greater self-reliance in marketing by individual primary producers.

The Industries Assistance Commission was established in 1974, replacing the Tariff Board which was a statutory authority in existence since 1921. The Commission was established to advise the Australian government on the nature and extent of assistance which should be given to industries in Australia. Its field of inquiry was much broader than that of the Tariff Board and covered all manufacturing, service and primary industries.

The first report of the Industries Assistance Commission (1974) placed agricultural industries in the category of requiring low levels of assistance in order to maintain efficient use of resources. The general philosophy of the Commission was to encourage the development of Australian industries which required little assistance from governments. One matter specifically mentioned as being a necessary encouragement to agriculture was for governments to be concerned with the scope of reducing fluctuations in the income of rural producers. These income fluctuations can adversely affect welfare, impair the efficiency with which resources are used and discourage development in the rural sector.

The Commission (now known as the Industries Commission) has since conducted numerous inquiries into different aspects of the rural industry, such as those on the superphosphate bounty, assistance to the apple and pear industry, and grain freight and handling costs.

Several categories of political influence are now discussed in more detail in order to illustrate the political influences stemming from the policy decisions of all tiers of government as well as from industry bodies. The ability of the Australian government to control or assist agricultural industries is limited by constitutional restraints, in many cases requiring approval of all state governments and often industry representatives before controls or assistance can be implemented by statute. These influences are discussed under five headings: Stabilisation, Legal, Capital Developments, Concessions and Miscellaneous.

Stabilisation

The reasons for farmers preferring to grow crops for which the price is relatively stable have been discussed earlier. Field crops generally have a far greater potential variability in regard to gross returns than the animal industries because of yield variations resulting from climatic variations and price variations caused by changes in market demand. It is natural to expect these industries to have received the earliest efforts in stabilised marketing and price support schemes.

Sugar

Government influence has had a long history of control in the sugar industry and has influenced its location, annual and total production, processing capacity and marketing for over 50 years.

A Royal Commission on the Sugar Industry in 1912 claimed the supreme justification for protection of the sugar industry was the part it could play in the development and settlement of northern Queensland. This had been accepted in all inquiries since, and the industry is still highly regulated.

Under a federal and state agreement, the federal government prohibits imports of sugar to Australia and in exchange the Queensland government agrees to:

(a) control production of cane sugar and accept any losses arising from export;

(b) acquire all raw sugar produced in Queensland and purchase all sugar produced in New South Wales. (This is done through the Sugar Board, only 5 per cent of total production being in New South Wales, the remaining 95 per cent in Queensland);

(c) make refined sugar available for the home market at fixed wholesale prices (approximately 25 per cent of the crop is sold on the home market).

Prices are fixed by annual proclamation under the Queensland Sugar Acquisition Act and related agreements with the New South Wales millers.

Production is controlled by a system of quotas to various mills and in turn to producers assigned to and supplying that mill. These quotas are determined initially by a Central Sugar Cane Prices Board and then locally by the Local Sugar Cane Prices Board. The quotas are based on the needs to meet home market requirements as well as regular export markets secured under international trade agreements.

As a result of these agreements and quota controls, the sugar industry has some areas producing sugar which are not as well suited agronomically to its production as areas where expansion has been restricted by quota allocations and the mill size.

The location of the cane producers and processing mills is directly related to the policies of the federal and state governments and will continue to be while these controls are in place. Proposals for tariff reform seem likely to change this long-standing arrangement.

Rice

The rice industry in Australia is almost a closed industry although no formal federal influence has determined this. The industry has developed through its own organisation and in co-operation with federal and state government bodies.

All paddy rice grown in New South Wales must be delivered to the Rice Marketing Board, a statutory body established in 1928 after a poll of growers. The Ricegrowers Cooperative Limited (owned by producers) is the sole enterprise for processing. The Cooperative handles most of the domestic and export sales of the rice.

The maximum area to be sown each year is determined annually after consultation between the Water Resources Commission of New South Wales, the Rice Marketing Board and representatives of rice growers. The Board is responsible for storage of rice received from growers.

The Board was formed in 1928 after success had been achieved in rice growing in the Murrumbidgee Irrigation Area. The initial trial work was started by the New South Wales Department of Agriculture at Yanco Experiment Farm in 1915–16. Had initial work and effort to develop the industry commenced elsewhere there may well have been a rice industry more diversified locationally than it is today.

Some rice is grown on the Burdekin River delta in Queensland, but attempts to produce rice commercially on the north coast of New South Wales were a failure. A commercial attempt at Humpty Doo on the Adelaide River near Darwin in the Northern Territory in 1958, where 2000 hectares were sown, resulted in low yields, eventually forcing the abandonment of the venture. Commercial rice-growing in the Ord River system has also ceased, although research continues.

The virtually closed nature of the rice industry in the Murrumbidgee Irrigation Area and along the northern side of the Murray resulted from government-sponsored action supporting research into crops suitable for that area. Other factors contributing to the closed nature of the industry were the formation of the Board, the introduction of production control, and the use of rice quotas in the settlement policies of the New South Wales government when expanding the irrigation areas. The industry is supported by research facilities of government origin and produces about 1 million tonnes from around 120 000 hectares throughout these irrigation areas.

Given the same initial stimulation of irrigation development and government assistance, other areas of Australia would have been agronomically suited to paddy rice production.

Tobacco

The tobacco producing industry has been a closed industry since 1965, when marketing problems experienced in the early 1960s caused the Australian government and the state governments of Victoria, New South Wales and Queensland, together with producer organisations, to

introduce the Tobacco Leaf Stabilisation Plan which is administered by
the Australian Tobacco Board.

Tobacco is sold by public auction in three centres: Melbourne, Brisbane
and Mareeba. State Tobacco Leaf Marketing Boards in the three states
arrange the sale of all tobacco produced.

Under the Tobacco Leaf Stabilisation Plan, the Australian Agricultural
Council sets a marketing quota which is allocated to three State Quota
Committees, who allocate the quota to individual producers. All sales of
leaf are made on a grade and minimum price basis set by the Australian
Tobacco Board. The industry is further protected by the Australian
government insisting that manufacturers using leaf must use a certain a
minimum of Australian tobacco in order to qualify for a concessional
rate of duty on imported tobacco.

The effect of this scheme is stabilisation but it now restricts expansion
of the industry to that determined by demand, and then only in those
areas where production now occurs (namely, Queensland, northern New
South Wales and northeast Victoria).

Wheat

Long-term price stability has existed in the wheat industry in Australia
although the control of the industry through production quotas is not as
rigid as the controls in the sugar, rice or tobacco industries. Wheat has
been stabilised principally through marketing and price policies.

The Australian Wheat Board has been in operation since 1939 as the
statutory marketing authority with the sole right, until 1990, to market
wheat in Australia and Australian wheat and flour overseas. Complementary state and federal legislation was required to ensure that all wheat
was marketed through the Board. It derives its authority from the *Wheat
Marketing Act 1989*. Further legislation in 1992 guaranteed the Board as
the sole exporter of Australian wheat, but it no longer has sole marketing
rights within Australia.

The broad objective of past marketing plans has been to provide the
industry with some security against price fluctuations without distorting
the underlying trend in market prices and to keep the cost to the Australian government within definite limits. The basis of most plans has
been to ensure a minimum stabilised price for wheat with a stabilisation
fund contributed to by growers in years when export prices were high and
to which the federal government contributes when export prices were low
and when growers' accumulated funds were insufficient to ensure the
stabilised price.

Details of the scheme are published annually by the Australian Wheat
Board. For wheat delivered to a pool, the Board makes a first advance to
growers against delivery to the various state handling authorities and this
payment is made soon after delivery. The funds are obtained by the
Australian Wheat Board from various financial institutions and by
various local and international mediums.

The various stabilisation schemes gave producers much confidence in the wheat industry, and this confidence was the main influencing factor in making wheat the stable crop on most mixed farms in Australia.

Because the grain has been readily marketable worldwide, production increased gradually up to the mid-1980s except during drought periods and a period in the early 1970s after quotas had been introduced.

The introduction of production quotas was an industry decision which was legislated for by both state and federal governments, after the Australian Wheat Growers Federation initiated talks. The decision occurred after a record harvest in 1968, which presented substantial handling and storage difficulties, associated with declining world prices and increasing world oversupply. Quotas were allocated for each state but had no effect on subsequent production after 1971. The quotas affected production efficiency in the industry and disadvantaged states which had expanded production rapidly in the years just prior to 1968–9.

The quotas also severely affected many individual wheat producers. They were based on historical production and favoured traditional wheat growers; people new to the industry were allocated very low quotas, and many of these producers withdrew or were virtually forced to withdraw from wheat production. This withdrawal certainly occurred in some areas admirably suited to wheat production. Should states need to impose production quotas on individual producers at some future time, some dislocation of production will inevitably occur as a result.

The world oversupply of wheat, which built up in the early 1980s, has presented extreme challenges to the Australian wheat industry to improve its efficiency in all areas, and governments have moved in several ways to assist in the industry's restructuring (e.g. inquiry into grain handling, inquiry into assistance for the wheat industry, counselling initiatives for farmers and further moves to deregulate the industry).

Other Industries

Through legislation, governments have ensured that various industries have been able to readily form marketing co-operatives and statutory boards with wide-ranging powers, and have allowed them access to storage and transport facilities to enable those organisations to pool and market a particular product. Producers have, in the main, taken advantage of these resources, and since the end of World War II a large number of marketing boards have been established throughout Australia.

Boards or authorities which have represented or are still representing some of the field crops not mentioned include the Australian Barley Board (statutory board), Barley Marketing Board (Queensland), New South Wales Oat Marketing Board (statutory board), Victorian Oatgrowers Pool and Marketing Co. Ltd, Atherton Tableland Maize Marketing Board (statutory), Grain Sorghum Marketing Board in New South Wales and Queensland, Tobacco Leaf Marketing Boards in Queensland, Victoria and New South Wales (statutory), Queensland Cotton Marketing

Board (statutory), Namoi Cotton Co-operative Ltd, Grain Pool of Western Australia (statutory), Oilseed Marketing Board of New South Wales (statutory), and Queensland Peanut Marketing Board (statutory).

This is not a complete list, but it illustrates the effect government encouragement has had on the establishment of boards designed to provide more orderly marketing of certain crops.

These boards provide a medium, particularly for export, for pooling individual growers' crops to the advantage of both producers and purchasers of exports. In many instances the existence of a well run and reputable board encourages the expansion of production of a crop in the area or state where the board operates (e.g. oats in Western Australia, barley in South Australia).

Whilst the large number of boards and co-operatives established would seem to indicate that they are a successful part of field crop production in Australia, this is not always the case. Stability of price can often encourage production to the point of overproduction at which point difficulties can be experienced in marketing and maintaining stability without production controls. The loss of free-market price indicators can often lead to poor or slow market intelligence reporting to the producer.

During the 1980s lower real commodity prices tended to expose the averaging effect of pooling and there has been both a government and an industry push to deregulate marketing authorities. This trend is expected to continue throughout the 1990s.

Legal

Whilst most industry assistance or stabilisation schemes are subject to statute by law, other government controls are aimed at protecting an industry from careless or unscrupulous producers or wholesalers of their products. Restrictions on factors such as moisture, insects, weeds and impurities in crop produce delivered are all aimed at protecting the whole industry and maintaining export markets. Such restrictions may lead to a crop losing favour in an agronomically suited area because of a build-up in a prohibited weed or insect (e.g. dodder in lucerne hay, black oats in wheat, saffron thistle in canola, weevils in wheat, *Helicoverpa* in cotton).

These restrictions can increase the cost of production in some areas, thus forcing down the production of the crop (e.g. artificial drying of sorghum or wheat, grading of wheat). They can also encourage the relocation of crop production into more suitable agronomic areas and thus increase industry efficiency through a reduction in production costs.

Prohibition, whilst only of minor importance, should be mentioned in relation to a crop such as marijuana which is agronomically adaptable, readily marketable and highly profitable.

Laws associated with land use or land settlement policies are designed to prevent or encourage land development for crop production. It is not generally appreciated that approximately 84 per cent of land in Australia is held under some licence or lease from the Crown, the remaining 16 per cent being freehold or in the process of becoming freehold. This apparent hold by the Crown has allowed state and federal governments to dictate land settlement policies. Many pastoral leases have restrictions on their stocking rates, clearing and cropping.

Large grazing properties which were situated in favourable cropping areas saw no need to engage in farming because of their size, so large areas were underutilised and in many cases would have remained so except for government action. The Closer Settlement Schemes implemented throughout most of Australia at various times have been instrumental in expanding crop production in areas which may not have been possible without changes in land ownership. Over five million hectares have been subjected to closer settlement by government in Australia.

Mentioned later is the effect of the large-scale settlement in the wheat belt of Western Australia, but more recent evidence of crop area expansion can be seen at Esperance in Western Australia, in irrigation crops along the Murrumbidgee River and in the more favoured brigalow areas of Queensland where sorghum and wheat production has expanded as a result of state and federal government land policies.

Capital Developments

Many areas of crop production have developed as a result of a significant capital investment on the part of government. Many such investments come as a result of economic and social pressures on government, but others may be seen as government initiatives designed to stimulate production or development. The former category includes some of the irrigation developments, the expansion of state railway systems, and road improvement schemes. The development of storage dams for irrigation and town water supply is often a result of pressures brought to bear on governments. These developments can favour the expansion or establishment of crop production in an area where development would otherwise not have occurred.

Associated with government development of major irrigation dams have been state government schemes to help producers develop irrigation land by providing finance at concessional rates. In New South Wales the attitude of the state government has helped with private irrigation developments in the cotton, coarse grains and oilseeds industries in the Macquarie, Gwydir and Namoi river systems as well as other developments throughout the state. The northern river systems are becoming large and expanding areas of production of these crops.

The addition of some branch lines to the state railway systems has occurred as a result of an expanding grain industry. Once enough producers were established on the frontier of the existing rail system and distant enough from the last delivery point, an additional line could be economically justified to carry the grain. Producers, faced with high road freight costs to the nearest railhead, were quick to apply pressure and reasoning to governments showing any willingness to listen. Once the extra branch line was built, settlement would move further out, necessitating continuation of the process. This expansion of the railway system allowed the expansion of the grain industries over most of Australia. More recently, contraction of the rail service has occurred as road transport has proven more efficient, especially from many more isolated areas.

Many examples of government initiatives which were designed to stimulate production and the expansion of specific industries are available. These illustrate the very large effect government policy can have on crop area development.

Glynn (1974) clearly indicates how government initiatives led to the very rapid expansion of wheat production in Western Australia. This expansion was more rapid than anywhere else in Australia and was primarily because of government investment and the availability of cheap land suited to crop production. The expansion principally took place during the period 1900–30 and was based on government assistance in land settlement schemes. Subsequently when war, drought and depression struck, generous and unprecedented government financial support was made available to keep producers on the land and viable. The major part of the railway development in Western Australia took place between 1909 and 1918. The crop area expanded from 350 000 hectares in 1910 to 730 000 and 1920 and to a phenomenal 1 940 000 in 1930, after the majority of the railway system was established.

The example of the Western Australian grain belt expansion is historical now and undoubtedly the area would have been settled and crop production expanded gradually over time, even without government assistance during those early years. This would have taken much longer, however, than was the actual case.

A major development, again in Western Australia, created by government initiative was the building of the Ord River Dam in the northeast corner of Western Australia. The Ord River rises in the Durack Range on the Western Australian and Northern Territory border and flows northward into the Joseph Bonaparte Gulf.

Since 1948, the CSIRO station in Kununurra on the Ord River had been experimenting with various crops and had shown that cotton, safflower, linseed, sugar cane and rice could be grown under irrigation in the Kimberleys. Economic analysis of their experiments, however, had shown that the crops, particularly cotton, would have to be subsidised for the area to be viable. This was due mainly to the isolation of the area from other agricultural regions, and market infrastructure.

The federal and Western Australian governments decided to give encouragement to northern development and during the 1960s a storage dam was built on the Ord River. The dam has some 60 000 hectares of alluvial flats below it suitable for irrigation. Farmers were encouraged to settle the area, a township was built at Kununurra and a cotton gin constructed for raw cotton processing. Little thought, other than this, was given to creating an infrastructure suitable for agriculture, particularly in the area of marketing of produce. This large investment by the federal and state governments came as a government initiative to encourage crop production in that area and to assist with the further development of the north of Australia. This development was being encouraged by goven-ments concerned with strategic defence of Australia and with a philosophy at the time based on populating the north of Australia. Subsequent to the initial investment, the federal government has withdrawn support for further expansion because of the many technical and economic problems encountered by farmers in the area.

Davidson (1965) critically examined the developments being encouraged by government action and concluded that such investments were eco-nomically unjustified.

On 1 May 1964 the *Western Australian* newspaper, in an editorial statement on criticism of the Ord River project, made the following comment:

> Doubtless their criticism is well founded on narrow economic grounds, but wider vision than that is needed to develop a nation. If our earlier political leaders had let such talk beguile them, much of the development for which we are now grateful would never have been started . . . All progress is a risk. It would have been difficult to show immediate economic justification for herding new chums into what is now a prosperous (W.A.) wheat belt . . . But the risks were taken and Australia is the better for it!

Whether government investment and support for the Ord River develop-ment will ever make the Ord scheme a viable development does not detract from the fact that government initiative and action was the prime reason for crop production expanding in this area.

Private investment in new avenues of agricultural production is often encouraged and supported by governments. Queensland State Lands authorities encouraged the British Food Corporation's investment in the Peak Downs area in 1949, New South Wales government authorities attempted to persuade American cotton growers to start their industry in the Murrumbidgee Irrigation Area rather than in the Namoi Region, the federal government encouraged the development of the ill-fated Humpty Doo rice project in the Northern Territory and the Chase Syndicates' Esperance Region development was also assisted by state government authorities.

Many other private large-scale agricultural developments can be found which have been encouraged and assisted by government policies.

Government Concessions and Control

Many actions of governments are designed to help the rural sector as a whole rather than a single industry. However, some policies are designed to help a specific industry by direct subsidy, price support or cost control.

Several categories of this type of assistance exist and some of the more common are discussed below.

Bounties and Subsidies

The use of direct payments to subsidise the price of a product or reduce the cost of an input is one means by which governments can encourage an industry to develop. Australia has minimal subsidies on agricultural commodities compared to Japan, the EC, the USA and Canada, where subsidies on production or market prices greatly distort good land use and efficient production. Reducing or changing these distorting support systems will take considerable time, as the agricultural infrastructures in those countries have become highly dependent on these subsidies.

A cotton bounty was used to encourage production in Australia so that the country would become self-sufficient in cotton production. In order to encourage the industry, a bounty was paid of up to 100 per cent of the commercial value of the crop. This support during the 1960s saw the development of the industry to the stage where cotton was actually exported, and by the end of 1971 the bounty had been phased out. As a result of this initiative, Australia now has a viable cotton industry centred on the areas found most suited to its production.

The fertiliser bounties on both nitrogenous and phosphatic fertilisers were a cost-cutting measure designed to stimulate investment in the grazing industries, particularly, but have aided most crop industries. The benefits are felt most where crops are particularly responsive to fertiliser and the effect, although slight, helps encourage production in more marginal areas.

The gradual replacement of tariffs on imported machinery and chemicals by bounties paid to Australian manufacturers is a more recent move to help reduce input costs for rural producers.

Direct assistance to an area may be given by governments in the form of relief for drought, flood or fire.

Australian government support for plague locust control in Queensland, New South Wales, Victoria and South Australia is designed to protect producers from the ravages of large-scale plagues.

Specific assistance may also be given to help producers in a particular area. An example was the direct help given by the Western Australian government to producers in the Ord River area to reduce the cost of spraying cotton crops.

During the period of high interest rates in the 1980s, a direct subsidy on interest payments was made to certain producers. These schemes were of short duration. Longer-term concessional finance was made available to a limited number of needy producers through the Rural Assistance Board in each state.

Tariffs

Major purchased inputs in most cropping industries include machinery and machine spare parts. For these and many other farm inputs, tariff protection has been implemented to help protect Australians manufacturing similar goods.

In many cases farmers suffer additional input costs as a result of this government action to protect Australian manufacturers. These cost increases as a result of tariff protection have been used by producer bodies in their demands for tariff compensation (Cobb 1983).

The tariff measure is not always disadvantageous to rural producers and does protect many rural products. For example, the tariff on many imported vegetable oils is designed to help protect the Australian oilseed industry. The tariff reduction policies of both major federal political parties indicates tariff protection is unlikely to be a major government initiative again in the foreseeable future.

Tax Concessions

All primary producers in Australia enjoy some concessions with respect to taxation. Measures vary and are generally designed to help with

(a) income variation;
(b) stimulation of capital investment;
(c) encouragement of production increases; and
(d) better conservation practices.

Income averaging allows a producer to average the taxable income over the previous five years to ascertain the tax payable in the current year. This ensures, to some extent, that the producer is not overtaxed in good seasons which helps compensate for poorer seasons or poorer prices. Averaging also allows investment decisions to be more rationally based rather than being a means of reducing tax in high income years.

Income Equalisation Deposits were introduced in 1975–6, allowing non-taxable deposits to be made from income in good years. Deposits were only taxed in the year of withdrawal. This scheme was later amended as the government believed it did not benefit those for whom it was intended. The scheme now offers slightly higher interest rates on deposits but these deposits are no longer tax deductible.

Various special investment allowances and accelerated depreciation rates on machinery have helped provide some compensation for tariff protection of local manufacturers. The allowances enable governments to stimulate investment in capital items. Governments tend to alter or withdraw these concessions regularly.

Most machinery and equipment for use in primary production is also exempt from state government sales tax. Estate duties and gift taxes, which were often responsible for the sale of family holdings or the reduction in their size to less viable units, have been virtually abolished on most asset transfers.

Freight

State governments have at times given concessions on freight for commodities such as grain and fuel to help in their economic production and distribution. Freights from more distant areas are often subsidised by those areas closer to centres of population. This ensures that distance from the seaport is not a major deterrent to crop production in agronomically favourable areas.

Miscellaneous

A large number of miscellaneous initiatives by governments produce various effects in various ways on the location and production of field crops. The effects of some of these initiatives are discussed below.

Research

The effect of research over time is a factor which can greatly influence where a particular field crop is grown. Industry support for research through levies on the sale of products is often matched by grants from governments.

Some industries, such as the wheat industry, have historically contributed considerable funds for research through a levy on growers. These funds have often been matched by grants from governments. The size of the wheat research contribution varied between 26 cents and 76 cents per tonne during the 1980s, providing a contribution from growers of up to $9.7 million. The size of these funds has allowed wheat breeders and researchers to develop varieties suited to most areas of Australia and to continue to develop varieties and production techniques which increase average wheat yields.

At the other end of the scale are the oilseeds and grain legumes, relatively new crops in Australia with the potential for further expansion. There is little government research into varieties and this, combined with lack of technical expertise and slowly developing markets, has possibly restricted expansion. With the introduction of plant variety rights in the mid-1980s, private research in variety development increased significantly for most crops, leading to continued expansion.

Governments have also initiated research into areas of production thought desirable by their advisers and annually spend considerable sums looking for new crops, evaluating production techniques and breeding and testing new varieties throughout Australia.

Extension

Both state and federal governments support their research efforts with widespread extension services to farmers. State governments have been particularly active in this area. Without this extension, farmers would be much slower to adopt new practices or introduce new crop varieties. A government can encourage the development of a crop in a particular area through the influence of these services.

Government extension services cover all areas of agronomy, animal husbandry, conservation, irrigation and farm management economics, as well as many other aspects of agriculture. They are complemented by a wide range of private extension services provided by industry in the areas of chemicals for weed and pest control, machinery services, and crop production, and by consultants in farm management, environmental management and agribusiness.

Education

Many crops are grown in areas completely unsuited to their production, often at great expense to the farmer involved. Lack of knowledge of the fundamentals of crop production and management are visibly evident throughout the country. Many producers, either through ignorance or stubbornness, do not see fit to use the extension services provided or to make themselves familiar with the requirements of certain crops. While this does not apply to the majority of producers, lack of education is a definite drawback to the efficient utilisation of the nation's resources in many areas.

Governments have realised this and have assisted in providing a range of agricultural technical and management courses for both full-time and part-time participation. The continuation and expansion of these courses combined with improved access will be necessary if the advances in technology and management are to be applied generally in future crop production.

Trade Agreements

Many international trade agreements are negotiated on a government to government basis in the interests of orderly world trade. The United States Sugar Act allocates quotas to various countries to help fill its shortfall and Australia shares in this quota. The International Wheat Agreement is designed to stabilise the world wheat trade and Australia is a party to this agreement. Other less specific agreements are in existence which subject Australian trade to certain requirements. Agreements under GATT and UNCTAD may influence overseas trade or price.[2] Protracted negotiations under GATT to resolve the agricultural trade

---

[2] GATT (General Agreement on Tariffs and Trade) is an organisation of over 70 nations which meets regularly in Geneva to negotiate jointly on matters of trade policy. Its basic principle is that of nondiscrimination between nations and the general agreement is that member nations will not give better treatment in trade to any single nations than it gives to other members of GATT.

UNCTAD (United Nations Conference on Trade and Development) was formed in 1964 with the main aim of discussing trade or tariff preferences for less-developed countries and of stabilisation of international commodity prices. The organisation was formed after realisation that organisations such as GATT had little relevance in solving the problems of less-developed countries.

crisis of the 1980s and 1990s illustrate how important such agreements may be.

The Closer Economic Relations (CER) Trade Agreement has removed most trade barriers between Australia and New Zealand and totally free trade will exist by 1995. Less formal arrangements such as those with Japan, China, the Middle East and Malaysia all influence trade in agricultural commodities.

Finance

While most financial concessions have occurred through welfare assistance to rural producers, investment in soil conservation with storage and irrigation development has often been assisted by loans at concessional interest rates.

A more recent development has been to use concessional finance to build up viable enterprises while assisting nonviable producers to leave primary production.

Quarantine and Export Inspection Services

Government controls on imports through quarantine regulations are designed to protect the nation's crop and livestock industries. These services have helped ensure that many diseases and pests which could have serious effects on crop production have not entered the country.

Conversely, government inspection of products for export helps to ensure that standards of these exports are maintained and quality is consistent with the description. This encourages the regular return of customers for Australian products.

ECONOMIC

Many economic factors affect crop production. Factors such as expected gross returns, total production costs, capital requirements and the producer's financial position all have a bearing on crop production and distribution.

Crop Profitability

The assessment of crop profitability by the use of gross margins analysis is covered in detail in Chapter 9. The production of any crop must be profitable to the producer and it is likely that an increase in production of a crop will be greatly dependent on the relative profitability of alternative land uses.

The effect of crop profitability on national production levels is exemplified by the relative production of wheat and canola since the 1960s. The introduction of wheat quotas after the 1968–9 record production year stimulated an increase in oilseed and coarse grain production (Fig. 1.1).

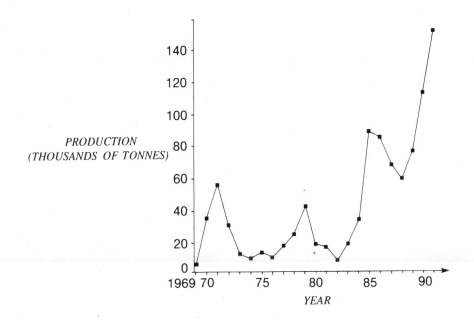

FIG. 1.1   Canola production in Australia, 1969–91 (ABARE, 1992)

This stimulation was brought about by a combination of circumstances, the overriding factor being economic. A poor outlook and world price for wheat, wheat quotas and lowering prices for wool and livestock all combined to stimulate producers to look for viable alternatives to wheat and livestock production.

Many crops, including rape (now canola), linseed, barley, sunflowers, lupins, field peas, maize and sorghum, were grown by many producers who had little or no experience in the culture of these crops. Some introduced varieties of oilseeds had not been well developed for oilseed, and disease cycles were not well understood. Many producers found that canola did not provide an easy alternative to wheat, and as soon as livestock and wheat prices started to recover, production, particularly of oilseeds, declined sharply. The downturn in wheat and grain prices generally in the mid-1980s again challenged the economics of wheat production. Wheat areas fell from record levels (Fig. 1.2) as alternative crops and livestock enterprises expanded. Newer improved varieties of canola, combined with relatively better prices, caused canola production to rise sharply in the late 1980s (Fig. 1.1).

It is likely, however, that wheat will remain the major crop in Australia

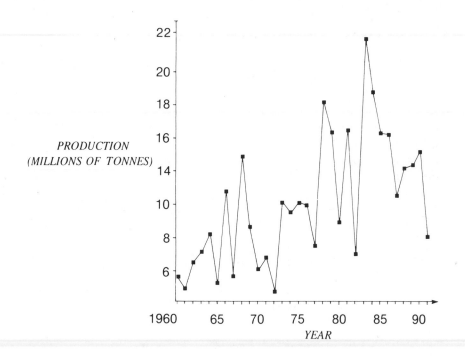

FIG. 1.2   Australian wheat production, 1960–91 (ABARE, 1992)

principally because of the experience in the industry, the relative stability of production and its adaptation to a wide variety of conditions.

Capital Requirements and Risk

The capital requirements for specific crop production may be high, but a high risk in crop production is usually balanced by a high financial return if the crop is successful.

Wheat, oats and barley probably have the lowest growing costs and capital requirement per hectare and this is a further factor in encouraging production of these and similar crops.

Cotton production has a very large capital requirement and a high degree of technology is required. This aspect tends to limit the type of investor and the number who are prepared to invest in the industry. Similarly, some broadacre vegetable crops require large investments in irrigation, harvesting and processing equipment and this has a limiting effect on expansion.

High-risk crops with large variations in yields and price have not been attractive in comparison to low-risk crops with government sponsored price stabilisation and as such tend not to be expanded in production.

Most producers do not have the reserves or the inclination to attempt crops with a higher than normal risk factor (e.g. broadacre tomatoes for the open market). Deregulation of markets is unlikely to change the general trend to low-risk, low-capital crops, although areas of dryland cotton have expanded significantly in recent years.

Market Access

Access to local markets assumes a critical role in the production and location of the more perishable field crops. Historically, most processors were established close to large centres of population with the processing crops being grown around the outskirts of these centres. However, as road and rail transport improved, the processors tended to move to centres where the crops could more favourably be grown and where capital investment in land was much lower. The gradual increase in beef feedlotting during the 1980s has provided a local market for many coarse grain producers.

Access to export markets is critical to Australian agriculture, as between 60 and 70 per cent of our total production in value is exported. Efficient transport and handling systems for grain, beef, wool and other products are essential to economic viability of Australian agricultural producers.

MARKETING INFLUENCES

Quality of Supply

The introduction and establishment of a new field crop in an area can be restricted by the requirements of the consumers of the crop. For example, the development of the lupin industry in Western Australia initially meant convincing feed mills of the value of the legume grain and then convincing growers to produce sufficient quantities to be of interest to the mills. Now that lupin production is well established, export markets can be developed.

The expansion of similar crops in the eastern states has been slower but is continuing as supply increases to a level of interest to more processors.

The establishment of lupins as a major crop in Western Australia occurred because the University of Western Australia bred new varieties suited to local soil types and because farmers were looking for alternatives to wheat to grow on the lighter soils which suited the lupin crop.

The same situation occurs in developing an export market, but it is generally necessary to have a board or large co-operative to market a crop if an export market is to be developed. The board can market in bulk, help to guarantee supply and organise storage and transport for the exported product.

Organised Marketing

The concept of marketing boards has been considered earlier in this chapter. The security offered to buyers and producers by some guarantee of supply and price creates a climate more favourable for the expansion of production of a crop than where no organised marketing exists.

A comparison between maize and sorghum production in eastern Australia helps to illustrate this point. Maize is produced throughout New South Wales and Queensland on a small scale, and production has been relatively stable in the 1980s. Sorghum production increased as a result of the downturn in wheat prices. Since 1972–3, sorghum has been marketed principally by boards, while most maize has been sold on an open market, little being exported.

The cotton industry in Australia has pursued an aggressive marketing program through co-operative organisations and this has led to forward selling, hedging and financing facilities being developed through these co-operatives.

As the Australian grain trade is predominantly for export, the influence of pooling through boards and co-operatives has enabled more stable marketing and supply to develop.

Producers' Attitude

Most primary producers centre their business and social life around their farm and local town community. This is a natural trend and typical of people involved in many other industries.

If the marketing of their crop involves significant organisation, travel and negotiation, then many producers would rather not become involved. They prefer to produce crops which can be grown and delivered locally under some organised marketing scheme, even if another crop is slightly more profitable and better suited to their area, but requires considerable effort in marketing after it is produced.

Export Markets

Marketing of the majority of field crops produced in Australia involves finding export outlets. These markets need to possess some degree of permanence and stability in order to encourage stability of supply. If a short-term export demand exists for a crop, even at attractive prices, an industry based on the crop is unlikely to develop with any permanence.

Competition from Imports

On occasions, crop products, which can be grown in ample supply in Australia, are subjected to competition from imports. For example, potatoes have suffered at the expense of imports from New Zealand because a free trade agreement exists between the two countries.

Where a particular crop is processed and marketed in Australia, competition from imports is likely to be greatest. This is due to the generally higher cost of processing in Australia. An edible bean industry has been established in the northern half of New South Wales, but it is subject to competition from similar canned and frozen imports.

Australian supermarkets have a wide range of imported processed crop products available, especially from areas where the cost of production and processing is much cheaper.

HISTORICAL INFLUENCES

Many of the influences on location and production of field crops already mentioned are of a certain historical nature (e.g. Western Australian railway development, rice industry in Murrumbidgee Irrigation Area, sugar industry in Queensland).

World War I and II had significant implications for agriculture, especially when combined with government policies during and after these events. Expansion of food production, although difficult, was encouraged during World War II. After World War II, land settlement policies of the federal government were aimed largely at helping resettle returning soldiers on to properties under the War Service Land Settlement Scheme. This policy was combined with a vigorous migration drive which brought many new settlers to the country. With these settlers came new dietary habits and a demand for production of crops new to the country. This effect was probably felt more in the vegetable and horticultural industries than in field crop areas, but many of these newer Australians helped to expand crop production in various ways (e.g. maize in New South Wales).

The dairy industry in New South Wales has been based principally on the coastal fringe, close to the bulk of the population. North coast dairy farms were small operations, and the industry became based traditionally on dairying and maize production. Because of their distance from market, most milk produced in the area was for processing. Small farm sizes and low prices for manufacturing milk caused the north coast area to become a poor agricultural region, so there has been a rationalisation period which reduced the extent of dairy farming in this region. Subsequently a reduction in maize production has occurred, and New South Wales maize-producing areas have moved further west. In other areas, the historical tie between dairying and maize or other grain production has gradually decreased as dairy farmers have found the cost of grain production did not warrant its production when supplementary grain based feeds could be bought in more cheaply, this grain being produced in other lower-cost areas.

SOCIAL INFLUENCES

The Australian population is urbanised: 60 per cent live in cities, 25 per cent in large towns, and only 14 per cent in the country on farms and in the smaller towns. The trend over the last 30 years has been for the number living in towns and cities to grow at the expense of those living in the country areas.

Many farmers' children, children of farm labourers, farm labourers themselves and people associated with redundant service industries are some of those making up this shift. The trend has been accelerated by the substitution of capital for labour. This is mostly associated with the mechanisation of the agricultural industries, so that more is produced annually from each work unit.

Casual Labour Availability

The shift in population has meant that there is now less casual labour available for employment on farms. This shortage of labour and increasing wage rates has accelerated the shift towards mechanisation.

The processing tomato industry provides an example of this effect. Most canning tomatoes were grown in blocks of 5 to 15 hectares and were reasonably close to relatively large towns so that labour for harvest was readily available. However, as reliable labour became harder to obtain and progressively more expensive, plant breeders gradually developed varieties suitable for mechanical harvesting. Effective and efficient mechanical harvesting requires larger areas and the areas of production are increasingly being located in the larger irrigation regions.

Location of Production Areas

The fact that the majority of the population lives in the city areas must influence the location of the production centres of the more perishable crops.

Horticultural production is centred principally around the major cities and towns, particularly production of products for sale in the fresh state. The continuing increase in the price of land close to large population centres, however, tends to force producers further away from these centres. Thus there is a balance between land price effect in production costs and cartage to the market.

Products required for processing tend to be grown close to the centre of processing (e.g. asparagus).

Most field crops however are less affected by distance. A reduction in land prices as distance from the major centres increases tends to reduce the cost of production from the more distant areas. With some freight equalisation by governments the location of the major grain industries is little affected by centres of population.

Governments have adopted policies which have encouraged the settling of new areas. There is little doubt that the Queensland sugar agreement was designed to help populate and stabilise the Queensland coastal areas. The initial support for the Ord River Scheme was obviously encouraged by social scientists as a means of developing and populating northern Australia.

Land Price Effect

The pressures of city business life have given many city dwellers the desire to escape from the urban life by owning a rural property either for commercial or recreational purposes. This increased demand for rural land in close proximity to suburban areas has caused prices to escalate beyond values capable of sustaining commercial agricultural production. This demand has also caused many originally viable properties to be subdivided into small hobby farmlets.

The trend has been for producers closer to the major cities to move further afield. This has been particularly noticeable in the dairy industry where many NSW dairy farmers have moved along coastal areas and to the inland as the suburban sprawl has increased along the coast. This trend will continue while quotas can be transferred.

The land price effect also tends to cause producers of the broadacre field crops to move further away from the centres of population to obtain land at prices which can be justified by the returns achievable.

References

ABARE (1992), Commodity Statistical Bulletin, Australian Government Publishing Service, Canberra.

Anon. (1986), *Australian Income Tax Legislation*, CCH Australia, Sydney.

Anon. (1987a), *The Australian Cotton Grower*, **8**(1), 6–7.

Anon. (1987b), *The Government Money Manual, Aug. 85 to Jan. 86*, Information Australia, 179–87.

Anon. (1987c), *Australian Hereford Quarterly*, **15**(1), 100.

Australian Bureau of Statistics (1986), *Year Book Australia, 1986*, ABS, Canberra.

Australian Wheat Board (1985), *Annual Report 1983–4*.

Bureau of Agricultural Economics (1987), *Quarterly Review of the Rural Economy*, **9**(1), Australian Government Publishing Service, Canberra.

Cobb, M. (1983), *The Great Tariff Debate*, Livestock and Grain Producers Association of New South Wales, Sydney.

Commonwealth of Australia (1986), *Economic and Rural Policy*, Australian Government Publishing Service, Canberra.

Department of Primary Industry (1985), *Rural Industry Directory*, Australian Government Publishing Service, Canberra.

Harris, S. *et al.* (1974), *Rural Policy in Australia*, report to the Prime Minister by a working group, May 1974, 2nd edn, Australian Government Publishing Service, Canberra.

Industries Assistance Commission (1974), *Annual Report 1973–4*, Australian Government Publishing Service, Canberra.

Glynn, S. (1974), *Government Policy and Agricultural Development*, University of Western Australia Press, Nedlands.

Lewis, J. N. (1971), 'Government Policy and the Location of Agricultural Industries' in G. J. R. Linge and P. J. Rimmer (eds), *Government Influence and the Location of Economic Activity*, Australian National University, Canberra, 161–4.

Miller, G. *et al.* (1987), *The Political Economy of International Agricultural Policy Reform*, Department of Primary Industry, Canberra.

Primary Industry Bank of Australia (1987), *Annual Report 1986*.

CHAPTER 2

CLIMATE AND CROP DISTRIBUTION

F. M. Kelleher

CLIMATE AND CROP DISTRIBUTION

Distribution of commercial crop production throughout the world is governed by many factors, principally climate, soils, topography, insect pests, plant diseases and economic conditions (Martin *et al.*, 1976). In Australia, as in most developed countries, segregation of crop types into particular production areas is largely governed by profit motives, the profitability of specific crops reflecting their adaptation and hence ability to grow and produce uniformly high yield of acceptable quality, in those areas. Crop adaptation is determined primarily by genotype–environment interaction, the suitability of a crop to a particular region depending largely on the climatic features of the region in relation to the requirements for normal growth and development of the crop (Janick *et al.*, 1974; Nix, 1975; Martin *et al.*, 1976).

Accepting climate as a dominant factor governing crop distribution, this chapter examines the distribution of commercial crop production in relation to climate in Australia, with emphasis on the climatic and physiological factors that confine the major crop species to current production areas. Where possible, the expansion of commercial crop production into currently unexploited regions is examined in relation to climate and possible alternative crop species and management systems.

AUSTRALIAN CLIMATE

The Australian Continent

Australia is a large, relatively flat island continent of 7 682 300 square kilometres, located between latitudes 10°41′S (Cape York) and 43°39′S (South East Cape, Tasmania) and longitudes 113°09′E (Steep Point) and 153°39′E (Cape Byron). Average surface altitude is only 300 m, with 87 and 99.5 per cent of the land mass below 500 and 1000 m respectively (Bureau of Meteorology, 1975). Three major landform features are evident—the western plateau comprising the western half of the continent,

26

of altitude 300 to 600 m; the interior lowlands of southwest Queensland and the Murray–Darling system of the south; and the easten highlands (Great Dividing Range) of tablelands, ranges and ridges with limited mountain areas above 1000 m (Mabbutt and Sullivan, 1970). The eastern highlands extend from north Queensland to Tasmania but with varying width, forming a significant barrier to the general westerly flow in the lower levels of the atmosphere (Bureau of Meteorology, 1975).

General Climatic Features

The predominant controlling factor in Australian climate is the descending air of the great subtropical high-pressure belt (average position centred at 30°S), individual anticyclones within this belt moving across the continent from west to east. During the winter semester (May to October) the anticlockwise circulation of these systems brings mild, dry southeast trade winds, carrying warm, dry, tropical continental air from the interior to the north of Australia, and westerly winds carrying cooler, moist Indian subtropical maritime air to the southwest of the continent (Gentilli, 1971). The westerly winds and frontal systems associated with extensive depressions travelling west to east over the Southern Ocean bring winter rainfall to southern Australia, with intense Southern Ocean depressions bringing strong onshore flow of very cold Pacific subtropical maritime and subpolar maritime air to southern, and particularly southeastern Australia (Gentilli, 1971; Bureau of Meteorology, 1975).

In the summer semester (November to April), the subtropical high pressure belt migrates 5 to 8° south of its winter position (Gentilli, 1971), bringing fine, warm weather to southern Australia with the passage of individual anticyclones. Heat waves may occur in southern and southeastern Australia if the eastward progression of anticyclones is halted by a stationary (blocking) anticyclone, causing winds to back northerly and then northwesterly and thus bringing intensely dry, hot air from the interior (Bureau of Meteorology, 1975).

The southward shift of the subtropical high-pressure belt at the same time allows the intertropical convergence zone to intrude from the equator to the northern tip of Australia, bringing warm, moist air of equatorial, Indian tropical maritime and Pacific tropical maritime origin to northern Australia (Gentilli, 1971), resulting in a hot rainy season. Tropical cyclones develop over the Coral Sea and Indian Ocean to the northeast and northwest respectively, bringing heavy rainfall, flooding and at times extensive wind damage.

The climate of eastern Australia is more complex, because of the combination of warm ocean currents offshore, onshore southeast trade winds and the eastern highlands. Tasmania is under the influence of the moist westerlies and Southern Ocean depressions throughout the year.

The climate of eastern and northern Australia is influenced by the Southern Oscillation (SO), a seesawing of atmospheric pressure between the northern Australian/Indonesian region and the central Pacific Ocean.

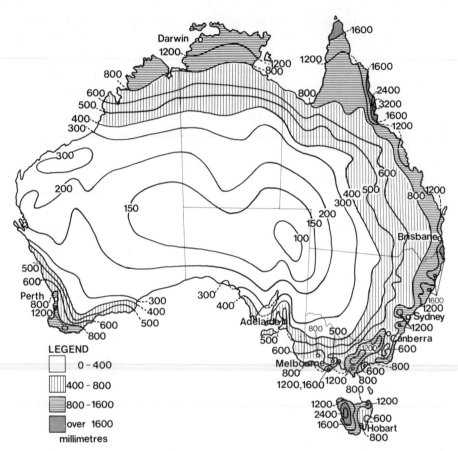

FIG. 2.1   Median (50 percentile) annual rainfall in Australia (after Australian Bureau of Statistics, *Year Book Australia*, 1974)

This oscillation is the second most important cause of climatic variation, after the annual season cycle, over eastern and northern Australia. The strength of the Southern Oscillation is determined by the Southern Oscillation Index (SOI) which is a measure of the difference in sea-level atmospheric pressure between Tahiti in the central Pacific and Darwin, northern Australia. At one extreme of the oscillation, the pressure is abnormally high at Darwin and abnormally low at Tahiti. Severe and widespread drought over eastern and northern Australia usually accompanies this extreme. These conditions generally commence early in the year, last for about 12 months, and recur every 2 to 7 years. This extreme is generally immediately preceded or followed by the opposite extreme, when pressures at Darwin are abnormally low and those at Tahiti are abnormally high. In this case, rainfall is generally above average over eastern and northern Australia.

The SO is linked to sea surface temperatures (SSTs) in the Pacific Ocean. Dry extreme SO years are accompanied by above-normal SSTs

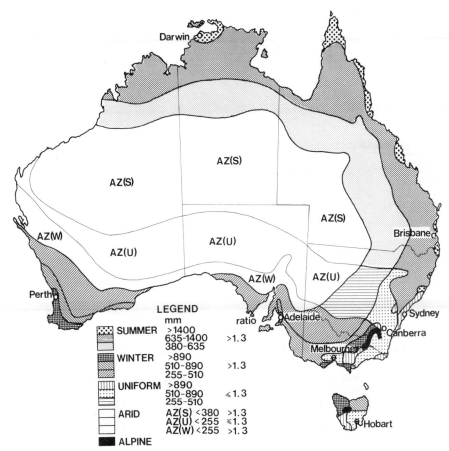

FIG. 2.2   Seasonal rainfall zones for Australia based on annual rainfall, seasonal incidence and altitude. Season incidence is based on the ratio of median summer (Nov. to Apr.) to median winter (May to Oct.) rainfall (after Australian Bureau of Statistics, *Year Book Australia, 1974*)

in the central and/or eastern equatorial Pacific and vice versa. Dry extreme years are called El Nino years. Wet extreme years are called La Nina years (Anon., 1992).

The Water Balance

PRECIPITATION

The dominance of rainfall over other climatic elements in controlling plant growth in Australia led Gaffney (1971) to develop a climatic classification based on median (50 percentile[1]) annual rainfall (Fig. 2.1) and seasonal rainfall incidence (Fig. 2.2).

---

[1] Fifty percentile (median) annual rainfall—amounts that are not exceeded by 50 per cent of all recordings.

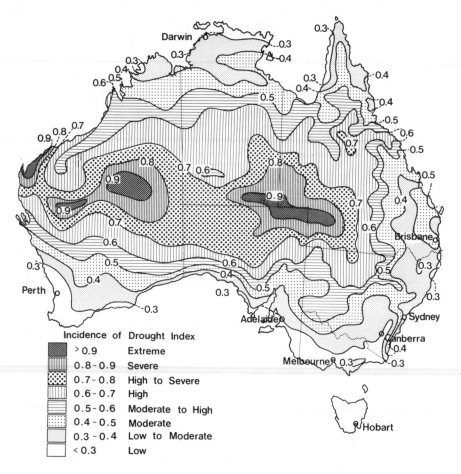

FIG. 2.3   Index of drought incidence in Australia based on 50, 30 and 10 rainfall percentiles where index $= \dfrac{50-10}{30}$ percentiles (after Australian Bureau of Statistics, *Year Book Australia, 1974*)

Median annual rainfall   This demonstrates the dry nature of the continent: 37, 57, 68 and 80 per cent (cumulative percentages) of the total area receive totals of less than 250, 375, 599 and 600 mm respectively. Fig. 2.1 shows a marked concentration of rainfall on the northern, eastern and southeastern continental fringe, Tasmania and southwestern Australia. Areas of greatest median annual rainfall concentration are the northeastern coast of Queensland, the western coast of Tasmania, and isolated mountainous areas of southeastern New South Wales and Victoria.

Seasonal rainfall incidence   Fig. 2.2 shows a steady transition from a wet summer – dry winter pattern (northern Australia, southeastern Queensland and northeastern New South Wales) through uniform distribution (central and southern New South Wales, eastern Victoria and

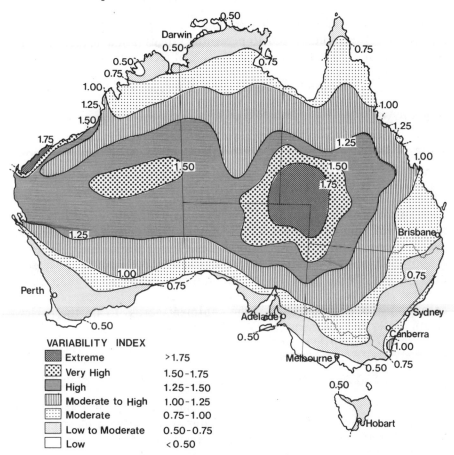

Fig. 2.4 Annual rainfall variability in Australia, based on the variability index $\frac{(90-10)}{50}$ rainfall percentiles (after Australian Bureau of Statistics, *Year Book Australia, 1974*)

southern Tasmania) to a marked wet winter–dry summer (southwestern Western Australia, the southern continental margin, and Tasmania).

Variability The overwhelming feature of Australian rainfall over most of the continent is the variability of both seasonal and annual totals. Non-seasonal droughts (Fig. 2.3) are very critical in Australia because so much of the country is near the threshold of usable rainfall, particularly with regard to crop production. Many monthly rainfall distributions are heavily skewed because of the combination of many low or zero totals with sporadic heavy falls (Maher, 1966). Annual rainfall and even seasonal rainfall totals may obscure this fact. Several indices of rainfall variability in Australia have been proposed since the first attempt by Taylor (1920) (for review see Gentilli, 1971), most of which are based on the acceptance of annual rainfall total following a normal distribution over time. One such index is that prepared by Gibbs and Maher (1967) in which the variability of annual rainfall is given by the ratio of the difference between the 90 and 10 percentile ranges (annual rainfall totals never reached in 90 and 10 per cent of years of record) to the 50 percentile (median) value, i.e.

$$\text{Variability index} = \frac{90 - 10}{50} \text{ percentiles.}$$

A high ratio indicates extreme variability (Fig. 2.4). Areas of extreme variability lie mostly in the summer rainfall sector of the arid zone, with those of lowest variability in the northern, eastern and southern continental fringe and in Tasmania.

EVAPORATION AND RAINFALL EFFECTIVENESS

Assessment of water availability for plant growth requires a knowledge not only of rainfall characteristics, but also of seasonal and annual evaporation and evapotranspiration losses. Seasonal and annual maps of evaporation in Australia have been published by the Bureau of Meteorology (1965, 1975). Features of the January evaporation map show greatest values of 350 to 400 mm in the arid centre, extending in a broad core from the western coast of the continent across to northern New South Wales and southeastern Queensland, with values declining to 150 mm around the northern, eastern and southern coasts. The July map is much simpler and evaporation rates much lower, showing an east – west zonation with values declining from 150 mm in the north to 50 mm in the south.

Early Australian attempts to translate rainfall and evaporation data into meaningful estimates of monthly water availability for plant growth (rainfall effectiveness) have been reviewed by Gentilli (1971). The methods used defined the growing season in terms of monthly rainfall excess over evaporation losses, using various ratios of monthly rainfall to either evaporation (e.g. Davidson, 1934, 1936; Prescott, 1938, 1949, 1956; Prescott and Thomas, 1949) or some closely related factor such as saturation deficit or temperature (e.g. Andrews and Maze, 1933). The major deficiency in these methods is that they delimit the growing season to months of 'effective rainfall', failing to take account of soil water storage, which can be a significant factor determining water availability at critical stages of crop development.

The Moisture Index proposed by Fitzpatrick and Nix (1970) is a refined water budgeting technique which overcomes this problem, accounting for changing soil moisture storage with rainfall additions and evapotranspiration withdrawals. The index ranges in value from 1.0, where water is not limiting to plant growth, to zero, where available soil moisture within the root zone is inadequate and plant growth has ceased. Plant growth is assumed to follow a linear increase as values increase from zero to 1.0. Although derived for a single soil type of medium texture under conditions of permanent vegetative cover, factors which could vary significantly with changes in soil type and extent of vegetative

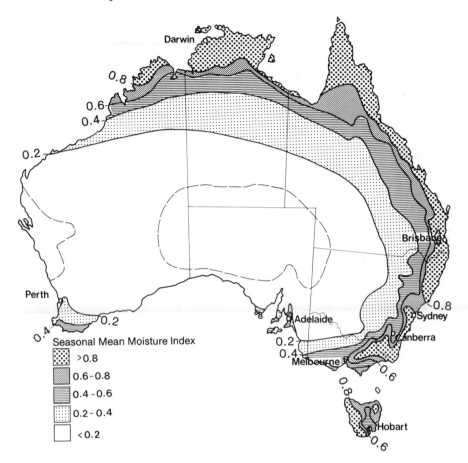

FIG. 2.5   Mean Moisture Index values for the summer semester (Nov.–Apr.)
(after Fitzpatrick and Nix, 1970)

cover under Australian crop production systems, the index provides an
excellent summary of moisture availability for plant growth. Summer
(November to April) and winter (May to October) maps of the Moisture
Index (Figs 2.5 and 2.6) show clearly the patterns of seasonal water
availability, with a pronounced summer dominance in the north and
winter dominance in the south. Despite the differences in assumptions
used to derive them, a comparison of these seasonal Moisture Index
patterns with bi-monthly hydric efficiency ($r/e^{0.7}$ where r = rainfall and
e = evaporation) (Prescott) ratios (Gentilli, 1971) shows close agreement,
indicating that both systems are applicable to a consideration of seasonal
water availability for plant growth in Australia.

MEAN TEMPERATURES

Because of its latitude and east–west extent (4000 km) in the vicinity of
the Tropic of Capricorn, Australia may be considered the hottest continental mass. Midsummer daily maxima exceed 35°C over much of the

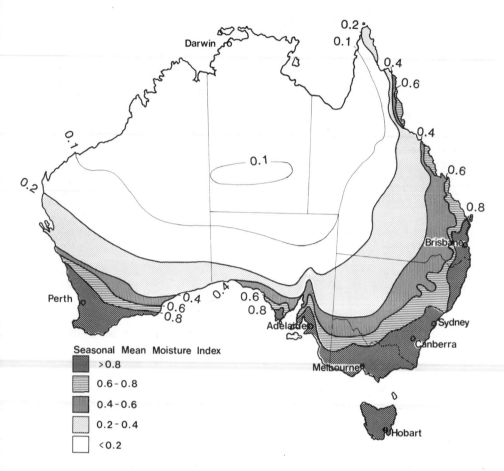

FIG. 2.6   Mean Moisture Index values for the winter semester (May–Oct.)
(after Fitzpatrick and Nix, 1970)

interior and 40°C in the northwest, with extremes up to 50°C. In
contrast, extreme daily minima are not as low as those recorded in other
continents, because of the absence of extensive highlands (Gentilli, 1971).
Highest July daily minima (18 to 21°C) occur along the northern coastal
fringe, declining with increasing latitude south of the Tropic to values of 3
to 6°C over much of the interior and 0 to 3°C in the eastern highlands
and Tasmania (Figs 2.7 and 2.8). Mean daily temperatures for January
and July show a strong latitudinal zonation, with highest values in the
north and northeast of the continent in both months. Temperatures in the
eastern and southern coastal fringes are modified by proximity to the
surrounding ocean. Much of the southeast and southern sector of the
continent exhibits mean July daily temperatures of less than 13°C,
particularly in the highlands where frequent frosts occur.

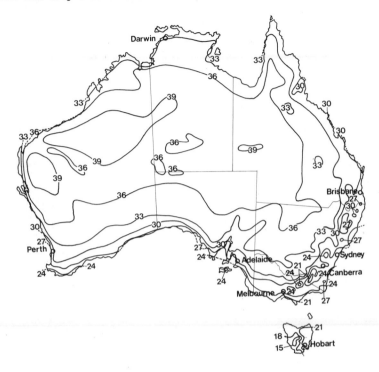

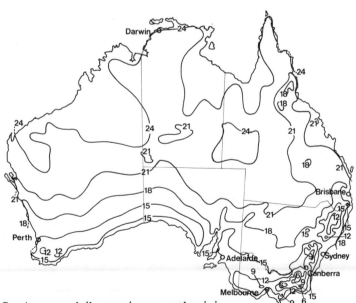

FIG. 2.7 Average daily maximum and minimum screen temperatures (°C) for January (after Australian Bureau of Statistics, *Year Book Australia, 1974*)

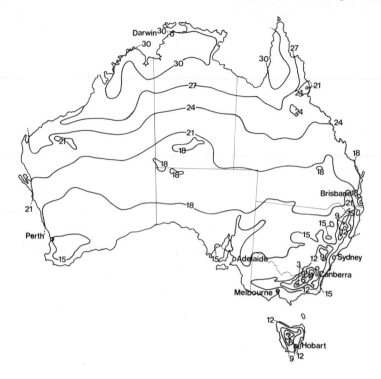

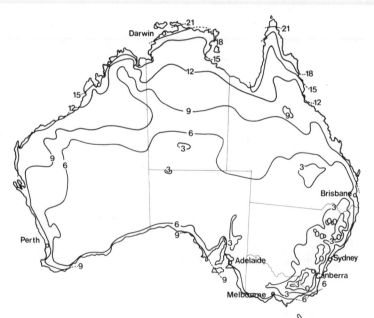

FIG. 2.8   Average daily maximum and minimum screen temperatures (°C) for July (after Australian Bureau of Statistics, *Year Book Australia, 1974*)

THE THERMAL REGIME FOR PLANT GROWTH

In terms of suitability for plant growth, the temperature environment of Australia was evaluated by Fitzpatrick and Nix (1970) for three broad groups of pasture plants, namely, tropical grasses (optimum temperature for growth >38°C), tropical legumes (26 to 30°C) and temperate legumes and grasses (21 to 22°C). Using a derived Thermal Index ranging in value from 0 (temperature completely limiting to plant growth) to 1.0 (no temperature restriction on growth), the following pattern was described for the summer (November to April) and winter (May to October) semesters:

Tropical grasses The most favourable environments extend in a broad arc across the continent north of 22°S latitude, where summer and winter Thermal Index values range from 0.70 to 0.80 and from 0.35 to 0.50 respectively. On the east coast, summer values decline southward to 0.20 at Bega on the New South Wales south coast, while the southern winter limit (0.10) is reached at Lismore. In the southern inland, summer and winter values average 0.20 to 0.50 and less than 0.10 respectively: killing winter frosts impose severe restrictions in the south.

Tropical legumes The pattern is broadly similar to tropical grasses, but extends further south along the east coast and to the southern inland. Summer eastern coastal values fall to 0.40 at Bega, while the southern winter limit (0.10) approximates to Port Macquarie. Southern inland areas have summer and winter values ranging from 0.40 to 0.90 and less than 0.10 respectively.

Temperate grasses and legumes The patterns for these crops are rather complex because of the growth restrictions imposed by both supra- and sub-optimal temperatures (>24°C and <4.4°C respectively). Supra-optimal temperatures over most of the north and much of the southern inland (including all the Australian wheat belt) limit summer Thermal Index values to between 0.60 and 0.90. Optimum values of 1.0 are attained in autumn and spring, with lower winter and summer values because of sub- and supra-optimal temperatures respectively. The most favourable year-round thermal environments occur along the central and north coast of New South Wales and at elevations above 300 and 600 metres in southern and northern Queensland respectively. Midwinter values are greatest in areas north of latitude 22°S.

Classifying the thermal environment using this index produces a useful description (based on dry matter production) of the overall suitability of the Australian environment for absolute plant growth for these groups but gives less information on a critical factor in crop production, i.e. phasic development of the crop. Different crop species, all of which could be classified into the broad groups outlined above, frequently exhibit different temperature requirements at different stages of development, so that some further refinement is needed to give a more thorough assessment of the suitability of the Australian thermal environment for these crops. Where appropriate, this is examined in more detail for individual crops later in this chapter.

FROST

Frost incidence in Australia imposes a major constraint on crop production, particularly in the southern inland and the highland areas. With winter crops such as wheat, management strategies aim at delaying the onset of frost-sensitive, critical phases of development (especially flowering (anthesis) until the spring frost danger is passed (Nix, 1971, 1975). For summer crops, practically all of the 'safe' growing season is dictated by the length of the frost-free period (i.e. average date of the last spring frost to average date of the first autumn frost in the following year). While winter crops are frost sensitive mainly in the later developmental stages (heading, anthesis and grain filling), summer crops are sensitive at all stages, with the exception of the early vegetative stages in some species such as sunflower. As dates of last spring and first autumn frost are averages taken over many years, some degree of frost risk must be accepted by the producer. If the crop under consideration is frost sensitive at all growth stages, then the producer may take the frost-free period as being that from the latest spring frost on record to the earliest autumn frost on record if he wishes to ensure absolute freedom from frost and accept no risk whatsoever. However, such an approach would considerably shorten the available growing season for a summer crop while considerably delaying flowering for a winter crop (if sown late so that flowering will not occur at a time of frost risk), and therefore under commercial conditions, some degree of risk must be accepted. An example of the average and extreme dates of frost occurrence is shown in Table 2.1 for Richmond, New South Wales, to illustrate the point. Here the length of the absolute frost period is from 16 April to 15 October, a period of 183 days, while that of the average frost period is from 16 May to 13 September, a period of 121 days. Median annual length of the frost period for Australia is shown in Fig. 2.9.

TABLE 2.1   Average dates of occurrence of first and last screen temperatures associated with light and severe frosts at Richmond, N.S.W. (1907−76)

	Average date of first occurrence	Earliest date recorded	Average date of last occurrence	Latest date recorded	Average frost period (days)
Light	16 May	16 April	13 Sept.	15 Oct.	121
Severe	28 May	5 May	22 Aug.	26 Sept.	87

GROWING DEGREE DAYS

For many crop species, the rates of growth and development have been shown to be closely related to the temperature environment in which they are grown. The length of individual phases of development (e.g. emergence to floral initiation, floral initiation to anthesis, etc.) and that of their sum, the entire growth period, have been shown to be largely dependent on temperature. For these crops, the concept of 'Growing Degree Days'

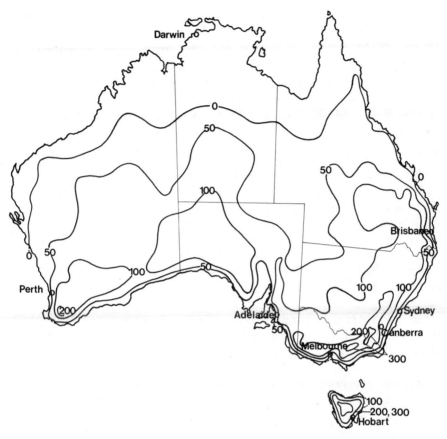

FIG. 2.9 Median annual length (days per year) of the frost period in Australia. Frost-day: minimum screen temperature $\leq 2°C$ (after Australian Bureau of Statistics, *Year Book Australia, 1974*)

(GDD) has proven extremely useful as a predictive tool to determine the likely length of successive developmental phases and hence that of the total growth period. The concept is based on the fact that, above a certain 'base' temperature, growth and development increases linearly with temperature, up to a certain point. A GDD is defined as a day in which the mean temperature for that day exceeds the base temperature by $1°C$. If the mean for a particular day exceeds the base temperature by $10°C$, then that day is scored as 10 GDD. Assuming that each phase of development requires a certain total number of GDD for its successful completion and hence initiation of the next phase, the actual number of calendar days required to achieve a given total will depend on ambient temperature, and hence season. Thus, the required GDD summation for a given crop may be achieved in a short time in a warm environment, but take much longer in a cold environment, resulting in a much shorter growing season in the warmer environment.

GDD has proven very useful as a predictive tool in determining the likely suitability of certain crops to particular environments where the length of the growing season is fixed by some other environmental variable, e.g. frost occurrence or heat and water stress in summer. In these situations where the growing period for a given summer crop *must* be completed during the frost-free period, successful production of that crop in that environment will depend on whether the crop receives the requisite GDD total *during* the frost-free period. If not, the crop will not successfully complete its life cycle before the end of the frost-free period. Such a technique has been used to establish the high risks associated with maize and sorghum production in southern New South Wales (Nix, 1971). Similarly, the technique of using GDD summations based on mean temperatures for a given environment can be used to predict the likely impact of variation in sowing date on the rate of development and hence the timing of critical growth phases in relation to environmental hazards such as water or heat stress (Nix, 1971, 1975).

A knowledge of the GDD requirements for particular crop species or cultivars therefore would prove invaluable as a predictive tool for determining the suitability of any environment for production of certain crops, and there is a pronounced need for much more data on the GDD requirements for phasic development of both crop species and cultivars (Nix, 1975). Some examples of the successful use of GDD for this purpose have been with wheat (Nix, 1971, 1975), winter Brassica oilseeds, maize and sorghum (Nix, 1971), sunflowers (Robinson, 1971; Doyle, 1975) and cotton (Basinski 1965; McMahon and Low, 1972).

Solar Radiation

Average daily global (direct and diffuse short-wave) radiation for Australia is shown in Fig. 2.10, while the variability of daily global radiation receipts at the earth surface is shown in Table 2.2 (Bureau of Meteorology, 1975). Average daily global radiation (Fig. 2.10) shows a concentric zonation, highest values occurring in the dry northwestern corner of the continent, declining to lower values in the north, east and south, reflecting a greater incidence of cloud cover in the coastal margins. Summer season radiation receipts are uniformly high and do not impose serious restrictions on the growth of most crop species at any site (Fitzpatrick and Nix, 1970). However, in winter, average daily radiation receipts are much lower, falling to values as low as 175 mWh/cm^2 in Tasmania and increasing northward in a distinct latitudinal banding pattern to 520 mWh/cm^2 in the northern extremes of the Northern Territory. Winter radiation receipts could therefore impose limitations on growth of many crop species, particularly in southern Australia and Tasmania, where the low values recorded reflect the high degree of cloud cover in this winter rainfall environment. Light Index values ranging in scale from zero (solar radiation completely limiting to plant growth) to 1.0 (no solar radiation limitation to growth) (Fitzpatrick and Nix, 1970) reach their

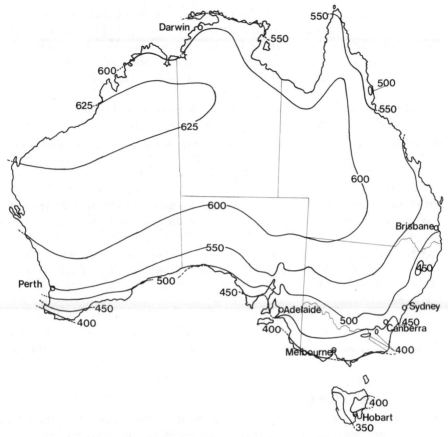

Fɪɢ. 2.10   Average daily global radiation in milliwatt hours per square centimetre (mWh/cm²) for Australia (after Australian Bureau of Statistics, *Year Book Australia, 1974*)

Tᴀʙʟᴇ 2.2   Global radiation: variability of daily amounts for June and December at selected centres (20, 50 and 80 percentile values in mWh/cm²) for the period 1964–68
Source: 'Climate and Physical Geography in Australia' in Australian Bureau of Statistics, *Year Book Australia, 1974*

Station	June percentiles			December percentiles		
	20	50	80	20	50	80
Alice Springs	360	450	480	580	760	810
Darwin	520	570	590	440	570	620
Melbourne	130	190	240	470	640	780
Perth	180	260	330	770	870	910
Townsville	360	490	510	550	710	760
Williamstown	210	270	330	490	650	780

lowest midwinter values (0.50) over western Tasmania, increasing to 0.60 to 0.75 over southern Australia south of the Tropic (23½°S latitude). Comparing these values to those for the Thermal and Moisture indices, it is evident that solar radiation constraints on plant growth are far less significant than those of temperature or moisture availability.

DAYLENGTH

Crop species can be divided into three major groups based on their daylength requirement to initiate flowering, namely, long-day plants, in which floral initiation occurs in response to increasing daylength or, more correctly, decreasing dark period (winter crops, floral initiation occurring in response to increasing daylength from winter to spring); short-day plants, in which floral initiation occurs in response to decreasing daylength or, more correctly, increasing dark period (summer crops, floral initiation occurring as daylength declines from summer to autumn); day-neutral plants (including both winter and summer crops) which show no response to daylength variation, flowering after a given period of growth during which a required sum of 'radiation' or 'heat units' (GDD) has been received.

With the first two groups, daylength is critical to successful flowering and ultimate grain production, latitudinal and seasonal variation in daylength (Table 2.3) governing their adaptation and potential value as commercial crops.

Classification of Climate

A number of systems of climate classification have been proposed by geographers over the last century, and their applications to the Australian continent, together with various special purpose climatic indices, have been summarised by Gentilli (1971, 1972). Of the systems available, the most widely used for agricultural purposes in defining the suitability of various climatic zones for plant growth have been those of Köppen and Thornthwaite, and more recently the Growth Index system of Fitzpatrick and Nix, which was first proposed in 1970. A brief description of these systems follows.

The Köppen System

The most widely known and used system, it is an empirical classification based on temperature, precipitation, seasonal characteristics and the fact that natural vegetation is the best ultimate expression of the climate of a region, so that many of the climatic boundaries have been selected with vegetation limits in mind (Janick *et al.*, 1974). Five main climatic groups are identified, of which four are based on temperature and the remaining one based on moisture. Within each, a number of further subgroups, based on temperature and seasonality of rainfall, are identified. A summary of the main groups and their subgroups for the Australian

TABLE 2.3 Duration of daylength at selected latitudes in Australia
Source: Gentilli, 1971, based on data from *Astronomical Ephemeris*, HMSO, London

Latitude (°S)	Length of astronomical day on the 21st day of each month											
	Jan.	Feb.	Mar.	Apr.	May	Jun.	Jul.	Aug.	Sep.	Oct.	Nov.	Dec.
	h min	h min	h min	h min	h min	h min	h min	h min	h min	h min	h min	h min
10	12 37	12 23	12 08	11 51	11 38	11 33	11 38	11 50	12 07	12 23	12 37	12 42
20	13 09	12 41	12 08	11 35	11 07	10 55	11 06	11 34	12 07	12 41	13 09	13 20
30	13 45	13 03	12 10	11 17	10 33	10 13	10 30	11 14	12 08	13 01	13 45	14 04
35	14 07	13 15	12 11	11 07	10 12	9 48	10 10	11 03	12 09	13 13	14 07	14 30
40	14 32	13 28	12 12	10 55	9 49	9 20	9 46	10 51	12 09	13 27	14 31	15 01

continent is outlined in Table 2.4, and their distribution in Australia in Fig. 2.11. Most of the continental inland is semi-arid to desert, with moist climate limited to a relatively narrow band extending inland from the northern, eastern and southern coasts to a distance of approximately 250 to 400 km. Boundaries shown in Fig. 2.11 can best be regarded as transition zones between climate types, with no sharp delineation except in isolated cases such as southern Victoria and Tasmania. The Köppen system has been criticised on the basis that many of its boundaries are arbitrary (Trewartha, 1968) and do not necessarily correspond consistently

TABLE 2.4   Major climatic groups and subgroups and their characteristics for the Australian continent, based on the Köppen system
Source: 'Climate' in *Atlas of Australian Resources*, 2nd edn, Department of National Resources, Canberra 1973, and Janick *et al.*, 1974

Main groups			Subgroups	
Symbol	Climate type	Symbol		Characteristics
A	Tropical rainy			All months with average temperature greater than 18°C
			Af	No distinct dry season
			Am	Monsoon—short dry season
			Aw	Marked winter dry season
B	Dry			Annual potential evaporation exceeds annual rainfall
		BS		Steppe or semi-arid, hot to warm
			BSwh	Winter drier than summer, hot
			BSfh	Uniform rainfall, hot
			BSfk	Uniform rainfall, cool to warm
			BSsh	Summer drier than winter, hot
			BSsk	Summer drier than winter, cool to warm
		BW		Desert
			BWh	Hot
			BWk	Cool to warm
C	Temperate			Coldest month with average temperature between 0° and 18°C
		CS		Mediterranean—dry summer, subtropical
			Csa	Hot dry summer, mild wet winter
			Csb	Warm dry summer, cool wet winter
		Cf		Humid subtropical, uniform rainfall
			Cfa	Hot summer, mild winter
			Cfb	Warm summer, cool winter
			Cfc	Short cool summer, cold winter
		Cw		Moist summer, subtropical
			Cwa	Mild dry winter

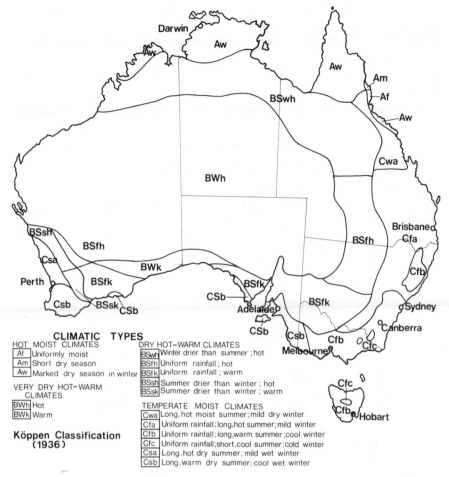

CLIMATIC TYPES

HOT MOIST CLIMATES
Af	Uniformly moist
Am	Short dry season
Aw	Marked dry season in winter

VERY DRY HOT-WARM CLIMATES
BWh	Hot
BWk	Warm

Köppen Classification (1936)

DRY HOT-WARM CLIMATES
BSwh	Winter drier than summer ; hot
BSfh	Uniform rainfall ; hot
BSfk	Uniform rainfall ; warm
BSsh	Summer drier than winter ; hot
BSsk	Summer drier than winter ; warm

TEMPERATE MOIST CLIMATES
Cwa	Long, hot moist summer; mild dry winter
Cfa	Uniform rainfall; long, hot summer; mild winter
Cfb	Uniform rainfall; long, warm summer; cool winter
Cfc	Uniform rainfall; short, cool summer; cold winter
Csa	Long, hot dry summer; mild wet winter
Csb	Long, warm dry summer; cool wet winter

FIG. 2.11 Australian climate classification after the Köppen system (after Division of National Mapping, *Atlas of Australian Resources*, Department of National Resources, Canberra 1973)

with natural vegetation distribution changes or in some cases, for Australia, changes in the distribution of crops and pastures (Hazlewood, 1973).

The Thornthwaite System

This system, proposed by C. W. Thornthwaite in 1931, resembles that of Köppen in so far as letter combinations are used to denote climatic types, and plant response is used to integrate climatic elements. It is based primarily on three factors:

precipitation effectiveness, represented by monthly P:E ratios, their summation yielding a P:E index; five humidity provinces are recognised, based on the index (Table 2.5);

seasonal distribution of precipitation—humidity provinces are further subdivided on the basis of seasonal distribution specifying seasonal deficiencies (Table 2.6);

temperature efficiency index—obtained by summing monthly mean Fahrenheit temperature, subtracting 32° F and dividing by 4, yielding six temperature provinces (Table 2.7).

A summary of the relationships between precipitation and temperature indices to classify climate groups is shown in Fig. 2.12, and climate of the Australian continent in Fig. 2.13. Of the 120 possible combinations, only 32 are recognised as major climatic types, and of these only 12 apply to Australia. The map of Australian climate based on the Thornthwaite system (Gentilli, 1972) shows good general agreement with that based on the Köppen system, but gives much finer and more useful delimitation based on moisture availability, particularly within Köppen's A and C climatic groups (Hazlewood, 1973). The Köppen system was strongly

TABLE 2.5   Humidity provinces of the Thornthwaite system (from Janick *et al.*, 1974)

Symbol	Humidity province	Vegetation	P:E index
A	Wet	Rainforest	≥ 128
B	Humid	Forest	64–127
C	Subhumid	Grassland	32–63
D	Semi-arid	Steppe	16–31
E	Arid	Desert	< 16

TABLE 2.6   Humidity province subdivisions of the Thornthwaite system (from Janick *et al.*, 1974)

Symbol	Seasonal precipitation distribution
r	abundant all seasons
s	sparse in summer
w	sparse in winter
d	sparse all seasons

TABLE 2.7   Temperature efficiency provinces of the Thornthwaite system (from Janick *et al.*, 1974)

Symbol	Temperature province	T–E index
A'	Tropical	≥ 128
B'	Mesothermal	64–127
C'	Microthermal	32–63
D'	Taiga	16–31
E'	Tundra	1–15
F'	Permanent frost	0

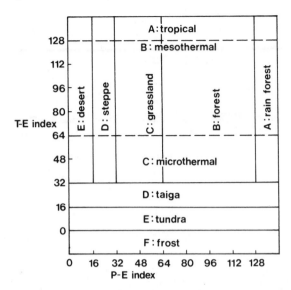

FIG. 2.12 Temperature and humidity provinces and their interrelationships in the Thornthwaite system of climate classification (after Thornthwaite, 1931, from Janick *et al.*, 1974)

criticised on this point by Gentilli (1971), who stated that while it gave significant regional boundaries in the tropical zone, it was inadequate to delimit subtropical regions if they were to correspond with vegetation distribution. Much of this criticism is based on the failure of the Köppen system to adequately separate regions within the subtropical zone on the basis of moisture availability.

Despite the limitations of the Thornthwaite and Köppen systems (see Gentilli, 1972), both have been utilised extensively by crop geographers in the analysis of both natural vegetation and crop and pasture distribution, and have proved a useful basis in most cases.

The Growth Index System

This system, developed by Fitzpatrick and Nix (1970), was aimed at simulating the basically non-linear functions of plant response to temperature, moisture and light and transforming these into linear functions or indices with a scale range from zero to unity. By combining the Moisture, Thermal and Light indices into a composite multi-factor Growth Index, these workers were able to characterise the climate of the Australian continent directly in terms of plant response to environment for three major groups of pasture plants, namely, tropical grasses, tropical legumes, and temperate legumes and grasses. While the Köppen and Thornthwaite systems were developed on the basis of temperature and precipitation characteristics, with vegetation distribution assumed to

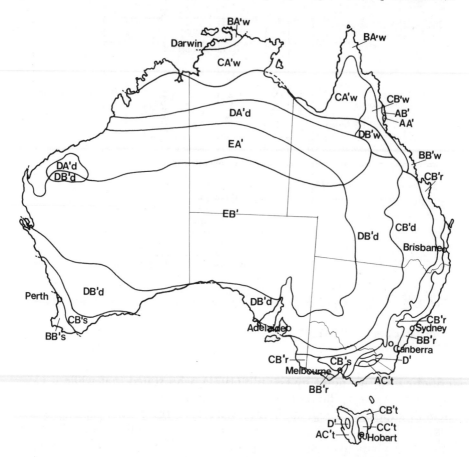

Fig. 2.13 Australian climate classification under the Thornthwaite system (after Gentilli, 1972)

closely follow identified climatic zones, the Growth Index system was derived directly on the basis of plant response to temperature, light and moisture, allowing a climatic classification based on the *suitability* of different areas for growth of the three groups of plants.

In particular, it allows easy identification of the major factor(s) limiting plant growth at any site for any time of year. The Growth Index itself cannot exceed the index value of the factor limiting growth. A favourable temperature and light regime, resulting in high Thermal and Light indices, may coincide with a low Growth Index value because moisture is limiting and the Moisture Index value low. While derived for pasture plants with permanent vegetative cover, the system has proved equally applicable in the analysis of the Australian environment for the production of wheat (Nix, 1975). Detailed maps of Thermal, Light, Moisture and Growth indices for the three groups of plants have been published (Fitzpatrick and Nix, 1970; *Atlas of Australian Resources*, 1973) and a general summary of the main features follows:

Tropical grasses have high temperature requirements, and the summer rainfall patterns of northern Australia are most suitable for these species; summer Growth Index values decline to the southern and central inland due to less favourable moisture regimes, and southward along the east coast as temperature regimes become less favourable. Winter values are low throughout the continent, primarily because of winter drought in the north and unfavourable temperatures in the south.

Tropical legumes have similar requirements to tropical grasses, but their lower temperature requirements allow the favourable area for summer production to extend further south along the eastern coastal margin. Winter values are similar to those for tropical grasses.

Temperate grasses and legumes are predominantly restricted to the southwestern corner of the continent from Carnarvon to Esperance, and in a broad band paralleling the coast from the Eyre Peninsula in South Australia through Victoria and New South Wales, to isolated areas in southeastern Queensland and elevated regions such as the Atherton Tableland in northeastern Queensland. On the inland boundaries, moisture regimes impose the major restrictions throughout the year, while in the most favourable winter rainfall areas of the wheat belt, thermal and light regimes are out of phase with the most favourable moisture regime. In essence, autumn and spring offer the best combination of thermal, moisture and light regimes (and hence highest Growth Index). In winter, Growth Index values are lowered by sub-optimal temperatures, while summer values are restricted by both moisture supply and supra-optimal temperatures. Successful production of temperate species in summer rainfall areas is achieved by management strategies which enable soil storage of summer rainfall (Nix, 1975).

THE GROWTH INDEX IN RELATION TO KÖPPEN AND THORNTHWAITE SYSTEMS
A comparison of Growth Index patterns for tropical grasses, tropical legumes and temperate legumes and grasses with Australian climatic zones under the Köppen and Thornthwaite systems (Figs 2.11 and 2.13) shows a general similarity in boundaries, and allows broad generalisations to be drawn in matching favourable areas for both tropical and temperate species under the Growth Index system with defined climatic zones under the Köppen and Thornthwaite classifications (Table 2.8). It must be emphasised that boundaries do not coincide exactly with one another, but the general similarity between the three allows reasonable comparisons to be made.

In general terms, most favourable areas for production of tropical legumes and grasses are the Aw and BSwh Köppen zones, while those for temperate species are the subtropical or temperate moist zones with uniform (Cfa, Cfb) or winter dominant (Csb) rainfall distribution and mild winters. For a comparison of the Köppen, Thornthwaite and Growth Index systems, see Table 2.8 and Figs 2.11 and 2.13.

TABLE 2.8   Approximate distribution of areas with favourable Growth Index values for tropical grasses, tropical legumes and temperate legumes and grasses in terms of Köppen and Thornthwaite climate zones

Species group	Growth index values	Köppen climates	Thornthwaite climates
Tropical grasses	0.4–0.6	BSwh	DA'w
		Cwa	CA'w
	0.6	Aw	BA'w
			Ca'w
Tropical legumes	0.4–0.6	BSwh	CA'w
		Cfa	DA'w
	0.6	Aw	AA'r
		Bswh	BA'w
			CA'w
Temperate legumes and grasses	0.4–0.6	Cwa	CA'w
		Cfa	CB'd
		Bsfk	CB's
			DB's
	0.6	Cfa	CB'r
		Cfb	BB'r
		Csb	

CROP DISTRIBUTION IN AUSTRALIA

Australian Field Crops

Major Australian field crops, their botanical classification, principal uses and average sown area for the period 1987 to 1991 are shown in Table 2.9. While seven families are listed, the overriding importance of the *Poaceae* or grass family is evident. The total area sown to crops varies from year to year on the basis of market movements, seasonal conditions and other factors, showing a steady increase over time to a peak of 21.7 million ha in 1984 but declining to 17.1 million ha by 1991 (Table 2.10). When viewed in the context of a total Australian continental area of 787 million ha, the total area utilised for crop production accounts for only 2.2–2.8 per cent.

PRODUCTION PATTERNS — WINTER AND SUMMER CROPS

Generalised seasonal patterns of winter and summer crop development in Australia are shown in Table 2.11. Considerable within-season variation may occur with some operations such as planting, where delays may arise from a variety of factors such as inadequate soil water, excessively wet soil and delays in land preparation.

TABLE 2.9   Major Australian field crops—classification, principal uses and average area sown for the period 1987 to 1991. Crop species within each family are divided into two major groups (temperate and tropical*) on the basis of climatic adaptation and production period under Australian conditions (autumn–winter–spring and spring–summer–autumn respectively)
*Includes tropical and subtropical species.
Source: Australian Bureau of Statistics, Summary of Crops Australia 1987–88; 1990–91. ABS Cat. No. 7330.0

Crop	Species	Principal uses	Area sown (thousand ha)
Family *Poaceae (Gramineae)*			
Temperate species			
Wheat	*Triticum aestivum* } *Triticum durum* }	Human consumption —grain and grain products	9438
		Hay and green forage	58
Barley	*Hordeum vulgare*	Grain for malting and stock feed	2335
		Green forage and hay	54
Oats	*Avena sativa* *Avena byzantina*	Grain—stock feed and human consumption	1171
		Hay and green forage	416
Rye	*Secale cereale*	Grain—stock feed and human consumption	46
		Land stabilisation, hay and green forage	19
Canary seed	*Phalaris canariensis*	Bird seed	12
Triticale	*Triticosecale* [a]	Stock feed grain and green forage	117
Tropical species			
Maize	*Zea mays*	Grain—stock feed and human consumption	53
		Green feed and silage	6
Grain sorghum	*Sorghum bicolor*	Stock feed grain	589
		Green feed and silage }	119
Forage sorghum	*Sorghum* spp. hybrids	Green feed and silage }	
Rice	*Oryza sativa*	Grain—human consumption	99
Millets			
White Panic	*Echinochloa frumentacea*		
White French Panicum Panorama Hungarian Siberian Setaria Foxtail Italian	*Panicum miliaceum* *Setaria italica*	Birdseed grain	35

TABLE 2.9 continued

Crop	Species	Principal uses	Area sown (thousand ha)
Japanese Shirohie	*Echinochloa utilis*		
Pearl	*Pennisetum americanum* (syn. *Pennisetum glaucum*) (open-pollinated and hybrids)	Hay	18 [b]
		Green forage	108 [c]
Sugar cane	*Saccharum officinarum*	Sugar production	393

Total, *Poaceae*			15,080

Family *Fabaceae (Leguminosae)*
Temperate species

Lupins	*Lupinus angustifolius* *Lupinus cosentinii* *Lupinus albus*	Stock feed—high protein grain	841
Field peas	*Pisum arvense* (syn. *Pisum sativum* var. *arvense*)	Grain—human consumption and stock feed. Green manure	371
Chickpeas	*Cicer arietinum*	Grain—human consumption and stock feed	
Faba beans	*Vicia faba*	Grain—stock feed	
Tropical species			
Cowpeas (Poona peas)	*Vigna sinensis* (syn. *Vigna unguiculata*)	Forage, green manure Grain—stock feed	190
Navy beans	*Phaseolus vulgaris*	Grain—human consumption	
Pigeonpea	*Cajanus cajan*	Grain—stock feed	
Peanuts	*Arachis hypogea*	Grain—confectionery, oil and protein meal	25
Soybeans	*Glycine max*	Grain—protein and oil	51

Total, *Fabaceae*			1478

Family *Asteraceae (Compositae)*
Temperate species

Safflower	*Carthamus tinctorius*	Grain—oil and protein meal	33
Tropical species			
Sunflower	*Helianthus annuus*	Grain—oil and protein	162

Total, *Asteraceae*			195

TABLE 2.9 continued

Crop	Species	Principal uses	Area sown (thousand ha)
Family *Malvaceae*			
Tropical species			
Cotton	*Gossypium hirsutum*	Cotton fibre, oil and protein meal from seed	223
Total, *Malvaceae*			223
Family *Brassicaceae* (*Cruciferae*)			
Temperate species			
Canola	*Brassica napus* / *Brassica campestris*	Oil and protein meal	58
Forage—canola	*Brassica campestris*		
—turnips	*Brassica oleracea*		
—swedes	*Brassica napus*	Forage crops	22
—kale	*Brassica oleracea*		
—chou mollier	*Brassica oleracea*		
Total, *Brassicaceae*			80
Family Linaceae			
Temperate species			
Linseed (flax)	*Linum usitatissimum*	Oil—industrial uses, stock feed, protein meal	5
Total, *Linaceae*			5
Family *Solanaceae*			
Tropical species			
Tobacco	*Nicotiana tabacum*	Tobacco products	5
Total, *Solanaceae*			5
TOTAL OF ALL CROPS			17066

[a]Taxonomy not resolved. *Triticosecale* commonly used for convenience in the literature.
[b]Comprises millets, sorghum, sorghum—sudan grass hybrids and rye harvested for hay.
[c]Includes millets, triticale, cow and poona peas grown for green forage.

TABLE 2.10 Area (million ha) of crops sown in Australia, 1977 to 1991 (year ending 31 March) excluding pastures, nurseries, fruit and vegetables
Source: Australian Bureau of Statistics, *Crops and Pastures Australia 1981–82; 1984–85: Crops Australia 1980–81: Summary of Crops Australia 1987–88; 1990–91.*

Year	1977	1979	1981	1983	1985	1987	1989	1991
Total area	14.8	17.2	18.1	19.1	20.9	19.5	17.2	17.1

TABLE 2.11   Generalised seasonal development patterns for winter and summer grain crops in Australia

	AUTUMN			WINTER			SPRING			SUMMER			AUTUMN		
	Mar	Apr	May	Jun	Jul	Aug	Sep	Oct	Nov	Dec	Jan	Feb	Mar	Apr	May
WINTER CROPS		PLANTING						HARVESTING			FALLOW			PLANTING	
		Establishment												Establishment	
			Vegetative Growth												
					Floral Initiation		Flowering								
								Grain Filling / Maturity							
SUMMER CROPS	HARVESTING						PLANTING						HARVESTING		
	Grain Filling						Establishment								
								Vegetative Growth				Flowering			
										Floral Initiation		Grain Filling			
												Maturity			

Winter Crops

The development of commercial production principles for winter (temperate) crops largely reflects strategies to capitalise on available moisture throughout the growth period, and to minimise both frost risk and exposure to high temperatures during the flowering and grain filling phases. All are sensitive to water stress throughout growth, with the establishment, flowering and grain filling stages critical to yield and grain quality. Frost sensitivity is limited during establishment and vegetative growth and probably has little yield effect, but all crops exhibit pronounced susceptibility during the flowering and early grain filling phases. The cereals are all determinate in flowering habit and even a single heavy frost during anthesis (flowering) can be catastrophic (Single, 1985). The indeterminate flowering habit of winter grain legume and oilseed crop species allows considerably greater yield tolerance to frost, as compensatory later flowering results in little yield penalty; as a result, strategies for avoiding frost are less critical than for cereals. All species are sensitive to high temperature during flowering and grain filling, the effects ranging from inhibition of grain filling due to induced water stress, accelerated senescence and inhibition of photosynthesis, to deleterious effects on grain composition, particularly in oilseeds (Lovett *et al.*, 1979; Armstrong *et al.*, 1985).

Production Strategies

The principal strategies for successful winter crop production based on climatic considerations are cultivar selection combined with planting time and rate. Other factors, particularly nitrogen availability with cereals and oilseeds, can also markedly affect crop performance and must be addressed as a prerequisite to successful production.

CULTIVAR SELECTION

All winter crop species produced in Australia, with the exception of faba beans and safflower, have a range of cultivars available with maturity ratings to cover most production environments. Commercially, selection is based on a combination of local experience and advice from government extension officers, private consultants, rural merchandisers and seed companies. Generally, cereal and grain legume producers retain their own seed from the previous year's crop or purchase from other producers, and tactical decisions to change cultivar sowing intentions are made on the basis of grower experience, current season crop performance, seed availability, and seasonal conditions in late summer and autumn of the new crop year. Delays in sowing caused by an unseasonally dry or wet autumn in most cases do not result in a change of cultivars, as seed purchases have already been finalised and the consequence is a yield penalty with late sowing. All species and cultivars exhibit accelerated maturity with late sowing due to a combination of increasing spring daylength and temperatures promoting rapid onset of flowering and thus a short vegetative phase. Individual plants thus exhibit limited branching/tillering, a low biomass at flowering and reduced yield potential.

PLANTING TIME

A number of factors determine the optimum planting time for winter crops, viz:

(i) *Cultivar maturity*

In any location in the temperate croplands of Australia, the optimum sowing date for any winter crop is determined by the combination of cultivar maturity, average date of last severe frost in spring, and the availability of adequate soil moisture for flowering and completion of grain filling. For determinate crops such as cereals, date of anthesis becomes the critical consideration (Fischer, 1979; Nix, 1971, 1975), planting strategies aiming at timing anthesis as soon as possible after the average date of last severe frost, to capitalise on favourable soil water availability and moderate temperatures for anthesis and grain filling in spring. In this situation, approximations of crop development can be made based on heat unit or radiation summations (or both), given average radiation and temperature data for the location, together with phenological data for the cultivar. A 'safe' anthesis date can thus be determined, and from that it is possible to extrapolate back to a 'safe' sowing date to achieve anthesis at the desired time. Such an approach has been used extensively by Nix (1971, 1975) in the analysis of crop adaptation and production systems for the temperate region. The approach assumes limited vernalisation or photoperiod response in the cultivars under examination, a characteristic of over 80 per cent of sown wheat cultivars in 1967 (Nix 1975) although a strong photoperiod response was observed in commercial hard spring wheat cultivars in northern New South Wales by McDonald *et al.* (1983).

Cultivar maturity rating is an important factor in selecting the sowing date, as early cultivars must be sown later than mid-season or late maturing ones to minimise frost risk at flowering. Indeterminate species, for which spring frost risk at flowering is less critical to yield prospects than for determinate ones, have greater flexibility in planting time. Again, however, cultivars should be sown according to maturity rating, with early cultivars sown last for a given situation. Topographic factors should also be considered, and frost-prone, low-lying areas subject to cold air drainage and accumulation should be sown last or to a late maturity cultivar.

(ii) *Soil water status*

For all crops, adequate soil water availability for germination and emergence is a precursor to successful crop establishment and ultimately yield performance. Simple approximations of available soil water reserves may be made using hand probes to establish the depth of wet soil in the profile. These probes have given useful indications of starting soil moisture and likely crop performance in wheat, particularly in summer rainfall regions of the northern New South Wales and southeastern Queensland wheatbelt (Fawcett *et al.*, 1976). The probes are cheap, commercially available and widely used as crop management tools. A more sophisticated assessment of soil water status may be obtained by direct measurement of volumetric soil moisture content using neutron probes (Cull, 1985, 1987), but their use is restricted largely to irrigated production systems. Computer prediction of safe sowing dates for winter crops in South Australia based on long-term rainfall records and the occurrence of specific autumn rainfall events (Huda *et al.*, 1992a) is emerging as a useful risk minimisation component of decision support systems for dryland crop production in that state. For example, statistics for Minnipa in South Australia show that if 10 mm of rain falls on any two consecutive days after April 1, sowing will be successful in 80 per cent of years if the crop is sown by the end of May, and in 95 per cent of years if it is sown by the end of June. Huda's analysis showed that early season rainfall was a useful predictor of subsequent seasonal conditions and could be used as a management support tool to both predict likely wheat yields and to vary the crop area sown depending on the level of risk accepted (Huda *et al.*, 1992a, 1992b; Taylor, 1992).

Other computer models such as WATBAL (Keig and McAlpine, 1974) can be used to generate simulated soil water balance for a crop period using rainfall and evaporation records, while the WHEATMAN model of Woodruff (1987) allows simulation of growth and yield performance in wheat under varying climatic conditions. These and other models are increasingly being used as extension tools to help farmers develop better risk management strategies in crop production systems and will in time contribute to the development of decision support systems for producers.

PLANTING RATE

Crop planting rates are generally well established for most production areas and planting guides are readily available from extension, retail and supply services in each state. Research over many years has shown that, if planting is delayed, the planting rate should be increased to compensate for reduced tillering or branching, smaller mature plant biomass and lower individual plant grain yield.

Summer Crops

Production principles for summer crops also focus on frost escape, water availability throughout the growth period, and minimisation of water and high-temperature stress during flowering and grain filling. As with winter crops, the principal production strategies to minimise climatic constraints on crop performance and yield are cultivar selection, planting time and, to a lesser extent, planting rate. The growing season for summer crops is generally delimited by temperature, the minimum soil temperature required for successful establishment of most being 15–16°C, a value attained in the earliest areas in Queensland during late September–early October. The end of the growing season is determined by the onset of autumn frosts and, provided grain has reached physiological maturity, subsequent frosts may actually assist in hastening grain drying and crop ripening for harvest. Some variation exists between species in their low-temperature tolerance; sunflower, for example, is a frost tolerant up to the eight leaf stage (Lovett *et al.*, 1979) while rice is much more low-temperature sensitive, having a minimum temperature tolerance of 13°C during the critical panicle initiation stage (Owen, 1971; McDonald, 1979) and a commercial production period considerably shorter than the frost-free period for the location. With the exception of soybeans, summer crop species are largely photoperiod insensitive, and the suitability of any environment for their production can be assessed by the length of the period of suitable temperatures for growth and development. The absolute length of the growing season is set by the frost-free period, but the actual length is usually considerably less than this due to the lag in reaching the minimum soil temperature for establishment. An analysis of southern New South Wales areas for their suitability for producing summer crops based on heat unit (degree day) accumulation showed little prospect for commercial production of any but the very earliest maize cultivars (Nix, 1971), even under irrigation. The production of summer crops for grain is thus, even under irrigation, generally restricted to areas north of the New South Wales–Victoria border.

CULTIVAR SELECTION
All commercially produced summer crop species have a range of cultivars which vary in maturity from early (typically 3 to 3.5 months) to late (5 to 6 months); and cultivar selection is based on average seasonal temperature patterns, expected water availability during critical stages of development, and other factors such as land preparation in multiple cropping systems. The availability of irrigation greatly increases flexibility in cultivar selection and in general allows the use of late maturing full-season cultivars with higher yield potential than shorter season ones. In these situations, the length of the growing season is defined by temperature conditions and cultivar selection is based on the timing of grain filling to avoid frost prior to physiological maturity. In temperate regions, the higher yield potential of full-season cultivars may be offset by harvesting difficulties due to late autumn – early winter rainfall, and earlier cultivars may be chosen to minimise this risk. Earlier cultivars may also be used in multiple cropping systems where the changeover between summer and winter crops constrains both planting and harvest times for each. Under dryland (rainfed) conditions, summer crops may often be sown as opportunity crops, particularly in the northern production areas of New South Wales and Queensland. In these areas, inadequate fallow-stored soil water for planting of winter crops may result in a decision to maintain the fallow through to late spring and sow a summer crop

instead. Cultivar selection for the summer crop then depends on soil water availability: if still inadequate in late spring – early summer (November–December), early to mid-season maturity cultivars would be needed for a successful crop finish before the onset of autumn frosts. In these dryland situations, cultivar selection reflects the availability of soil water and, if inadequate in early spring, the timing of effective planting rains and thus length of the potential growing season.

PLANTING TIME
The planting time for summer crops is determined by a combination of cultivar maturity, soil water availability, frost risk, soil temperature and ambient temperature.

(i) *Cultivar maturity*
Cultivar maturity is an important determinant of planting date as it establishes the length of the growing season required before low temperatures or frost limit or curtail growth. With full-season (late maturing) cultivars, planting must be done in the early part of the season to enable the crop to reach maturity before the onset of autumn frosts. Early maturing cultivars provide much greater flexibility in planting date and can be planted up until January if seasonal conditions dictate.

(ii) *Frost risk*
Most summer crops are frost sensitive at all growth stages, and in general terms the production period is delimited by the length of the frost-free period. Some species such as sunflower are tolerant of frost up to the eight leaf stage and can thus be planted 6 to 8 weeks before the average date of the last severe frost without yield penalty. Maize also shows some tolerance early in growth, provided the stem apex is not above ground level—leaf damage certainly occurs, but the apex appears insulated by the soil and pseudostem in the early stages of growth.

(iii) *Soil temperature*
With the exception of sunflower, which can germinate at soil temperatures as low as 8°C, all summer crops require minimum soil temperatures of 15–16°C for rapid germination and successful establishment. Planting may often be delayed well beyond the average date of the last severe frost due to the lag in soil temperature rise to acceptable levels. The planting of temperature-sensitive crops such as cotton is delayed until soil temperature reaches 16°C *and rising*—a brief period of hot weather may provide a 'false start' and a subsequent drop in soil temperature post-planting may result in patchy establishment and the necessity to replant. The standard for soil temperature measurement is the 9.00 am reading at 10 cm soil depth.

(iv) *Ambient temperature*
The length of the growing season for many summer crops is set by the length of the period of suitable temperatures for active growth, rather than the frost-free period. For the start of the season, this factor is

usually less important than soil temperature in determining the planting time, but can be a factor where cool weather follows a hot early season period and soil temperature allows planting; slow subsequent growth may result in patchy establishment and slow crop development.

(v) *Soil water status*

As with winter crops, starting soil water content is critical to successful crop establishment; under rainfed conditions, considerable delays in planting may result from inadequate soil water storage and erratic early season rainfall. Routine determinations of available soil water may be made with hand (Fawcett *et al.*, 1976) or neutron (Cull, 1992) probes and where significant delays occur, a change in cultivar may be necessary due to the shortened season. In many cases, a decision may be made to abandon the summer crop plan and retain the land under fallow for a following winter crop. Under irrigated conditions, pre-sowing irrigation is practised to establish sufficient water reserves for crop establishment. This practice is preferable to that of 'watering up' by post-sowing irrigation, particularly on crusting soils.

CROP DISTRIBUTION IN RELATION TO CLIMATE

Australian Croplands

The distribution of crop production (excluding pastures, fruit, vegetables and nurseries) is shown in Fig. 2.14. Four major regions dominate, namely:
1. The Atherton Tableland, coastal plains, southeastern downs and central highlands of Queensland.
2. The coastal plain, tablelands, slopes and the eastern margin of the Western Plains of New South Wales.
3. The coastal and sub-coastal plains and the Mallee country of western Victoria and southeastern South Australia.
4. The coastal and sub-coastal plains of southwestern Western Australia.

The great majority of production areas are located in the temperate region, based on the Growth Index patterns of Fitzpatrick and Nix (1975), reflecting the predominance of the temperate cereals wheat, oats, barley and triticale and the temperate grain legumes lupins and field peas. The inland margins of the Australian croplands reflect the overriding influence of rainfall and coincide with the arid inland margin for wheat production. Although the general production period for temperate crops coincides with the most favourable rainfall regime, a major constraint on growth and yield is imposed by low temperatures, with the most favourable rainfall and temperature periods being out of phase, resulting in a low Growth Index overall (Nix, 1975). This contributes to low overall crop yields when compared to those of other major producing countries.

FIG. 2.14 Australian croplands—distribution of commercial grain crop production in Australia, based on Australian Bureau of Statistics crop returns by local government areas

Cereal Crops

Winter Cereals

Winter cereals grown in Australia comprise wheat, oats, barley, triticale and cereal rye (ryecorn). Collectively they are temperate grasses, exhibiting varying degrees of vernalisation and long-day response for floral initiation. They are grown in a broad, crescent-shaped area extending from southeastern Queensland, through central New South Wales to northern Victoria and southern South Australia, with a further separate concentration in southwestern Western Australia (Figs 2.15 and 2.16). The southern production areas of Western Australia, South Australia, Victoria and southern New South Wales lie within the temperate moist and semiarid zones typified by uniformly distributed to winter dominant rainfall and a pronounced late spring and summer drought. The central New

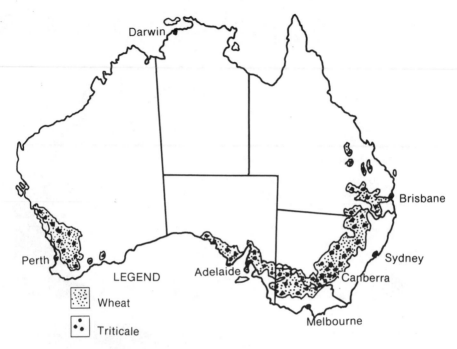

FIG. 2.15   Distribution of commercial wheat and triticale crop production in mainland Australia, based on Australian Bureau of Statistics crop returns by local government areas

South Wales sector experiences somewhat more uniformly distributed rainfall, although both annual receipts and reliability diminish rapidly with distance inland. The northern New South Wales and southeastern Queensland areas receive uniform to summer-dominant rainfall—a feature of these areas is the low and erratic nature of winter and early spring rainfall, making successful winter crop production heavily reliant on fallow storage of summer and autumn rainfall (Fawcett *et al.*, 1976; Holland *et al.*, 1987). In the winter-dominant rainfall areas, little grain yield advantage is achieved by fallowing in years of adequate crop season rainfall (Kohn *et al.*, 1966; Nix, 1975; Poole, 1987). As a result, fallowing has been discontinued in all southern areas except northern Victoria (Ridge, 1986) since the 1950s. In dry years, however, significant yield advantage results from greater water availability, as a result of fallowing, during grain filling (French, 1978). The role of fallowing in winter crop production in these southern areas has been reviewed by Ridge (1986).

With the exception of small areas of irrigated wheat and triticale for grain, and irrigated oats and barley for green forage, winter cereals are grown under dryland (rainfed) conditions.

Winter temperatures impose the second major constraint on growth and development within the winter cereal zone. In the northern sector warm days, which promote rapid vegetative growth, are combined with

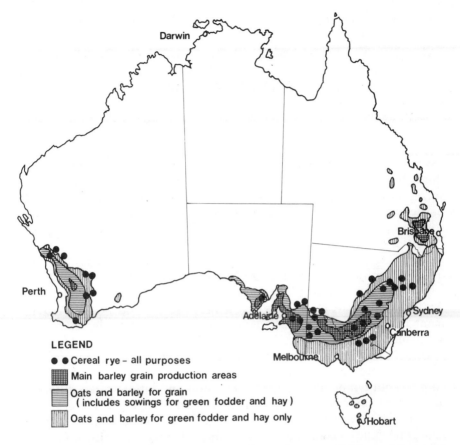

FIG. 2.16 Distribution of barley and oats for grain, green fodder and hay, and cereal rye for grain, green fodder, hay and land stabilisation in Australia, based on Australian Bureau of Statistics crop returns by local government areas

clear nights and relatively high frost incidence. In the southern, winter rainfall sector, days are cool to cold, while the relatively high incidence of cloud and greater air movement associated with the westerlies results in fewer frosts. In addition to greater frequency, frosts in the north are generally more severe than those in the south.

WHEAT

Australian commercial cultivars of wheat (*Triticum aestivum, T. durum*), particularly those grown in the northern sector, are predominantly short season spring types exhibiting limited photoperiod sensitivity. Early Australian wheats had a pronounced long-day requirement for floral initiation, but this has been greatly reduced through the use of Gabo and Mexican wheats of Gabo parentage in breeding programs. This has endowed both earlier maturity and a higher degree of adaptability on current cultivars, allowing them to be grown successfully over a wider latitudinal range than those with a strong photoperiod requirement. McDonald *et al.* (1983) showed photoperiod to be the most important factor governing pre-anthesis development in a range of spring cultivars grown in northern NSW. Their earlier maturity in particular makes them less prone to early summer drought damage during the critical phases of anthesis and grain filling.

Commercial release of modified winter cultivars with strong vernalisation but weak photoperiod requirement has enabled sowing with the autumn rainfall break, often four to six weeks earlier than the safe sowing time for spring cultivars. Their vernalisation requirement delays floral initiation and anthesis to much the same time as the later sown spring types (Martin, 1983). They are well suited to heavy clay soils, particularly in winter rainfall areas where late autumn – early winter sowing of spring cultivars is often delayed, if not prevented, by wet soil conditions. The high yield potential, flexible sowing time and grazing value of winter wheat, particularly for high rainfall areas, has been outlined by Johnston (1984) and Davidson *et al.* (1985).

In the colder regions of the southern wheat belt in Victoria and South Australia, commercial cultivars have a more pronounced long-day requirement and thus delayed anthesis, resulting in a greater degree of frost avoidance.

Under Australian conditions, wheat is sown from autumn to early winter, undergoing vegetative growth under conditions of short days and cool to cold temperatures. Floral initiation and inflorescence differentiation occur under conditions of increasing daylength, temperature and solar radiation during late winter – early spring, while heading, anthesis and grain filling occur under a combination of high temperatures and rapidly rising water stress during late spring and early summer (Evans *et al.*, 1975).

For the three development phases, sowing to floral initiation, floral initiation to anthesis and anthesis to maturity, growth and development during the first two is largely governed by the thermal and light regimes, particularly the thermal regime. A comparison of development rates for the photoperiod insensitive Gabo type under cold and mild winter conditions (Nix, 1975) revealed development times for the first two phases of 20 and 8 weeks respectively. In the final phase, anthesis to maturity, the moisture regime imposes the major constraint throughout the wheat belt, although the thermal regime can also impose a major constraint through incidence of post-heading spring frosts (Single, 1971; 1985). In the northern sector of the wheat belt, favourable thermal regimes, which stimulate rapid development through the first two phases, are typically associated with unfavourable moisture regimes in spring. This often results in severe water stress (Nix, 1975), which can impose serious limitations on grain yield and quality (Asana and Williams, 1965; Fischer, 1973; 1979); it is here that carryover soil-stored water from the previous summer fallow can be critical to successful production. Rapid development during the first two phases may also predispose the crop to severe frost damage. Hence sowing times must be determined by the need for anthesis to occur after significant frost risk has passed, yet early enough to allow adequate grain filling before moisture stress becomes severe. In contrast, development in the less favourable thermal regimes of the south is relatively slow, resulting in a delay of the final phase of grain filling into a period of both high temperatures and severe moisture stress during the late spring – early summer drought. In this environment, water availability for the final phase is largely dependent on soil water storage from crop-season rainfall (Nix, 1975; Ridge, 1986).

In addition to direct yield effects, low water availability during grain filling also markedly reduces grain quality. Reduced carbohydrate supply to the developing grain results in poor grain filling and small 'pinched' grain. Because of its small size and hence ability to pack densely, pinched grain has high weight per unit volume, while protein percentage is also high. The latter is a straight dilution effect, the grain protein component forming a higher percentage of total grain weight solely because of a deficit of carbohydrate.

Spring frosts impose a second major constraint during the final phase, severe frosts typically being associated with cool, still and dry conditions followed by a light southerly air stream, particularly in low-lying areas of the northern wheat belt (Single, 1971). Frost damage is most severe during the anthesis–grain filling period, after the developing ear has been extruded from the protective flag leaf base (boot), and if heading occurs during a period of high frost risk, extensive ear and floret damage can result (Single, 1971; Evans *et al.*, 1975; Nix, 1975). For photoperiod insensitive cultivars, frost avoidance during the critical post-heading phase is achieved by adjustment of sowing dates to delay heading until the frost danger is low. On the Liverpool Plains of the northern wheat belt, sowing is late because of high frost risk (Single, 1961, 1975) despite the fact that later sowings *per se* can result in significant yield reductions (Doyle and Marcellos, 1974; Fischer, 1979; McDonald *et al.*, 1982; 1983). Australian cultivars, while not as frost hardy as overseas ones, can develop 'cold hardiness' if subjected to low but not lethal temperatures from germination onwards (Single, 1971, 1975; Arnon, 1972) although 'cold hardiness' offers little protection once heading has occurred. Frost damage is most severe in northern New South Wales and Queensland, where mild winter temperatures may prevent development of cold hardiness, resulting in 'soft' growth and rapid development. Crops head during the spring frost danger period unless sown late to delay heading (Single, 1971; McDonald *et al.*, 1982; 1983). In contrast, winters in the southern zone of the wheat belt are uniformly cold, resulting in slow crop development, development of cold hardiness, and later ear emergence with lower frost risk. In these areas, early to mid-winter sowing into cold wet seedbeds can significantly delay germination and cause patchy establishment (Nix, 1975) while waterlogging may also cause serious problems. Early sowing of winter wheats in these areas is increasing steadily.

Although spring moisture supply is one of the most critical factors in successful production, the erratic nature of spring and early summer rainfall, particularly in the drier inland margins of the wheat belt and the summer rainfall zones of the northern sector, can frequently result in heavy falls during the final phase of development. Excessive rainfall in this phase can cause serious disease outbreaks, particularly of stemrust (*Puccinia graminis*), Septoria spot (*Septoria tritici*) and root rots (Moore, 1974), while serious yield and quality reductions can also be incurred through lodging, harvesting difficulties, delayed maturity, loss of grain weight through endosperm leaching, and pre-harvest sprouting (Barnes, 1973; Moore, 1974; Nix, 1975; Mares, 1984, 1985). Wet, humid conditions during grain ripening can cause serious grain quality decline by initiating α-amylase activity when sprouting occurs. Such conditions frequently occur in the northern wheat belt.

Sprouting accounts for annual losses as high as 35 per cent, the average annual loss being about 20 per cent for northern New South Wales and Queensland (Mares, 1984, 1985), with a much lower proportion in southern areas where the frequency of wet spring conditions is considerably less. Losses result from reduced crop value and in addition create substantial storage and marketing problems.

Sprouting resistance in white hard wheat for northern areas is a major breeding objective (Marshall, 1984). A number of resistant or tolerant lines have been identified through screening some 3500 lines, and subsequently used in breeding programs (Mares, 1984, 1985; Marshall, 1984). Derera (1973) promoted the introduction of sprouting-resistant hard red wheats to northern areas, but the probability of developing resistant white cultivars from sources identified in screening tests has obviated Derera's arguments for red wheats (Marshall, 1984). Two tolerant white cultivars (Suneca and Sunelg) have been released and further development of resistant lines is continuing (Mares, 1985).

In summary, wheat production in Australia is carried out successfully over a wide range of moisture, thermal and light regimes, much of the success being attributable to extrapolation of cultivar requirements and management strategies to environmental conditions. Apart from absolute limits to the timing and duration of the crop cycle set by seasonal water availability, the major constraints on Australian wheat crop systems were summarised by Nix (1975) as:

–the timing of sowing rains;

–the duration of the mid-winter depression in temperature and solar radiation;

–the timing of earliest safe ear emergence date as set by frost occurrence; and

–the rapid increase in temperature and evaporation rate during spring and early summer.

For a thorough analysis of the agro-climatic relationships of wheat production in Australia, refer to Nix (1975).

OATS

Oats are produced on a wider geographical scale than wheat, being grown as a dual purpose grazing and grain crop throughout the wheat belt. Major grain production areas are concentrated in the southern sector (Fig. 2.16), while their use for forage extends beyond both the drier inland and wetter coastward margins of the wheat belt. High rainfall (up to 1200 mm) and disease precludes their use as a grain crop in the coastal margins, where they are used primarily for forage and hay. In the dry inland areas, where pasture establishment carries high risk, oats are grown essentially as a forage crop and, in good seasons, an opportunity grain crop (Walkden Brown and Fitzsimmons, 1972). They are the most widely grown crop for hay and green forage, and are the most important New South Wales dual grazing/grain crop (Simmons, 1987). The main areas for grain production are concentrated in the winter-dominant and uniformly distributed rainfall areas of the wheat belt in Western Australia, South Australia, Victoria and southwestern and central New South Wales (Fig. 2.16).

Australian cultivars are primarily spring types with little or no vernalisation requirement, but with varying degrees of photoperiod sensitivity and a range of maturity types from very early to very late. When grown for grain only, sowing times are governed by maturity rating and spring frost risk, tempered by the timing of sowing rains, with most grain crops being sown in early to late winter. Under these conditions oats, with a similar but more rapid development pattern (and subject to the same constraints as wheat), undergo grain filling under less severe conditions of late spring water stress, although the crop is predisposed to a greater risk of post-heading frost damage (Walkden Brown, 1975). Oats are frost sensitive during both the early vegetative and post-heading phases, damage during the early vegetative stages being accentuated by grazing injury. Extension of oat grain production into the warmer regions of the northern wheat belt is precluded largely by its lower tolerance to high spring temperatures than wheat and the risk of crown rust outbreaks, both of which, when combined with premium prices for prime hard wheat from this zone, make it unattractive as a commercial crop. However, its greater cold tolerance allows it to be successfully grown for grain in the higher tableland margins of the northern New South Wales wheat belt where wheat production is limited (Walkden Brown, 1975).

BARLEY

Like oats, barley (*Hordeum vulgare*) is grown extensively throughout the wheat belt both as a grain crop and for green forage and hay. Intensive grain production is limited to restricted areas in southwestern Western Australia, the Eyre Peninsula and southeastern region of South Australia, northern Victoria and southwestern New South Wales (the latter centred on the Riverina and southwest slopes), and the Darling Downs in Queensland (Fig. 2.16). Barley as a grain crop was of minor importance in most states, with the exception of South Australia, until the imposition of short-lived wheat quotas in 1969, but has since expanded to be second only to wheat (Table 2.9). In addition to the main grain producing areas, barley is grown extensively as a green forage crop over a similar geographic spread to oats, although the area sown for both forage and hay is considerably less than that of oats. One of the most outstanding features of barley is its early maturity, the crop being able to mature in a shorter time than other winter cereals (French and Shultz, 1982; Cook, 1987), enabling it to be grown successfully for grain in areas where the season is cut short by low water – high temperature stress in spring. The fact that it is early maturing has led to it being regarded in many countries as drought resistant (Martin *et al.*, 1976), but this is really a reflection of drought escape rather than resistance *per se* (Klages, 1942).

The growth cycle of barley is very similar to that of wheat, and the same general considerations apply. In Australia, moisture supply imposes the major restriction on grain yield (Doolette, 1968), the yield potential of crops being generally correlated with water supply during the growing period. Even short periods of moisture stress, particularly during the post-heading phase in spring, can have serious effects on yield and grain quality (Sparrow and Doolette, 1975). Premium-quality barley for the malting trade characteristically exhibits large, plump grains with a high carbohydrate content and relatively low nitrogen content (Cook, 1987). As starch deposition in the grain largely occurs after most of the grain protein has been formed, water stress in the latter phases of grain filling can inhibit starch deposition, resulting in smaller grain of relatively high protein percentage, which would be rejected by the malting trade (Sparrow and Doolette, 1975). Producing high-quality malting grain requires both high humidity and mild temperatures, with adequate soil moisture during the post-heading phase, resulting in a long period of grain fill (Gilmour *et al.*, 1992). These conditions are typical of early spring in the southern grain-producing areas (Anon., 1975a). Sowing time is critical, late sowing reducing grain quality and suitability for malting (Wheeler and Nitschke, 1985). Malting barley in Australia is derived from two-row cultivars, while 'feed' barley is largely from six-row types. Grain quality standards for feed barley are quite different from those of the malting trade, higher protein levels being desirable. Nevertheless, grain quality standards for the export trade as feed grains are stringent, particularly for the Middle East market (Shawyer, 1984). Six-row barley can be grown successfully throughout the wheat belt, and is adaptable to a wider range of soil and climatic conditions than two-row cultivars, producing more dependable yields than either oats or wheat on low fertility soils, and producing profitable crops in areas either too dry or too hot over the ripening period for malting barley (Anon., 1975a). It does not, however, withstand wet conditions and is much inferior to wheat, oats and triticale in wet situations (Cook, 1987). Barley has proved a very useful alternative to wheat in much of the southern wheat belt, particularly in seasons with a delayed autumn break in rainfall necessitating late sowing. Under these conditions, its early maturity makes it a safer proposition than wheat, particularly in the spring-drought prone southern areas of the wheat belt.

CEREAL RYE (RYECORN)
The most winter-hardy and low-fertility tolerant of all the winter cereals (Klages, 1942), cereal rye (*Secale cereale*) is a multi-purpose crop grown throughout the drier and colder margins of the wheat belt (Fig. 2.16). Main crop uses by state vary quite significantly as follows (Lovett, 1975; Quinlan, 1984):

New South Wales —tablelands for forage and grain
—coast for green forage
South Australia and—primarily for sand dune stabilisation and land
Victoria   reclamation, forage and grain
Tasmania —green forage
Western Australia —erosion control and green forage
Queensland —grain (very small area)

Compared to the other winter cereals, cereal rye is a crop of relatively minor importance in Australia, the area of greatest concentration for grain being located in southeastern South Australia (Fig. 2.16), where some 70 per cent of the Australian crop is produced (Quinlan, 1984).

The most cold tolerant of all the winter cereals, cereal rye is produced in Europe in areas extending beyond the Arctic Circle, replacing wheat in areas of intense winter cold (Klages, 1942). In Australia, its cold tolerance has enabled its use as a substitute for oats for green forage in elevated tableland districts of the wheat belt, growing actively under conditions too cold for active growth in oats. Despite its extreme cold tolerance during vegetative growth, rye is quite sensitive to post-heading frost damage, and is usually grazed out in these elevated regions (Fitzsimmons, 1984).

In addition to cold tolerance, cereal rye is capable of withstanding hot dry conditions, its extensive root system imparting drought tolerance far in excess of that tolerated by wheat or oats (Klages, 1942). This drought resistance, in combination with the excellent soil-binding properties of its root system and its resistance to sand blast, make it suitable for vigorous growth on dry sand dunes where wheat, oats and barley would barely survive (Lovett, 1975; Fitzsimmons, 1984), and accounts for its widespread acceptance as a primary coloniser of sand drifts and dunes in the dry inland margins of the Victorian, South Australian and Western Australian wheat lands.

When grown for grain, cereal rye is extremely sensitive to hot dry conditions at anthesis. Unlike other winter cereals, it is largely wind pollinated, and a combination of high temperature and low humidity at anthesis proves lethal to pollen, thereby severely limiting potential grain set (Fitzsimmons, 1984). In the main grain-growing areas of South Australia, reliable winter rainfall combined with the early maturity of cereal rye enables it to escape the hot dry conditions typical of late spring.

There has been renewed interest in cereal rye as a green fodder and grain crop in acid soil areas of New South Wales with the release of Rysun, a cultivar with resistance to stem rust (Verrell and Gammie, 1992). However, the total area of cereal rye sown is unlikely to increase significantly in future, with production limited to existing areas. The advent of triticale may further reduce the importance of rye as a commercial crop (Fitzsimmons, 1984).

TRITICALE

Triticale is an artificial hybrid between wheat and cereal rye with most of the genetic material coming from the wheat parent (Reeves, 1980). It is a relatively new crop to Australia, being first produced commercially in 1976 with a single cultivar, Growquick (May, 1981). Yields were disappointing, being significantly inferior to wheat (Cook, 1983). Subsequent release of new cultivars in New South Wales, Victoria and South Australia led to substantial interest in the crop particularly on acid soils with high levels of available aluminium, where it out-yielded wheat significantly (Reeves, 1980). The crop had been promoted for grazing and grain, but forage yields have been disappointing and inferior to oats and barley except in harsh conditions of low winter temperatures and acid soils. Under these conditions, however, triticale has been inferior to ryecorn, and its use for forage is limited. All cultivars available are grain types (Reeves, 1980; Cook, 1983) and its future as a forage crop will depend on development of forage types (Duncan, 1983). Triticale is grown throughout the wheat belt (Fig. 2.15) and the annual sown area is increasing steadily (Table 2.10). Its future lies in its further expansion in acid soil areas with the development of better grain cultivars in all states. Its yield advantage over wheat on acid soils with high aluminium and manganese availability may, however, be limited with the future release of wheat cultivars tolerant to these conditions (Fisher and Scott, 1983). On average wheat soils, current wheat cultivars outyield the best triticales, and this, together with the price advantage of wheat over triticale, does not justify its substitution for wheat. Any future substitution on the average and better wheat soils will depend largely on yield and price considerations in the stock feed grain market. Concern about a growth inhibiting factor (King, 1980) which becomes manifest with high inclusion of triticale in animal diets has caused some reservation about its value as a stock feed grain, although conflicting results on its nutritive value have been reported (Castleman, 1982).

Under irrigation, triticale out-yields wheat by 10 to 15 per cent (Cook, 1983; Verrell and Gammie, 1992), and its use as a substitute for wheat as an irrigated disease-break crop in cotton rotations is increasing. In these situations the amount of irrigation water available is frequently limited and timing of its application is critical to grain yield, with maximum response when water is applied after rather than before anthesis. Further, under conditions of continuous low water availability, triticale yields significantly less than wheat (Sutton and Dubbelde, 1980). The latter condition is a feature of the inland margins of the wheat belt, thus limiting expansion of triticale into these areas.

Summer Cereals

Summer cereals grown in Australia encompass maize, sorghum, rice and millet. Production of these crops on a major scale is restricted almost entirely within the uniform to summer-dominant rainfall zones of north-eastern New South Wales and Queensland, with further small concentrations in the semi-arid regions of inland New South Wales (Fig. 2.17). In these latter areas, production is almost entirely under irrigation, the low and erratic summer rainfall precluding reliable dryland production. Although less important in a national sense than the winter cereals, they have significant regional importance. All four are annual summer-growing grasses of tropical origin and can generally be regarded as short-day plants, although photoperiod response varies among cultivated types of each species from photoperiod insensitivity to moderate short-day sensitivity.

MAIZE

Maize exhibits a great diversity of types, enabling its adaptation to a wide range of climatic conditions (Duncan, 1975), although major areas of production are typified by relatively high summer temperatures with cool to warm nights (Wilsie, 1962). Production in Australia is limited by high temperatures combined with low relative humidity rather than the length of the growing season. The crop is therefore confined in Australia to the eastern coastal and tableland areas of New South Wales and Queensland, and to irrigated inland areas (Colless, 1979; Crosthwaite, 1983) (Fig. 2.17).

The rate of development of maize from planting to anthesis is governed almost entirely by temperature experienced by the growing point rather than by total photosynthesis. As the growing point is beneath or near the soil surface for more than half the vegetative growth phase, development during the early stages of emergence, stem and leaf growth is governed largely by spring soil temperatures (Duncan, 1975). Maximum yield potential is attained where hot days of 30 to 33°C are combined with cool nights, resulting in a moderate rate of development and thus a longer growth period and large plant size. In contrast, where hot days are combined with warm to hot nights, plants develop rapidly, culminating in small plant size and early maturity with resultant low yield potential (Duncan, 1975). Where high temperatures (up to 38°C) are combined with low relative humidity during the critical tasselling and silking (anthesis) phase, seed set and yield can be substantially reduced by pollen

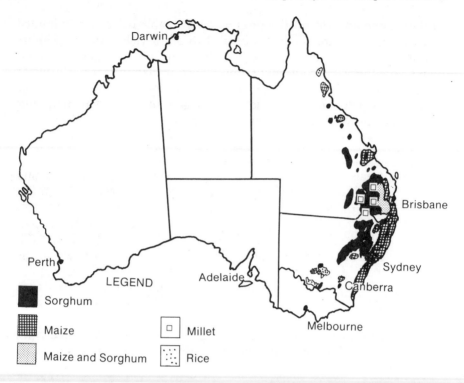

FIG. 2.17   Distribution of the summer cereals sorghum, maize, millet and rice
for grain, based on Australian Bureau of Statistics crop returns by local
government areas with greater than 200 and 400 ha for maize and grain sorghum
respectively but no minimum area limitation on rice and millet

and silk desiccation (blast) (Colless, 1982). As a consequence, sowing
strategies aim at avoidance of anthesis from late December to mid-
February, when these conditions are most likely to occur in the main
production areas. This can be achieved through either early sowing (soil
temperatures permitting) of early to mid-season hybrids which undergo
anthesis before mid-December, or later sowing of early to mid-season
hybrids which will undergo anthesis after mid-February, yet still mature
before autumn frosts. Full-season hybrids are sown in spring as early as
soil temperatures permit (Colless, 1979).

The length of the period from anthesis to harvest maturity varies
considerably among hybrids, with considerable variation between genotypes in the length of the period from anthesis to physiological maturity,
and thence from physiological maturity to harvest maturity (Colless,
1979). The final phase involves essentially grain drying, and its length
depends on temperature and humidity. Low temperatures combined with
moderate to high humidity can often significantly delay grain drying in
autumn, with the result that harvest is delayed into late autumn – early
winter, when wet soils may seriously hamper harvesting operations,

particularly in inland areas (Dale and Colless, 1990). While grain drying is weather dependent, there is also considerable genetic variation in the rate of grain dry-down under the same conditions (Purdy and Crane, 1967).

Maize production in Australia until the 1950s utilised tall, locally adapted open-pollinated cultivars with moderate yields but high yield stability. Introduction of the first hybrid from the U.S. in 1947 and subsequent establishment of both public and private hybrid breeding programs in New South Wales and Queensland led to the rapid transition to hybrids in all production areas, with little open-pollinated maize grown today. Two distinct hybrid types were developed, namely

  (i) tall, late maturing (24 to 28 weeks) types resistant to a range of diseases and insect pests endemic to the humid coastal production zone. These were developed from the adapted open-pollinated cultivars on which production was originally based;

  (ii) short, high yield potential, early maturing (20 weeks) hybrids of U.S. origin or developed in Australia from U.S. inbred parents. These are grown in the inland areas of New South Wales and Queensland where their early maturity matches the frost-free growing season. Production of these types in coastal regions is limited by their susceptibility to endemic diseases and pests (McWhirter, 1972; Colless, 1979; Henzell *et al.*, 1985).

The subsequent commercial release of single cross hybrids that combine the high yield potential of U.S. inbreds with the disease resistance of Australian inbreds (Colless, 1982; Henzell *et al.*, 1985) has resulted in a substantial swing to these types in the coastal regions in response to their excellent yield performance.

Production in coastal regions is predominantly rainfed, with some supplementary irrigation, while that in inland areas is almost entirely irrigated. Maize is the most widely grown fodder crop in the world (Pritchard and Moran, 1987) and its production as a silage crop using full-season hybrids under irrigation has become widely established in New South Wales, Queensland and northern Victoria, particularly for the dairy industry.

SORGHUM

Sorghum is a crop of tropical origin which has long been renowned for its tolerance of high temperature and water stress (Klages, 1942; Wilsie, 1962; Doggett, 1970) although the bulk of the crop is now produced in temperate regions from temperate-adapted hybrids developed by intensive breeding and selection programs (Anderson, 1979). Temperate adaptation has resulted in a significant loss of photoperiod (short day) sensitivity and resultant developmental response to temperature, high temperatures within the range 20° to 35°C leading to a reduction in length of each development phase and thus total growing season.

Commercial grain sorghum hybrids are sensitive to temperature extremes, low soil temperatures (<16°C) at sowing inhibiting germination and emergence (Anderson, 1979) while high temperatures (>38°C) at panicle emergence and anthesis, particularly when combined with low atmospheric humidity, result in reduced pollen viability (Pasternak and Wilson, 1969) and a substantial reduction in seed set and potential yield. Some degree of acclimatisation occurs in sorghum in continuously high temperatures such that the effects of short-term heat waves, which have severe effects on plants grown in temperate regions, are not so pronounced (Anderson, 1979). Considerable research into the eco-physiological adaptation of sorghum has been conducted in Australia, and an excellent overview is provided by Wilson and Eastin (1982).

Sorghum is more tolerant of water stress than maize, and in Australia is produced largely under rainfed conditions where water stress frequently occurs. Under these conditions sorghum develops an extensive root system which, depending on soil physical properties, may extract water down to a depth of one metre (Doggett, 1970). It has both sensitive stomatal control which enables effective internal control of transpiration (Slatyer, 1955; Glover, 1959) and high photosynthetic efficiency (Downes, 1970), and these, combined with its extensive root system, enable it to successfully exploit limited moisture under the rainfed conditions in which it is commercially produced in Australia (Anderson, 1979). As a consequence of its greater adaptation to high temperatures and particularly water limitations, sorghum is commercially produced on the hotter, drier inland margins of the main maize production areas, and additionally is produced dryland in the summer-dominant rainfall inland areas, where maize is produced only under irrigation. Dryland production is conducted in the semi-arid zones of northern New South Wales and southeastern Queensland (Fig. 2.17). Sorghum is also produced, although on a limited scale, under irrigation in inland areas, often as a disease break crop in cotton rotations.

It is frequently grown as an alternative opportunity crop to wheat in northern areas where seasonal conditions significantly delay wheat sowing. It is also grown as a summer rotation crop to control major weeds of wheat, notably wild oats. In recent years it has become the most important summer crop component of strip cropping systems which are rapidly substituting for broadacre production of wheat and other crops in erosion-prone summer-dominant rainfall areas.

Despite the high yield potential of U.S. hybrids, which were first introduced in the early 1960s and resulted in a rapid increase in sown area, particularly in Queensland, average yields in Australia, at less than 2 tonnes per ha, are erratic and low. This has been attributed to expansion into previously unsuitable marginal areas with low yield potential (Anderson, 1979; Henzell *et al.*, 1985) and to low moisture availability.

Within the extensive maturity range of hybrids available in Australia, early maturing lines have proven most successful, probably because of limitations on the length of growing season imposed by water, temperature, disease and insect pests (Anderson, 1979).

Besides its production as a grain crop, forage sorghums and sorghum-sudan grass hybrids (e.g. Sudax) are an extremely important source of forage and are widely grown beyond the grain production areas for forage, silage and hay, particularly in the coastal areas of New South Wales and Queensland.

MILLETS

Millets are grown on a fairly small scale in Australia for the production of grain and forage. The term 'millets' embraces a range of species (Table 2.9) which feature considerable tolerance to heat and water stress when compared to other cereals (Klages, 1942; Wilsie, 1962). Pearl millet (syn. bulrush millet) in particular produces good grain yield in regions too hot and dry for other crops (Burton *et al.*, 1972). Australian production for grain in a range of species (Table 2.9) is essentially for the birdseed market, and is unlikely to expand significantly beyond current production areas (Fig. 2.17).

Forage species, particularly Japanese millet and Shirohie, a selection from Japanese, have been grown as an early spring-sown crop in coastal districts for many years. Acceptance of pearl millet as a forage crop was limited until the recent release of semi-dwarf types which combine high productivity and easy grazing management. The latter are beginning to make significant inroads as an alternative to the highly productive but prussic acid containing forage sorghums and sorghum-sudan grass hybrids through the main New South Wales and Queensland crop lands.

RICE

Essentially a tropical crop, rice (*Oryza sativa*) is grown over a wide latitudinal range throughout the world, ranging from 49°N to 35°S, the latter extreme in southern New South Wales (Arnon, 1972; McDonald, 1978). Rice in Australia is produced in two distinct zones, the erratic rainfall semi-arid Murrumbidgee and Murray river valleys of southern New South Wales and the tropical wet summer Burdekin River delta in northeastern Queensland (Figure 2.17). Attempts to establish a commercial rice industry at other sites in the tropical and semi-arid region of northern Australia, notably on the coastal plains east of Darwin and the Fitzroy and Ord river systems of northwestern Western Australia, have failed for a number of reasons, largely agronomic.

Commercial production in the two discrete zones of New South Wales and Queensland is entirely under irrigation, utilising high yield potential japonica types well adapted to intensive cultivation (Arnon, 1972; Boerema, 1973; McDonald, 1979).

Highest rice grain yields are typically obtained in temperate zones, where dry summers with little cloud cover and resultant high solar radiation levels associated with lower temperatures during ripening occur (Arnon, 1972). Record and average yields attained in southern New South Wales are considerably greater than those of the tropics, where summer rainfall and associated heavy cloud cover reduce solar radiation

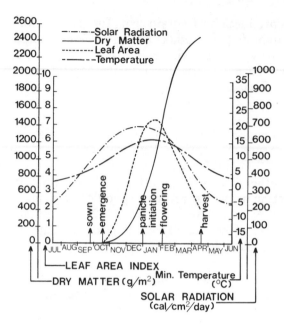

FIG. 2.18   The growth of rice in relation to minimum temperature and global solar radiation in the Murrumbidgee Irrigation Area (after Boerema, 1973)

receipts (Boerema, 1969; McDonald, 1979). Although the high radiation levels attained are conducive to high yields, the major control of growth and development and ultimately grain yield in southern New South Wales is the length of the growing season, delimited by low spring and autumn temperatures (Boerema, 1973). This problem is common to other temperate zone rice production areas of the world (Owen, 1971; McDonald, 1979). An excellent summary of growth and development in relation to solar radiation and minimum temperature in the Murrumbidgee Irrigation Area (Boerema, 1973) is shown in Fig. 2.18.

   Rice requires relatively high (25° to 30°C) temperatures for growth and development, although temperature requirements vary between growth stages (Owen, 1971). Low spring temperatures common to southern New South Wales are conducive to slow germination and distorted seedling growth (Boerema, 1973), although excessively high temperatures (>30°C) at this time can result in poor root penetration and anchorage (Owen, 1971). Perhaps the most important temperature effect on grain yield is the occurrence of low temperatures during the panicle initiation, panicle differentiation and flowering phases (McDonald, 1972) with a critical minimum temperature of 13°C (McDonald, 1972; Owen, 1971). Minimum night temperatures of 15°C or less, particularly if extended over 2 to 3 days during the panicle initiation to flowering period (December to early February in southern New South Wales) result in reduced panicle branching and excessive floret sterility (flatheads), the type and extent of damage depending on the actual timing of the low temperatures (McDonald, 1972). Low temperatures during grain ripening are conducive to high grain yields through extension of the grain filling period (Boerema, 1973), although evidence on this point is conflicting (Owen, 1971).

The average length of the growing season, as defined by low temperature, in southern New South Wales is 180 to 200 days, although that in the southern region of the Murray Valley is even shorter (Boerema, 1969). In these areas, management strategies (in particular aerial seeding, which allows more rapid establishment and 10 to 14 days earlier maturity) have been used to shorten the crop growing season. Intensive plant breeding efforts are being devoted to the development of very early (130 to 140 days) maturity cultivars, which will prove a safer proposition in the Murray Valley, and the entire southern rice growing area where the season is further limited by erratic spring and autumn rainfall. Although classified as a uniform rainfall zone the erratic nature of rainfall has frequently resulted in significant delays to ground preparation and sowing in spring, while autumn rainfall also imposes significant constraints on machinery access for harvest. Consequently, any factors conducive to a shorter duration of the crop cycle without detriment to yield will impart greater flexibility and allow more reliable production.

Oilseed Crops

Winter Oilseeds

Winter-growing oilseed crops produced in Australia are canola (rapeseed), linseed and safflower. Each is valued as a rotation crop in cereal production systems, as a disease-break (buffer) crop and as a means of significant improvement in soil physical conditions. All, particularly safflower, have deep and vigorous tap-roots which break up hard soil layers and improve both subsequent crop performance and soil structure. Their distribution in Australia is largely confined to the temperate moist and semi-arid zones of uniform to winter-dominant rainfall (Fig. 2.19). The area sown to each varies markedly from year to year due to fluctuations in climatic and market conditions, pests and diseases, and the outlook for wheat, the major competitor for land and resources. Winter oilseed growing on a large scale followed the short-lived quota on wheat production imposed in 1969. The removal of the quota, together with poor yield performance through limitations imposed by climate, disease, pests and lack of grower expertise, led to marked fluctuations in area sown since then (Fig. 2.20).

CANOLA (RAPESEED)

Rapeseed, now known as canola, has been produced commercially on a significant scale only since 1971 (Fig. 2.20), as an alternative to wheat

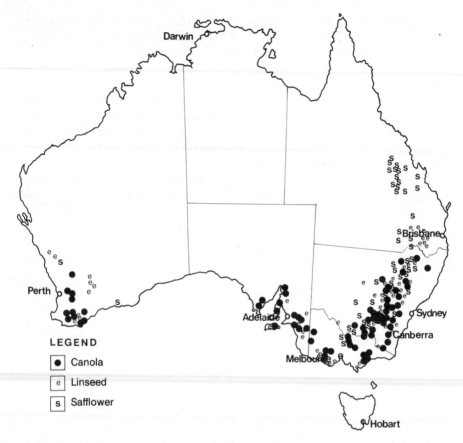

FIG. 2.19 Distribution of the winter-growing oilseeds safflower, linseed and canola in Australia, based on Australian Bureau of Statistics crop returns by local government areas

following the 1969 introduction of wheat production quotas. Area initially expanded rapidly, but fluctuated markedly from year to year as a result of poor grower experience, removal of wheat quotas, seasonal conditions and the impact of blackleg disease, which decimated Western Australian crops in the mid-1970s. Production since then has stabilised in the better rainfall temperate moist regions of the New South Wales, Victorian, South Australian and Western Australian wheat belt. In New South Wales it has extended into the higher rainfall areas of the Tablelands and, more recently, into the northwest Plains and the northern coastal regions (Colton, 1985). The recent release of suitable hybrids has also extended production into the better rainfall wheat districts of southeast Queensland (Chudleigh, 1991). Although produced almost entirely as a rainfed crop to date, irrigated production is increasing steadily along the western river systems of New South Wales and in flood years, on western lakebeds and river flood plains as flood waters recede (Colton and Jones, 1985).

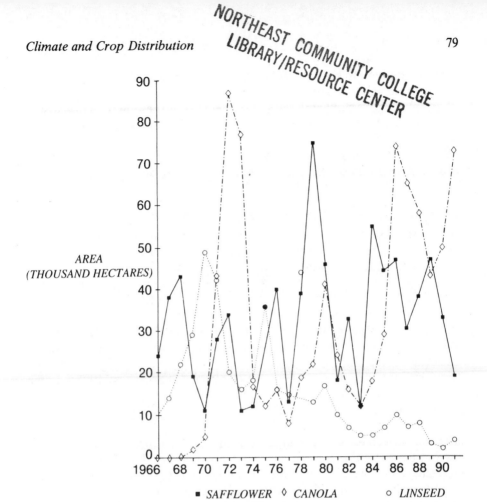

FIG. 2.20 Area of winter oilseeds (thousand ha) grown in Australia 1966–91
Source: Australian Bureau of Statistics crop returns by local government areas

Distribution in the southern areas of Western Australia, South Australia, Victoria and southern New South Wales has been markedly affected by the incidence of blackleg and, despite the release of cultivars with some resistance through the 1980s, the disease remains a major factor limiting production. Canola production in Western Australia, the largest producing state in the mid to late 1970s, is now down to almost negligible levels. In addition to blackleg resistance, breeding programs have focused on developing cultivars with low levels of erucic acid and glucosinolates ('double low' lines) which meet the international double low standards of canola. The result has been the release of cultivars with high seed and oil yields, canola seed and oil quality, and resistance to blackleg (Wratten *et al.*, 1987), with the first cultivars commercially released in New South Wales in 1988. Hybrids have also been released by commercial seed companies, but their use has been restricted to northern New South Wales and southern Queensland where blackleg incidence has been negligible (Sykes, 1989). New blackleg resistant hybrids in New South Wales should enable substantial expansion into southern areas where blackleg is now endemic—the yield from these new hybrids is at least 20 per cent higher than that from existing open-pollinated cultivars (Chudleigh, 1991).

Canola has a large water demand in comparison with winter cereals, and both seed and oil yield are severely depressed by water stress during flowering (Buzza, 1979; Colton, 1985; Sykes, 1992), a major factor in poor yield performance in drier areas of the wheat belt where canola was widely grown in the 1970s. It is also frost sensitive to a minor extent during the rosette stage, although damage is generally negligible. Frost can cause severe yield reductions if early cultivars are planted too early, or if mild winter conditions promote rapid development, resulting in flowering throughout the frost danger period in spring (Colton, 1985; Sykes, 1989; 1992). In general terms, early planting (in low rainfall areas, as early as late March), given normal seasonal conditions and crop development, results in higher yields (Sykes, 1989). Normal planting times for rainfed crops in the main wheatbelt areas are mid-April to mid-May. If late autumn breaks occur, planting can be delayed until late June without serious yield penalty (Colton, 1985; Sykes, 1992). Later plantings suffer a yield penalty of 10 per cent per week (Sykes, 1992). High temperatures during the flowering and grain filling periods result in significant yield reductions. During grain filling, oil quality is also adversely affected by high temperatures (Canvin, 1965; Buzza, 1979).

Commercial cultivars are derived from two species, *Brassica napus* (Argentine or Swedish rape) and *B. campestris* (Polish or turnip rape). Within each, both annual and biennial *and* spring and winter types occur, the latter with a pronounced vernalisation requirement, necessitating early planting to ensure floral initiation (Buzza, 1979). Commercial Australian cultivars are all annual, spring types, and *B. napus* cultivars account for the great majority of plantings. *B. campestris* cultivars (80 to 86 days) mature earlier than *B. napus* ones (100 to 110 days) (Matheson, 1976), and are useful in areas with unreliable spring rainfall or where planting has been delayed significantly by seasonal or other conditions. Recent cultivar releases are almost all of the higher-yielding *B. napus* type, and this species and hybrids will continue to dominate the industry in future years. Higher yielding early lines of *B. napus* are available which offer the same advantages of spring drought escape, and this factor is being incorporated into breeding programs (Armstrong *et al.*, 1985; Mendham and Russell, 1982).

LINSEED

Linseed is the seed of the flax plant (*Linum usitatissimum*) and is produced for its oil, which is used for industrial purposes as its high (45—60 per cent) linolenic acid content makes it unsuitable for human consumption. Recent research at CSIRO has utilised induced mutation to reduce linolenic acid to only 1 to 2 per cent, with an accompanying rise in linoleic acid from 20 to 60–70 per cent, in the process creating a new crop, Linola, with a high-quality polyunsaturated edible oil similar to sunflower (Green 1992).

Linseed has been produced commercially in Australia since 1947. Before then the crop was grown on a fairly large scale for flax between 1938 and 1945, the total area reaching over 25 000 ha in 1945. Production for flax utilised tall cultivars with little branching and seed production, and continued on a small scale, in parallel with that for linseed, until 1964. Production for linseed began in the 500 to 700 mm rainfall zones of the northern New South Wales and Queensland wheatbelt, and quickly expanded to over 20 000 ha. Subsequently the principal area of production shifted south into the central western and southern Slopes of New South Wales, the western district of Victoria, and Western Australia (Matheson, 1976; Sykes and Green, 1988). Area and production fluctuated widely from year to year (Fig. 2.20) with varying demand for linseed oil, and since 1975 worldwide demand has fallen markedly through competition from synthetic substitutes. The crop had a resurgence in the early 1970s with the onset of wheat quotas, but has declined to negligible levels since 1983.

Linseed can be planted from late autumn to early spring, depending on cultivar and seasonal incidence of frost and rainfall. The plant is particularly sensitive to frost during flowering and seed setting, and planting times have to be adjusted to take this into account. Higher temperatures result in a growing period for spring-planted crops of 120 days, compared to late autumn plantings which take 210 days but which result in larger plants and higher yield potential (Sykes and Green, 1988). Flowering and capsule maturation occur over a protracted period, and if soil moisture levels are adequate the crop may undergo a renewed period of flowering, leading to uneven maturity but potentially high yields. The lack of an extensive root system predisposes linseed to water stress, although it can withstand relatively dry conditions during vegetative growth (Matheson, 1976). Water stress during flowering, seed set and seed filling can result in severe yield reduction, so the crop does not perform well in areas where spring rainfall is low or unreliable (Sykes and Green 1988). The crop is intolerant of waterlogging for prolonged periods and performs poorly in wet seasons; irrigated plantings should be restricted to well-drained land (Wightman, 1984). Dry conditions are essential at harvest because the crop has tough stems, and secondary growth and flowering make harvesting difficult; in this situation a desiccant may be necessary (Sykes and Green, 1988). The crop is particularly sensitive to high temperatures (particularly if associated with low humidity) from flowering to seed maturation, which reduce seed and oil yield and adversely affect oil quality (Arnon, 1972; Price, 1968). Dry, hot conditions after maturity, however, facilitate harvesting.

SAFFLOWER

The production of safflower (*Carthamus tinctorius*) in Australia is largely restricted to the northwest slopes and plains and upper-central western plains of New South Wales (Colton, 1988), northern wheat belt of Victoria (Wightman, 1985) and the Darling Downs and Central Highlands of Queensland (Beech, 1979; Jackson and Berthelsen, 1986) (Fig. 2.19). Successful production in northern, summer rainfall areas depends on fallow water storage, a full moisture profile to one metre depth being a prerequisite for avoiding spring water stress during the critical flowering–grain filling period (Colton, 1988). Some opportunity cropping is conducted on drying floodplains and lake beds of far western New South Wales after seasonal floodwaters recede.

Commercial safflower production began in Queensland in the 1950s and quickly spread to New South Wales, which soon became the largest producing state. Poor yield performance due to unfavourable seasonal conditions, disease and lack of suitable cultivars (Beech, 1969; 1979) resulted in a decline in New South Wales, and Queensland again became the major producer in the 1980s. Commercial prospects have been severely hampered by the availability of a single, late-maturing cultivar, Gila, which matures some 4 to 8 weeks later than for wheat for the same environment. It has a significantly greater water requirement than wheat for growth, development and the production of satisfactory grain yields (Matheson, 1976). It is grown almost entirely as a rainfed crop, although production under irrigation as a rotation crop to break subsoil hardpans in cotton is common in northern New South Wales (Colton, 1988). The crop is sensitive to water availability, and both low and high extremes of water supply can impose restrictions on yield. Drought stress during the early stage of vegetative growth has little influence on subsequent growth and yield (Basinski and Beech, 1972), but subsequent stresses before and after flowering significantly reduce the seed and oil yield (Cutting, 1974; Colton, 1988). Under excess water conditions, particularly if waterlogging occurs at any stage on poorly drained soils, significant yield reduction can result from *Phytophthora* root rot, while rainfall during or after flowering can stimulate germination in the head and initiate severe outbreaks of *Alternaria carthami* with substantial yield loss (Matheson, 1976; Harrigan and Sykes, 1987; Colton, 1988). The crop is also temperature sensitive, low soil temperatures being blamed for emergence failure and poor establishment (Basinski and Beech, 1972), while frost damage during stem elongation, flowering and grain filling can severely reduce yields (Matheson, 1976; Jackson and Berthelsen, 1981). Frost risk necessitates mid-winter (May–June) planting in New South Wales, but is associated with late maturity under high temperature and water stress in summer (December–January). High summer temperatures and rainfall in northern New South Wales and Queensland predispose the crop to severe outbreaks of *Phytophthora cinnamomi* and *Alternaria carthami* and to seed germination in the head (Jackson and Berthelsen, 1986;

Harrigan and Sykes, 1987). High but not excessive temperatures (26–29° C) at and after flowering have also been shown to have deleterious effects on seed yield and oil content (Basinski *et al.*, 1961; Beech and Norman, 1963; Beech, 1969).

As a commercial crop, safflower will continue to be restricted to current production areas until the many problems associated with limited cultivar choice are resolved. Although traditionally regarded as day-neutral, significant long-day responses have been recorded with a range of lines by Horowitz and Beech (1974), who proposed that earlier maturity and drought escape during flowering and grain filling in southern areas would require photoperiod-insensitive cultivars which mature at the same time as wheat. Commercial production to date has been based on a single cultivar, Gila, and until new cultivars which are early maturing and disease resistant become available, future prospects for the crop are limited. The 1989 release of Sironaria and Sirothora (CSIRO-developed cultivars resistant to *Alternaria* and *Phytophthora* respectively) is a major advance, but both are agronomically similar to, and slightly later maturing than, Gila, and their major contribution is to enable more reliable production in existing dryland production areas where safflower is traditionally grown (Harrigan and Sykes, 1987).

Summer Oilseeds

The summer-growing oilseed crops produced in Australia are soybeans and sunflower. Two other summer crops, cotton and peanuts, are important sources of vegetable oils, although grown primarily for the cotton fibre and confectionery trades respectively. For convenience, cotton and peanuts have been included in this section as summer oilseeds.

The production of summer oilseeds is restricted almost entirely to eastern Australia (Fig. 2.21), in rainfall zones ranging from uniformly distributed to temperate moist or subtropical humid-summer dominant, with sunflower plantings extending into the semi-arid inland margins.

SOYBEANS

Soybeans (*Glycine max*) were a relatively minor crop until 1968, being restricted to dryland production in the South Burnett and Darling Downs areas of Queensland, where sowings totalled less than 2000 ha per annum. Subsequent production expanded rapidly in both Queensland and New South Wales, peaking at 71 000 ha in 1989, but by 1991 had fallen to 40 000 ha. Queensland is the major producer, with 1989 to 1991 plantings approximately double that of New South Wales. Until 1985, expansion in New South Wales occurred almost solely under irrigation. In 1971, 96 per cent of plantings in New South Wales and 61 per cent of those in Queensland were irrigated (Laing and Byth, 1972). Since the 1983 season there have been steady increases in dryland sowings in the summer-dominant rainfall areas of northern coastal New South Wales

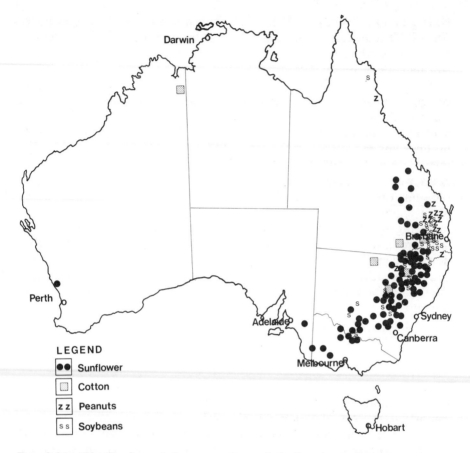

FIG. 2.21   Distribution of the summer-growing oilseeds sunflower, soybeans, cotton and peanuts in Australia, based on Australian Bureau of Statistics crop returns by local government areas with greater than 100 ha of peanuts and greater than 200 ha of sunflower, soybeans and cotton

and southeastern Queensland, although irrigated production still accounted for almost the entire crop by 1985. Since then, dryland production has also extended to the Northern Tablelands and Northwest Slopes of New South Wales (Colton, 1987). Here fallow stored soil water, with a full profile at planting, is necessary to offset unreliable autumn rainfall when crops are in the critical flowering–pod-filling phases (Colton, 1987; Rose *et al.* 1989).

New South Wales production is now spread throughout the inland irrigation areas, extending as far south as the Riverina in southwestern New South Wales (Fig. 2.21). The industry in New South Wales was founded almost entirely on cultivars of U.S. origin. In Queensland, production was initially based on the locally selected lines Wills and Semstar (Laing, 1974) and later on U.S. cultivars (Lawn and Byth, 1979).

Intensive soybean breeding programs have been conducted since the early 1960s in Queensland, and since the mid-1970s in New South Wales. A major focus of the New South Wales program has been the development of early maturity, indeterminate cultivars for dryland production (Rose *et al*. 1987).

The soybean plant is particularly sensitive to photoperiod, commercial cultivars being separated into 10 groups on the basis of their relative short-day requirement. In general, the photoperiod requirement is the critical factor in cultivar selection, requiring correct matching of cultivar daylength requirements with latitude and sowing date, modified by both temperature and planting arrangement (Carter, 1970; Laing and Byth, 1972; Lawn and Byth, 1979; Colton, 1987). Initial testing in Australia was largely based on the extrapolation of introduced commercial U.S. cultivars with known daylength requirements to similar latitude zones in Australia, but subsequent experience has shown that direct extrapolation is not entirely successful because of the complicating effects of temperature–daylength interactions in continental Australian production areas, compared to the more typical coastal plains locations where they were developed in the southeastern U.S. (Laing and Byth, 1972). In addition, the U.S. system of maturity groups has shown major deficiencies in latitudes north of 27°S in Australia where considerably less daylength variation occurs (Byth, 1968), and would have to be expanded by the addition of at least three groups to encompass cultivars adapted to lower latitudes such as far northern (15°S) Australia (Lawn and Byth, 1979). Successful cultivar adaptation occurs where flowering begins approximately 60 days after sowing. When sown north of its adapted region, a cultivar will flower early, producing a small plant with low-set pods that have a low potential and harvestable yield. If sown south of its adapted region, it will flower late in autumn, produce excessive vegetative growth, and have a low yield recovery if subjected to killing frosts prior to maturity, as well as being very difficult to harvest because of lodging (Lawn and Byth, 1979). Cultivar and daylength requirements are also subject to considerable modification by sowing date, planting arrangement and temperature (Laing and Byth, 1972), an interaction investigated on a wide latitudinal basis in eastern Australia (Laing, 1974; Lawn and Byth, 1979).

Soybeans show high sensitivity to water stress, particularly during flowering, pod development and pod filling (Shaw and Laing, 1966; Colton, 1987; Rose *et al.*, 1987). Even with adequate soil moisture supply, high temperature – low humidity conditions can cause severe flower and pod abortion. Temperature also imposes strict controls on growth and development, with minimum soil temperatures for germination and emergence of 13 to 15°C, although values of 19°C are desirable (Matheson, 1976; Colton, 1987). The optimum temperatures for growth are 27 to 32°C (Arnon, 1972), while the temperature requirements during initiation, flowering and pod filling are complicated by interactions with daylength (Lawn and Byth, 1979). Yields attained under irrigation have been satisfactory based on U.S. standards, and it is likely that further expansion will largely be at the expense of rice, maize and sorghum in inland irrigation areas. With careful attention to cultivar adaptation and sowing date, soybeans are well adapted to current areas of production, ranging from 26°S to 33°S. Additional areas with potential are the Murray–Murrumbidgee region of northern Victoria – southern New South Wales, the Queensland highlands around Emerald, and the Burdekin and Ord River irrigation areas. The crop performs well in zero-tillage situations, enabling opportunity dryland cropping as an alternative to fallow in favourable summer seasons in northern New South Wales (Wheatley *et al.*, 1992).

SUNFLOWERS

Like soybeans, sunflowers (*Helianthus annuus*) are a relatively recent crop of commercial importance in Australia, with little interest shown until 1969 when wheat production quotas were first imposed. Since then, however, sunflowers have become the major Australian oilseed crop. Plantings rose to a record 295 000 ha in 1972, fluctuating annually thence to 166 000 ha in 1991. Initial production was confined to dryland culture on the heavy wheatbelt soils of northwestern New South Wales and southern Queensland, the north-central Tablelands of New South Wales, and smaller irrigated plantings on the inland central and southern Plains of New South Wales (Matheson, 1976; Dale, 1984). Further expansion resulted in plantings extending throughout the wheat belt of Western Victoria and New South Wales, the western margin of the Darling Downs and into the Central Highlands of Queensland (Fig. 2.21). These areas have uniform to summer-dominant rainfall, and long, hot summers. Limited plantings are also located in northern Victoria. The initial cultivars grown in Australia were largely of Russian and Eastern European origin, poorly adapted to Australian conditions and grown with little agronomic expertise (Lovett *et al.*, 1979). These open-pollinated cultivars were grown across a wide range of climatic and edaphic environments, with disappointing yield results. The introduction of U.S. hybrids in the early 1970s, together with subsequent development of Australian hybrids, has markedly increased the range of cultivars available and enabled selection for specific environments. By 1982, practically all crops were of hybrid cultivars.

The length of the growing period ranges from 80 to 160 days, and is almost solely dependent on temperature (provided adequate water is available). Variation in total length of the growing period has been shown to be closely related to the total Growing Degree Days (GDD) experienced. Summer plantings, where daily mean temperatures are considerably in excess of base temperature, receive the requisite number of GDD in a short period, in contrast to winter plantings (Robinson, 1971; Doyle, 1975). Hence the length of the growing season is much shorter for a summer than a spring-grown crop. Individual cultivars exhibit relatively constant GDD totals for development from sowing to maturity, total requirement representing the sum of those for each of the five developmental stages: sowing to emergence, emergence to head visible, head visible to first anther, first anther to last anther, and last anther to maturity (Robinson, 1971; Anderson *et al.*, 1978; Dale, 1984). The longer growing period of early planted crops has been shown to be a result of an extended emergence to head visible stage, such that very early (late winter) sown crops do not flower very much in advance of later (spring) sown crops.

Base temperatures used in GDD summations were 7.2°C (Robinson, 1971) while Doyle (1975), after testing base temperatures of −2 to +8°C, obtained best correlation between development and GDD during the first three phases with a base temperature of 1°C. Base temperatures have also been shown to vary between development stages, increasing from emergence to flowering, and decreasing to 0.5–1°C for the seed maturation stage (Lovett *et al.*, 1979). Though they are traditionally regarded as a day-neutral plant, some research suggests that sunflowers are sensitive to photoperiod (Woodruff, 1973) but evidence on this point is conflicting (Robinson, 1971). Sunflowers will germinate at relatively low (5–8°C) soil temperatures, although emergence is retarded and may take up to 30 days. In contrast, emergence and establishment may take as little as four days under high (24–27°C) temperatures (Matheson, 1976; Shaw, 1991). Newly emerged seedlings are frost tolerant, withstanding temperatures as low as −6°C at ground level. Frost tolerance is retained until the six to eight-leaf stage, although some leaf damage may occur. The ripening seeds are also frost tolerant, but between this stage and the young seedling stage the crop is frost sensitive (Anon, 1973; Lovett *et al.*, 1979; Dale, 1984). High temperatures at and after flowering can adversely affect pollination, oil yield and oil quality (Matheson, 1976; Lovett *et al.*, 1979; Ralph, 1982).

Prior to flowering, the crop is tolerant of high temperatures during vegetative growth, followed by mild temperatures post-flowering (Martin *et al.*, 1976; Dale, 1983). Crops maturing in midsummer under high temperature conditions may exhibit both lower oil yield and oil quality, the latter through a marked reduction in linoleic acid content of the oil. Most crops are medium to late maturity hybrids, planted either early (October) or late (December–January) to avoid flowering and grain filling during the high risk heat stress period (mid-January to late February). The greater drought tolerance of sunflowers than of maize during vegetative growth (Anon., 1973) allows production in dryland areas unsuitable for maize, and in these areas the crop is in direct competition with sorghum. However, the crop is particularly sensitive to water stress during anthesis and early grain development, severe stress resulting in significant reduction in grain yield (Lovett *et al.*, 1979). The plant exhibits lax stomatal control, and has a low water-use efficiency under restricted water conditions (Rawson, 1979). Yield is closely correlated with photosynthetic leaf area during seed filling, and is adversely affected by water stress during the period 2 to 3 weeks before flowering, as upper leaf expansion, and ultimately photosynthetic area during seed filling, is severely inhibited. Earlier water stress, in contrast, had little effect on yield provided establishment was not affected (Takami *et al.*, 1981). Successful dryland production thus depends on adequate water availability from 2 to 3 weeks before flowering through to physiological maturity of the seed. A fully charged, deep soil profile wet to 150 cm depth is an essential pre-requisite for planting, particularly in areas with unreliable late summer – early autumn rainfall (Dale, 1984).

The combination of low-temperature tolerance at each end of the growing season, relative insensitivity to photoperiod and tolerance to a wide temperature range during vegetative growth make sunflowers adaptable to a wide range of environments, provided water availability is adequate. Current areas appear very suitable but yields are generally low, particularly in New South Wales, and probably reflect water stress at critical stages. Better agronomic practices, which focus on more efficient water utilisation, are essential for better yield performance to ensure the long-term future of dryland production. A simulation of potential areas for expansion (Smith *et al.*, 1978), based on a maximum fallow water reserve of 100 mm, with a minimum of 75 mm for planting, suggested extension of production into the eastern wheatbelt of Victoria, New South Wales and Queensland. Further extension, but with lower yield potential, was also indicated for South Australia and central Queensland. To date, extension into these potential production areas has not occurred and appears unlikely.

COTTON

Although cotton (*Gossypium hirsutum*) is grown primarily as a fibre crop, the seed after ginning (lint removal) is also a source of vegetable oil. On a world basis, cotton ranks second to soybeans as an oilseed. It has been produced in Australia since the early 1920s, initially only as a dryland crop in Queensland, where it reached a peak planting of over 30 000 ha in 1932. Since the early 1960s, its rapid expansion as an irrigated crop in Queensland and New South Wales has resulted in an increase in sown area to 279 000 ha in 1991. Although the annual area fluctuates due to market conditions and irrigation water availability, total Australian area planted has averaged over 220 000 ha since 1987 (Table 2.9). New South Wales is the largest producer, with over two-thirds of total plantings. Production is restricted to the Namoi, Gwydir, Macquarie, and Darling River valleys of New South Wales and to the Condamine River, St George, Darling Downs, Dawson, Callide and Emerald districts of Queensland. The largest production areas are in the Namoi, Gwydir and Macquarie River valleys of New South Wales. Earlier cotton industries in the Murray–Murrumbidgee Irrigation Area of southern New South Wales and the Ord River valley of northwest Western Australia ceased production in the mid-1970s. The short growing season in southern New South Wales restricted production to short staple lines which had limited market demand, while high costs and insect pressure led to closure of the industry in the Ord (Thomson, 1979). The high capital costs of land development for irrigated production led to renewed interest in dryland production in northern New South Wales during the 1980s, and the dryland area sown has increased slowly but steadily. Excellent yields of over 5 bales per ha in the 1990/91 season stimulated renewed confidence, following indifferent performance due to erratic rainfall in earlier years, and expansion appears likely to continue (Kay *et al.*, 1992).

Cotton originated in hot, arid regions and these conditions are conducive to high yields, provided adequate water is available. Provided the water needs of the crop are met, growth and development are controlled almost entirely by temperature (Basinski, 1963; McMahon and Low, 1972; Thomson, 1979; Constable and Shaw, 1988). Within limits, the higher the average seasonal temperature, the higher the growth rate; the longer and hotter the season, the higher the potential yield. The absolute length of the growing season is delimited by the average dates of last spring and first autumn frosts, but the major spring factor determining start of the season is soil temperature. Successful establishment requires a minimum soil temperature (10 cm soil depth, 8.00 am) of 14°C maintained for at least 3 days. At this level, complete emergence at Narrabri took 17 days and only 73 per cent of seed produced surviving seedlings. At 18°C, results improved sharply to 5 days and 90 per cent survival (Constable and Shaw, 1988). Growth and development in the absence of water stress is governed by temperature, with all growth and development ceasing at 12°C. The development of the crop can be predicted from temperature data by GDD summations, using a base temperature of 12°C (Table 2.12).

McMahon and Low (1972) reported a high degree of correlation between length of the growing season and GDD accumulation, using a base temperature of 10°C, with most commercial cultivars requiring 900 GDD to flowering and 3000 GDD to final harvest. Allowing for the difference in base temperature used, their results agree well with those of Constable and Shaw (Table 2.12), at least up to flowering. However,

TABLE 2.12   Total Degree Days for some phases of cotton development: typical values (after Constable and Shaw, 1988)

Phase	Total degree days	Days at 28/20°C*
Sowing to final emergence	80	7
Sowing to squaring	505	42
Sowing to flowering	777	65
Flower to open boil	750	63

*Day/night temperatures

10°C appears too low in practice, as McMahon and Low predicted successful production in the Murray–Murrumbidgee area of southern New South Wales with GDD summations as low as 2250, given adequate water and radiation receipts. Experience has shown, however, that seasons of this length are too short for commercial production, an infant industry in that region failing in the early 1970s. In general terms, length of the growing season and thus yield potential increases with distance north from the current southern limit (Trangie, in the Macquarie Valley of New South Wales), to the Queensland production areas. Planting time is critical in New South Wales, with a substantial and progressive yield penalty for crops planted beyond mid-October. Safe planting time then becomes a trade-off between waiting for favourable soil temperatures and planting before mid-October, with most New South Wales cotton plantings beginning in mid-September. Low temperatures during the season can also lead to severe setbacks in development through cold shock (Thomson, 1979). For every day with a minimum temperature at or below 11°C, growth on the following day is retarded, regardless of maximum temperature experienced. Development is thus delayed and yield potential significantly reduced (Constable and Shaw, 1988). The probability of this occurring increases from north to south, southern areas having a much higher incidence of cool nights during the season.

Besides the benefit of high temperatures and radiation receipts, semi-arid areas offer the best production environment, as water supply to the crop can be controlled without waterlogging or production program interruptions through rainfall. Irrigation management aims to avoid severe water stress at any stage of development, particularly during squaring (flower bud formation), flowering and boll filling. Erratic rainfall in the Australian production areas can be a major problem in some seasons, creating difficulties such as hasty land preparation, delayed planting, interference with weed and pest control operations, water-logging, lint damage and delayed harvest due to secondary growth, inaccessibility to machinery and retarded maturity (Basinski, 1975; Thomson, 1979). Floods in some years have resulted in total crop loss in some regions such as the Namoi Valley. A lack of irrigation water has been a major problem in New South Wales over the last decade, especially in the Namoi and Gwydir River Valleys, where the expansion of irrigated crop production has increased demand for water and a succession of dry years has prevented recharge of the major supply storages such as the Keepit (Namoi) and Copeton (Gwydir) Dams. Projected allocations in some years have been notified to growers at levels as low as 5 per cent of the quota. This (together with fluctuating prices) has limited recent industry expansion in these areas and has forced growers to aim for efficient water use rather than maximum production per hectare (Browne, 1984) as well as developing major on-farm storages to take advantage of high river flows in wet periods.

Water requirements of the crop are high, ranging from 600 to 900 mm over the growth season, depending on cultivar maturity, radiation, temperature and evapotranspiration during the season. Daily water use increases over the season from 1 to 2 mm in early growth to a peak as high as 8 mm during late flowering and early boll fill, then declining to 4 mm during late boll filling through to maturity. Cotton is most sensitive to water stress during the peak demand period, which in New South Wales occurs in late January – early February. If attempting to grow a crop with restricted water supply, limiting water and nitrogen availability during early growth will restrict plant and canopy size and thus peak demand. Available water should be saved until this stage to maximise water-use efficiency and resulting lint yield. Under normal conditions, where adequate water is available for full irrigation, growers should aim to limit soil moisture depletion to no more than 50 per cent before irrigating again (Browne, 1984). Water monitoring using neutron probes and irrigation scheduling software is now routine in the industry, and has allowed substantial improvement in irrigation efficiency and cost-effectiveness (Cull 1987, 1992). A recent analysis of water use and yield records for several major enterprises in northern New South Wales produced an overall efficiency of 1.3 bales of lint per megalitre of water (including rainfall), which is higher than the industry planning standard of 1.1 bales per megalitre (D. A. Robson 1992, pers. comm.).

The future for the industry appears bright in the long term, provided the problems of irrigation water supply and pesticide resistance can be overcome. Both have provided the stimulus for a substantial change in production practices over the last decade, and production systems are continually evolving to cope with change.

PEANUTS

The production of peanuts (*Arachis hypogea*) in Australia is confined to restricted areas of red kraznozem soils in southeastern Queensland (Fig. 2.21), with the major production area centred on the Kingaroy district. A small industry near Lismore in northern New South Wales closed down in the mid 1970s.

Peanuts are completely intolerant of frost at any growth stage (Martin *et al.*, 1976) and require relatively high temperatures throughout the growing season (Arnon, 1972). They are drought sensitive at all growth stages, having rather lax stomatal control (Slatyer, 1955), and require crop season rainfall of at least 508 to 635 mm during the growing season. This restricts production to areas of high summer rainfall (Matheson, 1976). Despite a high moisture demand, rainfall requirements for optimum yield are rather specific, short periods of moisture stress followed by rain inducing flushes of flowering and aiding production of seeds of uniform size and quality (Matheson, 1976). Consistent rainfall throughout the growing period stimulates continuous flowering and subsequent production of pods of a wide range of maturity, making prediction of harvesting time for maximum yield of mature pods difficult. Dry weather is essential during ripening and harvesting. The ideal rainfall environment for peanut production has been defined by Saint-Smith *et al.* (1969) as having low rainfall during early flowering, followed by a short period of heavy rainfall to stimulate rapid flowering and a single crop of pods, and culminating in dry weather at harvest, the latter to facilitate pod separation from soil and allow adequate pod curing in open air stacks. The crop is regarded as day neutral, the length of the growing period being determined primarily by temperature (Anon., 1967b; Martin *et al.*, 1976), and appears ideally suited to current production areas in Australia. Further expansion of production under dryland conditions would appear most promising in the summer-dominant rainfall zone of eastern Queensland, although soil requirements may impose a major controlling influence.

Grain Legume Crops

Winter Grain Legumes

This group comprises temperate large-seeded legumes grown for their high protein grain. Four species are of commercial importance: lupins, field peas, chickpeas and faba beans. Both lupins and field peas have been grown on an extensive scale throughout the southern croplands of southern New South Wales, Victoria, South Australia and Western Australia, while chickpeas and faba beans have only developed into commercial-scale production since 1985. Interest in winter grain legumes as alternatives to or as rotation crops with the winter cereals, particularly wheat, has accelerated since 1984 with declining market prospects for wheat, their recognition as effective disease-break crops, their significant contribution to soil nitrogen status, and a favourable market outlook for legume grains. Lupins, field peas and faba beans are grown primarily for the stockfeed trade, while chickpeas, which are produced mainly for culinary purposes, are also suitable for stock feed. Future prospects for all are closely linked with export markets.

LUPINS

Although grown extensively in the southern wheat belt (especially in Western Australia) for many years as green manure and forage crops, development as a grain crop began only in 1969 with the release in Western Australia of Uniwhite, a bitter alkaloid-free cultivar of *Lupinus angustifolius* (narrow-leaved lupin). This was quickly followed by Uniharvest and Unicrop, and subsequently by a number of non-shattering, disease-resistant cultivars covering a range of maturity. These were the world's first true crop cultivars of narrow-leaved lupin. Further breeding with sandplain lupin (*Lupinus cosentinii*) has resulted in registration of the first low-alkaloid crop cultivar of this species (Gladstones, 1982).

    Reduced brancing, early maturing types which have out-yielded current lines by 30 per cent in drier areas show promise for development of

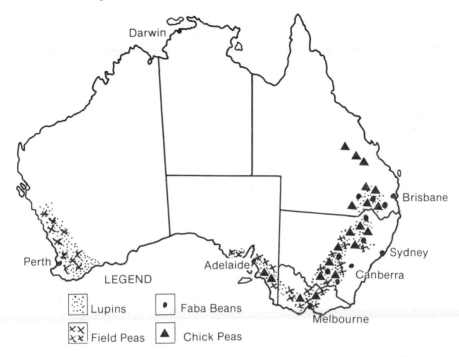

FIG. 2.22 Distribution of winter legumes for grain in mainland Australia. Lupin and field pea areas based on Australian Bureau of Statistics crop returns by local government areas. Faba beans and chickpea areas based on 1985–91 commercial sowings

cultivars suitable for dry regions, where low yields have been a problem (Delane *et al.*, 1986). Crop cultivars of white (*L. albus*) and yellow (*L. luteus*) lupins have also been released and are now in commercial production. Current research with a range of other *Lupinus* species shows promise for the release of a number of new cultivars (Buirchell and Cowling, 1992).

The development of lupins as a crop has been spectacular: the area sown since the release of Uniwhite has risen from nil to over 1 million ha in 1988, 876 000 ha of which were in Western Australia. Although down to 793 000 ha in 1991, further increases are expected. Smaller but significant production areas extend throughout the southern croplands of South Australia, Victoria and New South Wales (Fig. 2.22). With the exception of New South Wales, where production extends into the semi-arid zone, lupin production for grain is limited to the dry summer, temperate areas with reliable winter rainfall. They are well adapted to low fertility, lighter soils which are frequently acid (Gladstone, 1969, 1970), but perform poorly on heavier clay soils, particularly those that are poorly drained (Walton, 1982) or alkaline (Walton, 1982; Badawy *et al.*, 1985). They are best adapted to mild temperate Mediterranean climates with reliable winter and spring rainfall (Southwood and Scott, 1972), with water availability during the reproductive phase critical to grain yield (Farrington, 1974; Biddiscombe, 1975; Nelson *et al.*, 1986). Yields of 1.5 tonnes per hectare may be expected in rainfall districts of 450 mm per annum or higher in Western Australia (Walton, 1983). Average yields of 1.1 tonnes per ha have been reported for low rainfall (340 mm) areas in Victoria (Mock, 1984).

The crop is sensitive to temperature, particularly during flowering, where temperatures outside the range 11–25°C will cause flower and seed loss and reduced yield (Walton, 1982). Early April–May sowing of non-vernalised cultivars will result in early flowering and low-temperature (particularly frost) damage (Gladstones, 1970; Walton, 1982; Boundy *et al.*, 1982). This may be offset to some extent by the indeterminate flowering habit of lupins: Unicrop flowers for up to 100 days (Anon., 1972) although later compensatory flowers may be subjected to spring water stress. Successful grain production requires a growing period of at least five months free of moisture stress, coinciding with rainfall areas of 500 mm per annum or more in New South Wales (Southwood and Scott, 1972); early maturing cultivars are recommended for the drier production areas (Smith and Walker, 1984; Walton, 1982).

Further expansion of lupins on a wider range of lighter soils in northern New South Wales and southeastern Queensland is feasible, but has to date been limited. Breeding programs focusing on developing cultivars with higher yield potential, greater resistance to lodging, disease resistance, reduced branching and pods borne higher from the ground will play a major role in the extent of crop expansion (Marcellos, 1984). Cultivars resistant to *Phomopsis* stem blight, the cause of lupinosis (a mycotoxicosis of sheep that ingest lupin forage infected with *Phomopsis leptostromiformis*), are close to release (Cowling *et al.*, 1986) with one commercial cultivar (Ultra) of *L. albus* showing marked resistance (Wood and Allen, 1980).

FIELD PEAS

Field peas are the most adaptable of the winter grain legumes and have for many years been grown as a rotation crop with cereals in the main winter cereal belt of Western Australia, South Australia, Victoria and Tasmania. In recent years they have been grown on a rapidly increasing scale, extending in 1983–85 into the southern and central wheat belt of New South Wales on a substantial scale. The area sown peaked at 456 000 ha in 1989, Victoria and South Australia being the major producers. This area declined to 318 000 ha in 1991, coinciding with an increase in area planted to wheat.

Field peas may be grown as a specialty grain crop, but are most frequently grown as a rotation crop with cereals, providing both soil nitrogen build-up and a break against soil-borne cereal diseases, particularly cereal cyst nematode and take-all.

They are grown as an alternative to lupins in cereal rotations, either where lupins have performed poorly due to unsuitable soils or where there is a reluctance to grow lupins because of lupinosis. The advantages of field peas over lupins are suitability for later sowing, adaptation to a wider range of soils (particularly heavy soils), better winter growth, availability of a better range of registered herbicides, and their greater suitability for hay (Simmons and Sykes, 1984; Simmons, 1989). They are adapted to cool temperate winter rainfall areas with a long, cool growing season (Arnon, 1972), making better cool season growth than other legumes (Anon., 1986b). Annual rainfall requirements for good yields range from 500 mm in warmer areas to 400 mm in cooler regions such as the tablelands and slopes of southern New South Wales (Simmons and Sykes, 1984). The timing of rainfall is more important than total amount received (Farrington, 1974; Simmons, 1989), yields of 2 tonnes per hectare having been achieved in the Mallee region of Victoria (300 mm annual rainfall) when rainfall coincides with flowering and grain filling (Boundy and Smith, 1981). Field peas are adapted to a wider range of soil types than lupins, and are more suited to heavier alkaline soils. However, they cannot tolerate poorly drained conditions. Production areas (Fig. 2.22) have undergone significant expansion since 1983. The crop is frost tolerant during vegetative growth, but frost during flowering or pod development can cause heavy loss of pods and yield (Boundy and Smith, 1981; Simmons, 1989). High temperatures, particularly during flowering and pod development, can also cause a significant reduction in yield, particularly when combined with low water availability. Field peas are also susceptible to damage from hail after flowering (Simmons and Sykes, 1984) and from rain on a ripe crop, causing a substantial reduction in quality (Simmons, 1989). The most recent extension of field peas has been into the southern and central New South Wales wheat belt, increasing from a negligible area in 1982 to 41 000 ha in 1988; seasonal conditions limited 1991 plantings to 28 000 ha. Current production is constrained by the limited adaptation of the major cultivars available and their susceptibility to diseases such as *Mycosphaerella*, *Ascochyta* and bacterial blights, all of which are favoured by cold wet conditions and poor drainage.

Earlier maturing lines are continually evaluated for yield and disease resistance, and improved cultivars may initiate wider interest. A recent development of leafless and semi-leafless types, in which the normal leaf is replaced by tendrils, allowing greater standability for harvest without yield penalty over existing lines (Berry, 1985; Anon., 1986b), may result in release of cultivars of this type in the near future. Such cultivars, with reduced leaf area, may also enable extension in drier areas. Future expansion will reflect market outlook, the current utilisation of the Australian crop being 55 per cent for domestic stock feed, 20 per cent exported for stock feed and 25 per cent exported for human consumption (Simmons *et al.*, 1992).

CHICKPEAS

Chickpeas (*Cicer arietinum*) are a very recent commercial crop in Australia, and there is substantial interest in their role as a rotation crop with cereals. Interest centres particularly on their role as a substitute for field peas on alkaline soils unsuitable for lupins (Pye, 1982), particularly where high temperatures occur at flowering (Simmons, 1989). Commercial production began on a small scale in 1980 in Victoria, and has since expanded rapidly into New South Wales and Queensland, where 1987 plantings were 18 000 ha and 52 000 ha (Hooke, 1987) respectively. New South Wales plantings in 1991 had increased to 66 000 ha, a substantial increase over the 1989 area of 22 000 ha (Dale *et al.*, 1992). Recent rapid expansion in Queensland has occurred on alkaline black earths of the Darling Downs and Central Highlands (Brinsmead, 1992). The total Australian area in 1991 was 215 000 ha, up from 94 000 in 1989 (Dale *et al.*, 1992). On a world scale, chickpeas are the most important grain legume crop of semi-arid regions. The crop is adapted to dry environments and sets seed successfully under moderately high temperatures (Simmons, 1982). It appears best suited to the heavier soils of the wheat belt provided they are well drained (Brinsmead *et al.*, 1984; Marcellos, 1984; Pye, 1984; Mullen, 1986). The crop is particularly sensitive to diseases caused by *Phytophthora megasperma* and *Botrytis cineria* under wet conditions (Brinsmead *et al.*, 1984; Marcellos, 1984), both of which can result in complete crop failure. *Botrytis* infection also causes severe yield reduction in early planted crops with excessive vegetative growth (Brinsmead, 1992). Initial expansion was limited by lack of adapted cultivars—the only two available commercially were the *desi* type Tyson, recommended for dryland sowing in all production areas, and the large-seeded *kabuli* type Opal, which is late maturing and recommended only under irrigation (Knights, 1982). Both have been superseded by the release of Barwon, the first *Phytophthora*-resistant commercial cultivar (Simmons, 1991).

Reduced branching may also offer considerable yield advantages over existing types which have a low harvest index (Siddique and Sedgley, 1985). Coloured seed coats offer some resistance to seedling damping-off caused by *Botrytis*, and selection for this character, together with double-podding, may substantially improve crop prospects (Knights, 1982).

FABA BEANS

Also known as horse or tick bean, the faba bean (*Vicia faba*) is a small-seeded broad bean which is moderately tall, erect, easy to grow and harvest and well adapted to heavier, neutral to alkaline soil types (Marcellos, 1984), where lupins perform poorly (Marcellos and Constable, 1986). Initial interest in the crop centred on heavy alkaline soils in the South Australian wheat belt where it demonstrated high yield potential. This led to the release in South Australia of the cultivar Fiord, the only commercial cultivar currently available (Marcellos and Constable, 1986).

In both 1989 and 1990 the area planted to faba beans in Australia was 44 000 ha. In 1991 it had risen to 58 000 ha, producing an estimated 64 000 tonnes. The area in New South Wales increased from 3500 ha in 1990 to 8800 ha in 1991, reflecting poor wheat prices and planting cutbacks. With yields around two-thirds that of wheat, future planting will be closely determined by relative prices (Benson and Dale, 1992). The crop has shown high yield potential on alkaline clays of the northern New South Wales wheat belt (Marcellos, 1984). Current production centres on these soils, together with heavy alkaline clays in the southern Riverina region. Breeding programs underway will result in the release of alternative cultivars to Fiord, and extend the range of commercial adoption of the crop (Marcellos, 1984). A feature of the crop is its tolerance to frost, cold wet weather and waterlogging when compared to field peas or chickpeas (Mahoney, 1984) and it appears the logical substitute for them under those conditions.

Summer Grain Legumes

Crops included in this group are cowpeas (Poona peas), mung beans, navy beans and pigeonpeas. The first three have been produced on a relatively small scale for some years, while pigeonpea came into commercial production in 1982. All are grown as grain crops in the summer rainfall areas of northern New South Wales and southeastern Queensland. Their commercial distribution is shown in Fig. 2.23.

COWPEAS

Cowpeas (*Vigna sinensis*) are grown as a dryland summer crop in northwestern New South Wales and southeastern Queensland. The grain produced is used mainly as seed for forage crop sowing (Bott, 1970). Substantial production increases are unlikely; further expansion would be at the expense of sunflower and sorghum in these northern areas, and the lack of suitable grain cultivars resistant to stem rot (*Phytophthora vignae*) is a major obstacle. Expanded grain production would be largely reliant on development of the high protein grain market in the stock feed trade. Release of cultivars suitable for the culinary trade (Banjo, Blackeye) may result in some expansion, although the extent will depend entirely on market prospects. They are adapted to a wide variety of soils from light to heavy texture, but will not tolerate waterlogging.

Cowpeas require a soil temperature of 18°C for successful establishment, and planting in the main production areas is delayed until late November – early December. The crop is well adapted to semi-arid regions and can withstand prolonged water stress better than most other crop species.

MUNG BEANS

Mung beans (*Vigna radiata* ssp. *aureus*, *V. radiata* ssp. *mungo*) are short-season (60 to 120 days), summer-growing legumes yielding high protein

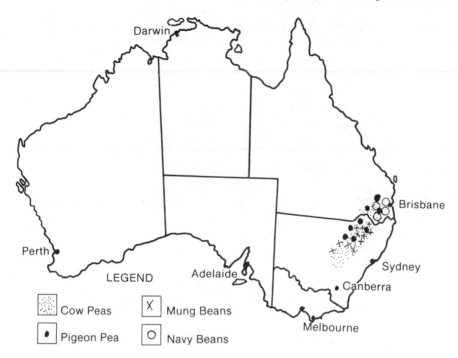

FIG. 2.23   Estimated commercial distribution of summer grain legumes, based
on Australian Bureau of Statistics returns for the 1985–91 seasons

grain, with the additional advantages of soil nitrogen build-up and value
of the stubble for forage. Mung beans appear best adapted to tropical
and subtropical areas where summer rainfall is too low for other crop
legumes (Lawn and Russell, 1978), with growing season rainfall totals of
300 to 400 mm. Potential areas of production encompass existing cowpea
areas, extending to the north into high rainfall areas of central and
coastal Queensland, and to the south into the irrigation areas of northern,
central and southern New South Wales where they would compete
directly with soybeans, cotton, maize, sorghum and rice. The domestic
market is limited to approximately 1000 tonnes and is unlikely to expand
rapidly, making further expansion dependent on export markets. Export
markets of considerable scope for this purpose exist in Asia.

   The mung bean's short growing season enables production as a complementary crop to existing winter crops in suitable areas, extending the
cropping phase and at the same time making a valuable contribution to
soil nitrogen status.

NAVY BEANS
Navy beans (*Phaseolus vulgaris*) are produced in the South Burnett,
Darling Downs and Atherton Tablelands regions of Queensland. Navy
beans became economically important in the early 1970s with commercial

introduction of new cultivars (Gallagher, 1972). With favourable market conditions, production expansion would be at the expense of maize, sorghum and dryland soybeans in these areas, while its short growing season (8 to 14 weeks) may enable its production during summer as a double crop in association with wheat and barley. The beans are grown for the edible dry bean canning (baked beans) and package trade. Some interest has been shown in their production as edible dry beans in New South Wales, but the lack of an organized marketing body, together with the high risks of weather damage and downgrading through autumn rainfall, has resulted in very few growers persisting with the crop (Strong, 1981). New, early maturing cultivars have shown a high yield potential in Tasmania, and the crop is in demand for processing in that state. A small but growing industry is now under development there (Thanomsub and Mendham, 1992).

PIGEONPEAS

The pigeonpea (*Cajanas cajan*) is a frost-sensitive, short-lived perennial legume grown widely as a grain crop in the tropics and subtropics, where the grain is used largely for human consumption (Brinsmead and Wallis, 1984). The plant is drought tolerant, and cropping is possible in areas currently producing sorghum (Holland and Byrnes, 1986). Production in higher rainfall areas is possible only if a dry period during seed maturation in autumn is normal. Severe water stress causes significant delay in flowering and reduction in yield. The plant requires warm to hot temperatures for active growth, while frost will defoliate and even kill it (Brinsmead and Wallis, 1984). It is adapted to a wide range of soil types, but will not tolerate poor drainage. Successful establishment requires soil temperature at planting of 19°C or more (Holland and Byrnes, 1986).

The crop is of very recent commercial importance in Australia, significant sowings not being recorded until the 1985–6 season. Initial success led to rapid expansion in both Queensland and New South Wales in 1986–7, preliminary estimates for the latter (R. R. Komoll, 1987, pers. comm.) being over 9000 ha, with almost 8000 of this in the Moree and Narrabri districts of northern New South Wales. The crop is basically an export prospect, and market development will be the key to its future.

MISCELLANEOUS CROPS

Summer Crops

The important crops in this category are sugar cane and tobacco, both of which are grown in reasonably restricted areas of Australia. Production areas are shown in Fig. 2.24.

SUGAR CANE

Sugar cane (*Saccharum officinarum*), typically a crop of the tropical and subtropical regions, is confined to a narrow eastern coastal strip of

FIG. 2.24   Distribution of tobacco and sugar cane production in Australia, based on Australian Bureau of Statistics crop returns by local government areas

Australia in the wet sub-tropical and tropical climatic zones (Fig. 2.24). A perennial, tropical grass requiring high temperatures (21° to 31°C) throughout the growth period, it requires from 8 to 24 months to produce a crop (Bull and Glasziou, 1975; Williams, 1975). Under Australian conditions, cane is planted in Queensland in autumn and harvested during the dry winter of the following year, production entailing a 15-month cycle (one-year cane). In the southern cane producing areas of northern New South Wales, the production period is much longer, ranging from 20 to 24 months (two-year cane). Here cold winter temperatures limit sowing to spring, the crop subsequently growing through two summers and being harvested in winter of the second year. The crop is frequently allowed to regrow following harvest, producing a second or ratoon crop some 18 to 20 months later, although ratoon yields are typically lower than those of the initial sown crop (Williams, 1975). Moisture requirements are very high, approximating and even exceeding pan evaporation rates during active growth. Irrigated crops in Queensland require up to 2000 mm of water per year (Ham, 1970; Bull and Glasziou, 1979), while successful dryland production throughout the world is limited to areas with a summer rainfall of 1250 to 1625 mm (Klages,

1942). Rapid summer growth at high temperatures is followed by a period of cool temperatures in the following autumn and winter, during which growth rate slows down considerably and sucrose is stored in the stem (Martin *et al.*, 1976). Moisture deficits during this period can result in reduced sucrose storage (Bull and Glasziou, 1975; Williams, 1975) and lower crop value. Supplementary irrigation during the drier winter period of northern Australia is utilised to offset this problem. Intense summer cyclone activity in northeastern Queensland is a significant hazard to crop production in some seasons, high winds causing extensive mechanical damage.

TOBACCO

The production of tobacco (*Nicotiana tabacum*) is a highly specialised industry in Australia, and is carried out over a fairly wide latitudinal range from the Atherton Tableland in northeast Queensland to northern Victoria (Fig. 2.24), reflecting in part the wide climatic adaptation of the crop (Martin *et al.*, 1976). Current distribution largely reflects the availability of suitable irrigation water (particularly in Victoria) combined with suitable soil types, the best-quality leaf being obtained on chemically poor soils where close control can be maintained over crop nutrition. Successful production requires a hot, humid environment with a frost-free period of 100 to 150 days and, particularly, freedom from strong winds and hail (Chippendale, 1961). Temperature requirements desired during the initial 6 to 10 weeks seedbed period are 24° to 26°C, followed by field temperatures after transplanting averaging 26° to 27°C (Martin *et al.*, 1976). The crop is sensitive to water stress at all growth stages, growth checks induced by water shortage having deleterious effects on leaf quality (Chippendale, 1961). At the same time, excessive water availability and waterlogging can cause root damage (Williamson, 1970). Precise irrigation control is essential to minimise both water deficiency and excess throughout the growth cycle. Both low humidity and excessively high temperatures are detrimental to leaf quality during ripening in autumn, the finest tobaccos in the world being produced in areas with climates subject to oceanic influences (Klages, 1942). At the same time, fine weather is required for harvesting, picking of mature leaves generally being delayed for two to three days after rain to ensure adequate carbohydrate levels in the leaf (Chippendale, 1961). Some degree of cloud cover throughout the growing season also appears desirable, since excessively high solar radiation levels cause leaf blemishes and thus a lower-quality crop.

In general, production is closely controlled by water availability during the growing season, and it is likely that production will continue to be restricted to irrigation areas that have an acceptable water quality. Summer temperatures throughout the areas where production is now concentrated appear quite suitable.

POTENTIAL EXPANSION OF CROP PRODUCTION
IN AUSTRALIA

The potential for future expansion of crop production in Australia hinges on many factors, not the least being industrial, urban, socio-economic and pollution pressures, which combined could modify or inhibit expansion of all agricultural activities. The development of new crop industries, particularly in remote regions, could be severely hindered by the high cost of populating an area to service the needs of the local industry, combined with the high costs of industrial development to process certain crops (e.g. cotton, oilseeds, rice and sugar cane). These cost factors alone may prohibit development of new crop industries in regions otherwise well suited in terms of climate, natural water resources and arable land. These issues are discussed in more detail in Chapter 1.

Accepting these limitations, expansion or modification of the Australian cropping industry from the current situation appears to revolve around three major questions:

(a) What areas, currently unexploited although climatically and edaphically suited to either current or potentially important crop species, are available?

(b) What alternative crop species to those of current commercial importance are suitable for either existing or potential croplands in Australia?

(c) What economic advantages, in a regional and national sense, can be gained through: expanded production of currently important crops into new areas; substitution of new for traditional crops in current production areas; or foundation of new cropping industries utilising new species in currently unexploited areas?

In the final analysis, it is the third factor, that of economic advantages, which is most likely to influence decision making and hence the actual expansion or modification of the current crop industries (Marshall, 1992). Basic economics may indicate that the greatest economic advantages could in fact ensue through raising the productivity of current crop systems, or by the substitution or addition of alternative crops of greater economic potential, in existing cropping areas. The remainder of this chapter is concerned with the first two factors mentioned, i.e. potential new areas for crop production, and alternative crop species to those of current commercial importance. Where possible, the economic implications associated with these factors will be considered briefly, particularly in situations where economic considerations are of primary importance.

Potential Croplands

Any assessment of potential croplands in Australia hinges basically on four factors: moisture regimes, temperature regimes, soil characteristics and topography. Topography imposes a major constraint, precluding cultivation in steep or rugged areas that are inaccessible to machinery, highly susceptible to erosion, or characterised by shallow skeletal soils unsuitable for cultivation on a commercial scale. In Australia, this excludes some 29.2 million ha, of which 16.5 million ha are located in tropical areas and 12.7 million ha in temperate areas (Davidson, 1961). Using the Prescott ratio to delimit suitable cropping areas based on moisture availability, and subtracting topographically unsuitable areas, Davidson concluded that potential arable land area in Australia totalled 124.4 million ha, comprising 16.6 million ha in tropical and 107.8 million ha in temperate Australia. Excluding land areas under pastures and horticultural crops, the total cropped area in Australia in 1991 was only 17 million ha (Table 2.10). Using more rigorous analysis based on the Moisture Index, Nix (1976, pers. comm.) estimated that total area of land in Australia suitable for cropping (minimum Moisture Index of 0.4 during the crop cycle) based on moisture regimes alone was considerably greater than the estimates of Davidson, and amounted to 237 million ha, leaving a remaining area of 531 million ha that was too dry for arable agriculture. Applying terrain and topographical constraints reduced this to 132 million ha, within which the total area with suitable soils for cropping was only 77 million ha. The greatest proportional reductions based on terrain and soil constraints were in northern Australia. In considering the potential for arable land development, Nix divided the Australian continent into four zones: Arid, North, East and South. In the Köppen classification (Fig. 2.11) these zones could be categorized as follows:

Arid   :   BWh, BSwh, BSfh
North :   Aw, Af
East   :   Cwa, Cfa, Cfb
South :   BSfk, BSsk, BSsh, Cfb, Csa, Csb

Of the four, Nix saw little potential for further arable land development in the Arid or Northern zones, with greatest potential for expansion in the Eastern zone of warm temperate to subtropical climate, particularly in the inland sector characterised by high variability of seasonal rainfall. In the South, where the greatest development has occurred to date, an extension of crop production is most likely to occur with changes in land-use patterns, substituting crop production for present pasture, livestock and forestry industries. Concomitant with this development would be increased management expertise to minimise erosion risk.

Potential Alternative Crops

Potential alternative crops to those outlined in this chapter have been reviewed by Matheson (1979) and Marshall (1992). They include jojoba, buckwheat, sesame, cassava, kenaf, pyrethrum, duboisia and essential-oil crops such as tea-tree, boronia, lavender and peppermint. (Refer to the two reviews mentioned for further information.) Brief comments on some of these crops follow.

SESAME

Sesame (*Sesamum indicum*) requires a high temperature, being essentially adapted to the tropics and subtropics where summer temperatures during the growing season average 27° to 32°C. Dryland production would be most successful in Queensland and northern Australia, although severe disease and harvesting problems may be encountered under high temperature – high humidity conditions. Irrigated production is most likely to be successful in areas currently devoted to cotton production in New South Wales, Queensland and northwestern Western Australia. In all these areas it would come into direct competition with cotton, and with soybeans in the Queensland and New South Wales areas. Serious problems of uneven maturity and seed shattering would have to be overcome before it assumed a role as a commercial crop. Seed is utilised both for oil and the confectionery trade.

BUCKWHEAT

Buckwheat (*Fagopyron esculentum*) is a summer-growing, short-season (70 to 84 days) plant best adapted to cool moist temperate climates. It is very susceptible to temperature extremes, being completely intolerant of frost at all growth stages and extremely sensitive to high temperatures at flowering (Cutting, 1971). At the other extreme, it is tolerant of low soil fertility and poor seedbed preparation, although yields are considerably enhanced by fertiliser application and thorough seedbed preparation. At present it appears best adapted to spring plantings on the higher tablelands of central and southern New South Wales. Market prospects are limited on the domestic scene, but a reasonable Japanese market exists which would allow some development as a commercial crop. Grain is used both for drug extraction and as a source of specialty flour for 'buckwheat cookies' in the U.S. and Japan. Trial commercial plantings in the Orange district of New South Wales have been made under contract for the Japanese market.

CASSAVA

Cassava (*Manihot esculenta*) is a tropical species confined to areas in a broad belt between 30°N and S latitudes. The crop appears limited to Australian regions in which the lowest winter (July) temperature is 13°C, confining potential production areas to a narrow strip in northeast Queensland and extending across northern Australia to the west coast of Western Australia (De Boer and Forno, 1975). For commercial yields to be attainable, rainfall of 1000 mm is required, with a distribution spread ideally over 9 to 10 months of the year. Areas defined as suitable for cassava production were outlined by De Boer and Forno as having a summer season Moisture Index of 0.80 or more. The principal commercial product is the roots, which when dried and pelleted provide a high energy carbohydrate source for stock feed. Markets exist in the EEC. Additionally, high-protein meal produced from the leaves and stems would be competitive with fishmeal in the domestic stock feed market. Starch manufacture from the roots for domestic markets is limited, major competitors being maize, wheat and potatoes. Major restraints at present are the isolated nature of suitable production areas, the high costs of mechanisation for production, and the need to establish expensive drying and pelleting plants to process roots for the EEC market (De Boer and Forno, 1975).

KENAF

Kenaf (*Hibiscus cannabinus*) is a tropical species cultivated presently in the U.S., Nigeria, South Africa and New Guinea for pulp production for paper manufacture. It is a tall (3 to 6 metres) annual species which at present appears most suitable for the summer rainfall areas of northern Australia. In addition to the use of stems for paper pulp, the plant tops and leaves are suited as a stock feed protein source, while the seed is useful for oil and protein extraction. Current prospects appear limited by the demand for pulp, which at present is restricted by its suitability for production of high-quality paper only, the market for which is restricted in size.

JOJOBA

Jojoba (*Simmondsia californica*) is a bushy, evergreen perennial shrub which grows to three metres and is native to desert, as well as coastal, regions of Central and South America. It usually grows on hillsides and is well adapted to arid situations with an annual rainfall of 100 to 300 mm. Growth occurs in the warmer months, with flowering in spring and fruit development and ripening in early to mid-summer. Winter flowering may occur, but buds and developing fruit are sensitive to frost damage. There are separate male and female plants; one in ten plants must be male for successful pollination. Plants mature sexually at 2 to 3 years of age, but may not mature fully until 8 years. They prefer deep, well-drained, fertile soils of sandy to medium texture and pH of 5.5 or above. The ripe seed contains about 50 per cent by weight of a liquid wax which has similar properties to sperm whale oil and is in demand for high-grade cosmetics and lubricants. Consistent trial yields in central New South Wales of 0.5 tonne per ha would produce gross returns of $750 per ha at current prices. The release of three cultivars in 1992, together with development of an agronomic management package, should ensure the crop a sound future as a new industrial crop (Milthorpe, 1992).

Acknowledgements

Grateful acknowledgement is extended to the federal and the various state offices of the Australian Bureau of Statistics for supply of statistical data utilised in preparation of the various maps of crop distribution, and to the New South Wales, Victorian, Tasmanian and Western Australian Departments of Agriculture, the Queensland Department of Primary Industry, and the South Australian Department of Agriculture and Fisheries for supplying crop estimates and unpublished material.

References

Anderson, W. K. (1979), 'Sorghum' in Lovett and Lazenby (1979), 37–69.

Anderson, W. K. (1980), 'Some Water Use Responses of Barley, Lupin and Rapeseed', *Australian Journal of Experimental Agriculture and Animal Husbandry*, **20**, 202.

Anderson, W. K., Smith, R. C. G. and McWilliam, J. R. (1978), 'A Systems Approach to the Adaptation of Sunflower to New Environments. I. Phenology and Development,' *Field Crops Research*, **1**, 141.

Andrews, J. and Maze, W. H. (1933), 'Some Climatological Aspects of Aridity in their Application to Australia', *Proceedings of the Linnaean Society of New South Wales*, **58**, 105.

Anon. (1965), *Climate and Meteorology of Australia*, Bureau of Meteorology Bulletin No. 1, Melbourne.

Anon. (1967a), *A Guide to Cotton Growing in Southwestern New South Wales*, Irrigation Research and Extension Committee, Griffith.

Anon. (1967b), *Pest Control in Groundnuts*, PANS Manual No. 2, Ministry of Overseas Development, London.

Anon. (1973), *Oilseed Sunflowers*, New South Wales Department of Agriculture Division of Plant Industry Bulletin P456.

Anon. (1975a), *Barley Growing in New South Wales*, New South Wales Department of Agriculture Division of Plant Industry Bulletin P317.

Anon. (1975b), *Rapeseed Newsletter*, Oilseeds Marketing Board of New South Wales Bulletin No. 1.

Anon. (1976), 'Hybrid *v.* Open Pollinated Sunflowers', *Agnote* No. 40/76, New South Wales Department of Agriculture.

Anon. (1982), *Crops Australia, 1980–81*, Australian Bureau of Statistics Catalogue No. 7321.0.

Anon. (1983), *Crops and Pastures Australia 1981–82*, Australian Bureau of Statistics Catalogue No. 7321.0.

Anon. (1986a), *Crops and Pastures Australia 1984–85*, Australian Bureau of Statistics Catalogue No. 7321.0.

Anon. (1986b), *Growing Field Peas in Western Australia*, Department of Agriculture, Western Australia, Farmnote 28/86, Agdex 166/20.

Anon. (1992), 'Geography and Climate', in *Year Book Australia 1992*, Australian Bureau of Statistics, No. 75, Catalogue No. 1301.0.

Anon. (1989), *Summary of Crops Australia 1987–88*, Australian Bureau of Statistics Catalogue No. 7330.0.

Anon. (1992), *Summary of Crops Australia 1990–91*, Australian Bureau of Statistics Catalogue No. 7330.0.

Armstrong, E. L. (1985), 'Sowing Time Effects on Yield, Growth and Development of Rapeseed in Central Western New South Wales', in *Proceedings of the 3rd Australian Agronomy Conference*, Hobart, 317.

Armstrong, E. L., Bernardi, A. L., Banks, L. W. and Drew, T. P. (1985), 'Flowering Date and Yield of Rapeseed', in *Proceedings of the 3rd Australian Agronomy Conference*, Hobart, 308.

Arnon, I. (1972), *Crop Production in Dry Regions.* Vol. 2: *Systematic Treatment of the Principal Crops*, Plant Science Monographs (ed. N. Polunin), Leonard Hill, London.

Asana, R. D. and Williams, R. F. (1965), 'The Effect of Temperature Stress on

Grain Development in Wheat', *Australian Journal of Agricultural Research*, **16**, 1.

Badawy, N. S., Mahoney, J. E. and Jessop, R. S. (1985), 'The Performance of Lupins on the Alkaline Grey Clays of the Wimmera Region of Victoria' in *Proceedings of the 3rd Australian Agronomy Conference*, Hobart, 330.

Barnes, W. C. (1973), 'Rain Damage in Wheat', *Agricultural Gazette of New South Wales*, **84**, 268.

Basinski, J. J. (1963), *Cotton Growing in Australia—an Agronomic Survey*, CSIRO Division of Land Research and Regional Survey, Canberra.

Basinski, J. J. (1965), 'The Cotton Growing Industry in Australia', *Journal of the Australian Institute of Agricultural Science*, **31**, 206.

Basinski, J. J. and Beech, D. F. (1972), 'Prospect and Problems of Safflower in Australia' in *Proceedings of Australian Specialist Conference on Crops of Potential Economic Importance*, Manly, 3–27.

Basinski, J. J., Beech, D. F. and Lee, L. C. (1961), 'Effect of Time of Planting on Yields of Safflower in Northern Australia', *Journal of the Australian Institute of Agricultural Science*, **27**, 156.

Beech, D. F. (1969), 'Safflower—a review article', *Field Crop Abstracts*, **22**, 107.

Beech, D. F. (1979), 'Safflower' in Lovett and Lazenby (1979), 161.

Beech, D. F. and Norman, M. J. T. (1963), 'The Effect of Time of Planting on Yield Attributes of Safflower', *Australian Journal of Experimental Agriculture and Animal Husbandry*, **3**, 140.

Benson, R. J. and Dale, A. B. (1992), *Faba Bean Update February 1992*, New South Wales Agriculture and Fisheries Agnote 2/056, Agdex 168/00.

Berry, G. J. (1985), 'Performance of New Field Pea Plant Types' in *Proceedings of the 3rd Australian Agronomy Conference*, Hobart, 332.

Biddiscombe, E. F. (1975), 'Effect of Moisture Stress on Flower Drop and Seed Yield of Narrow-leafed Lupin (*Lupinus angustifolius* L. cv Unicrop)', *Journal of the Australian Institute of Agricultural Science*, **41**, 70.

Boerema, E. B. (1969), 'Rice Growing in New South Wales', *Agricultural Gazette of New South Wales*, **80**, 450.

Boerema, E. B. (1973), 'Rice Cultivation in Australia', *Il Riso*, **22**, 131.

Bott, W. (1970), 'Cowpea Seed Growing on Darling Down', *Queensland Agricultural Journal*, **96**, 722.

Boundy, K. and Smith, I. (1981), *Growing Field Peas*, Department of Agriculture, Victoria, Agnote 1639/81, Agdex 140/10.

Boundy, K. A., Reeves, T. G. and Brooke, H. D. (1982), 'Growth and Yield Studies of *Lupinus angustifolius* and *L. albus* in Victoria', *Australian Journal of Experimental Agriculture and Animal Husbandry*, **22**, 76.

Brinsmead, R. B. (1992), 'Chickpea Cultivar by Planting Time Studies in Queensland', *Proceedings of the 6th Australian Agronomy Conference*, Armidale, 244.

Brinsmead, R. B., Rettke, M. L. and Keys, P. J. (1984), *Chickpeas in Queensland*, Department of Primary Industries, Queensland, Farmnote F29/84, Agdex 143/20.

Brinsmead, R. B. and Wallis, E. S. (1984), *Pigeonpeas in Queensland*, Queensland Department of Primary Industries Farmnote F133/84, Agdex 164/20.

Browne, R. L. (1984), *Irrigation Management of Cotton*. Department of Agriculture, New South Wales, Agfact P5.3.2, Agdex 151/560.

Buirchell, B. and Cowling, W. (1992), 'Domestication of Rough Seeded Lupins', *Journal of Agriculture, Western Australia.* **33/4**. 131.

Bull, T. A. and Glasziou, K. T. (1975), 'Sugar Cane' in Evans (1975), 51–72.

Bull, T. A. and Glasziou, K. T (1979), 'Sugarcane' in Lovett and Lazenby (1979), 95–112.

Bureau of Meteorology (1975), 'Climate and Physical Geography of Australia' in *Year Book Australia 1974*, Government Printer, Canberra, 25–75.

Burton, G. W., Wallace, A. T. and Rachie, K. O. (1972), 'Chemical Composition and Nutritive Value of Pearl Millet (*Pennisetum typhoides* (Burm.) Stapf and E. C. Hubbard) Grain', *Crop Science*, **12**, 187.

Buzza, G. (1979), 'Rapeseed' in Lovett and Lazenby (1979), 183–97.

Byth, D. E. (1968), *The Soybean in Australia—Past, Present and Future*, Australian Institute of Agriculture Science Symposium, Sydney.

Canvin, D. T. (1965), 'The Effect of Temperature on the Oil Content and Fatty Acid Composition of the Oils from Several Oilseed Crops', *Canadian Journal of Botany*, **43**, 63.

Carter, O. G. (1970), 'Soybean as a World Cup and as a Crop for Australian Conditions', *Journal of the Australian Institute of Agricultural Science*, **36**, 316.

Castleman, G. H. (1982), 'Realisation of the Potential of Triticale' in *Proceedings of the 2nd Australian Agronomy Conference*, Wagga Wagga, 243.

Cawood, R. J. (1982), 'Crop Evapotranspiration and Soil Water Depletion of Dryland Sunflowers' in *Proceedings of the 2nd Australian Agronomy Conference*, Wagga Wagga, 219.

Chippendale, F. (1961), 'Flue-cured Tobacco Growing' in *Tobacco Growing in Queensland*, Queensland Department of Agriculture and Stock, 1–67.

Chudleigh, P. D. (1991), 'Breeding and Quality Analysis of Canola (Rapeseed)', ACIAR Economic Assessment Series No. 6, Australian Centre for International Agricultural Research, Canberra.

Claxton, R. A. (1972), 'Rapeseed—a New Oil Crop for New South Wales', *Agricultural Gazette of New South Wales*, **83**, 194.

Colless, J. M. (1979), 'Maize' in Lovett and Lazenby (1979), 1–36.

Colless, J. M. (1982), *Maize Growing*, Department of Agriculture, New South Wales, Agfact P3.3.3, Agdex 111.

Colton, R. T. (1981), *Rapeseed Growing*, Department of Agriculture, New South Wales, Agfact P5.2.1, Agdex 144.

Colton, R. T. (1983), *Safflower Growing*, Department of Agriculture, New South Wales, Agfact P5.2.2, Agdex 145/20.

Colton, R. T. (1985), *Rapeseed Growing*, Department of Agriculture, New South Wales, Agfact P5.2.1, Agdex 145/20.

Constable, G. A. and Shaw, A. J. (1988), *Temperature Requirements for Cotton*. New South Wales Agriculture and Fisheries, Agfact P5.3.5, Agdex 151.022.

Cook, L. J. (1983), *Triticale Recommendations for 1983*. Department of Agriculture, New South Wales, Agdex 117/32.

Cook, L. J. (1987), *Barley Growing*, Department of Agriculture, New South Wales, Agfact P3.2.3, Agdex 114/10.

Cornish, P. S. and Pratley, J. E. (eds) (1987), *Tillage, New Directions in Australian Agriculture*, Australian Society of Agronomy, Australian Land Series, Inkata Press, Melbourne and Sydney.

Cowling, W. A., Allen, J. G., Wood, P. McR. and Hamblin, J. (1986), 'Phomopsis Resistant Lupins—Breakthrough Towards the Control of Lupinosis', *Journal of Agriculture, Western Australia*, **27**, 43.

Crosthwaite, I. C. (1983), 'Maize Growing on the Atherton Tableland', *Queensland Agricultural Journal*, **109**, 40.

Cull, P. O. (1985), 'Case Studies of Irrigation Scheduling and On Farm Water Management in Eastern Australia', *Proceedings of the 3rd Australian Agronomy Conference*, Hobart, 366.

Cull, P. O. (1987), 'Software for Irrigation Scheduling Using the Neutron Probe', *Proceedings of the 4th Australian Agronomy Conference*, Melbourne, 317.

Cull, P. O. (1992), 'Irrigation Scheduling and Crop Management System', *Proceedings of the 6th Australian Agronomy Conference*, Armidale, 1992, 561.

Cutting, F. W. (1971), 'Technical Aspects of Oilseed Crops' in *Proceedings of the Workshop on Production and Marketing on the Southern Tablelands, N.S.W.*, Southern Tablelands Research—Extension Liaison Committee, CSIRO, Canberra, 61–77.

Cutting, F. W. (1974), 'Safflower Not Yet a Success Story', *Agricultural Gazette of New South Wales*, **85**, 2.

Dale, A. B. (1983), *Sunflower Variety and Planting Guide 1983–84*, Department of Agriculture, New South Wales, Agfact P5.1.2, Agdex 145/30.

Dale, A. B. (1984), *Sunflower Growing*, Department of Agriculture, New South Wales, Agfact P5.2.3, Agdex 145/20.

Dale, A. B., Benson, R. J., Knights, E. J. and Schwinghamer, M. W. (1992), *Chickpea Update February 1992*, New South Wales Agriculture and Fisheries, Agnote Reg 2/059, Agdex 168/00.

Dale, A. B. and Colless, J. M. (1990), *Maize Hybrid and Planting Guide 1990/91*, New South Wales Agriculture and Fisheries, Agdex 111/30.

Davidson, B. R. (1961), 'The Distribution of Agricultural Land in Australia', *Journal of the Australian Institute of Agricultural Science*, **27**, 203.

Davidson, J. (1934), 'The Monthly Precipitation–Evaporation Ratio in Australia as Determined by Saturation Deficit', *Transactions of the Royal Society of South Australia*, **58**, 33.

Davidson, J. (1936), 'Climate in Relation to Insect Ecology in Australia, 3. Bioclimatic Zones in Australia', *Transactions of the Royal Society of South Australia*, **60**, 88.

Davidson, J. L., Christian, K. R. and Bremner, P. M. (1985), 'Cereals for the High Rainfall Zone of Temperate Australia' in *Proceedings of the 3rd Australian Agronomy Conference*, Hobart, 112.

De Boer, A. J. and Forno, D. A. (1975), 'Cassava: Potential Agro-industrial Crop for Tropical Australia', *Journal of the Australian Institute of Agricultural Science*, **41**, 241.

Delane, R. J., Hamblin, J. and Gladstones, J. S. (1986), 'Reduced-branching Lupins', *Journal of Agriculture, Western Australia*, **27**, 47.

Derera, N. F. (1973), 'Should Red Wheat be Introduced into the Northern Wheat Belt?', *Journal of the Australian Institute of Agricultural Science*, **39**, 48.

Doggett, H. (1970), *Sorghum*, Tropical Agriculture Series, Longmans, Green and Co., London.

Doolette, J. B. (1968), 'Some Factors Which Influence Barley Yield and Quality in South Australia' in *Proceedings of the 10th Convention of Australian and New Zealand Section of the Institute of Brewers*, 103.

Doyle, A. D. (1975), 'Influence of Temperature and Daylength on Phenology of Sunflowers in the Field', *Australian Journal of Experimental Agriculture and Animal Husbandry*, **15**, 88.

Doyle, A. D. and Marcellos, H. (1974), 'Time of Sowing and Wheat Yield in Northern New South Wales', *Australian Journal of Experimental Agriculture and Animal Husbandry*, **14**, 93.

Duncan, M. (1983), *Winter Fodder Crops: Northern Tablelands*, Department of Agriculture, New South Wales, Agfact P2.3.3, Agdex 120/20.

Duncan, W. G. (1975), 'Maize' in Evans (1975), 23–50.

Evans, L. T. (ed.) (1975), *Crop Physiology*, Cambridge University Press, London.

Evans, L. T., Wardlaw, I. F. and Fischer, R. A. (1975), 'Wheat' in Evans (1975), 101–49.

Farrington, P. (1974), 'Grain Legume Crops: Their Role on the World Scene and in Australian Agriculture', *Journal of the Australian Institute of Agricultural Science*, **40**, 99.

Fawcett, R. G., Gidley, V. N. and Doyle, A. D. (1976), 'Fallow Moisture and Wheat Yields in the Northwest', *Agricultural Gazette of New South Wales*, **87/2**, 28.

Fischer, R. A. (1973), 'The Effect of Water Stress at Various Stages of Development on Yield Processes in Wheat' in *Plant Response to Climatic Factors*, UNESCO, Paris, 233–41.

Fischer, R. A. (1979), 'Growth and Water Limitations to Dryland Wheat Yields in Australia: A Physiological Framework', *Journal of the Australian Institute of Agricultural Science*, **45**, 83.

Fisher, J. and Scott, B. (1983), 'Wheat for Acid Soils', *Bulk Wheat*, **17**, 74.

Fitzpatrick, E. A. and Nix, H. A. (1975), 'The Climatic Factor in Australian Grassland Ecology' in R. M. Moore (ed.), *Australian Grasslands*, Australian National University Press, Canberra, 3–26.

Fitzsimmons, R. W. (1984), *Rye Growing in New South Wales*, Department of Agriculture, New South Wales, Agfact P3.2.5, Agdex 117.

French, R. J. (1978), 'The Effect of Fallowing on the Yield of Wheat. I. The Effect of Soil Water Storage and Nitrate Supply', *Australian Journal of Agricultural Research*, **29**, 653.

French, R. J. and Schultz, J. E. (1982), 'The Phenology of Eight Cereal, Grain Legume, and Oilseed Crops in South Australia', *Australian Journal of Experimental Agriculture and Animal Husbandry*, **22**, 67.

Gaffney, D. O. (1971), *Seasonal Rainfall Zones in Australia*, Bureau of Meteorology Working Paper 141, Melbourne.

Gaffney, D. O. (1973), 'Climate' in *Atlas of Australian Resources*, 2nd edn, Government Printer, Canberra, 3–9.

Gallagher, E. C. (1972), 'Navy Beans, Part 1', *Queensland Agricultural Journal*, **98**, 562.

Gentilli, J. (1971), 'Climates of Australia and New Zealand' in *World Survey of Climatology*. Vol. 13: *Climates of Australia and New Zealand*, Elsevier, Amsterdam, 35–210.

Gentilli, J. (1972), *Australian Climate Patterns*, Nelson, Melbourne.

Gibbs, W. J. and Maher, J. V. (1967), *Rainfall Deciles as Drought Indicators*, Bureau of Meteorology Bulletin 48, Melbourne.

Gilmour, R., Tarr, A. and Harasymow, S. (1992), 'Breeding Better Malting Barleys', *Journal of Agriculture, Western Australia*, **33**, 114.

Gladstones, J. S. (1967), 'Uniwhite, a New Lupin Variety', *Journal of Agriculture', Western Australia*, **8**, 190.

Gladstones, J. S. (1969), 'Lupins in Western Australia—1. Species and Varieties: *Journal of Agriculture, Western Australia*, **10**, 318.

Gladstones, J. S. (1970), 'Lupins as Crop Plants', *Field Crop Abstracts*, **23**, 123.

Gladstones, J. S. (1976), 'The Mediterranean White Lupin', *Journal of Agriculture, Western Australia,* **17**, 70.

Gladstones, J. S. (1982a), 'Breeding the First Modern Crop Lupins', *Journal of Agriculture, Western Australia*, **23**, 70.

Gladstones, J. S. (1982b), 'Breeding Lupins in Western Australia', *Journal of Agriculture, Western Australia*, **23**, 73.

Green, A. G. (1992), 'The Evaluation of Linola as a New Oilseed Crop for Australia', *Proceedings of the 6th Australian Agronomy Conference*, Armidale, 471.

Ham, G. J. (1970), *Water Requirements of Sugar Cane*, Report of Water Research Foundation of Australia No. 32.

Harrigan, E. K. S. and Sykes, J. D. (1987), *New Safflower Varieties*, Department of Agriculture, New South Wales, Agfact P5.1.14, Agdex 145/30.

Hazlewood, L. K. (1973), 'Climatic Classification' in *Atlas of Australian Resources*, 2nd edn, Government Printer, Canberra, 9–12.

Henzell, R. G., Martin, I. F. and Cox, M. C. (1985), 'The Contribution of Science to Australian Tropical Agriculture. 5. Tropical Coarse Grains', *Journal of the Australian Institute of Agricultural Science*, **51**, 42.

Hesketh, J. D. and Low, A. (1968), 'Effect of Temperature on Components of Yield and Fibre Quality of Cotton Varieties of Diverse Origin', *Cotton Growing Review*, **45**, 243.

Holland, J. F. and Byrnes, P. J. (1986), *Pigeonpea Recommendations*, Department of Agriculture, New South Wales, Agdex 168/00.

Holland, J. F., Doyle, A. D. and Marley, J. M. (1987), 'Tillage Practices for Crop Production in Summer Rainfall Areas' in Cornish and Pratley (1987), 48–71.

Hooke, M. (1986). 'The Chickpea Crop in Queensland', *The Pulse*, **2**, 6.

Horowitz, B. and Beech, D. F. (1974), 'Photoperiodism in Safflower (*Carthamus tinctorius* L.)', *Journal of the Australian Institute of Agricultural Science*, **40**, 154.

Huda, A. K. S., Holloway, R. E. and Rahman, M. S. (1992a), 'Simulating the Probabilities of Early Seeding, Water Use and Wheat Yield on Eyre Peninsula' in *Abstracts of Poster Presentations*, *42*. National Soils Conference, Australian Society of Soil Science Inc., Adelaide, 19–23 April 1992.

Huda, A. K. S., Lymn, A. S., Holloway, R. E. and Rahman, M. S. (1992b), 'Agroclimatic Analysis for Assessing Cropping Risk' in *Proceedings of Agronomy Technical Conference*, 116–23. Department of Agriculture, South Australia, Adelaide, 24–25 March 1992.

Jackson, K. J. and Berthelsen, J. E. (1982), 'Yield Compensation in Safflower

Following Hand Pruning to Simulate Frost Damage' in *Proceedings of the 2nd Australian Agronomy Conference*, Wagga Wagga, 231.

Jackson, K. J. and Berthelsen, J. E. (1986), 'Production of Safflower, *Carthamus tinctorius* L., in Queensland', *Journal of the Australian Institute of Agricultural Science*, **52**, 63.

Janick, J., Schery, R. W., Woods, F. W. and Ruttan, V. W. (1974), *Plant Science: An Introduction to World Crops*, 2nd edn, Freeman and Co., San Francisco.

Johnston, B. G. (1984), 'Expansion of Wheat-Growing into High-Rainfall Areas' in J. L. Davidson and D. R. de Kantzow (eds), *Directions for Australia's Wheat Industry* IV/9. Australian Institute of Agricultural Science Occasional Paper No. 17, 1–15.

Kay, G., Shaw, A. J. and Stewart, J. (1992), *Dryland Cotton Growing*, New South Wales Agriculture, Agdex 151.

Keig, B. and McAlpine, J. R. (1974), 'WATBAL: A Computer Program for the Estimation and Analysis of Soil Moisture from Simple Climatic Data', 2nd edition, Technical Memorandum, CSIRO Division of Land Use Research, Canberra, **74/4**, 45.

King, R. H. (1980), 'The Nutritive Value of Triticale for Growing Pigs', *Proceedings of the Australian Society of Animal Production*, **13**, 381.

Klages, K. H. W. (1942), *Ecological Crop Geography*, MacMillan, New York.

Knights, E. J. (1982), 'Some Basic Breeding Objectives for Dryland Chickpeas' in *Proceedings of the 2nd Australian Agronomy Conference*, Wagga Wagga, 336.

Kohn, G. D., Storrier, R. R. and Cuthbertson, E. G. (1966), 'Fallowing and Wheat Production in Southern New South Wales', *Australian Journal of Experimental Agriculture and Animal Husbandry*, **6**, 233–41.

Laing, D. R. (1974), *Co-operative Soybean Variety Evaluation Programme 1970–1972*, University of Sydney, Faculty of Agriculture Report No. 8.

Laing, D. R. and Byth, D. E. (1972), 'Soybeans in Australia' in *Proceedings of Australian Specialist Conference on Crops of Potential Economic Importance*, Manly, 4–23.

Launders, T. E. (1971), 'The Effects of Early Season Soil Temperatures on Emergence of Summer Crops in the North Western Plains of New South Wales', *Australian Journal of Experimental Agriculture and Animal Husbandry*, **11**, 39.

Lawn, R. J. and Russell, J. S. (1978), 'Mungbean: A Grain Legume for Summer Rainfall Cropping Areas of Australia', *Journal of the Australian Institute of Agricultural Science*, **44**, 28.

Lazenby, A. and Matheson, E. M. (eds) (1975), *Australian Field Crops. 1 — Wheat and Other Temperate Cereals*, Angus and Robertson, Sydney.

Lovett, J. V. (1975), 'Rye' in Lazenby and Matheson (1975), 508–37.

Lovett, J. V., Harris, H. C. and McWilliam, J. R. (1979), 'Sunflower' in Lovett and Lazenby (1979), 137–60.

Lovett, J. V. and Lazenby, A. (eds) (1979), *Australian Field Crops. 2 — Tropical Cereals, Oilseeds, Grain Legumes and Other Crops*, Angus and Robertson, Sydney.

Mabbutt, J. A. and Sullivan, M. E. (1970), 'Landforms and Structure' in R. M. Moore (ed.), *Australian Grasslands*, Australian National University Press, Canberra, 27–43.

McDonald, D. J. (1972), 'Low Temperature and Rice Sterility', *Agricultural Gazette of New South Wales*, **83**, 377.

McDonald, D. J. (1978), 'Rice and its Adaptation to World Environments', *Journal of the Australian Institute of Agricultural Science*, **44**, 3.

McDonald, D. J. (1979), 'Rice' in Lovett and Lazenby (1979), 70–94.

McDonald, G. K., Sutton, B. G. and Ellison, F. W. (1982), 'Yield Limitations of Irrigated Wheat in the Lower Namoi Valley' in *Proceedings of the 2nd Australian Agronomy Conference*, Wagga Wagga, 249.

McDonald, G. K., Sutton, B. G. and Ellison, F. W. (1983), 'The Effect of Time of Sowing on the Grain Yield of Irrigated Wheat in the Namoi Valley, New South Wales', *Australian Journal of Agricultural Research*, **34**, 229.

McMahon, J. and Low, A. (1972), 'Growing Degree Days as a Measure of Temperature Effects on Cotton', *Cotton Growing Review*, **49**, 39.

McWhirter, K. S. (1972), 'An Appraisal of Present and Future Production and Breeding Problems of Maize, Sorghum and Millet' in *Proceedings of Australian Specialist Conference on Crops of Potential Economic Importance*, Sydney, 5–10.

Maher, J. V. (1966), 'Climatological Normals', *Australian Meteorological Magazine*, **14**, 30.

Marcellos, H. (1984), 'Winter Grain Legumes for Northern New South Wales', *Bulk Wheat*, **18**, 76.

Marcellos, H. (1985), 'Developmental Attributes of Faba Beans for Northern New South Wales' in *Proceedings of the 3rd Australian Agronomy Conference*, Hobart, 328.

Marcellos, H. and Constable, G. A. (1986), 'Effects of Plant Density and Sowing Date on Grain Yield of Faba Beans (*Vicia faba* L.) in Northern New South Wales', *Australian Journal of Experimental Agriculture*, **26**, 493.

Mares, D. (1984), 'Pre-harvest Sprouting in Wheat—the Problem and its Solution', *Bulk Wheat*, **18**, 46.

Mares, D. (1985), 'New Horizons for Tolerance to Pre-harvest Sprouting in Wheat', *Bulk Wheat*, **19**, 79.

Marshall, D. R. (1984), 'Red Wheats and Feed Wheats' in J. L. Davidson and D. R. de Kantzow (eds), *Directions for Australia's Wheat Industry* III/7. Australian Institute of Agricultural Science Occasional Paper No. 17, 1–6.

Marshall, D. R. (1992), 'Developments—New Kinds of Plants', *Proceedings of the 6th Australian Agronomy Conference*, Armidale, 104.

Martin, R. H. (1983), 'Improved Winter Wheats for New South Wales', *Bulk Wheat*, **17**, 20.

Martin, J. H., Leonard, W. H. and Stamp, D. L. (1976), *Principles of Field Crop Production*, 3rd edn, McMillan Publishing Co., New York.

Matheson, E. M. (1976), *Vegetable Oil Seed Crops in Australia*, Holt, Rinehart and Winston, Sydney.

Matheson, E. M. (1979), 'Other Crops of Potential Importance' in Lovett and Lazenby (1979), 263–86.

May, C. E. (1981), 'Triticales in New South Wales', *Agricultural Gazette of New South Wales*, **92/6**, 28.

Mendham, N. J. and Russell, J. (1982), 'The Yield Potential of New Oilseed Rape Cultivars in Tasmania' in *Proceedings of the 2nd Australian Agronomy Conference*, Wagga Wagga, 230.

Milthorpe, P. J. (1992), 'Jojoba: An Opportunity Too Good To Miss', *Australian Grain*, **2**, 22.

Mock, I. (1984). *Growing Lupins in Low Rainfall Districts.* Department of Agriculture, Victoria, Agnote 2536/84, Agdex 161/20.

Moore, K. J. (1974), 'Cereal Diseases and Frost Damage in 1973–74', *Agricultural Gazette of New South Wales*, **85**, 9.

Mullen, C. (1986), *Chickpea Performance in the Dubbo–Gilgandra–Narromine Districts*, Department of Agriculture, New South Wales, Agdex 168/34.

Nelson, P., Smart, W. and Walton, G. (1986), *Growing Lupins in Low Rainfall Areas*, Department of Agriculture, Western Australia, Farmnote 23/86, Agdex 161/20.

Nix, H. A. (1971), 'Environment in Relation to Crop Production' in *Proceedings of the Workshop on Crop Production and Marketing on the Southern Tablelands, N.S.W.*, Canberra, 1.

Nix, H. A. (1975), 'The Australian Climate and its Effects on Grain Yield and Quality' in Lazenby and Matheson (1975), 183–226.

Nix, H. A. (1976), 'Resource Limitations: Land and Water' in Proceedings of *Limitations to Growth: Options for Action*, Australian Institute of Agricultural Science National Conference, Canberra, 18–27.

Owen, P. C. (1971), 'The Effects of Temperature on the Growth and Development of Rice', *Field Crop Abstracts*, **24**, 1.

Pasternak, D. and Wilson, G. L. (1969), 'Effects of Heat Waves on Grain Sorghum at the Stage of Head Emergence', *Australian Journal of Experimental Agriculture and Animal Husbandry*, **9**, 636.

Poole, M. L. (1972), 'Economic Potential of Rapeseed in Australia—a critical review' in *Proceedings of Australian Specialist Conference on Crops of Potential Economic Importance*, Sydney, 3–9.

Poole, M. L. (1987), 'Tillage Practices for Crop Production in Winter Rainfall Areas' in Cornish and Pratley (1987), 24–47.

Prescott, J. A. (1938), 'Indices in Agricultural Climatology', *Journal of the Australian Institute of Agricultural Science*, **4**, 33.

Prescott, J. A. (1949), 'A Climatic Index for the Leaching Factor in Soil Formation', *Journal of Soil Science*, **1**, 9.

Prescott, J. A. (1956), 'Climatic Indices in Relation to the Water Balance, Australia', *UNESCO Symposium on Arid Zone Climatology*, Canberra, 5a–5g.

Prescott, J. A. and Thomas, J. A. (1949), 'The Length of the Growing Season in Australia as Determined by the Effectiveness of the Rainfall: a revision', *Proceedings of the Royal Geographic Society of Australia, South Australian Branch*, **50**, 42.

Price, M. (1968), 'Fatty Acid Composition of Linseed in Northern and Southern Queensland', *Queensland Journal of Agricultural and Animal Sciences*, **22**, 129.

Pritchard, K. and Moran, J. (1987), *Maize for Fodder. A Guide to Growing, Conserving and Feeding Irrigated Maize in Northern Victoria*, Department of Agriculture and Rural Affairs, Victoria, Technical Report Series No. 146, Agdex 111/00.

Purdy, J. L. and Crane, P. L. (1967), 'Influence of Pericarp on Differential Drying Rate in "Mature Corn" (*Zea mays* L.)', *Crop Science*, **7**, 379–81.

Pye, D. L. (1982), 'The Development of Chickpeas as a Viable Grain Legume

Crop for the Victorian Mallee' in *Proceedings of the 2nd Australian Agronomy Conference*, Wagga Wagga, 226.

Pye, D. L. (1984), *Growing Chickpeas*, Department of Agriculture, Victoria, Agnote 2600/84, Agdex 143/20.

Quinlan, J. (1984), *Growing Ryecorn*, Department of Agriculture, Victoria, Agnote 2594/84, Agdex 117/20.

Ralph, W. (1982), 'Towards Improved Sunflower Yields', *Rural Research in CSIRO*, **115**, Winter 1982, 4.

Rawson, H. M. (1979), 'Vertical Wilting and Photosynthesis, Transpiration and Water Use Efficiency of Sunflower Leaves', *Australian Journal of Plant Physiology*, **6**, 109.

Reeves, J. T. (1980), 'Triticale—a New Grain Crop?', *Journal of Agriculture, Western Australia*, **21**, 120.

Ridge, P. E. (1986), 'A Review of Long Fallows for Dryland Wheat Production in Southern Australia', *Journal of the Australian Institute of Agricultural Science*, **52**, 37.

Robinson, R. G. (1971), 'Sunflower Phenology. Year, Variety and Date of Planting Effects on Day and Growing Degree Day Summations', *Crop Science*, **11**, 635.

Rose, I. A., Holland, J. F. and Shaw, A. J. (1987), *Dryland Soybean Guide for Northern New South Wales*, Department of Agriculture, New South Wales, Agfact P5.1.12, Agdex 162/32.

Saint-Smith, J. H., Rawson, J. E. and McCarthy, G. P. (1969), 'Peanut Growing in Queensland—1', *Queensland Agricultural Journal*, **95**, 296.

Shaw, A. J. (1991), *Sunflower Variety and Sowing Guide 1990–91*, New South Wales Agriculture and Fisheries, Agnote Reg 2/002, Agdex 145/32.

Shaw, R. H. and Laing, D. R. (1966), 'Moisture Stress and Plant Response' in *Plant Environment and Efficient Water Use*, American Society of Agronomy and Soil Science, Madison, 73–94.

Shawyer, E. (1984), 'Barley—the Export Scene', *Bulk Wheat*, **18**, 22.

Siddique, K. H. M. and Sedgley, R. H. (1985), 'An Ideotype for Chickpea (*Cicer arietinum* L.) in a Dry Mediterranean Environment, in *Proceedings of the 3rd Australian Agronomy Conference*, Hobart, 335.

Simmons, K. V. (1982), *Chickpea Growing*, Department of Agriculture, New South Wales, Agfact, Agdex 133/20.

Simmons, K. V. (1987), *Oats*, Department of Agriculture, New South Wales, Agfact P3.2.2, Agdex 133/010.

Simmons, K. V. (1989), *Field Peas*, New South Wales Agriculture and Fisheries, Agfact P4.2.9, Agdex 166/10.

Simmons, K. V. (1991), 'Chickpeas Update February 1991', *The Pulse*, February/March 1991, 12.

Simmons, K. V., Armstrong, E., Walker, S. and Nickandrow, A. (1992), *Field Pea Update February 1992*, New South Wales Agriculture and Fisheries, Agnote Reg 5/110, Agdex 166/10.

Simmons, K. V. and Sykes, J. A. (1984), *Growing Field Peas for Grain*, Department of Agriculture, New South Wales, Agfact, Agdex 160/10.

Single, W. V. (1961), 'Studies on Frost Injury in Wheat, 1: Laboratory Freezing Tests in Relation to the Behaviour of Varieties in the Field', *Australian Journal of Agricultural Research*, **12**, 767.

Single, W. V. (1971), 'Frost Damage in Wheat Crops', *Agricultural Gazette of New South Wales*, **82**, 211.

Single, W. V. (1985), 'Frost Injury and the Physiology of the Wheat Plant', *Journal of the Australian Institute of Agricultural Science*, **51**, 128.

Slatyer, R. O. (1955), 'Studies of the Water Relations of Crop Plants Grown under Natural Rainfall in Northern Australia', *Australian Journal of Agricultural Research*, **6**, 365.

Smith, I. and Walker, B. (1984), *Grain Lupin and Field Pea Varieties*, Department of Agriculture, Victoria, Agnote 2507/84, Agdex 140/32.

Smith, R. C. G., Anderson, W. K. and Harris, H. C. (1978), 'A Systems Approach to the Adaptation of Sunflower to New Environments. III. Yield Predictions for Continental Australia', *Field Crops Research*, **1**, 215.

Southwood, O. R. and Scott, R. C. (1972), 'Sweet Lupins—a Potential High Protein Grain Crop for N.S.W.', *Agricultural Gazette of New South Wales*, **83**, 99.

Sparrow, D. H. B. and Doolette, J. B. (1975), 'Barley' in Lazenby and Matheson (1975), 430−80.

Strang, J. (1978), *Dry Beans*, Department of Agriculture, New South Wales, Agfact, Agdex 143.

Sutton, B. G. and Dubbelde, E. A. (1980), 'Effects of Water Deficit on Yield of Wheat and Triticale', *Australian Journal of Experimental Agriculture and Animal Husbandry*, **20**, 594.

Sykes, J. D. (1989), *Canola (Rapeseed) Variety and Management Guide 1989*, New South Wales Agriculture and Fisheries, Agfact P5.1.11, Agdex 144/30.

Sykes, J. D. (1992), *Canola Variety and Management Guide 1992*, New South Wales Agriculture and Fisheries, Agnote Reg 3/46, Agdex 144/30.

Sykes, J. D. and Green, A. G. (1988), *Linseed Growing*, New South Wales Agriculture and Fisheries Agfact P5.2.7.

Takami, S., Rawson, H. M. and Turner, N. C. (1981), 'Leaf Expension of Four Sunflower (*Helianthus annuus* L.) Cultivars in Relation to Water Deficits. I. Patterns During Plant Development', *Plant, Cell and Environment*, **4**, 399.

Taylor, G. (1920), *Australian Meteorology*, Oxford University Press, Melbourne.

Taylor, G. (1992), 'Best Bet Sowing Options for Risky Environments', *Australian Grain*, **2/2**, 14.

Thanomsub, W. and Mendham, N. J. (1992), 'Effects of Plant Density Arrangement and Nitrogen Nutrition in Navy Beans', *Proceedings of the 6th Australian Agronomy Conference*, Armidale, 508.

Thomson, N. J. (1979), 'Cotton' in Lovett and Lazenby (1979), 113−36.

Thornthwaita, C. W. (1931), 'The Climate of North America According to a New Classification', *Geographical Review*, **21**, 55.

Trewartha, G. T. (1968), *An Introduction to Climate*, 4th edn, McGraw-Hill, New York.

Verrell, A. G. and Gammie, R. L. (1992), *Winter Cereal Variety Sowing Guide 1992*, New South Wales Agriculture, Agdex 110/10.

Walkden Brown, C. (1975), 'Oats' in Lazenby and Matheson (1975), 481−507.

Walkden Brown, C. and Fitzsimmons, R. W. (1972), 'Oat Varities for 1972', *Agricultural Gazette of New South Wales*, **83**, 2.

Walton, G. H. (1982), 'Better Practices for Higher Production—Lupins', *Journal of Agriculture, Western Australia*, **23**, 77.

Walton, G. H. (1983), *Lupin Growing in Western Australia*. Department of Agriculture, Western Australia, Farmnote 3/83, Agdex 161/20.

Wheatley, D. M., McLeod, D. A. and Jessop, R. S. (1992), 'Effects of Zero Tillage Practices on Soybean Yield and Soil Cover', *Proceedings of the 6th Australian Agronomy Conference*, Armidale, 302.

Wheeler, R. D. and Nitschke, R. A. (1985), 'Management Techniques to Improve Barley Yield and Malting Quality in S.A.' in *Proceedings of the 3rd Australian Agronomy Conference*, Hobart, 270.

Wightman, B. (1984), *Growing Linseed*, Department of Agriculture, Victoria, Agnote 2457/84, Agdex 148/10.

Wightman, B. (1985), *Growing Safflower*, Department of Agriculture, Victoria, Agnote 2673/85, Agdex 145/10.

Williams, C. N. (1975), *The Agronomy of the Major Tropical Crops*, Oxford University Press, London.

Williamson, R. E. (1970), 'Effect of Soil Gas Composition and Flooding on Growth of *Nicotiana tabacum* L.', *Agronomy Journal*, **61**, 80.

Wilsie, C. P. (1962), *Crop Adaptation and Distribution*, Freeman and Co., San Francisco.

Wilson, G. L. and Eastin, J. D. (1982), 'The Plant and its Environment' in *Sorghum in the Eighties. Proceedings of the International Symposium on Sorghum*, ICRISAT, Patancheru, India.

Wood, I. M. W. (1969), *Peanuts in the Northern Territory: a Guide to Production Practices*, CSIRO Division of Land Use Research, Technical Memorandum 69/18.

Wood, P. McR. and Allen, J. G. (1980), 'Control of Ovine Lupinosis: Use of a Resistant Cultivar of *Lupinus albus*—cv. Ultra', *Australian Journal of Experimental Agriculture and Animal Husbandry*, **20**, 316.

Woodruff, D. R. (1973), *Preliminary Systems Analysis of Sunflowers, December 1972*, Queensland Department of Primary Industries, Agriculture Branch Technical Report No. 12.

Woodruff, D. R. (1987), *WHEATMAN*, Queensland Department of Primary Industries Project Report Q087014.

Wratten, N., Mailer, R. J. and Sykes, J. D. (1987), *Rapeseed Variety Testing 1986–1987*, Department of Agriculture, New South Wales, Agfact P5.1.13, Agdex 144/34.

CHAPTER 3

CROP IMPROVEMENT

G. M. Halloran and D. J. Luckett

ORIGIN AND VARIATION OF CROP PLANTS

Knowledge of the geography and history of the origin of crops and of variability within crop species is important to the crop breeder. Because present-day crop breeding is in effect a continuation of the domestication process, this knowledge is useful in designing efficient and appropriate breeding programs, and for obtaining appropriate sources of genetic variability for crop improvement.

Early History

Prior to the beginnings of agriculture, human beings were nomadic food gatherers. They ate fruits and berries and the seeds of grasses and of a range of dicotyledonous species. Archaeological evidence indicates that among the plants supplying them with seed were the ancestors of present-day crop plants and, in addition, many of the present weeds of agriculture (Helbaek, 1954). With increasing demands on food supply, due most likely to population increases, our forebears were forced to gradually adopt a system of deliberate sowing and harvesting of food plants, and in areas such as the Near East they also used irrigation.

In crop domestication human cultivators were selective for those species that were consistently productive, whose seed could be stored and which were able to provide people with a food that satisfied both their quantitative needs and their qualitative preferences for such characteristics as taste, colour and texture. These were the species that became successful domesticates and that are the crops of present-day agriculture and horticulture. Those species not chosen or retained but which were able to adapt to the changed conditions of cultivation became the weeds of agriculture.

Some of these weeds are close relatives, and even ancestors, of crop plants and, throughout the history of domestication, many exchanged genes with the crop plant through repeated hybridisation with it. In fields of cultivated rice (*Oryza sativa*) in Southeast Asia, for example, two weed species, *O. rufipogon*, an annual, and *O. nivara*, a perennial, occur and hybridise with the cultivated species. Similarly, in and around fields of cultivated maize (*Zea mays*) in Central America, two weed species, teosinte (*Zea mexicana*) and *Tripsacum* are still to be found. These species have contributed substantially to genetic changes in maize under domestication through repeated hybridisation with it (Mangelsdorf, 1965). The process of gradual 'infusion' of characters of wild and weed species into the cultivated species through hybridisation, coupled with natural selection favouring the character incorporated from the wild or weed parent, is termed introgression.

Crop Plant Domestication

Most present-day crop plants have had long histories of domestication during which time, in conjunction with increased productivity, quite marked changes have taken place in many morphological and physiological characters. These changes have been in the direction of increased seed size and number and the reduction or loss of adaptive characters of seed, such as dormancy and dispersal mechanisms. A significant difference between the wild and weed relatives and cultivated forms of wheat and barley, for example, is the change from the fragile to the non-fragile rachis. This change has occurred under domestication and has been most likely strongly selected for by humans. Genetic changes of domesticated forms of the crop plant from that of the wild and weed forms have been those most likely favoured by the changed selection pressures of agriculture. These adaptations in time have rendered the crop plant unable to survive in the wild environment. Not only have humans been dependent on the crop for their survival, but the crop has become dependent on humans for its continued existence.

The changes taking place under domestication from the ancestral forms in most crop plants have been much more substantial than the changes effected since the advent of plant breeding approximately 80 years ago. The adoption of plant species into agriculture took place in the areas of natural distribution of the wild forms. Many of these wild and subsequent weed forms are still in existence today and knowledge of their geographical distributions has been used to indicate the likely geographical origins of present-day crop species. Extensive studies have been made of these distributions, together with those of the domesticated forms (De Candolle, 1909; Vavilov, 1926) from which the concepts of primary and secondary centres of origin of crop plants were proposed (Vavilov, 1926). A primary centre of origin was defined as an area of the world where the wild relatives of the crop plant occurred together with the domesticated forms, in what appeared to be the area of original distribution of wild species. An example is the restriction of the distribution of wild diploid wheat, *Triticum monococcum*, and wild tetraploid wheat *Triticum dicoccoides* to the Fertile Crescent (Northern Iraq, Southeastern Turkey, Lebanon and Israel), indicating that this was the likely area where these wheat forms were first domesticated. A secondary centre of origin, or diversity, was regarded as an area where the crop plant occurred in a wide range of types but without the wild forms. It was likely to have been taken from the primary centre early in the history of cultivation of the crop. Vavilov (1926) proposed eight primary centres of origin of crop plants, six of them in the Old World and two in the New World. Examples of some of the cultivated plant species occurring there are shown in Table 3.1. While these concepts proposed by Vavilov have had a profound impact on evolutionary studies of crop plants and plant collecting activities, it is now believed that plant domestication was much more complex in evolutionary terms and of wider geographical extent than previously believed. Some crops originated in more than one centre and with some crops the area of origin is not known, while others appear to have been domesticated over vast areas and not in more restricted geographical areas as the centres-of-origin theory of Vavilov implied. Harlan (1976) proposed the idea of diffuse origins of many crop plants. The areas of origin proposed by him for most crop plants are geographically less localised than the areas proposed by Vavilov.

Knowledge of the geographical origins and evolution of crop plants and their wild and weed relatives is fundamental for genetic and evolutionary study, and for conserving a wide range of genetic variability for crop breeding needs (Simmonds, 1976).

TABLE 3.1   Centres of original domestication of some important crop plants as proposed by Vavilov (1926)

Centre	Species believed to have been first domesticated in that area of the world
1. Chinese	Soybean, apricot, peach, orange
2. Indian	Rice, chickpea, cucumber
2a. Indo-Malayan	Banana, coconut
3. Central Asiatic (Afghanistan, Tibet, Iran)	Bread wheat, cereal rye, peas, pear, apple, walnut
4. Near Eastern (Transcaucasia, Turkey, Syria, Southern Russia)	Diploid wheat, barley, lucerne
5. Mediterranean	Durum wheat, oats, broad bean, lettuce, cabbage, olive
6. Abyssinian	Durum wheat, barley, peas, flax
7. South and Central American	Maize, common bean, pepper, cotton (upland), squash, pumpkin
8. South American (Peru, Ecuador, Bolivia)	Sweet potato, potato, Lima bean, tobacco, tomato
8a. Chile	Potato
8b. Brazilian-Paraguayan	Potato

Gene Pool of a Species

Present-day crop breeding draws on useful genetic variability, not only within the crop species but also from related species and genera. An example of the use of distantly related genetic variability in crop breeding is the incorporation of wheat of a gene for resistance to stem rust (*Puccinia graminis tritici*) from *Thinopyrum elongatum* (*Agropyron elongatum*) (Knott, 1963). This species is a distant relative of wheat but it is able to be hybridised with wheat and possesses genes that can usefully contribute to the genetic improvement of the crop species. The concept of the gene pool of the species embraces genetic variability within the cultivated forms, its wild and weed relatives and more distantly related species which can be hybridised with the commercial species and can contribute genes for its improvement. Harlan *et al.* (1973) recognised the gene pool of a species as having primary and secondary levels related to the degree of accessibility of genetic variability for crop breeding within closely and more distantly related species, involving considerations such as cross-compatibility, hybrid viability and chromosome pairing in hybrids with the crop species.

The concept of the gene pool of the species is important, not only for securing the most appropriate sources of variability for present breeding programs but also for devising strategies for conserving the maximum range of potentially useful variability for future crop breeding needs.

Conservation of Genetic Variability

The life blood of crop breeding for further improvements in yield, disease resistance, quality and other characters is the genetic variability available within the gene pool of the species. Therefore to retain maximum flexibility in breeding it is necessary to have available as wide a range of genetic variability as possible in the crop species. Because many of the long-term future crop breeding needs are unspecifiable it is necessary that genetic variability, regardless of its immediate value, be conserved for future crop improvement.

A growing awareness is being shown by crop breeders and geneticists of the need to conserve genetic variability for future crop breeding (Frankel and Bennett, 1970; Frankel and Hawkes, 1975). The Food and Agriculture Organisation (FAO) of the United Nations has established an International Board for Plant Genetic Resources to organise the collection and maintenance of genetic variability of crop plants and related species.

One important activity in genetic conservation has been the collection of the wild, weed and primitive cultivated forms of crop plants. Most of the areas of distribution of this variability throughout the world were, until recently, areas of primitive agriculture using mainly primitive cultivars which generally contained large amounts of genetic variability. However, owing to the implementation of agricultural improvement programs in these areas over the last 30 years much of the variability of the primitive cultivars and associated weed species has been lost due to the widespread cultivation of advanced cultivars from overseas.

Gene Banks for Crop Breeding

A number of countries maintain gene banks of many of the world's crop plants. These are large collections of advanced cultivars and crossbreds, primitive cultivars, wild and weed relatives and other related species of the crop plant. The United States Department of Agriculture maintains a large gene bank for crop plants at Beltsville, Maryland, and Russia has a similar collection at the N.I. Vavilov All-Union Institute of Plant Breeding at St Petersberg and at other centres throughout the country. Crop breeders worldwide can obtain samples of seed from these collections for use in crop breeding. In Australia a national winter cereals collection has been established at the Agricultural Research Centre, Tamworth, New South Wales, for use by Australian breeders. Germplasm collections of other crop plants (and their wild relatives) important to Australian Agriculture are held at CSIRO Canberra, Biloela Queensland, and Horsham Victoria.

In the gene bank, storage is mainly as seed which is housed under conditions of low humidity and low temperature to ensure its viability for 10 to 20 years. Interest has been shown in the possibility of storing pollen of crop plants in gene banks but it generally has a much shorter viability than stored seed. Pollen collections are also much more difficult to regenerate than seed collections, a factor which is a serious limitation to the facility of maintaining the viability of lines in a gene bank. Before the viability declines to a low level the seed collections are sown and harvested to produce fresh seed and in this way the gene bank can be perpetuated. Some of these collections are being extended to accommodate vegetative propagules, such as tissue or single-cell cultures, of those species whose seed or pollen is not easily stored due to rapid loss of viability. They have been referred to as 'recalcitrant' species (Roberts, 1973).

The FAO is active in promoting activities for collecting and conserving genetic variability for crop breeding. It is also active in disseminating information on the location and content of breeders' collections and gene banks within a wide range of crops. It publishes a world catalogue of genetic stocks describing potentially useful attributes of the lines in these collections. Information on such collections is also being computerised to facilitate the dissemination of these data, and hence improve the efficiency in the use of appropriate forms of genetic variability for crop breeding.

GENETICS AND CROP BREEDING

Present-day crop breeding and selection methods are based on the genetic principles propounded in Mendel's laws of inheritance. Many characters of crop plants exhibit simple inheritance; that is, their expressions are controlled by one gene, or a small number of genes, of large effect whose expressions are little influenced by environment. They are called major genes and can be bred and selected for on the basis of segregation frequencies expected from simple Mendelian inheritance. Other characters owe their expression to a large number of genes, each of small effect. These characters are said to exhibit quantitative inheritance, and the field of their genetic study and selection is called quantitative genetics. A number of characters exhibit a combination of both types of genetic control; major genes determine wide differences in their expression, and genes of smaller effect have an influence within, and sometimes beyond, the range of the major genes.

Simple Inheritance

Characters of simple inheritance in crop plants segregate similarly to those obtained by Mendel for seed shape and flower colour in peas, and can be bred and selected for with relative ease. They include many forms of disease resistance, morphological and colour variation of the plant or seed and, in some crops, flowering time and its component processes. Pugsley (1965, 1971) has demonstrated that the component processes of flowering time in wheat—photoperiodic (or daylength) response and vernalisation (or cold) response—are each controlled by one, and one to four, major genes respectively. However, segregation ratios for some characters under the control of a small number of major genes quite often depart from simple Mendelian inheritance. This can be due to interactions between major genes if at least two genes are involved, linkage with other genes, alteration of their expressions by modifier genes, or environmental influences.

Quantitative Inheritance

Many characters of crop plants exhibit inheritance patterns that are not divisible into classes but show a continuous gradation in expression from that of one parent to the other. A character which segregates into discrete classes in inheritance and which can be visually assessed is referred to as a qualitative character. Such characters are usually under simple genetic control. The character which segregates to give continuous gradation in expression is referred to as a quantitative character and is specified accurately only in terms of measurable quantities such as length, time, weight or proportion.

While quantitative characters do not exhibit discrete Mendelian class segregation, the principles of Mendelian genetics, with some elaboration, are used to explain their inheritance. One of the first demonstrations of the relevance of Mendelian genetics in explaining inheritance of quantitative characters was that given by East (1916). He proposed the multiple factor hypothesis which explained the inheritance of corolla length in *Nicotiana langsdorfii*, a quantitative character, on the basis of the segregation of a definite number of genes. However, most quantitative characters in crop plants can only be analysed as the combined effect of all the genes governing their expressions.

The distinction between qualitative and quantitative character expression depends not so much on the magnitude of the effect of individual genes but on the relative magnitude of the influence of heredity versus environment on the phenotype.

Genotype–Environment Interaction

The phenotypic expression of most characters of crop plants is the consequence of interaction between the genes it carries for that character and the environment in which it has been grown. Seasonal, or locality, variation in the phenotypic appearance, yield or quality of grain can be used to obtain an estimate of the environmental versus the genotypic component of phenotypic expression. Quantitative characters, such as yield and grain quality, whose expressions are very dependent on the very variable environmental factors of moisture, nutrient availability and temperature, have very large environmental components. Consequently they are of low heritability.

In breeding and selecting for crop improvement it is necessary to partition phenotypic variance for a character into that due to environmental causes and that due to genetic effects. An important statistic in this analysis of segregating populations is the variance, which is the sum of the squared deviations of individuals (plants or lines) in the population from the population mean for the character being studied. This is written as

$$V = \frac{d^2}{n-1}$$

where V is the variance, d^2 is the sum of the squared deviations and n is the number of observations.

The total variability of a population for a particular character is called the phenotypic variance (V_P), which is the summation of the genetic variance (V_G) and the environmental variance (V_E) (Mather and Jinks, 1971; Falconer, 1981) such that

$$V_P = V_G + V_E$$

Estimates of these genotypic and environmental components of variability can be obtained by growing the hybrid population as families in a number of environments or seasons, or both. Methods of this type offer a powerful means of analysing phenotypic variance, and of accurately estimating heritability for selection.

The crop breeder needs a detailed knowledge of the nature and magnitude of genotype-by-environment interaction ($G \times E$) for any new cultivar approaching commercial release (see 'Cultivar Evaluation and Release' below). The level of $G \times E$ will influence the growing areas for which the cultivar will be recommended, and will provide information

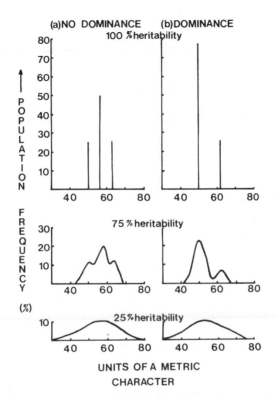

FIG. 3.1 F_2 distributions for a monogenic character by which the parents differ by twelve metric units in situations of (a) no dominance and (b) complete dominance with different levels of heritability (100, 75 and 25 per cent) for the character

regarding situations where the cultivar will not perform well, e.g. 'avoid late sowings' or 'only suitable for southern New South Wales'.

Heritability

Heritability is an estimate of the genetic variance. The phenotypic variance is caused basically by the difference between individuals in the genes they carry, but the environment acting on each individual modifies the outward expression of its genes. This environmental influence can range from little or no effect to a very pronounced effect that masks the genetic differences between individuals. A representation of the likely effect of different degrees of environmental influence on the second-generation (F_2) segregation patterns of a single gene is given in Fig. 3.1. With 100 per cent heritability the F_2 segregation ratios are 1:2:1 (dominant:heterozygote:recessive) or 3:1 (dominant:recessive) according to whether the character exhibits no, or complete, dominance respectively. As heritability decreases to 75 per cent, i.e. as the environmental influence increases, each of the F_2 segregation classes exhibits increased variance. When heritability is only 25 per cent, the segregation of the F_2 population is as one continuous distribution whether the gene exhibits no, or complete, dominance.

The magnitude of the heritable component of the phenotypic variance of a character is of interest to the crop breeder for selection. It is expressed as the ratio of the genetic variance to the total variance and is usually written in the form

$$H = \frac{V_G}{V_G + V_E}$$

where H is the heritability value, V_G is the genetic variance and V_E is the environmental variance.

Heritability estimates can be conducted in two main ways, as a broad-sense value and as a narrow-sense value. The broad-sense value is calculated using the total genetic variance for the character being selected and is expressed as a percentage. It is written in the following way:

$$H = \frac{V_G}{V_P} \times 100$$

where V_P is the phenotypic variance and

$$V_P = V_G + V_E$$

The narrow sense estimate is more restricted and gives a slightly lower value than the broad sense estimate. It has the following formula:

$$H = \frac{V_A}{V_P} \times 100$$

where V_A is the additive genetic variance.

Methods are available for calculating V_A, the additive genetic variance, for which the reader is referred to an authoritative text on the topic (Mather and Jinks, 1971).

Gene Action

Selection for improved performance in genetically variable populations is made more efficient if the breeder can characterise and quantify the type of gene action involved in terms of hereditary parameters. Fisher (1918) partitioned hereditary variance into three components: additive, which is the difference in expression between homozygotes at a single locus; dominance, the result of interactions between alleles at a locus; and epistatic, which results from interactions between non-alleles, also called interallelic interaction.

The hereditary, or genetic, variance V_G can be expressed as a summation of the three components as follows:

$$V_G = V_A + V_D + V_I$$

where V_A is the additive variance, V_D is the dominance variance and V_I is the epistatic variance.

Additive genetic variance is contributed by those alleles that exhibit linear quantitative effects and accounts for the resemblance of offspring with parents for the character under study. The dominance component of genetic variance is, statistically, the deviation of the heterozygote from the mean value for the parents for the character being selected. This component is usually small in comparison with the additive variance. Non-allelic gene interactions, or epistasis, are a summation of the interactions between additive components, interactions between dominance components, and interactions between additive and dominance components. The components of this portion of the total genetic variance are difficult to estimate, but it is believed that they are small in comparison with the additive and dominance components. Because of this, the epistatic component of total genetic variance is often omitted in calculating heritability.

For further information on methods of calculating the components of genetic variance, see the detailed discussion of this topic in Mather and Jinks (1971).

Linkage

A problem confronting the crop breeder is that of linkage of favourable genes with unfavourable ones in segregating populations. Unfavourable genes are those which influence expressions of characters which detract from agronomic performance. Included are genes promoting lodging, too late or too early a flowering time, shattering of the inflorescence, or disease susceptibility, or those which detract from commercial or industrial quality such as lower than optimal chemical composition of grain or low physical extractability of substances from grain. Linkage in crop breeding, when one is concerned with improving the genetic performance of highly improved cultivars, is a major problem limiting selection advance. It restricts independent assortment of genes in segregation and produces an abundance of parental combinations and a deficiency of recombinants. This limits the chance of obtaining the desired recombinants.

An indication of the effect of the degree of linkage on recombination value can be gained from considering a theoretical case of linkage of a gene for disease resistance (R) with an undesirable gene (l) conferring lax ear in wheat. It can be presumed that the genotype $RRll$ was being used as a parent in a wheat breeding program to incorporate the disease resistance gene, R, in a commercial cultivar of the genotype $rrLL$. The cross would be made between these two parents and the F_1 and F_2 genotypes and phenotypes would be as shown in Fig. 3.2. With independent segregation between genes and R and l, the genotype classes and their proportions in the F_2 would be as shown in the two-way table (Table 3.2).

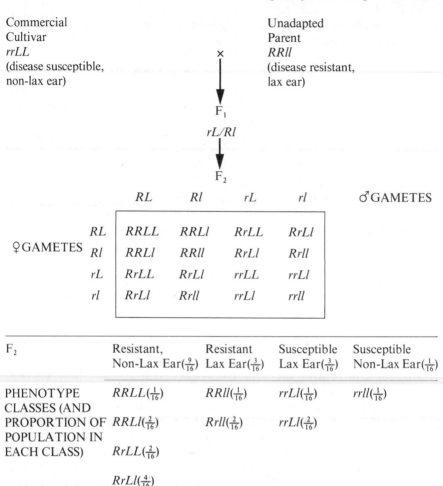

FIG. 3.2   Phenotypic and genotypic segregation classes for two genes (disease resistance, *R*, vs susceptibility, *r*, and non-lax, *L*, vs lax, *l*, ears), in the F₂ of a wheat cross whose parents differ for these characters

The desired recombinant (*RRLL*) would have a frequency in the F₂ of 1:16 or 6.25 per cent.

However, if the genes *R* and *l* exhibit linkage the recombination fraction of the desired genotype (*RRLL*) in the F₂ would be less than with independent assortment and its frequency will vary according to the degree of linkage, i.e. the closeness of these gene loci on a particular chromosome. The closer they are together the lower will be the recombination frequency, i.e. the lower the chance for a crossover occurring between genes *R* and *l* to separate them in segregation. When the two desired genes are present on different homologous chromosomes in the F₁ (as here) they are said to be in repulsion linkage. The influence of the

TABLE 3.2 Influence of linkage on the recombination fraction of genotype *RRLL* in the F_2 from the two types of F_1 double heterozygote

Recombination value (linkage)	Percentage of *RRLL* genotypes in F_2 if the F_1 is:	
	Rl/rL Repulsion	*RL/rl* Coupling
0.50 (Independent assortment of *R* and *l*)	6.25	6.25
0.25	1.56	14.06
0.10	0.25	20.25
0.02	0.01	24.01
0.01	0.0025	24.50
p	$\frac{1}{4}p^2$	$\frac{1}{4}(1-p)^2$

recombination value, or linkage, on the frequency of the recombinant genotype (*RRLL*) in the F_2 is shown in Table 3.2.

In the converse situation, a cross might be made between two parents, *RRLL* and *rrll* in order to manipulate some other character. In this case, the *RL* gene combination is already present on the same homologous chromosome (coupling linkage). Among the F_2 segregants the *RL* gene combination will tend to be held together and will appear at greater-than-expected frequency depending on the recombination value (Table 3.2).

A very useful means of breaking a linkage in crop breeding between desirable and undesirable genes is the backcross method. Its effectiveness in breaking linkage is discussed under 'Breeding and Selection Methods in Self-Pollinated Crops'.

Genetic Advance Under Selection

The rate at which the crop breeder can achieve genetic advance in selecting for a particular character will depend on three main factors. First, the amount of genetic variability in the original population for the character being selected must provide adequate potential to achieve the level of improvement envisaged. Second, the masking effect of the environment on the genes for the character must not be so high as to limit the variability of its expression for selection, i.e. the character must exhibit a moderately high level of heritability. Third, and closely related to the other two considerations, the intensity of selection must be such that it does not restrict the rate of genetic advance.

Crop breeders are interested in calculating the likely genetic advance they can expect from selection, whether it be a cross-fertilising population, a segregating population, or a set of homozygous lines. Among populations of different genetic origins they are interested in estimating the relative selection advance likely to be obtained from each population.

Consider the use of these predictive estimates for a situation where a breeder wishes to ascertain the likely gains in yield from selection amongst a set of *n* self-pollinated lines; *n* lines are grown in a number of

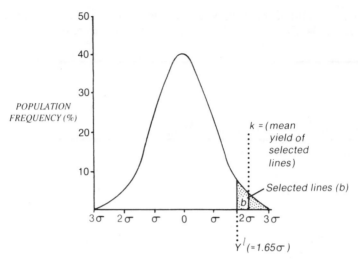

FIG. 3.3   Theoretical distribution of mean yields of lines from replicated yield trials. Assuming the breeder selects the upper 5 per cent of lines, i.e. with yields greater than $Y^l (=1.65\sigma)$, the mean of the b selected lines is expected to be k, the selection differential

locations in a number of seasons and evaluated for yield of each line (Y). Assuming the values of Y amongst the lines are normally distributed with a standard deviation of σ_Y the frequency distribution for yield will be as shown in Fig. 3.3. If the breeder sets the yield selection value at a level Y^l, the selected families are those falling in the stippled area below the yield distribution curve and represent a proportion (b) of the population. By calculus it can be shown that the expected genetic gain represented by the selected lines is

$$G_s = k \times \sigma_Y \times H$$

where G_s is the expected genetic gain in yield, σ_Y is the standard deviation of the mean yields of the total n lines, H is the heritability coefficient and k is the selection differential. The value G_s measures the difference between the mean genotypic value of the l selected lines, i.e. $\bar{a}_s$ and the means genotypic value of the n original lines, $\bar{a}$. Therefore $G_s = \bar{a}_s - \bar{a}$.

The selection differential value, k, is assessed from the mean phenotypic value of the l selected lines ($\bar{Y}_s$), the mean phenotypic value for the total number (n) of lines tested, the phenotypic standard deviation (σ_Y), and the intensity of selection l/n. The value, k, is expressed in standard deviation units and varies only according to variation in selection intensity. If selection for yield retains 5 per cent of the lines, k can be shown to have a value of 2.06 and if 20 per cent of the lines are retained k has a value of 1.40. The formula for calculating the rate of genetic advance under selection evaluates comparatively the three components, genetic variability, environmental effects and selection intensity for the character under study.

REPRODUCTIVE SYSTEMS IN CROP PLANTS

Differences in the mode of reproduction amongst crop plants influence the level of genetic variability present in the crop population and therefore breeding and selection methods vary considerably amongst crops according to their breeding systems. In addition, the particular type of reproduction can impose practical limits on the efficiency of certain breeding and selection procedures for the attainment of higher commercial yield. For this reason breeders have sought to genetically alter the breeding systems of crops. The use of cytoplasmic male sterility and the modification of self-incompatibility are examples of alterations that have been effected in breeding systems to produce new methods for crop improvement.

Crop plants can be divided into two classes on the basis of their common modes of reproduction, sexual and asexual. Sexually reproducing species can be divided into two groups—those that are predominantly self-pollinating and those that are predominantly cross-pollinating. The modes of reproduction of the most common crop plants are shown in Table 3.3.

TABLE 3.3   Modes of reproduction of the major crop plants

Crop type	Self-pollinated[a]	Cross-pollinated
	Mode of reproduction	
Cereals	Barley	Maize
	Oats	Rye
	Wheat	
	Rice	
	Sorghum	
	Foxtail millet	
Legumes	Broad bean	Scarlet runner bean
	Chickpea	
	Cowpea	
	Field pea	
	Lupin	
Oilseeds	Linseed	Safflower
	Canola (*Brassica napus*)	Sunflower
Fibre crops	Cotton	Hemp
	Flax	
Other crops	Tobacco	Sugar beet
	Tomato	Sugar cane
		Sweet potato

[a]Most self-pollinated crops have been shown to exhibit small amounts of cross-pollination (1 to 10 per cent)

Self-Pollinating Species

The self-pollinating character in plants ensures that fertilisation is between male and female gametes of individual flowers. The chance for cross-pollination is low because the flower remains enclosed by floral parts during anthesis. This is called cleistogamy. In self-pollinating cereals the glumes enclose individual flowers, while in self-pollinating legumes it is the corolla. In the cultivated tomato the anthers form a protective cone over the stigma until after anthesis. Other important adaptations of the self-pollinating character are synchronous production of female and male gametes in individual flowers and genetic self-compatibility.

Self-pollinating species can occasionally exhibit low to moderately high levels of cross-pollination. One cause is believed to be extremes of environmental stress such as heat, drought or cold which can cause partial male sterility and simultaneous opening of the flower, which provides the possibility for cross-pollination. The occurrence of occasional cross-pollination in self-pollinating crop species can be significant to the problem of maintaining purity of breeding lines and nucleus seed of commercial cultivars. The incidence of cross-pollination, even of very low frequency, has probably been of high significance in the evolution and domestication of self-pollinated crop plants in providing new sources of variability for natural and human-influenced selection.

Cross-Pollinating Species

Cross-pollination confers on plant populations a far greater degree of genetic flexibility than does self-pollination. Under domestication, in-breeders appear to have arisen from outbreeders, but rarely the reverse. Amongst crop plants the self-pollinating character is much more common than cross-pollination (Table 3.3). The self-pollinating character gives the species a greater immediate fitness than does the cross-pollinating character and therefore may be more favoured under the stable conditions of agriculture than cross-pollination. Cross-pollinating species have a number of mechanisms which either minimise the level of selfing or totally preclude it. One type of mechanism is concerned with certain morphological and developmental characters which are uniform for the species. One such mechanism, monoecism, the separation of male and female flowers on the plant, increases the chance for cross-pollination. The asynchronous maturation of anthers and stigma of flowers, whether they be monoecious or hermaphroditic (i.e., having bisexual flowers) is another method of ensuring mainly cross-pollination. When pollen is shed first in such flowers the mechanism is called protandry; when the stigma is receptive first it is called protogyny. The combination of monoecism and protandry in maize can ensure almost complete out-crossing. The mechanisms which preclude selfing in cross-pollinated species are dioecism and self-incompatibility. Dioecism, i.e. separate male and female individuals, prohibits selfing but does not prevent brother–sister mating. It is not common in higher plants and is the breeding system of only a few crop plants such as hemp, hops and jojoba. Self-incompatibility is much more widespread in higher plants and is found in most cross-pollinating species. It is significant in the evolution of cross-pollinating species and is encountered when breeding and selecting for their improvement. Another form of variation in reproductive systems of plants is sterility, which occurs in low frequency in most plant species. Systems of sterility have been studied in detail for their use in crop breeding. One form, cytoplasmic male sterility, has contributed substantially to crop improvement and especially maize improvement. Another form, genetic male sterility, is of limited use, but current research on its use in breeding holds promise for its wider application.

Self-Incompatibility

Cross-pollinating plant populations have a genetic structure in which the heterozygote is favoured and the individual which arises from selfing is adaptively disadvantaged. The reason for this is believed to be that selfing of heterozygotes produces individuals with many genes which are double recessive, in which condition they are deleterious to normal vigour, growth and fertility, whereas in the heterozygous condition the recessive allele is masked by the more favourable dominants. Self-incompatibility is an efficient mechanism of either preventing or minimising the amount of selfing in cross-pollinated plants thereby keeping inbreeding to a minimum. It also enables fuller exploitation of the productive and regenerative potential of the species. The incompatibility reaction is determined in the stigma where mutual recognition between pollen and stigma is necessary for pollen tube growth down the style to effect fertilisation. The recognition is between pollen-wall proteins and glycoproteins interacting with proteins on the surface of the stigma.

Two types of incompatibility are present in cross-pollinating species, gametophytic and sporophytic incompatibility. They are both genetically controlled and in most cases by a single gene which, in some species, has a small number of alleles but in others a very large number.

Gametophytic Incompatibility

This system of incompatibility was first described by East and Mangelsdorf (1925) in *Nicotiana sanderae*. It is usually controlled by a single gene, S, which has a number of different alleles in most systems studied.

Incompatibility is brought about by inhibition, or gross retardation, of the growth of pollen tubes on styles whose tissue contains the same allele of S. As a result, plants are always heterozygous at the S locus. It would not be possible to have two alleles with gametophytic control and no dominance for S alleles because all plants would be incompatible and the species sterile.

Gametophytic incompatibility produces three main types of pollination pattern depending on the allelic constitution of pollen and style at the S

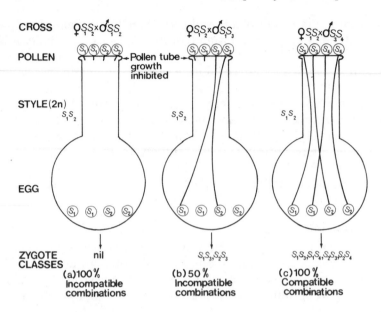

FIG. 3.4 Gametophytic incompatibility showing pollen–style interactions according to their genetic constitution for S alleles

locus (Fig. 3.4). In the cross where the allelic constitution of the parents at the S locus is $♀S_1 S_2 \times ♂S_1 S_2$ the stylar tissue, which is diploid, will be $S_1 S_2$. It will inhibit pollen tube growth of both types of pollen combinations, i.e., S_1 and S_2, and gives 100 per cent incompatibility. In the cross where the parent allelic constitution is $♀S_1 S_2 \times ♂S_1 S_3$ the stylar tissue will inhibit only S_1 pollen. S_3 pollen tubes will pass down the style to the ovule and fertilise both S_1 or S_2 eggs. In this cross there will be 50 per cent incompatibility in the pollen–style reaction. In the cross where the allelic constitution of the parents is $♀S_1 S_2 \times ♂S_3 S_4$ there will be 100 per cent compatible combinations because the stylar tissue does not possess an allele in common with either of those in the pollen.

Certain plant species such as cereal rye (*Secale cereale*) possess a two locus system of incompatibility (Lundquist, 1956).

Sporophytic Incompatibility

This system of incompatibility was discovered by Hughes and Babcock (1950) in *Crepis foetida* and by Gerstel (1950) in *Parthenium argentatum* (guayule). It is similar to the gametophytic system in that control is exercised by a single gene with multiple alleles. However, it differs from the gametophytic system in a number of ways. Firstly, the incompatibility reaction is determined by the genotype of the plant in which the pollen is borne. Another difference from the gametophytic type of compatibility is that its alleles may exhibit dominance, independent action or competition in either pollen or styles depending on the type of alleles present at the S

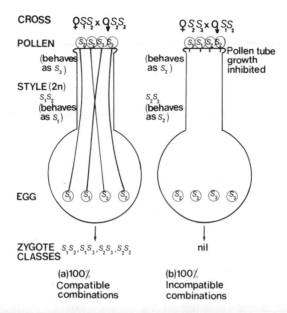

FIG. 3.5 Differences in sporophytic compatibility between the crosses ($♀ S_1 S_2 × ♂ S_2 S_3$) and its reciprocal when the order of dominance of S alleles in the style is $S_1 > S_2 > S_3$ but is reversed in the pollen, i.e. $S_3 > S_2 > S_1$

locus. It is genetically much more complex than the gametophytic system and in consequence leads to a larger number of compatibility reactions.

It is possible in the sporophytic system, because of reversal of the order of dominance of alleles in the pollen as against their action in the style, for reciprocal crosses to exhibit complete compatibility and complete incompatibility. For instance if with the three alleles S_1, S_2 and S_3 the order of dominance in stylar tissue was $S_1 > S_2 > S_3$ but in the pollen reactions $S_3 > S_2 > S_1$, this reversal of dominance would lead to reciprocal differences in compatibility vs. incompatibility as shown in the cross $S_1 S_2 × S_2 S_3$ and its reciprocal (Fig. 3.5).

In the selfing of species that possess sporophytic incompatibility it is often possible to obtain a very small amount of seed set. One reason is that an occasional pollen tube can penetrate the length of the style and fertilise the egg despite the oppositional allelic content of the pollen with that of the style. Self compatibility can also be conferred by a self fertility allele (Sf) which occurs in many species with sporophytic incompatibility. It can overcome the influence of the incompatibility alleles. Being different from gametophytic incompatibility, the sporophytic system can produce homozygotes at the S locus because of dominance or reversal of dominance of alleles in the style and pollen, as shown in Fig. 3.5. Another difference between the two systems is that, because of dominance amongst S alleles in the sporophytic system, two different genotypes can be in the same compatibility class. For example, $S_1 S_2$ and $S_1 S_3$ have the same compatibility reaction if S_1 is dominant to S_2 and S_3, whereas in the gametophytic system the alleles exhibit independent action in pollen and styles.

Sterility

One difference between sterility and incompatibility in plants is that one manifestation of sterility is non-viable gametes, whereas with incompatibility, gametes are viable but failure of fertilisation is due to inhibition in germination and growth of pollen tubes down the style. Sterility can arise from chromosomal aberration, mutant genes or cytoplasmic influences, whereas incompatibility reactions are determined by the action of one gene or, in a few cases, two genes. Sterility expressions are wider ranging than incompatibility, such as non-viable gametes, failure of dehiscence of anthers, failure of the embryo sac to develop, and even abnormal embryo or endosperm development.

Sterility occurs in most, if not all, crop plants at a very low frequency. The two types of sterility of interest to the crop breeder are genetic male sterility and cytoplasmic male sterility. The interest of the crop breeder in male sterility is for devising breeding methods for exploiting hybrid vigour for commercial production of F_1 populations of crop plants, both cross-pollinated and self-pollinated. These forms of sterility enable the breeder to render the female parent of a cross fully sterile in unlimited numbers. It allows large amounts of F_1 seed to be produced cheaply. This is done by planting the male parent in close proximity to the female (male sterile) parent, allowing the male parent to pollinate the male sterile parent and then harvesting the F_1 seed from the male sterile parent. Commercial production of hybrid (F_1) seed has only become practicable since the use of these male sterility-inducing mechanisms. It would not be practically or economically possible to do this if the breeder had to rely on the laborious method of hand emasculation to achieve hybrid seed in the amounts necessary for commercial production. The possibility is particularly low for those crops which, for each emasculation and pollination, yield only one seed such as the cereals, or only a few seeds, such as the legumes.

Genetic Male Sterility

Genetic male sterility occurs in low frequency in most crop plants and is conferred by the action of a single recessive gene, often denoted *ms*. It is of interest to crop breeders in that it offers a genetically simple control over fertility and its action is complete. This means that in the homozygous recessive condition, *ms ms*, plants are completely male sterile. Its use in breeding involves incorporating the male sterile gene in the female parent of a proposed cross, usually by backcrossing. The male sterile female parent is then sown, usually in alternating rows with the male parent. The seed harvested from the female rows is F_1 seed. One problem with the use of genetic male sterility is that the genetic male sterile condition has to be maintained by crossing the male sterile line with the same line of normal fertility. Therefore in using the male sterile line in a breeding scheme the maintainer line will contain male sterile and male fertile individuals. A considerable amount of work has been carried out to find genes for vegetative characters which are very closely linked to, or are a pleiotropic expression of, the male sterile locus so that before flowering, the homozygous dominants, *Ms Ms*, and heterozygotes, *Ms ms*, can be rogued out of the female parent rows leaving only *ms ms* plants for pollination by the male parent (Fig. 3.6).

One problem with the use of male sterility in breeding, particularly with regard to self-pollinating species which have not been selected for cross-pollination, either by wind or insects, is the low level of cross-pollination able to be achieved on to the male sterile line. This might be improved if it were possible to alter floral morphology towards a more exposed stigma or to choose environments where climatic conditions might favour higher levels of cross-pollination.

Cytoplasmic Male Sterility

This form of male sterility is conferred by the cytoplasm and its expression is dependent on the action of a gene which in the homozygous recessive condition allows male sterility to be expressed, but in the heterozygous, or homozygous dominant, condition the plant maintains normal fertility. This mechanism of sterility induction has been found in a large number of crop plants, both cross-pollinated and self-pollinated, and has been of significance in the large-scale production of hybrid seed for the commercial exploitation of heterosis in the F_1 plant. Cytoplasmic male sterility was first used for the production of hybrid seed in onion by Jones and Davis (1944). They found a single male sterile plant in the cultivar Italian Red which, when hybridised with a number of plants of normal fertility, gave three types of breeding behaviour in the progeny of the crosses. One type of progeny gave all male sterile plants, another male fertiles: male steriles in the ratio 1:1, and the third type all male fertiles. These results were accounted for by the presence in the original male sterile plant of a sterility-inducing cytoplasm, denoted *S*, whose influence was expressed by virtue of a double recessive gene for sterility, *rr*, on a chromosome of this plant. Its cytoplasm–gene designation was *S rr*. The plants of normal fertility which in hybridisation with the original male sterile plant gave the three types of progeny—all male steriles, male sterile:male fertile (1:1) and all male fertiles—had cytoplasm–gene constitutions of *S rr*, *F Rr* and *F RR* respectively. The presence of the fertile cytoplasm *F* gave male fertility regardless of the type of alleles at the fertility/sterility locus on the chromosome. Thus the possible cytoplasm–gene combinations for fertility or sterility are as follows:

F RR–male fertile	*S RR*–male fertile
F Rr–male fertile	*S Rr*–male fertile
F rr–male fertile	*S rr*–male sterile

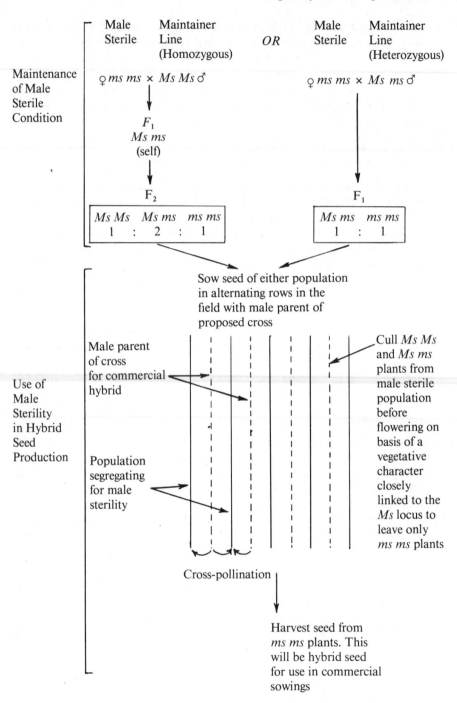

Fig. 3.6   Outline of a scheme for maintaining genetic male sterility and using it in a hybrid breeding program

The *R* type genes within maize which have the capacity to restore fertility are called restorers. One common source of cytoplasmic male sterility in maize is that derived from the cultivar Mexican June and is referred to as the Texas type of male sterility (Rogers and Edwardson, 1952; Duvick, 1965). Fertility can be restored to this type of sterility by two dominant genes, Rf_1 and Rf_2. The gene Rf_2 is found in nearly all forms of maize, and the two genes in combination will completely restore fertility. The mechanisms of cytoplasmic male sterility and fertility restoration that occur in maize have been found in a number of other crop plants, including wheat, linseed, tobacco, sorghum and millet.

METHODS OF HYBRIDISING CROP PLANTS

Hybridisation and selection, being different from selection alone which was practised prior to the discovery of Mendel's laws of heredity, provide more objective methods of crop improvement. They can bring together genes of specially chosen parents from diverse geographical sources to produce genetic recombinations that are likely to give much greater potential for genetic improvement than that from selection alone. Hybridisation is therefore a key activity of present-day breeding for crop improvement.

Techniques for Hybridisation

In self-pollinated plants the technique of hybridisation involves the following steps:
1. If flowers are borne in inflorescences, they are examined to find those in which the earliest flowers are 2 to 3 days off flowering.
2. Those parts of the inflorescence bearing flowers which are too small or too late in maturity for hybridising are removed.
3. The anthers of each flower are removed using forceps. This is called emasculation. If flowers are very small this operation is carried out under a dissecting microscope. The emasculated flowers are enclosed in a glassine bag fastened so that foreign pollen cannot enter. If flowers are borne singly in axils of leaves they can be covered individually with cotton-wool wrapping. The emasculated flowers are left for 1 to 2 days, the time for the stigma to become receptive, at which time anthers from the male parent are chosen immediately prior to anthesis. This is judged when the anthers are deep yellow and have not dehisced. In this condition they will dehisce soon after removal from the flower. Upon dehiscence, the pollen is dusted on the stigma with the anther held in forceps. If the flowers are very small, this operation should be carried out under a dissecting microscope where it can be verified that pollen grains are on the stigma.
4. After pollination, the inflorescence or the single flowers are again tightly bagged, or wrapped in cotton wool.

5. The bags can be removed from the flowers about 10 to 12 days after pollination and the hybrid seed allowed to ripen. The seed so obtained is called F_1 seed.

For cross-pollinating species which are monoecious, and those hermaphroditic (i.e. having bisexual flowers) species which are strongly self-incompatible, emasculation is not necessary prior to hybridisation. The hybridisation procedure for monoecious and self-incompatible dioecious species is somewhat different from that for self-incompatible, hermaphroditic species and is as follows:

1. Female flowers of the female parent of the cross are enclosed in glassine or paper bags well before the stigma(s) becomes receptive. This prevents self-pollination or within-parent pollination for the monoecious and dioecious parents respectively, and also prevents cross-pollination with other adjacent plants.
2. When the stigma(s) becomes receptive, pollen is collected from the male flowers of the male parent of the cross. The pollen is collected when the male flowers have just reached anthesis; if the male inflorescence is many-flowered, the pollen can be collected by shaking the inflorescence into a paper bag.
3. Pollen is then placed, or dusted out of the bag, onto the stigma(s) of flowers of the female parent. The pollinated flowers are again bagged to prevent pollination from foreign pollen. The seed which develops in these flowers is the F_1, or hybrid, seed.

For hermaphroditic species which are strongly self-incompatible the flowers or inflorescences which are at a similar stage of maturity are bagged together just prior to anthesis of the earliest flowers. It may be necessary to agitate the bags a few times daily during anthesis to ensure that there is pollen movement between the flowers of the two parents. The seed which develops in the bagged heads will be hybrid, or F_1, seed.

For hermaphroditic cross-pollinating species which are moderately self-compatible it is necessary to emasculate the female parent well prior to anthesis. The female flower or inflorescence can then be bagged with a normal flower or inflorescence at a similar stage of maturity. Cross-pollination will therefore take place inside the bag and the seed harvested from the emasculated head will all be hybrid, or F_1, seed.

SELECTION IN CROP IMPROVEMENT

From the beginnings of agriculture human beings have been, in a sense, plant breeders, in that they have exerted a selection pressure on the crop for both improved and more assured productivity. Their selection has also been qualitative in that they have found certain variants of the crop species especially suitable for particular uses. This is evident in a crop like maize where types especially suited to boiling, cooking, popping and colour variants for ceremonial use have been selected and retained as discrete types. Another example of the influence of human selective preference in crop domestication is the choice of one particular crop over others for main use. The pre-eminence of wheat as the main cereal, over barley, oats and cereal rye, which had somewhat similar ranges of adaptiveness and yield potential as wheat, is an example of human discernment in selection. It is likely that the choice of wheat was that of palate preference in that the wheat grain possessed a protein complex, called gluten, which was unique in physical properties such as extensibility and resilience and in consequence produced a bread much more pleasing to the palate than the other cereals.

During crop domestication, selection by humans, together with natural selection, produced a wide range of variability in each crop plant which is essentially the variability that present crop breeders are using for further improvement. In cross-pollinated crops, genetic variability for the advances made by selection under domestication have come from new forms of variability arising from continual hybridisation and recombination within crop populations. Both mutation and hybridisation with the 'wild' and weed relatives of the crop plant have been significant sources of new variability in the areas of original domestication of the crop species. In self-pollinated crops, variability for selection has come from mutation and the occasional outcrossing between component varieties in the crop populations which were likely to have been, in the main, mixtures of a large number of different varieties. Small frequencies of hybridisation between the domesticated and wild and weed forms of the crop species are likely to have been very significant in contributing new variability for advances made from selection under domestication.

There are two main methods of selection in both self-pollinated and cross-pollinated plants. One method, called mass selection, is similar in procedure in these two groups of plants; it has been of significance in crop plant improvement, both throughout the history of crop domestication and in recent plant breeding programs. The other method of selection for self-pollinated plants is called pure line selection and for cross-pollinated plants is called progeny selection and line breeding. It involves selecting single plants from crop populations and evaluating their progenies for superior performance. From this selection, lines or cultivars can be produced which are genetically improved for the character(s) being selected. Owing to the different population structures in self-pollinated and cross-pollinated crops, the implications and procedures in selection are somewhat different. Because the homozygous condition is achieved through selfing in self-pollinated crop plants, single plant selection can be practised from which homozygous pure line cultivars can be produced. In cross-pollinated crops, the heterozygous nature of all the individual plants in the population means that single plant selection would lead to inbreeding and homozygosity and a consequent loss in vigour and fertility. Selection in these latter crops therefore has to be based on a population of plants where out-breeding and hence heterozygosity is maintained at a high level.

Mass Selection

Mass selection involves the selecting of a group of individuals from a population on the basis of their similar phenotype in an attempt to improve the performance of the population for that character or characters. A knowledge of the genetic basis and the extent of the population variability for the character indicates to the crop breeder the potential offered for improvement through selection. In the mass selection procedure there is no progeny test, as with pure line selection and progeny selection and line breeding, but the selected population is usually evaluated against the original unselected population to gauge the effectiveness of selection.

In self-pollinated crops, mass selection has two main uses. Firstly, it is an efficient method of bringing rapid improvement to populations that have a proportion of individuals that are obviously (i.e. visually) unfit, such as too early or late in maturing, disease susceptibility or poor standing ability. Mass selection is used in this way to 'purify' populations for a narrower range of character value(s), a practice which is believed to confer closer to optimum fitness for that character, or characters, in commercial production. It is also used in pure seed production in self-pollinated crop cultivars where selection is based on discarding phenotypic variants that diverge from the standard type, or range, for the cultivar.

In cross-pollinated crops, mass selection has contributed substantially to their improvement, particularly for characters of high heritability such as disease resistance, date of maturity and protein or oil content of grain, and there are numerous examples of its successful application. The increase in sugar content in sugar beet (*Beta vulgaris*) from approximately 7.5 to 18 per cent, from early in the nineteenth century until the present day, is an example of the contribution of this method to crop improvement.

Mass selection is, however, not an efficient method for improving characters of low heritability such as yield. Further discussion of this matter is given under 'Breeding and Selection Methods in Cross-Pollinated Crops'.

Pure Line Selection

The pure line theory of selection was established by Johannsen (1903, 1926) who showed, using a seed lot of beans of different sizes, that selection for seed size was effective in isolating large and small-seeded lines which bred true for this character. Through continued selection he showed that there was no significant change in seed size within the large and small seeded lines. He concluded that the lines were homozygous for seed size and that the original seed lot was a mixture of pure lines. A pure line can be defined as the progeny of a single self-fertilised homozygous individual.

The pure line selection method depends on the knowledge that continued selfing of a heterozygous individual or population of a self-pollinated plant species will result in an increasing proportion of the subsequent population becoming homozygous. Eventually all individuals will be homozygous. In crop breeding the parents of a cross usually differ by a large number of gene pairs and this number will influence the number of generations of self-fertilisation necessary for the hybrid population to reach homozygosity. At this point the population would be heterogeneous and composed of a large number of different homozygous individuals. The degree of homozygosity of the hybrid population depends on the number of independent gene pairs by which the parents differ and the number of generations of selfing since hybridisation and is expressed as follows:

$$\text{Proportion of the population which is homozygous} = \left(\frac{2^m - 1}{2^m}\right)^n$$

where m = the number of generations of self-fertilisation, and
n = the number of independent gene pairs by which the parents differ

Thus if the parents of a cross differ by 12 independent gene pairs the proportion of homozygosity of the hybrid population after 6 generations of self-fertilisation would be 83 per cent.

In practice, segregating populations of self-fertilising plants often do not conform exactly to the expectation of the preceding formula for two main reasons. First, selection, artificial or natural, can favour the heterozygote giving it a higher survival value than the homozygous individuals and hence its rate of decline, as a proportion of the population, is slower than expected from theory. Second, linkage can cause an increase in the proportion of homozygous individuals in any generation over that expected on the basis of independent assortment of genes.

The pure line method is the basis of selection in breeding programs for improvement in self-pollinated crops.

Progeny Selection and Line Breeding

Progeny selection and line breeding has been an important method of improving the performance of cross-pollinated crops over the last 70 years. It involves selecting plants on phenotypic appearance from open-pollinated populations, selfing them and evaluating their progenies, usually as small rows or plots of 10 to 50 plants, in the following year. The inferior progenies are eliminated and the selected progenies are incorporated in a composite population which should be genetically superior to the original population for the character, or characters, being selected.

The procedure of incorporating the selected superior progenies in a composite population is referred to as line breeding. The risks of inbreeding can be minimised by firstly incorporating a large number of progenies, usually greater than 50, in the composite population and secondly ensuring that they are genetically dissimilar.

The source population in which progeny selection is carried out, besides being an open-pollinated population, can be an inbred or hybrid population. An outline of the progeny selection and line breeding method is shown in Fig. 3.7.

This method of selection has proved useful for improving the performance of cross-pollinated crops for qualitative characters and also quantitative characters such as yield in unadapted populations. However, in highly adapted populations it is not an efficient method for improving yield, due both to its low heritability and to the complex genetic make-up of yield in such populations.

Natural Selection

While most breeding programs in self-pollinated and cross-pollinated crops depend mainly on artificial selection for character improvement, there has been some interest in the possibility of using natural selection to achieve improvement in crop breeding.

Its use applies particularly to adaptive characters such as cold or drought tolerance and disease resistance or tolerance. Natural selection is allowed to act on segregating populations of self-pollinated crops, e.g. 'composite cross' populations of barley, or normal populations of cross-pollinating crops, when grown for a number of seasons in a particular environment. Under natural selection, significant shifts can be obtained in gene frequency of the population towards greater adaptiveness, and hence increased or more assured yield of the crop in that environment.

BREEDING AND SELECTION METHODS IN SELF-POLLINATED CROPS

Breeding and selection methods for self-pollinated crops are based on the knowledge that the genetic variability produced through hybridisation and recombination between carefully selected parents provides scope for obtaining more favourable recombinations of characters, and that it is possible to obtain homozygous lines containing these recombinants through selfing and selection. There are a number of methods for breeding and selecting self-pollinated plants. In the choice of a particular method the breeder considers the genetic control of the character, i.e. whether simple or complex in inheritance, whether it is of high or low heritability and the degree of linkage with undesirable characters, and the time, labour and space available for the proposed breeding program.

The main methods used for self-pollinated crops, as outlined here, are the pedigree, multiple cross, bulk, composite cross, backcross and single seed descent methods.

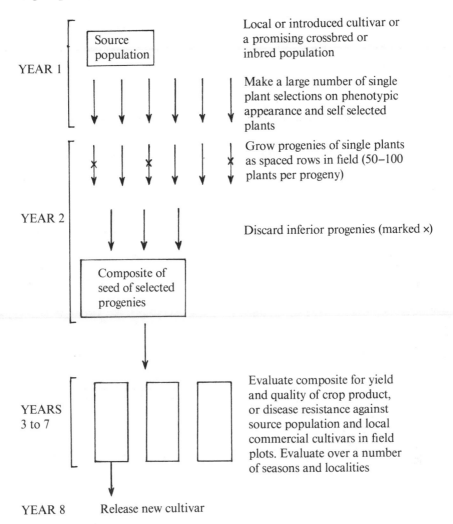

FIG. 3.7   Progeny selection and line breeding procedure

Pedigree Method

The pedigree method has found wide acceptance amongst breeders of self-pollinated crops (Fig. 3.8). It involves the hybridisation of two parents, one usually a commercial cultivar and the other chosen because of a particular superior attribute, or attributes, it possesses. The breeder aims to incorporate this attribute(s) in a genetic combination that is at least equal to the commercial cultivar in all characters and possessing the attribute(s) of the donor parent. The pedigree method involves selecting single plants from segregating populations of a cross and evaluating the performance of their progeny with repeated selection within the better progeny for single plants followed again by further progeny evaluation.

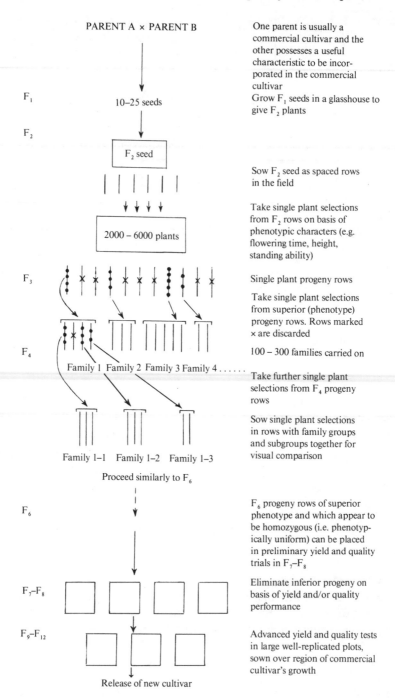

PARENT A × PARENT B

One parent is usually a commercial cultivar and the other possesses a useful characteristic to be incorporated in the commercial cultivar

F_1 — 10–25 seeds

Grow F_1 seeds in a glasshouse to give F_2 plants

F_2

F_2 seed

Sow F_2 seed as spaced rows in the field

2000 – 6000 plants

Take single plant selections from F_2 rows on basis of phenotypic characters (e.g. flowering time, height, standing ability)

F_3

Single plant progeny rows

Take single plant selections from superior (phenotype) progeny rows. Rows marked × are discarded

F_4

100 – 300 families carried on

Family 1 Family 2 Family 3 Family 4 . . . . . .

Take further single plant selections from F_4 progeny rows

Sow single plant selections in rows with family groups and subgroups together for visual comparison

Family 1–1   Family 1–2   Family 1–3

Proceed similarly to F_6

F_6

F_6 progeny rows of superior phenotype and which appear to be homozygous (i.e. phenotypically uniform) can be placed in preliminary yield and quality trials in F_7–F_8

F_7–F_8

Eliminate inferior progeny on basis of yield and/or quality performance

F_9–F_{12}

Advanced yield and quality tests in large well-replicated plots, sown over region of commercial cultivar's growth

Release of new cultivar

FIG. 3.8   Pedigree method of breeding

Single plant selection can be commenced as early as the F_2 population or it can be delayed until the F_3 or F_4 generation from the bulk segregation population. Single plant selection followed by progeny evaluation is followed until the F_6 or F_7 generation, by which time most of the lines are homozygous. This is judged visually for phenotypic uniformity of the lines and for a character like grain quality, when there is little or no variation for it amongst selections from within a line. At the F_7 the homozygous selected lines can be placed in replicated yield trials for further selection on the basis of yield performance.

The pedigree method is useful for handling large numbers of recombinants with comparative ease. The crop breeder usually selects in early generations, on a visual basis, for qualitative characteristics such as height, standing ability, morphology of the inflorescence, disease resistance and seed characters such as shape, colour and size. Because most of these characters are of reasonably high heritability and under simple genetic control this selection is very efficient in improving the overall agronomic suitability of the resultant population. Selection in later generations can be commenced for quantitative characters such as yield and quality. Because there are fewer selected lines and greater amounts of seed of the selections, small replicated trials are practicable, and necessary, for reliable evaluation of such characters. Many studies have been conducted on the efficiency of selecting for yield on a single plant, or row, basis in early segregating generations but the general finding is that it is not a reliable method.

The superior lines obtained from preliminary yield and quality trials are further evaluated for yield and quality in larger plots, with greater replication and often in a range of environments representative of the agricultural area for which the breeding is being conducted.

The theory behind the use of the pedigree method of selection bears upon its practicability in enabling the breeder to handle a very large number of recombinants in early segregating generations. When the breeder is using parents which differ by a large number of gene pairs, even when the character being selected is under simple genetic control, the possible number of recombinant types for the overall parental differences is very large. If the parents differ by n allelic pairs of genes the kinds of possible phenotypes in the F_2 generation are 2^n, if all genes exhibit full dominance, or 3^n if they exhibit no epistasis and no dominance. Thus for a difference between the parents of 20 allelic pairs of genes the number of possible phenotypes in the F_2 is 2^{20} (1 048 576) with full dominance, or 3^{20} (3 486 784 401) with no epistasis and no dominance. It is for this reason that the pedigree method is popular in that it enables the crop breeder to handle large numbers of recombinants. While the upper limit to the number he can handle in any one cross does not, in most situations, embrace the desired number on theoretical expectations of phenotype numbers in the F_2, it appears to offer a reasonably good chance of selecting favourable recombinants. However, much research needs to be conducted on improving the efficiency of pedigree selection, particularly in breeding for yield improvement.

Multiple Cross

This method of breeding is similar to the pedigree method except that it involves the hybridisation of more than two parents, sometimes from 4 to 8. It has been used in attempts to combine the attributes for a particular quantitative character, such as yield or grain quality, from a number of adapted parents of divergent genetic origins. An outline of the method is shown in Fig. 3.9. One severe limitation in the use of the method is the large number of crosses which would have to be performed in the four parent cross, i.e. in hybridising two F_1s, each of different parentage, and the extremely high number necessary in the eight parent cross, i.e. in hybridising the two double F_1s, to have a good chance of adequately sampling the possible range of recombinants in the two parent and four parent crosses respectively. The subsequent selection procedure could be that shown for either the pedigree or the bulk selection method.

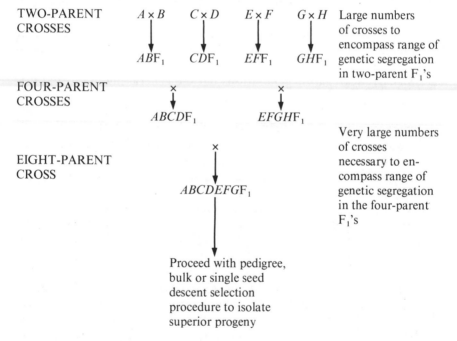

FIG. 3.9   Multiple cross breeding procedure

Bulk Method

The bulk method of breeding involves hybridising two chosen parents and growing the F_2 and subsequent generations until the F_6 generation in bulk populations as field plots in successive years (Fig. 3.10). During this

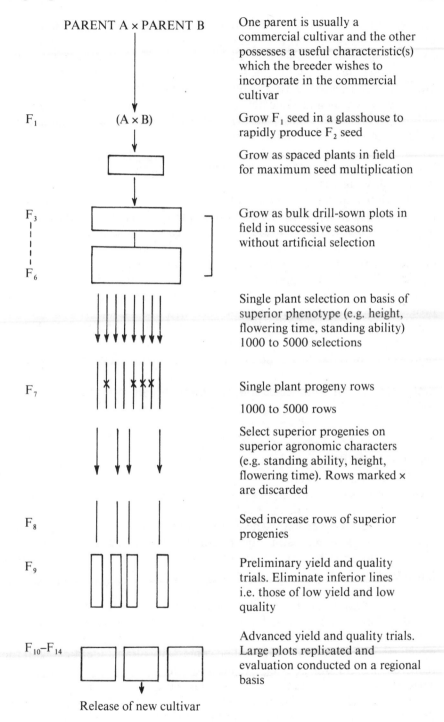

PARENT A × PARENT B — One parent is usually a commercial cultivar and the other possesses a useful characteristic(s) which the breeder wishes to incorporate in the commercial cultivar

F_1 (A × B) — Grow F_1 seed in a glasshouse to rapidly produce F_2 seed

Grow as spaced plants in field for maximum seed multiplication

F_3 ... F_6 — Grow as bulk drill-sown plots in field in successive seasons without artificial selection

Single plant selection on basis of superior phenotype (e.g. height, flowering time, standing ability) 1000 to 5000 selections

F_7 — Single plant progeny rows

1000 to 5000 rows

Select superior progenies on superior agronomic characters (e.g. standing ability, height, flowering time). Rows marked × are discarded

F_8 — Seed increase rows of superior progenies

F_9 — Preliminary yield and quality trials. Eliminate inferior lines i.e. those of low yield and low quality

F_{10}–F_{14} — Advanced yield and quality trials. Large plots replicated and evaluation conducted on a regional basis

Release of new cultivar

FIG. 3.10   Bulk population breeding method

time selection is usually not practised but in the F_5 or F_6 a large number of single plant selections are taken and their progeny sown as individual rows in the following season. Progenies can be culled on agronomic performance, including height, standing ability, threshing characteristics and such characters as flowering time and grain protein content. The progeny rows which are selected and which appear to be homozygous, as judged on the phenotypic uniformity of the row, can be evaluated for yield in small-plot replicated trials in the following year. Progeny rows which are of superior performance but which are still segregating can be single plant selected and progenies of individual plants evaluated in rows in the following year.

As with the pedigree method, extensive seasonal and locality evaluation of yield and quality of the homozygous lines is carried out. Elimination of inferior progenies for these characters early in the evaluation stages enables more elaborate evaluation, in larger plots and at a large number of testing sites, to be carried out on the superior progenies.

The bulk method of breeding relieves the breeder of the large amount of work in sowing, selecting and harvesting single plant selections and progeny rows from F_2 to F_6 as is the case with the pedigree method. It has the advantage of enabling the breeder to handle a larger number of crosses because of the relative ease in bulk sowing and harvesting the segregating populations. The comparative evaluation of bulk yields of the segregating populations of different crosses has been used to indicate those crosses wherein selection for yield might provide the highest yielding segregates.

Backcross Method

The backcross method of breeding was developed by Briggs (1938) and it is used for incorporating a gene, or genes, from a donor parent (called the non-recurrent parent) in the background of an adapted line or cultivar. This is achieved by crossing the adapted line, or cultivar, with the donor parent and again crossing the adapted line, or cultivar (called the recurrent parent) on to the F_1 of this cross. This is called the first backcross.

Segregates of the first backcross generation carrying the desired character of the donor parent are hybridised with the recurrent parent. This is called the second backcross. This procedure of selection and hybridisation with the recurrent parent can be continued until the sixth backcross at which stage the backcross line will be genetically almost identical with the recurrent parent but bearing the desired character of the donor parent. The selection procedure during backcrossing will differ according to whether the character being transferred is controlled by one or more dominant or recessive genes. An outline of the two different procedures is shown in Fig. 3.11.

The backcross method is efficient in the saving of labour and space

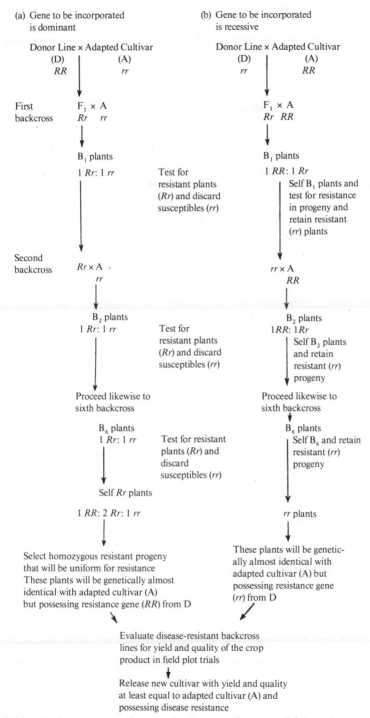

FIG. 3.11 Backcross breeding method according to whether the gene to be incorporated is (a) dominant or (b) recessive

when compared with the pedigree method. While the time to reach homozygosis is no faster than with the pedigree method it narrows segregation to those individuals which converge on the genotype of the recurrent parent. The relative gene contents of the recurrent and non-recurrent parents in successive backcross generations is shown in Table 3.4. After six backcrosses the backcross line should be approximately 99 per cent of the gene content of the recurrent parent.

The backcross method of breeding is also more efficient than the pedigree method when there is close linkage between the desired gene and undesirable genes in the donor parent. Repeated backcrossing with the recurrent parent increases the chance of breaking such linkages. The effectiveness of backcrossing and selfing in breaking linkage, compared with continued selfing as in the pedigree method, depends on the degree of linkage and on the number of backcrosses performed. The probability through backcrossing of eliminating an undesirable gene which is linked to the desired gene in the donor parent is expressed in the formula

$$1 - (1 - P)^{n+1}$$

where P is the recombination fraction between the desirable and the undesirable gene and n is the number of backcrosses.

TABLE 3.4   Relative constitution of the backcross population for the gene content of the recurrent and non-recurrent parents with successive backcrosses

Cross	Donor line (D) × adapted cultivar (A) (non-recurrent × recurrent parent)	Relative gene content (%) of backcross generation D : A
First cross—F_1	$(D \times A)$	50 : 50
First backcross—B_1	$(D \times A) \times A$	25 : 75
Second backcross—B_2	$(D \times A^2) \times A$	12.5 : 87.5
Third backcross—B_3	$(D \times A^3) \times A$	6.25 : 93.75
Fourth backcross—B_4	$(D \times A^4) \times A$	3.125 : 96.875

The effect of varying degrees of linkage (recombination fractions) on the probability of eliminating an undesirable gene linked to a desired gene, after five backcrosses, is shown in Table 3.5. It indicates that backcrossing is more efficient in breaking linkage than selfing and its efficiency over selfing is greater when there is very close linkage between the desired and undesired genes.

Crop breeders have often used the backcross method in a modified form by proceeding with, say, two or three backcrosses, after which they select for both the retention of the specific character being transferred from the donor parent and the possibility also of increased yield over the recurrent parent. It is possible at this stage of backcrossing, when the

TABLE 3.5 Comparative effectiveness of backcrossing (five backcrosses) versus selfing in a cross in eliminating an undesirable gene linked to a desirable gene

	Probability that undesirable gene will be eliminated	
Recombination fraction	With five backcrosses	With selfing
0.05	0.98	0.50
0.20	0.74	0.20
0.10	0.47	0.10
0.02	0.11	0.02
0.01	0.06	0.01
0.001	0.006	0.001

populations are still reasonably heterozygous, that transgressive segregates for a character like yield may be obtained from selection within such populations.

Single Seed Descent

Single seed descent is a recent method of selection in the breeding of self-pollinated plants. The method adopts the procedure of advancing lines from generation to generation through the agency of a single seed of the progeny of each plant rather than single plant and progeny row selection as in the pedigree method. Like the pedigree method, however, selection can be practised for qualitative characters throughout the segregating generations to homozygosity provided the populations are grown in the normal season in the field. In practical terms it enables the breeder to handle much larger numbers of initial recombinants than the pedigree method, and larger numbers of lines throughout the segregating generations. There is a great saving of time and labour in the sowing, maintenance and harvest of the lines compared with the pedigree method. When selection for qualitative characters is not being practised during the segregating generations the method, because of the compactness of each sowing, enables line to be sown under artificial conditions such as a glasshouse or growth room. It is therefore possible to grow two to three generations per year and greatly shorten the time to reach homozygosity compared with the pedigree method. An outline of the single-seed descent procedure is shown in Fig. 3.12.

BREEDING AND SELECTION METHODS IN
CROSS-POLLINATED CROPS

Earlier selection methods in cross-pollinated crops, mass selection and progeny selection and line breeding, have certain limitations, particularly when used for improving highly-adapted cultivars for characters of low heritability such as yield. Mass selection is of limited use for such characters, firstly because of the failure of phenotypic selection to isolate superior genotypes for the desired character and secondly because pollination is uncontrolled in that superior genotypes hybridise with inferior genotypes in the population being selected. Progeny selection and line breeding is a more efficient method in that it embodies progeny evaluation

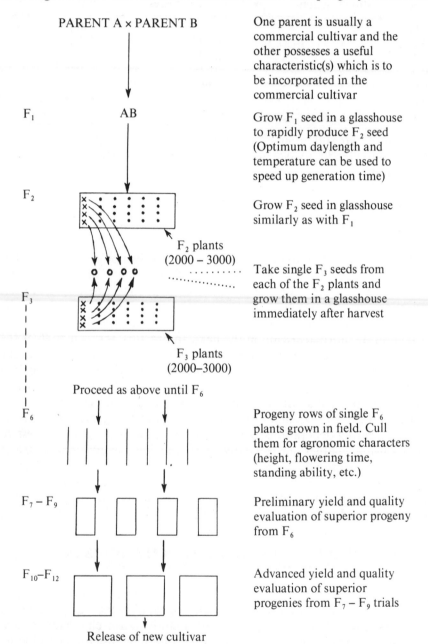

PARENT A × PARENT B — One parent is usually a commercial cultivar and the other possesses a useful characteristic(s) which is to be incorporated in the commercial cultivar

F_1 AB — Grow F_1 seed in a glasshouse to rapidly produce F_2 seed (Optimum daylength and temperature can be used to speed up generation time)

F_2 — Grow F_2 seed in glasshouse similarly as with F_1

F_2 plants (2000 – 3000)

Take single F_3 seeds from each of the F_2 plants and grow them in a glasshouse immediately after harvest

F_3

F_3 plants (2000–3000)

Proceed as above until F_6

F_6 — Progeny rows of single F_6 plants grown in field. Cull them for agronomic characters (height, flowering time, standing ability, etc.)

$F_7 - F_9$ — Preliminary yield and quality evaluation of superior progeny from F_6

$F_{10}-F_{12}$ — Advanced yield and quality evaluation of superior progenies from $F_7 - F_9$ trials

Release of new cultivar

FIG. 3.12 Single-seed descent breeding procedure

of selections and, particularly if done in replication, provides substantiation for the retention of superior genotypes.

However, these methods of selection have been found to be ineffective in improving yield in highly adapted genotypes. This is mainly due to the low frequency of favourable genes for yield in cross-pollinated populations, and the ineffectiveness of these methods in retaining genetic variability and allowing recombination to produce more favourable recombinants for yield within the population under selection.

In about the 1940s it became apparent that the low frequency of favourable genes for yield in cross-pollinated populations demanded a type of selection which acted to increase their frequency by allowing recombination to take place between the best performing selections. Another requirement of such a method was that it maintained genetic variability in the population being selected even through a number of cycles of selection.

The systems of selection which appeared to cater for these requirements are referred to generally as methods of recurrent selection and in broad principle they involve the following steps:

1. Individual plants are selected from an open-pollinated source population and selfed. These plants are evaluated at the same time for the character, or characters, being selected.
2. Plants of inferior performance (as adjudged either visually or on the behaviour of selfed progeny or test-cross progeny) for the desired character, or characters, are discarded.
3. Selfed plants in Step 1 with superior progenies are propagated from the retained portion of selfed seed.
4. Superior progenies are placed in a crossing block to obtain as many intercrosses between progenies as possible.
5. The intercross population is sown as a basis for further cycles of selection and intercrossing as in Steps 1 to 4.

There are four main methods of recurrent selection—simple recurrent selection, recurrent selection for general combining ability, recurrent selection for specific combining ability and reciprocal recurrent selection. The essential features of these four methods of selection are now discussed.

Simple Recurrent Selection

This method of selection involves visually selecting single plants from an open-pollinated source population and selfing them. The progenies of the selections are sown in a crossing block in the following year and allowed to freely intercross. The intercross population is sown in the following year and single plant selection and selfing carried out again. The progenies of the selfed plants are sown again in the following year in an intercross block. The second cycle of selection and sowing of progenies is called the first recurrent selection cycle. The cycle can be continued until there appears to be no further significant advance being obtained for the character or characters under selection. Because the method does not

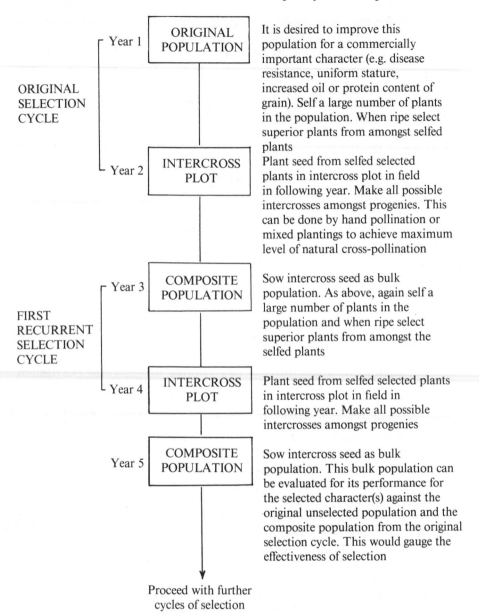

FIG. 3.13   Simple recurrent selection

involve progeny evaluation it is suitable only for characters of high heritability. An outline of the simple recurrent selection procedure is shown in Fig. 3.13.

Recurrent Selection for General Combining Ability

This method of recurrent selection involves selecting single plants (called S_0 plants) from a source population, selfing them and, at the same time, crossing them to a tester. This tester is a heterozygous population, and the progeny of the test crosses to it are evaluated comparatively for general combining ability for the character being selected. Selfed seed from those S_0 plants which are good combiners is then sown in a crossing block in the field, and the progenies are either intercrossed by hand or allowed to intercross naturally. The harvested seed is then sown, and a further cycle of selection and test-crossing can be carried out.

Recurrent selection for general combining ability has been shown to be a useful method for improving yield and adaptability of a number of cross-pollinated crops.

Recurrent Selection for Specific Combining Ability

This method of selection is basically similar in outline to recurrent selection for general combining ability except that a homozygous inbred tester is used rather than a heterozygous tester.

The aim of the method is to produce selections from an open-pollinated source population which possesses high specific combining ability with an inbred tester for the character being selected. Specific combining ability is the deviation in performance of a particular cross, for the character under selection, from the value predicted from its general combining ability. Through cycles of recurrent selection, combinations can be produced which have increased specific combining ability with the inbred tester. The selections can then be used in crosses with the tester to produce commercial hybrids.

Reciprocal Recurrent Selection

In this scheme of selection it is possible to select simultaneously for both general combining ability and specific combining ability (Fig. 3.14). The procedure involves the use of two genetically unrelated open-pollinated populations, A and B. Random single-plant selection (S_0 plants), and their selfing, is carried out in population A and at the same time they are crossed with a random sampling of plants from population B. The same procedure is also carried out in population B. The progenies of these reciprocal crosses are evaluated in replicated trials for the character or characters being selected. Selfed seed of S_0 plants of the A and B populations that gives superior progenies in these crosses is then sown as separate A and B populations and allowed to intercross within each population.

The bulk progenies of the intercrosses are then sown as separate A and B populations, and a further cycle of selection is carried out. It may be decided after the second cycle either to proceed to the third cycle or to use the selected populations to produce commercial hybrids. This is done by making crosses between the selected A and B populations.

Synthetics

Synthetic varieties have been produced in a number of cross-pollinated crops as a means of exploiting the heterosis of combinations amongst a

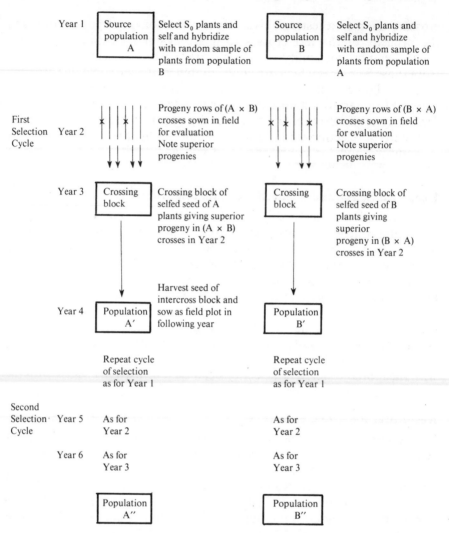

FIG. 3.14   A procedure for reciprocal recurrent selection

set of specially chosen parents. An important condition for constituting a synthetic variety is that the parents must exhibit good combining ability for yield in all pairwise combinations.

The level of performance of advanced generations of a synthetic variety depends on three important factors: the number of parents entering into the synthetic; the mean performance of these lines for the character; and the mean performance of these lines for the character or characters for which it is being constituted, and thereby the mean performance of all combinations arising from hybridisation amongst the parental lines. These requirements can be expressed in the formula:

$$Syn_2 = Syn_1 - \frac{(Syn_1 - Syn_0)}{n}$$

where Syn_2 is the estimated performance of the second generation, or F_2, of the synthetic after its constitution, Syn_1 is the mean performance of all the hybrids, or F_1s, of the parental components, Syn_0 is the mean performance of all the parental lines, and n is the number of parents in the synthetic.

This formula can be used to predict the performance of the second generation (Syn_2) of a synthetic. The value $Syn_1 - Syn_0$ represents the excess in performance of the Syn_1 over its parents. A two-parent combination will therefore have a second generation (Syn_2) performance equal to

$$Syn_1 - \frac{(Syn_1 - Syn_0)}{2}$$

i.e. it will have lost half of the excess vigour of the Syn_1 over the parents. With three, four, and five parents this decline in vigour will be $1/3$, $1/4$, and $1/5$ respectively that of the Syn_1, over the mean parental value. According to the Hardy–Weinberg rule, there should be no further decline in vigour in following generations, i.e. Syn_3, Syn_4, if mating in the synthetic population is completely at random and there is no differential selection.

parents in the synthetic is made large, the loss in vigour of the Syn_2 over the Syn_1 could be made very small. However, there is a limit to the number of parental lines which can be chosen to give high combining ability for yield. The optimum parent number has been shown to be approximately five to six for maize. Beyond this number the average combining ability declines.

The predictability of the Syn_2 performance from Syn_1 and Syn_0 performance has been shown to be very close and is used as a routine method in constituting synthetic varieties. It has been shown (Lonnquist and McGill, 1956) that yields of advanced generations of synthetic varieties in maize are maintained by mass selection, and there is evidence for yield being increased by selection.

Synthetic varieties are important in the commercial production of a number of cross-pollinated forage species like lucerne, and a number of pasture grasses such as perennial ryegrass, timothy grass and orchard grass. Similarly, as with crop plants, combining ability tests must be made amongst potential parents to choose those that will genetically constitute the synthetic. Combining ability measurements are usually made on hybrids between the parents as a result of natural pollination between them. An important method of assessing combining ability is the polycross test which is an evaluation of progeny of crosses obtained from random hybridisation amongst the parental lines when planted together in the field.

Hybrid Varieties

A hybrid variety is an F_1 population that is grown commercially to exploit heterosis, usually for yielding ability. Hybrid varieties were first developed commercially in onions and maize, and they have contributed significantly to increased yields in those crops. In fact the double cross F_1 (i.e. four-parent hybrid) in maize with its increased yield over open-pollinated varieties has been one of the most significant plant breeding achievements in modern times.

The concept of hybrid varieties was first developed by Shull (1909) in maize. Shull produced maize inbreds and crossed them in pairs. Those that gave superior F_1 performance were chosen for commercial production of hybrid seed. The advantages seen in the use of inbreds over open-pollinated varieties were that they were homozygous and were predictable in performance and they enabled the hybrid to be accurately reconstituted year after year. However, the single cross method of hybrid production did not succeed because the hybrids did not significantly surpass the yield of open-pollinated varieties. The seed was costly to produce because of the low yield of the inbred female parents and the low vigour and viability of the hybrid seed.

In 1918, Jones introduced the use of the double cross in maize, which is the hybrid between two F_1s of four parents, i.e. $(A \times B) F_1 \times (C \times D) F_1$, where A, B, C and D are four inbred parents. This method of hybrid production was more successful than the single cross hybrid. The single cross parents of the double cross were much more vigorous and higher yielding than the inbred parents of the single cross, and the hybrid seed was more vigorous and viable than the single cross seed.

The present-day procedures used in producing hybrid maize are as follows:

  (i) select desirable plants from open-pollinated populations;
  (ii) self the selections for several generations until homozygous to produce inbred lines;
  (iii) hybridise chosen inbreds to produce single cross F_1s;
  (iv) select those single crosses exhibiting the highest combining ability for the character(s) to be improved for use in the double cross hybrids;
  (v) produce double cross hybrids from the best-performing single crosses.

In the production of single-cross and double-cross seed, cytoplasmic male sterility is used to minimise the labour otherwise necessary in detasseling (emasculating) female parents to produce the single and double crosses. Fertility-restoring genes are also used to improve the

(i) One inbred male sterile and either one or two inbreds with dominant fertility-restoring genes

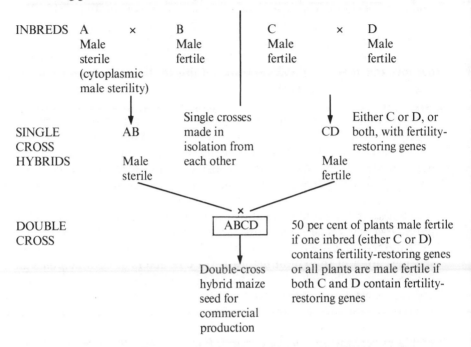

INBREDS   A   ×   B   C   ×   D
          Male   Male   Male   Male
          sterile   fertile   fertile   fertile
          (cytoplasmic
          male sterility)

                                     Single crosses   Either C or D, or
SINGLE   AB   made in   CD   both, with fertility-
CROSS   isolation from   restoring genes
HYBRIDS   Male   each other   Male
                    sterile   fertile

                                        ×

DOUBLE   ABCD   50 per cent of plants male fertile
CROSS   if one inbred (either C or D)
                                                contains fertility-restoring genes
                             Double-cross   or all plants are male fertile if
                             hybrid maize   both C and D contain fertility-
                             seed for   restoring genes
                             commercial
                             production

(ii) Two inbreds male sterile and one inbred with dominant fertility-restoring genes

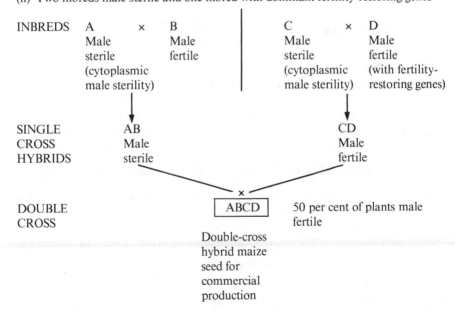

INBREDS   A   ×   B   C   ×   D
          Male   Male   Male   Male
          sterile   fertile   sterile   fertile
          (cytoplasmic   (cytoplasmic   (with fertility-
          male sterility)   male sterility)   restoring genes)

SINGLE   AB   CD
CROSS   Male   Male
HYBRIDS   sterile   fertile

                                        ×

DOUBLE   ABCD   50 per cent of plants male
CROSS   fertile
                             Double-cross
                             hybrid maize
                             seed for
                             commercial
                             production

FIG. 3.15   Two methods of producing double-cross hybrid maize seeds using cytoplasmic male sterility and fertility-restoring genes

efficiency of producing the double-cross hybrid. An outline of two methods of producing double-cross hybrid maize using cytoplasmic male-sterility and fertility-restoring genes is shown in Fig. 3.15. Four selected inbreds are hybridised through cross-pollination in pairs by growing the two pairs of parents together in the field and one pair in isolation from the other. A parent of one of the pairs, if it possesses cytoplasmic male sterility, will transmit this to the single cross to make it male-sterile. One or both parents of the other single cross contain fertility-restoring genes and each is male-fertile so that their hybrid will also be male fertile. The two single-cross hybrids are then planted together in the field and cross-pollination takes place between the male-sterile parent and the male-fertile plant. The seed which develops on the male-sterile parent will be double-cross hybrid seed. The presence of fertility-restoring genes in the double-cross hybrid from one of the single-cross parents restores male-fertility in the double-cross population to 50 per cent or 100 per cent depending on whether one or both of the original parents of the single cross possesses restoring genes. The double-cross hybrid seed is grown commercially as hybrid maize. Hybrid maize schemes attempt to exploit heterosis for characters of commercial significance, especially yield. In this context heterosis is the increased vigour, growth or yield of a hybrid population over that of its parents.

Hybrid cultivars are preferred by commercial breeders (as distinct from government or public breeders) because heterosis breaks down in F_2 and later generations by segregation. Farmers must therefore buy new F_1 planting seed from the breeder each season.

BREEDING FOR SPECIFIC CHARACTER IMPROVEMENT IN CROP PLANTS

Breeding programs for improving particular characters in crop plants require a detailed knowledge, not only of the genetic control of the character but also of the type and magnitude of environmental influences on its expression. Environmental influences include climatic factors such as moisture availability and temperature, and edaphic factors such as nutrient availability and soil pH which act to influence the growth, productivity and quality of the crop product. Modern crop breeding also makes use of knowledge of certain physiological and biochemical processes, influencing yield and quality of crop plants, in attempting to improve the efficiency of breeding for improvement in these characters.

Crop breeding activities are broadly concerned with improving the yield and quality of the crop product and improving or maintaining the resistance or tolerance of the crop to diseases and insect pests.

Yield Breeding

Breeding and selecting for yield improvement in crop plants is made difficult because of its low heritability and its genetic complexity. Because yield is the endpoint of the interaction of a large number of physiological and biochemical processes in the plant, its genetic control must be complex. Three main approaches may be discerned in breeding for increased yield in crop plants. First, yield can be regarded as a quantitative character, in which case use is made of quantitative genetics in both the choice of potentially useful parents for yield breeding and in selection procedures for increased yield. Second, certain approaches attempt to resolve yield into either its visible components (e.g. for wheat, grain size or weight, grain number per spikelet, number of spikelets per head and number of fertile tillers or shoots per plant), or physiological or biochemical processes, which are likely to be under simpler genetic control than total yield. This type of approach may enable the crop breeder to alter yield by means of simple genetic principles through genetic modification of one or more of its visible components or physiological or biochemical processes. The third approach aims at increasing yield through selecting for increased adaptability, either generally as high yield performance of hybrid lines and parents over a range of seasons and localities (Finlay and Wilkinson, 1963), or specifically as improved cold or drought resistance or escapism. Another approach to improving adaptability is selecting for greater tolerance to edaphic factors such as soil salinity or to high or low soil pH.

A number of methods for the genetic analysis of yield have been devised for predicting the potential contribution to increased yield of parental lines in hybrid combination. An important method is the diallel analysis (Jinks, 1954) which involves the hybridisation of a set of n parental lines in all possible pair-wise combinations to give $\frac{1}{2}n(n-1)$ single crosses or $n(n-1)$ reciprocal crosses. The diallel analysis uses the character expression of F_1 and F_2 and later generations of all the crosses, plus that of the parental lines, to interpret the genetic control of yield and to indicate those crosses likely to give the highest yielding segregates.

Selection for yield involves methods which partition the influences of environment, as season to season or locality to locality variation, against that of genotype on the expression of yield. These studies involve estimating heritability of yield and the selection intensity necessary to provide a reasonable chance of isolating high yielding segregates from hybrid populations. It has generally been found that visual selection for yield using a single plant or row in segregating populations of self-pollinated crops, and in populations of cross-pollinated crops, is not a very effective means of improving that character. To overcome these limitations, certain breeding methods (e.g. bulk selection and single seed descent in self-pollinated crops and recurrent selection schemes in cross-pollinated crops) enable very large numbers of random selections to be handled in breeding for increased yield.

The physical or physiological/biochemical component approach in breeding for increased yield is an area of great interest to crop breeders, physiologists and biochemists. Some of the visible components of yield (e.g. grain size and spikelet number in wheat) in crop plants have been shown to have higher heritability than total yield and hence provide more scope for improvement through breeding and selection than total yield.

Breeding for Improved Quality

The term quality in crop production has different meanings according to the various demands made of the range of crop products in industry and commerce. It can mean the suitability of the crop product to the technological demands made of it in its extraction or processing (e.g. the milling quality of wheat which is an evaluation of the ease of extraction, and the total amount of flour able to be extracted from the grain). Quality can also mean the extent to which the crop product meets the specifications of commercial demand (e.g. baking quality of wheat, malting quality of barley; taste, size and shape of fruit; fibre length, diameter and strength of cotton). The term nutritive quality is used to denote the dietary value of the crop product for human or animal consumption. It usually embodies an evaluation of the total carbohydrate, protein and fibre content and the available minerals. It often also includes estimates of the amino acid composition of the protein.

The quality of the crop product can usually be resolved into a small number of components whose variation is usually closely correlated with variation in the overall quality of the product. These components are used to measure quality both in its commercial evaluation and in breeding and selection studies for quality improvement. Breeders use small-scale tests for evaluating these quality components on small samples from their large numbers of breeding lines. This type of evaluation enables them to select with reasonable accuracy for quality improvement.

The level of quality is usually strongly influenced by environmental factors during the growth and ripening of the crop. Often, variation in specific environmental factors can be attributed to variation in the quality of the plant product. Soil nitrogen levels, for example, strongly influence the level of protein in the grain of wheat which in turn influences its baking quality. High humidity during the ripening of grain and fibre crops often has a detrimental influence on the quality of the crop product.

However, quality generally exhibits a moderate level of heritability and therefore selection for improved quality is usually successful in bringing about significant improvement, though often not of substantial magnitude. It has been shown, for instance, that breeding and selection have resulted in substantial improvement in oil and protein content of the maize grain and there are instances of increases in protein content having been achieved in the wheat grain from breeding and selection. In breeding for improved quality crop breeders choose high quality parental lines on the basis of their processing one, or a number, of superior quality components, and hybridise them with the commercial cultivar for quality improvement. In the segregating generations from these crosses, from the F_2 to the F_6, selection is carried out for high levels of quality in crossbreds which also bear all the desirable attributes of the commercial cultivar. When a small number of highly selected, high-quality lines of good agronomic performance have been produced they are multiplied and further evaluated for quality using methods relevant to commercial practice. In this way a new cultivar can be produced which is at least equal to the commercial cultivar in all of its desirable attributes (e.g. high yield, appropriate flowering and ripening times, disease resistance) but superior to it in quality.

Breeding for Pest Resistance and Tolerance

Breeding for pest resistance and tolerance is a very important activity in crop improvement. Its purpose is to ensure the continued productivity of the crop in situations where pest attack is significant in reducing yields or where the expansion of the crop into new environments is severely limited by the likelihood of pest attack.

The term pest resistance embraces both disease and insect resistance. Crop plants are subject to attack by a wide range of disease organisms such as fungi, bacteria, viruses and mycoplasma. Within each group of organisms there are a large number of genetic types, each of which causes a different identifiable disease which can be recognised on the basis of characteristic symptoms. Within particular disease-causing organisms there is usually wide variation in the capacity for, and degree of, infection of the host plant. This type of variation in fungi is attributed to different physiologic races of the pathogen and in bacteria, viruses and mycoplasma to different strains of the particular disease organism.

Fungal diseases usually attack only a particular plant part (inflorescence, leaf, stem or root) while bacteria, viruses and mycoplasma usually attack the plant by penetrating the vascular tissue and producing systemic infections in the plant.

Most crop plants are subject to attack from a wide range of insect pests whose feeding activities on the plant are usually restricted to a specific part of the plant, i.e. root, stem, leaf, or fruit and seed. There are three main ways in which insect infestation of crop plants can cause losses in productivity, and sometimes lowered quality, of the crop product: disease transmission, limitations to normal metabolic activity of the plant, and physical damage leading to death or lowered physiological efficiency of the plant. A large number of insects (e.g. aphids) transmit diseases (particularly diseases induced by viruses and mycoplasma and to a lesser extent by bacteria and fungi) to crop plants in conjunction with their dispersal and feeding activities. Many insect pests of crop plants release toxins into the plant, altering its normal metabolism. This may be expressed by reduced growth and vigour, reduced fertility, proliferation of shoots or even death of the plant. The lucerne bud mite (*Eriophyes medicaginis*), found on lucerne in many parts of Australia, releases a toxin into the plant which causes stunting and proliferation of shoots, resulting in greatly reduced yields in infested stands. Insects which feed on leaves, stems or roots of crop plants and destroy or damage significant amounts of plant tissue can cause severe losses in plant productivity due to reductions in normal physiological activity.

Disease Resistance

Crop plants have a genetic capacity to minimise the effects of disease attack which is manifested either as resistance or tolerance of the plant to the disease. Resistance includes those mechanisms which prevent or restrict the growth of the disease on the plant. The opposite condition to resistance is susceptibility, but the disease–plant relationship can also show varying degrees of resistance or susceptibility. Tolerance includes those mechanisms or that condition of the plant which allows the normal growth and reproduction of the disease on the plant without a significant depression in the yield of the plant.

The genetic basis of the plant–disease relationship (usually referred to as the host–pathogen relationship) can be of two general kinds—that controlled by major genes for resistance in the host and that controlled by minor genes for resistance in the host. However, with fungal pathogens the genetic basis of host–pathogen relationships also involves a corresponding system of genes in the pathogen which interact with the resistance/susceptibility loci in the host in the determination of resistance or susceptibility. This genetic interrelationship between the host and pathogen is quite specific. That is, for every resistance gene in the host there is a corresponding gene in the pathogen which interacts to determine the resistance or susceptibility of the host to the pathogen.

The capacity of the fungal pathogen to overcome the influence of a resistance gene in the host is termed *virulence*; the condition of its inability to infect the host is termed *avirulence*. A fungal pathogen possesses a range of different types with differing capacities to overcome resistance in the crop plant. These types are called physiologic races of the pathogen; most pathogens occur in nature as a few to numerous physiologic races.

An example of the relationship of virulence and avirulence genes in the pathogen with resistance and susceptibility genes in the host in determining the reaction of the crop plant to infection is shown in Table 3.6. Unless a race of the pathogen bears all the virulence genes corresponding to the number of resistance genes in the host, it is unable to attack the host.

Some fungal pathogens have a capacity to rapidly produce new physiologic races with the appropriate virulence gene or combination of virulence genes necessary to overcome the resistance gene (R) in the host. These virulence genes are either already present in the pathogen population, or arise by mutation, and are recombined by hybridisation or by asexual means into new virulent physiologic races. This type of active pathogen creates problems for the crop breeder because new cultivars have to be continually bred bearing new sources of resistance to the pathogen. One approach to this problem has been to incorporate two or more different genes for resistance to the pathogen in the one cultivar, it being reasoned that it would take the pathogen much longer to produce races with virulence genes for two or more resistance genes in the host than for a single resistance gene. However, while this approach generally delays the breakdown of resistance over that of the single resistance gene situation, it does not ensure indefinite breakdown of resistance. Wheat breeders incorporate multiple sources of resistance genes to wheat stem rust (*Puccinia graminis tritici*) to provide rust-resistant cultivars for Australian farmers.

TABLE 3.6 An example of the genetic relationships of host and pathogen in determining host disease reaction for two resistance loci in the host and the two corresponding virulence/avirulence loci in the pathogen

Genotype of pathogen	Genotype of host			
	$R_1^c R_1 R_2 R_2$	$R_1 R_1 r_2^d r_2$	$r_1 r_1 R_2 R_2$	$r_1 r_1 r_2 r_2$
$v_1^a v_1 v_2 v_2$	susceptible	susceptible	susceptible	susceptible
$V_1^b V_1 v_2 v_2$	resistant	resistant	susceptible	susceptible
$v_1 v_1 V_1 V_2$	resistant	susceptible	resistant	susceptible
$V_1 V_1 V_2 V_2$	resistant	resistant	resistant	susceptible

[a] v = virulence allele
[b] V = avirulence allele
[c] R = resistance allele
[d] r = susceptibility

The crop breeder uses a number of different methods for incorporating disease resistance and tolerance into a cultivar. The backcross method is commonly used to breed for major gene resistance. With the use of controlled environment facilities in glasshouses or growth rooms, two to three generations can be grown per year. After each backcross the segregating populations are inoculated with the disease and the resistant segregates are selected and crossed again with the recurrent parent. The recurrent parent is usually a commercial cultivar in which the resistance gene(s) is to be incorporated. In this way new sources of disease resistance can be rapidly incorporated to produce new cultivars possessing resistance to current races of the pathogen.

Major genes for resistance and tolerance can also be incorporated into new cultivars using the single seed descent or pedigree selection methods for self-pollinated crops and recurrent selection methods for cross-pollinated crops. When resistance or tolerance is controlled by minor genes these methods can also be used in breeding for resistance; the backcross method is not used for this type of resistance.

Studies of the genetics of resistance and breeding for resistance or tolerance to virus attack in crop plants are more recent than those for fungal pathogens. However, a number of cases of resistance or tolerance of crop plants to virus have been reported. An example is the tolerance of barley (*Hordeum vulgare*) to barley yellow dwarf virus which is conferred by the action of a major gene. An important approach in breeding to minimise losses in productivity due to virus disease is that of increasing the level of tolerance of crop cultivars to viruses. The scope for breeding for resistance, either to the penetration of the virus of the host or to its insect vectors, appears to be much more limited than that offered in breeding for resistance to fungal pathogens.

Certain crops are subject to severe attack and losses in productivity from bacterial pathogens, such as bacterial wilt (*Corynebacterium insidiosum*) of lucerne (*Medicago sativa*), and bacterial blight (*Xanthomonas campestris* var. *malvacearum*) of cotton (*Gossypium* spp.). Resistance to bacterial infection is genetically controlled and resistance breeding usually involves the incorporation of major genes for resistance against prevalent strains of the pathogen.

Insect Resistance

Methods of breeding for resistance and tolerance of crop plants to insect attack employ approaches similar to those used for disease resistance and tolerance.

Recent studies on the genetic basis of insect resistance and tolerance in crop plants indicate a generally similar genetic relationship between the host plant and the insect to that described for fungal pathogens and their hosts. There appear to be both major and minor genes for resistance to insect attack and the existence also of genetic variation for tolerance of the host to insect attack. There is also genetic variation within an insect population for its capacity to overcome resistance in the host. This variation is denoted as different biotypes within the insect species.

There are a number of different mechanisms which breeders are attempting to modify to minimise disease losses from insect attack in crop plants. Resistance breeding involves three main approaches— antibiosis, physical resistance and non-preference. The use of antibiosis involves the incorporation of genes which modify the metabolism of the plant such that the insect through its feeding is adversely affected in growth and/or reproduction. In this way the size of the insect population is restricted and hence losses in productivity are minimised. Breeding for increased physical resistance of plants (e.g. leaves, stems) to attack by sucking or boring insects has been effective in producing cultivars with higher resistance to such insects. Non-preference breeding involves the use of genes which modify the morphology or palatability of the plant tissue to deter the insect from feeding and reproducing on the plant. Pubescence of leaves or stems is a character which is often incorporated to confer non-preference in the host to the insect species.

Breeding for insect resistance and tolerance is likely to assume a role of increasing significance in future crop production. The deleterious environmental effects of many insecticides plus the capacity of many insect pests to produce resistant biotypes to chemical insecticides demands that more attention be given to breeding for insect resistance and tolerance in crop production. Crops which are particularly attractive to insects and therefore sprayed with insecticide at high frequency, such as cotton, are already the subject of breeding efforts for increased host-plant resistance, either by conventional means or with the aid of biotechnology.

TECHNIQUES WHICH AID THE CROP BREEDER

While new sources of genetic variability in crop breeding are usually obtained from hybridisation between cultivars and crossbreds of the crop species (and more rarely from more distantly related, perhaps wild or weedy, species), there are situations when the crop breeder has to resort to the use of more divergent sources of variability from species related to the crop plant. A number of methods are available for this purpose to enable the crop breeder to produce genetic combinations beyond those able to be obtained from normal hybridisation and selection methods. Certain of these methods also enable the crop breeder to gain closer insight into the genetic and evolutionary make-up of commercially important characters. This knowledge can lead to more efficient methods of breeding for improvement in such characters.

Induced Mutation

Methods for inducing mutation have the potential of producing new sources of genetic variability for crop breeding. These methods can be employed when it appears there is little, or no, variability for the character to be improved available within the gene pool of the species. In addition, when a desired character is so closely linked to an undesirable character that there would be little likelihood of obtaining recombinants between them, induced mutation could be used to produce mutants of more favourable expression for the undesirable character, making such genotypes more useful for crop breeding.

The effectiveness of using induced mutation in crop plants depends on the breeding system of the plant. Its use in self-pollinated plants is likely to be more successful than in cross-pollinated plants. Populations of cross-pollinated plants usually possess stores of genetic variability in the recessive condition and it would not be likely that induced mutation would produce significant amounts of new variability. Induced mutation is potentially useful in the improvement of asexually propagated crop plants. Much of the genetic improvement in present-day horticultural crops can be attributed to the occurrence, selection and propagation of 'sports', or naturally occurring mutations. This indicates that induced mutation could provide further useful variants in these species.

Mutation, in the broadest sense, implies a spontaneous, heritable change in the genetic material at the level of the gene, chromosome or genome. Chromosomal mutation involves such structural changes to chromosomes as translocations, deletions, inversions and duplications, while genome mutation involves loss or addition of one or more chromosomes or chromosome sets (genomes). It also includes the loss or addition of chromosome arms. Crop breeders are interested in the use of induced mutation for producing gene mutation and chromosomal translocations (see under 'Cytogenetics').

There are three broad categories of mutation-inducing agents (called mutagens)—ionising radiation, ultraviolet radiation, and chemical mutagens. Ionising radiation (e.g. X-rays, gamma rays and alpha, beta and other fast-moving particles) cause both gene mutation and chromosomal breakage. Their mutagenic effect is probably due both to direct 'damage' of the chromosome and to ionisation which produces chemical changes in the cell. Ultraviolet radiation has low penetrability and is used mainly for irradiation of pollen prior to its use in fertilisation. Chemical mutagens generally produce a greater ratio of point (gene) to structural mutations than radiations. The most commonly used chemical mutagens, hydroxylamine, nitrous acid, and alkylating agents (e.g. ethylmethanesulphonate, or EMS) modify the chemical composition of DNA to produce a wide range of mutations in higher plants.

Induced mutation techniques usually involve treating the seed with radiation or chemical mutagens under specified conditions of mutagen dose or concentration together with close control of the experimental conditions during treatment. The rate of induced mutation is dependent on the maintenance of precise conditions of temperature, oxygen concentration, pH, and moisture content of seed during the treatment period. Pollen irradiation is carried out under precise conditions of temperature and humidity. The irradiated pollen is then used in hybridisation, and mutants are sought in progeny arising from this hybridisation.

Because most of the induced mutations in plants are recessive, they can be detected in the second generation after that of the treated seed or pollen as segregating characters. A scheme for using induced mutation and selecting for mutants in crop breeding is shown in Fig. 3.16. The terminology for generation number in mutation studies is as follows: M_1 denotes the generation of mutagen treatment (i.e. the plants arising from treated seed or from hybridisation using irradiated pollen); M_2 denotes the following generation in which (recessive) mutations would segregate in the double recessive condition and M_3, M_4 ... subsequent generations.

While there are a number of examples of induced mutation having produced improvements of commercial significance in certain crop plants (e.g. stiff-strawed, lodging-resistant mutants in barley, and increased yield in peanuts), its contribution to crop improvement has, as yet, not been substantial. It appears that within the gene pool of most crop species there are still adequate amounts of genetic variability for crop

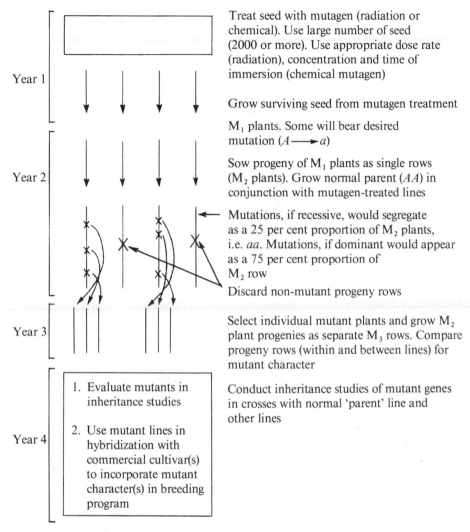

Treat seed with mutagen (radiation or chemical). Use large number of seed (2000 or more). Use appropriate dose rate (radiation), concentration and time of immersion (chemical mutagen)

Grow surviving seed from mutagen treatment

M_1 plants. Some will bear desired mutation ($A \longrightarrow a$)

Sow progeny of M_1 plants as single rows (M_2 plants). Grow normal parent (AA) in conjunction with mutagen-treated lines

Mutations, if recessive, would segregate as a 25 per cent proportion of M_2 plants, i.e. *aa*. Mutations, if dominant would appear as a 75 per cent proportion of M_2 row

Discard non-mutant progeny rows

Select individual mutant plants and grow M_2 plant progenies as separate M_3 rows. Compare progeny rows (within and between lines) for mutant character

Conduct inheritance studies of mutant genes in crosses with normal 'parent' line and other lines

Year 1

Year 2

Year 3

Year 4

1. Evaluate mutants in inheritance studies

2. Use mutant lines in hybridization with commercial cultivar(s) to incorporate mutant character(s) in breeding program

FIG. 3.16   Outline of a scheme for producing mutants for crop breeding

improvement for characters of commercial significance. Induced mutation would therefore be of significance only when genetic variability becomes limited from natural sources, or when potentially useful genes are very closely linked to deleterious genes in the crop plant or its related species.

Induced Ploidy Changes

Crop breeders can alter the ploidy level (the number of chromosome sets) of a crop plant, either to increase it (polyploidy) or to decrease it to half that of normal (haploidy). While induced ploidy changes have been of no great significance in the direct production of new crop types, the techniques are of indirect use in certain crop improvement programs.

Induced Polyploidy

Many crop plants are polyploids (e.g. wheat, oats, potato and cotton). The natural occurrence of polyploidy has been of significance to the evolution and successful domestication of these species. Its presence amongst crop plants has created speculation as to its possible value in increasing yield and adaptability of crop species of lower ploidy levels, especially diploids.

Induced polyploidy, however, has not made a substantial contribution to the production of higher yielding and better adapted crop types. The artificially induced polyploids—triploid watermelons, which are high yielding and seedless, triploid sugar beet (*Beta vulgaris*) which is generally of higher sugar yield than the diploid, and tetraploid red clover (*Trifolium pratense*), which has higher forage yield than the diploid—are three recent examples of commercially successful induced polyploids.

The technique most commonly used in inducing polyploidy involves immersion of germinating seeds or saturation of apical meristems with colchicine, an alkylating agent derived from the bulb of the autumn crocus (*Colchicum autumnale*). Colchicine is used in concentrations ranging from 0.1 to 0.5 per cent and treatment times ranging from 4 to 24 hours. The action of colchicine is to inhibit cell wall formation between the anaphase groups of chromosomes at mitosis and instead of two daughter cells with the diploid (2n) chromosome number the undivided cell will have the tetraploid (4n) number of chromosomes. Cells of this constitution, if they occur as a significant proportion of the cells in the meristematic region of the shoot, can give rise to tetraploid tissue and, in turn, to tetraploid shoots. Verification of the induction of tetraploidy can be carried out from chromosome counts at metaphase of meiosis of pollen mother cells in the developing flowers of these shoots or at metaphase of mitosis in root tips of germinated seeds from the next generation.

The most important role of induced polyploidy in crop breeding is as a technique in distant hybridisation programs between certain crop plants and their related species. It has been used as a means of increasing the frequency of successful hybridisations between the crop plant and a related species. For instance the frequency of successful hybridisations between cultivated tomato (*Lycopersicon esculentum*) and one of its related wild species (*Lycopersicon peruvianum*) is significantly increased if the cultivated tomato, as the female parent, is tetraploid rather than diploid.

Induced polyploidy is also an important technique for conferring fertility on those hybrids between the crop plant and related species, whereas the undoubled hybrid is almost completely or fully sterile due to a lack of significant amounts of, or any, pairing between the chromosomes of the two species.

Cytogenetics

Plant cytogenetics is a branch of science concerned with the study of the physical appearance (e.g. morphology, number) and behaviour (e.g. pairing, movement) of chromosomes in conjunction with genetic interpretation of the causes and consequences of variation in their appearance and behaviour.

Cytogenetic methods are of value in certain situations in crop breeding in providing both access to new sources of genetic variability for breeding and methods for genetic and evolutionary study of characters of commercial significance. This knowledge facilitates the use of genetic variability in crop breeding programs.

Present-day crop plants can be hybridised with the wild and weed relatives from which they were derived, and in these hybrids genetic recombination takes place between the chromosomes of the crop plant and the wild or weed species. However, many crop plants originate from more distantly related species which can be hybridised with them but whose chromosomes show little or no pairing with those of the crop plant. Cytogenetic methods can provide access to this source of genetic variability. Its possible incorporation in the genotype of crop plants can be considered at the different levels of genetic organisation—the whole genome, whole chromosome, chromosome segment and recombinational (chromosome pairing).

Whole Genome Addition

The crop plant can be hybridised with the related species and if there is no pairing between the two chromosome sets the resultant hybrid can be treated with colchicine to double the chromosome number giving a fertile hybrid. The genome(s) of the related species have, in effect, been added to those of the crop plant. The scope is limited for adding whole genomes of related species to the crop plant to produce commercially acceptable crop types. An example of a hybrid of this type which has achieved some commercial acceptance is triticale, which is the addition of the genome of cereal rye to hexaploid wheat to produce octoploid triticale or to tetraploid wheat to produce hexaploid triticale. Hexaploid triticale is grown commercially in a number of countries throughout the world and its grain is used as an animal feed.

Whole Chromosome Transfer

Individual whole chromosome transfer can be made from the related species to the crop plant by backcrossing to the hybrid between the crop plant and the related species, using the crop plant parent. The aim of selection is to retain a particular chromosome from the related species on the basis of chromosome number and morphology and its particular modification of the phenotype of the crop plant. In this way a chromosome from a related species can be added to the chromosome complement of the crop plant. As a single chromosome addition it is referred to as a *monosomic addition* line; as the homologous pair it is a *disomic addition* line. The chromosome addition line can then be used for substituting a pair of homologous chromosomes of the crop plant with the added chromosome pair from the related species and is referred to as a *chromosome substitution* line. However, the potential value of whole chromosome transfer from related species to the chromosome complement of the crop plant appears to be mainly the basis it provides for detecting genes borne by the added chromosome that may have value in crop improvement.

Chromosome addition and substitution lines between the crop plant and the related species are generally lower in fertility and productivity than the crop plant. Hence these methods do not hold much potential for direct use in crop improvement. Their value is in providing material for the more specific incorporation of useful characters in the crop plant from the related species by the transfer of small chromosome segments, carrying the desired gene(s), on to the chromosomes of the crop plant. Two methods, induced translocation and induced chromosome pairing, have been used for this purpose. Induced chromosome pairing has been demonstrated in wheat where, in hybrids between wheat and certain of its related species, the removal or suppression of pairing-inhibitor genes allows pairing to take place between the chromosomes of wheat and the related species. Cytogenetic techniques are also available for transferring homologous chromosome pairs between crop cultivars. In bread wheat a number of sets of lines have been produced between pairs of cultivars in which, in each line, a homologous chromosome pair in one cultivar is substituted with its homologous partner from the other cultivar. This substitution is carried out for each chromosome in turn and the number of lines, called a set of *intervarietal substitution* lines, equals the number of homologous chromosome pairs. In wheat which has 21 homologous chromosome pairs, there are 21 possible intervarietal chromosome substitution lines that can be made between two cultivars. While this method of chromosome transfer has not been of direct significance in crop breeding, it provides essential material for the genetic analysis of both simply and quantitatively inherited characters in the crop plant.

Chromosome Segment Transfer

Considerable success has been achieved in crop improvement through the use of methods which enable transfer to be made of segments of chromosomes of related species onto chromosomes of the crop plant. These methods have been used mainly for transferring major genes for disease resistance, from the related species to the crop plant. The transfer is achieved by irradiating the hybrid (or the chromosome addition or substitution line) between the related species and the crop plant with X-rays or gamma rays to produce random breakage and rejoining of its chromosomes. In the progeny of the irradiated plant it is possible to select genotypes which contain the full chromosome complement of the crop plant but with one chromosome bearing a segment of the related species' chromosome possessing the desired gene. This method of effecting genetic exchange between chromosomes through breakage and rejoining is termed *induced translocation*. The wheat cultivar Eagle released by the Agricultural Research Institute, Wagga Wagga, New South Wales, in 1969 possessed a translocation from the related species, *Thinopyrum elongatum* (*Agropyron elongatum*), which carried a gene conferring resistance to wheat stem rust (*Puccinia graminis tritici*).

Recombinational

Another possible approach to transferring useful genes from chromosomes of related species to those of the crop plant, when there is no pairing between their chromosomes, is the manipulation of genes restricting this pairing behaviour. In wheat it has been demonstrated that certain genes regulate the specificity of chromosome pairing within the plant to prevent pairing between genetically similar chromosomes in its three genomes. These genes also inhibit pairing between wheat chromosomes and those of certain of its related species in hybrids with them. However, by the removal or suppression of pairing-inhibitor genes it is possible to effect pairing and genetic recombination between wheat chromosomes and those of many of its related species. A large potential is offered by the use of these techniques for transferring desirable genes from the chromosomes of related species to those of wheat.

These techniques will most likely achieve wider significance for the future genetic improvement of other crop plants.

Distant Hybridisation

The wild and weed ancestors, related species and primitive cultivars of a crop plant constitute an important part of the gene pool of a crop species. Often because of the absence of useful genetic variability within modern cultivars and crossbreds of the crop plant, the breeder has to use more distantly related genotypes of the above-mentioned types to obtain this variability. Distant hybridisation involves the use of genetically divergent parents from such sources, in crosses with the crop plant. In such hybrids chromosome pairing, and hence the potential for genetic exchange, can vary from a high level to zero. When there is pairing between the chromosomes of the crop plant and the divergent parent, the breeder usually backcrosses the hybrid with the crop plant parent to obtain the desired character free from the large number of deleterious genes carried by the divergent parent.

When there is little or no pairing between the chromosomes of the crop plant and those of the divergent parent in these hybrids, the techniques for induced translocation are used for specific incorporation of the desired gene in the chromosome complement of the crop plant (see under 'Cytogenetics').

Many of the wild and weed ancestors of crop plants, their related species and primitive cultivars still occur in the areas of the world where the crop evolved and was first domesticated. These species have evolved or have been cultivated under conditions of continual disease and insect attack and thus contain potentially valuable sources of pest resistance for crop breeding. And because they have usually evolved or have been cultivated in a wide diversity of habitats, they also contain potentially useful variation for adaptiveness to extremes of environmental stress (e.g. drought and cold resistance). It is important that this part of the gene pool of crop plants be conserved and evaluated for future crop breeding needs. The incorporation of potentially useful genetic variability from these sources in the breeding of new crop cultivars presents many problems in obtaining well-adapted crop types. However, its use may provide higher levels of adaptation in the crop species.

Techniques Used in Plant Biotechnology

Modern biotechnology has become possible due to an increased under-standing of gene structure and function at the cellular and molecular level—an understanding that is advancing at a rapid rate (Wilke-Douglas et al. 1986, Buck, 1989; Gasser and Fraley 1989, Nijkamp *et al.* 1990). The main techniques associated with biotechnology (or 'genetic engineer-ing') in plant breeding are: recombinant DNA manipulation, tissue culture, protoplast manipulation, *in vitro* selection, genetic transform-ation, and the use of monoclonal antibodies. Undoubtedly others will be added as knowledge increases.

Meristem Culture, Clonal Multiplication and Disease Elimination

These processes are aspects of the same technology—micropropagation. Tissue culture is widely used to multiply a score of cultivated species (e.g. sugar cane and bananas); it is particularly useful for overcoming inter-national quarantine restrictions, and for speeding up the breeding of perennial and woody plants. The culturing of 'clean' meristem tissue allows disease-free (particularly virus-free) cultivars to be rapidly pro-duced, especially in clonally propagated crops (e.g. grapes, citrus and strawberries). These applications are most effective when genetic changes induced during culture can be kept to a minimum. New species are being added to the list of successes at a rapid rate; but problems with rooting and browning-off can be major impediments (Smith and Drew 1990).

Somaclonal Variation

Somaclonal variation is genetic change that occurs as a direct result of the stress of tissue culture. It is diverse in character and varies dramatically in frequency and type with species and genotype (Evans and Sharp 1986). Somaclonal variation may offer a different spectrum of genetic changes to spontaneous or induced mutations (for example, from the effects of ionising radiation). However, unless variant production is linked to some efficient selection procedure (preferably at the cellular level) then the system may be too unwieldy. Somaclonal variation can occur simultaneously in several characters. Consequently, a good deal of conventional clean-up breeding is required to remove undesirable gene combinations.

Induced Haploidy

Haploidy occurs at a low frequency in all crop plant populations. Haploids are individuals with half the chromosome number of the normal diploid or polyploid from which they are derived. They possess only one chromosome of each homologous pair of their parental source. In polyploid species the haploid is usually viable but sterile, while in diploids it often dies very early in its development.

Haploidy has been induced in many crop plants, both monocotyledons and dicotyledons, through the culture of immature anthers on specialised growth media under sterile conditions. In higher plants the young microspore, which usually gives rise to the pollen grain, also has the potency to develop callus tissue or plantlets, according to its nutritional environment. Because of the haploid condition of the microspore, the vegetative growth arising from it is also haploid and the induced plantlet will give rise to a haploid plant. In some species the fertile diploid condition is spontaneously restored in some of the regenerant plantlets produced from tissue culture, removing the need for induced chromosome doubling.

The technique of induced haploidy used in conjunction with that of induced chromosome doubling of haploid plants offers the crop breeder scope for greatly reducing the time normally taken to reach homozygosity in crop breeding programs. The induction of haploidy in the pollen of heterozygous plants and the induction of a doubling of chromosome number in the haploids enables homozygosis in the segregates to be reached in one generation rather than approximately six generations (F_2 to F_7) with normal reproduction and genetic segregation.

The combined methods of haploidy induction and that of chromosome doubling have application in breeding both self-pollinated and cross-pollinated crops for the rapid production of homozygous lines (see Fig. 3.17). While the use of induced haploidy holds considerable potential in shortening the length of time for the breeding of new crop cultivars, its impact must await the development of techniques which ensure a more consistent and higher frequency of (a) haploid induction from pollen microspores of a broad range of genotypes, and (b) chromosome doubling from colchicine treatment of haploid plants. Notwithstanding these difficulties, anther culture is an excellent tool for speeding up gene transfer by conventional hybridisation and backcrossing, where only a few fertile regenerants are required at each generation. The technique becomes even more powerful when coupled with the use of linked DNA markers (described later) because of the increased speed and precision of the selection process.

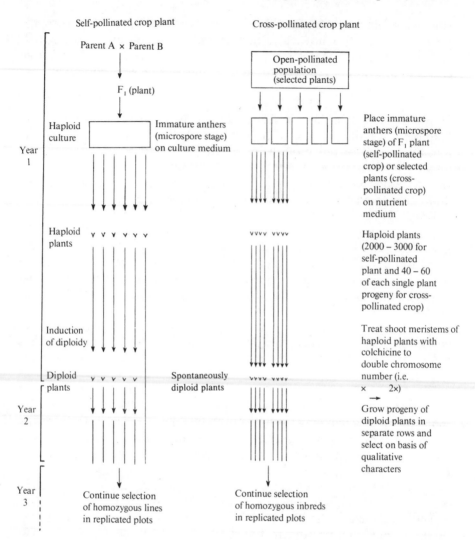

FIG. 3.17   Procedure for using induced haploidy in crop breeding to achieve
(a) early selection of homozygous lines in a cross in a self-pollinated crop, and
(b) rapid production of inbreds as parents for breeding in a cross-pollinated crop

In Vitro Selection

This technology allows the selection of cells in tissue culture that are
resistant to, or tolerant of, specific chemical hazards, such as fungal
phytotoxins, herbicides, acidity, salinity, or heavy-metal toxicity. Once
these cell lines are produced, there remains the problem of whole-plant
regeneration (common to much biotechnology) and whether the observed
resistance is genetically determined and therefore inherited.

In Vitro Hydridisation

Somatic cells (usually protoplasts) of a crop plant and a distantly related species can often be induced to fuse in tissue culture. This method is a useful tool for the transfer of nuclear or cytoplasmic genetic material between genotypes that are not normally sexually compatible. Desirable traits can be transferred even when the precise genes remain undefined. It has been used widely in the *Solanaceae* and *Cruciferae*, and also recently with lucerne, medics and rice.

DNA Manipulation

The manipulation of DNA is central to the genetic engineering aspect of biotechnology. There are a number of ways that geneticists identify and then isolate a gene of interest. The primary route is to identify a protein (or its precursor, an RNA) which is associated with a certain plant characteristic of interest. The protein is first isolated and purified. The amino-acid sequence allows several possible DNA coding sequences to be proposed, due to the redundancy of the triplet code. One of the DNA sequences is then synthesised, or more usually a portion of it. The resultant oligonucleotide is then used to 'probe' (i.e. test by DNA hybridisation) a genomic library of the organism's DNA maintained in some microorganism (usually *Escherichia coli*, but yeast is also used) to find which fragment in the library has the full original gene. The gene can then be located in the genome, cloned (multiplied in *E. coli*) and ultimately sequenced to understand its structure and function. With appropriate modifications the gene is then available to be inserted into another, usually unrelated, plant species requiring improvement.

Another way to track down a gene is to use a transposon ('jumping gene') of known DNA sequence to allow 'transposon tagging'. Essentially, the transposon is allowed to cause mutations by moving about the genome until the plant character of interest is affected. The transposon is then assumed to have inserted itself in a gene associated with that character and to have disrupted its normal function. A DNA library is then made of the genome, and the fragments which hybridise with the transposon sequence will be flanked by the gene of interest. Once isolated, investigation of the gene can continue as outlined above.

Biotechnology has stimulated a massive increase in our knowledge of gene structure and function. It is now routine to couple specific protein coding regions of DNA with control sequences from different species to allow tissue-specific or stimulus-specific gene expression. However, many fundamental problems remain. Most of the characters of plant breeding relevance are multigenic traits, with the genes of small effect scattered throughout the genome and exhibiting a myriad of pleiotropic or epistatic effects. Simply finding the genes is a major problem.

Once a foreign gene is inserted, its location, copy number, and orientation will affect its performance, and evidence is mounting that this may be a major problem. The inserted gene is disrupting a complex interconnected system of thousands of genes, so the results can be unpredictable.

Gene Shears and Antisense RNA

Gene shears is a recently perfected technology that has attracted much publicity in Australia due to the involvement of CSIRO (Maddox, 1989). It adopts a different approach—to remove the effect of an undesirable gene rather than to insert a new one. The technology depends upon RNA molecules (ribozymes) that themselves function as 'enzymes' and which are capable of cutting or splicing other RNA molecules and thereby disrupting their function. The ribozyme can be engineered to cut any m-RNA of known sequence; it can therefore be designed to interfere with virus replication, for example. There is a huge number of potential uses for this technique and commercial developments are occurring.

A similar approach to gene shears is being used with an undesirable gene that is responsible for fruit softening in ripe tomatoes. The effect of the gene can be overcome by chemical treatment with a polyamine after harvest but the more elegant solution is to identify and disable the gene responsible. One approach is to produce what is called antisense DNA to the gene to be disabled. The normal gene and its inserted antisense cousin both produce m-RNA but because the two strands are exactly complementary they bind together to produce non-functional molecules. The normal gene function is then largely removed (Day, 1989). The gene shears system outlined above could also be used to achieve a similar result. The outcome is that the neutralisation of this gene results in ripe tomatoes that stay firm for much longer and have a greatly increased shelf life. Conventional tomato breeding has made progress in this area by the use of non-ripening and ripening-inhibitor genes found in cultivated genotypes.

Genetic Transformation

Once a gene has been isolated, cloned and modified, the task is then to get it stably incorporated into the genome of a suitable recipient species (initially usually a 'model' plant species such as tobacco). This transformation can be attempted in a number of ways (Perani *et al.*, 1986; Potrykus, 1991):

1. Using a biological vector such as *Agrobacterium* or cauliflower mosaic virus (CMV) to carry the 'hitch-hiking' foriegn DNA into the plant. This technology is well understood for dicotyledons, but the two species of *Agrobacterium* will not generally infect monocots. However, extensive research has induced *Agrobacterium* into infecting monocots and success has been reported in at least asparagus and onion.

2. Naked DNA uptake by protoplasts or some other cell without a cell wall, such as a pollen tube, which may be facilitated by temporary

membrane disruption caused by treatment with polyethylene glycol (PEG), the application of a rapid electric current (electroporation), or sonication.

3. By physically inserting the DNA. One method is to use a very fine glass needle to inject the nucleus of protoplasts with a DNA suspension, using either pressure or electrophoresis to move the DNA (micro-injection). Another method inserts the DNA ballistically on the surface of very small, inert carrier particles, such as tungsten pellets (biolistics).

The major food crops (mainly monocotyledons) are not as amenable to these techniques as many dicotyledons, since fertile plants cannot easily be regenerated from protoplasts. However, progress has been made with rice, wheat and barley.

Antibodies

Highly specific monoclonal antibodies can now be produced by tumour cells growing in culture. The lifespan and production capacity of such cultures is virtually unlimited. The antibodies bind to certain specific protein molecules, thereby providing qualitative (and possibly quanti-tative) measurements. This powerful technique, coupled with its visuali-sation technology ELISA (enzyme-linked immunosorbent assay), can be used to detect, in plants, protein from such foreign sources as viruses, fungi, bacteria and mycoplasms and, more recently, mycotoxins and hormones (Reddy *et al.*, 1988). In addition, antibodies can reveal the presence (or absence) of specific genes by illuminating the gene's indirect protein product.

RFLPs

Restriction fragment length polymorphisms (RFLPs) are DNA markers based on differences in the actual DNA sequence. The piece of DNA in question may or may not be part of a gene with a recognised product (such as an enzyme, a structural protein, or a change in a morphological character). When DNA is 'cut' with restriction enzymes, the cut sites occur in different places and consequently the fragments of DNA produced are of different lengths. These pieces can be separated by electrophoresis, visualised, sized and quantified. The fragment sizes vary between species, cultivars and individuals (especially in outbreeding species or in hybrids). Once a number of RFLPs are identified, they can be located on chromosomes in relation to themselves and other known genes, such as morphological markers, isozymes and disease resistance genes. The RFLPs are then available as a fine-detail map of the plant's genome—the more RFLPs, the finer the map. This map can then be used to locate other major genes and, perhaps more importantly for the future, to locate genes that contribute to quantitative traits such as yield. These quantitative trait loci (QTLs) can then be tracked by monitoring a closely-linked RFLP, for example during breeding. The result is that selection procedures become unambiguous, rapid and highly efficient (Tanksley, 1983; Beckman and Soller, 1986).

RFLP maps are available for wheat, barley, maize, brassica, rice, lettuce, soybean, lentils, tomato, and potato. Progress is rapid in the *Triticeae* because of the availability of numerous cytogenetic stocks and because of the homoeology of chromosomes from different species. Homoeology results in a high level of cross-hybridisation; for example, some RFLPs from barley and a wild relative of wheat, *Triticum tauschii*, are also found in wheat.

The tomato RFLP map is well advanced and its value was dramatically demonstrated when a gene for resistance to a virus (*Tm-2*) from a wild tomato species was backcrossed into a cultivated line. By monitoring closely flanking RFLPs, the introgressed chromosome segment was so small after only two generations that to achieve the same result conventionally would have taken about 100 generations. RFLPs have been mapped close to very important QTLs in tomato, such as water-use efficiency, fruit soluble solids, fruit mass and pH.

RFLPs are very powerful. Suppose that one RFLP were closely linked to a QTL that was responsible for, say, 15% of yield in the major current cultivars of wheat: being able to guarantee the presence of that gene in all selected progeny from a cross by using the RFLP marker would boost the efficiency of the breeding program immensely. The use of RFLPs will make breeding more rapid and efficient and will be one of the major impacts of biotechnology on agriculture, irrespective of other developments.

More recently a technology to detect DNA markers has been developed based on the polymerase chain reaction (PCR). Randomly amplified polymorphic DNA (or RAPDs) are sections of DNA in the plant genome, the two ends of which are detected by molecular hybridisation with short DNA molecules of known sequence. The DNA section is copied and then multiplied rapidly in a doubling reaction on the laboratory bench. The resultant DNA molecules may vary in size (length) between genotypes, and genetic linkages may be detected between a particular RAPD and a gene of interest, e.g. for disease resistance. The RAPD can then be used, like an RFLP, to assist in selecting desirable segregants from breeding populations.

DNA Diagnostic Probes

These probes are useful in a similar way to antibodies. They hybridise with DNA of a corresponding sequence, and then when coupled with appropriate indicator molecules that are radioactive or produce a coloured reaction, the hybridisation can be visualised. This allows a qualitative (if not quantitative) test for the presence of particular DNA, be it from a closely related species or genotype with which hybridisation has occurred, or from a pathogenic organism such as a virus, bacterium or fungus. The polymerase chain reaction (PCR) is often employed in this context because of its ability to chemically amplify very small quantities of DNA.

Examples of Biotechnology Application

A wide range of genes have been cloned by DNA manipulation from a host of different organisms. Many of them could eventually be of agricultural significance once modified, equipped with the necessary controls, and stably inserted in an appropriate location in a recipient species. The list will continue to expand for a bewildering range of proteins and enzymes.

A range of gene control sequences have been isolated and characterised, and they can be coupled to structural genes to give a useful functional unit. Controlling sequences respond to stimuli such as light, zinc concentration, wounding, tissue location, and low oxygen concentration. This allows novel genes to be constructed; for example, insect toxin genes that are only expressed when tissue damage occurs from insect feeding, or a herbicide resistance gene which is activated by a topical zinc spray.

The modification of plant incompatibility systems has great potential, as the following example illustrates. A gene promoter specific to anther tissue was discovered and joined to a ribonuclease gene. The new gene was then inserted into tomato, tobacco and canola using *Agrobacterium*. It prevented normal anther development, and the transgenic plants were male-sterile. Such manipulation allows the economic production of hybrid F_1 seed in a normally self-pollinating crop (Peacock, 1990).

The prevention of plant disease has been addressed by biotechnology in at least two ways. The first is to engineer a 'disabled' viral genome to trigger a response in the 'infected' plant but cause no damage. This preconditioning allows the plant to resist subsequent infection much more effectively. The second approach is to produce synthetic resistance genes which employ a decoy satellite RNA to mop up the RNA polymerase produced by a virus, without which it cannot replicate. This technique is well developed in Australia at CSIRO (Courtice, 1987).

The nutritional quality of crops can be improved by biotechnology so as to benefit human health and animal production. An increased sulfur-containing amino acid concentration in cereal protein would be beneficial to humans and stock; for example, by inserting pea storage-protein genes into wheat.

The first products of plant biotechnology are being tested on farms. Cotton plants resistant to the herbicide 2,4-D, and others engineered to contain *Bt* toxin genes from the bacterium *Bacillus thuringiensis* for insect resistance, have been developed by CSIRO and are close to field testing. The protein produced by the *Bt* gene is toxic to *Lepidoptera* insect pests that graze on the plants (Cousins *et al.*, 1991).

More fundamental work on basic plant processes will be required to support long-term progress in the application of biotechnology to plant breeding, particularly in biochemistry, physiology and pathology. These basic processes include: the control of vernalisation, factors limiting photosynthesis, nitrogen fixation, differentiation, incompatibility systems, photoregulation, stress response, wound response, and response to pathogen invasion.

The relationship between conventional and biotechnological methods

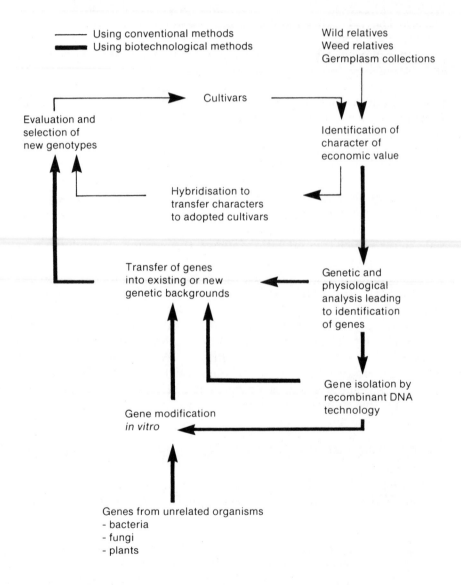

FIG. 3.18   The flow of hereditary material in plant breeding

in plant breeding in terms of the flow of hereditary materials is shown in Figure 3.18. The roles of these methods in natural science and production research are shown in Figure 3.19.

Cultivar Evaluation and Release

Before an advanced crossbreed of a crop can be released for commercial production it must undergo rigorous evaluation for yield, quality, and agronomic characters (e.g. ease of harvesting, non-shattering, lodging resistance, and disease resistance). The evaluation of the commercial potential of advanced crossbreeds of a crop is usually performed in replicated experimental plots over a range of sites representing the environments for which the cultivar is being bred. It usually involves growing them in plots of approximately 0.5 ha and at a sowing rate similar to commercial practice so that they are assessed under conditions as close as possible to commercial production. This evaluation is usually carried out over four to five years, or longer, to gain a more accurate

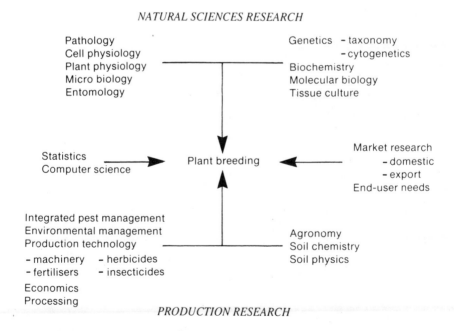

FIG. 3.19   The central nature of plant breeding

estimate of average seasonal performance. In conjunction with yield evaluation it is usual to conduct quality evaluation of the crop product using small-scale commercial evaluation methods closely related to commercial practice.

When a crossbreed has performed satisfactorily and sufficient seed is available, it is named and released for commercial production. While in commercial production it is important that the cultivar continues to perform according to the specifications determined at its release, hence the need to maintain pure seed supplies of the cultivar true to its identity, from which commerical stocks for planting are continually being derived.

In Australia, the release of a new cereal cultivar (wheat, barley, oats, rye and tritcale) entails a detailed description being made of its morphological characters, for both the plant and the harvestable product (e.g. grain, and the characteristics of commercial significance such as quality). In the case of disease resistance the description embodies a statement of the diseases, and the particular races of the disease, to which it is resistant. When it is known, the particular gene designations for resistance possessed by the cultivar are also stated.

The cultivar description also includes a statement on the pedigree of the cultivar. At release the cultivar is given a name which is carefully chosen so that it cannot be confused with existing or earlier cultivars of the crop.

Plant Variety Rights

In 1987 the Australian Government enacted Plant Variety Rights (PVR) legislation and Australia is now a member of UPOV (Union pour la Protection des Obtentions Végétales). PVR is an extension of the concept of allowing ownership of an invention, and is similar to copyright legislation. PVR is a limited form of proprietary ownership for up to 20 years to prevent others using a new plant cultivar without the owner's permission, except for their own use or for further research. However, under PVR, growers are able to retain seed from inbreeding crops for future sowings. The aim of PVR is to protect breeders' intellectual property, stimulate private-sector breeding in Australia, and to ensure that Australian farmers have access to suitable cultivars that are registered overseas.

New cultivars submitted for PVR registration must arise from selective breeding, must be uniform and stable over generations, and be distinguishable from all other cultivars. Seed, cuttings, or other propagules of a cultivar protected by PVR may not be freely sold. In this way, the PVR holder (which may, or may not, be the original breeder or their employer) is the only person permitted to distribute the cultivar and charge royalties. PVR has been enthusiastically received by the horticultural industry but, as yet, has had little impact in the crops of broadacre agriculture. There is an increasing world-wide trend to attempt to patent biological material, including plant cultivars, and especially the products of biotechnology. Patents provide much tighter ownership rights than PVR, but there is considerable concern about the moral difficulties and practical consequences of such a course.

Pure Seed Production

It is important that the genetic identity of the cultivar be maintained after release and throughout its commercial use so that its performance in all characters accords with that specified as justification for its release as an improved cultivar. For this reason the production and maintenance of pure seed stocks of the cultivar as a continuing source of seed for commercial sowing is an important activity in crop production.

Pure seed production is organised at a number of levels concerned with both the ensurance of genetic identity and providing sufficient volumes of seed for commercial plantings.

The initial level of seed multiplication is *breeder's seed* which is seed or vegetatively propagated material produced by the breeder, or his institution, which is exactly true to the cultivar type. Breeder's seed gives rise to *foundation seed* which is usually produced by the research organisation concerned with breeding the cultivar. In its production, strict precautions are taken to ensure that the foundation seed is true to its genetic identity. This seed is usually distributed to commercial seed growers who grow it to produce *registered seed*. Registered seed must also meet strict purity standards. Besides being genetically pure, it must be free of disease and weed seeds. The seed produced for commercial planting, having met the requirements for purity, is called *certified seed*. The activities of producing certified seed from breeder's seed are continuous; they ensure continued purity of the cultivar in its commercial production.

CROP PLANT INTRODUCTION AND BREEDING IN AUSTRALIA

Crop Plant Introduction

Despite the presence within the Australian flora of ancestral species of certain of the domesticated crops of Europe, Africa and Asia, it is evident that the Aboriginal people had not domesticated them. Such native species as wild rice (*Oryza sativa*), wild cotton (*Gossypium* sp.), wild flax (*Linum marginale*) and wild tobacco (*Nicotiana suaveolens*), were not deliberately cultivated and selected by them as were their related species in agricultural development elsewhere in the world. The only evidence for cultivation of plant species by Australian Aborigines is the planting of certain of the yam species (e.g. *Discorea hastifolia*).

Thus the European settler was confronted with a local flora with virtually no new domesticated species. The development of agriculture and horticulture in Australia was therefore dependant on the introduction and cultivation of the domestic crop plants of Europe. These were the crops the settlers were used to growing, and their palate preferences were for the tastes of the products of these crops.

The introduction of plant species in Australia with White settlement was both deliberate and accidental. Agricultural and horticultural crop plants were brought here by deliberate introduction because, as they can survive only under established cultural practices, their introduction and use had to be an objective process. However, most of the present pasture grasses and legumes came by way of accidental introduction in the bedding of animals, in ballast and packaging and most likely as seed contaminants in seed samples of introduced crop species.

The problems facing the adaptation of the introduced agricultural and horticultural crop species to the new environments of Australia were many and difficult. The first introduced cereal species, wheat, oats and barley, were European types and, being adapted to the long photoperiods of the European growing season, were ill-adapted to the shorter days of the Australian growing season. This rendered them too late in ripening in Australia with the consequence that, first, they often suffered from severe drought and produced either no grain at all or very low yields of grain, and second, their late ripening made them very vulnerable to disease attack, especially the leaf and stem rusts of wheat and oats.

Rather fortuitously, the early cereals were most likely mixtures of different types, quite different from the pure line cultivars now used in crop production. They afforded scope for selection of the early-maturing and disease-resistant types whose superior yielding ability would have been obvious in the late-maturing and disease-ridden mixtures from which they were selected. Thus the first phase of improvement in the temperate cereals was by selection, usually through the eye of the observant farmer.

The next phase of improvement was breeding and selection. In wheat, this was done by such notable breeders as William Farrer at Lambrigg in New South Wales and Hugh Pye at Dookie in Victoria. They were concerned with hybridisation and selection for earliness, disease resistance and improved baking quality. This work, which was commenced in the 1880s, laid the foundation for the development of the Australian wheat industry. The wheat cultivar Federation, bred by William Farrer and released for commercial use in 1900, combined the earliness of an Indian wheat with the productiveness and baking quality of a North American wheat. Because of its earliness, Federation could grow in drier, shorter growing season environments than those of the coastal areas where the first agricultural developments took place. Wheats of this type enabled the wheat industry to expand to inland areas of the continent with much lower annual rainfalls.

Although the cereals wheat, barley and oats were the first crops to be successfully adapted through breeding and selection to the agricultural environments of Australia, a range of other crops were introduced early in Australia's agricultural development. These included oilseeds, grain legumes, tropical or summer grown cereals such as maize and rice, and fibre crops such as cotton and flax. Breeding for closer adaptation of these crops to Australian conditions is less advanced than for the temperate cereals, but rapid progress has been made in the last decade. There appears to be a great potential with these crops for increased and more assured levels of yield and improved quality of the crop product.

Crop Breeding in Australia

Crop breeding in Australia is carried out mainly by state government organisations and universities (Downes, 1990; Lazenby, 1986). All state agricultural departments have plant improvement activities in which crop breeding is usually the major activity. A number of Australian universities conduct crop breeding programs and are also involved in research training in this area. The CSIRO has a limited involvement in the breeding of temperate and tropical crops in Australia and is more concerned with basic research on the physiology, biochemistry, genetics, tissue culture and molecular biology of crop plants relating to crop yield, product quality and pest resistance. This information is used in improving the efficiency of breeding and selection for crop improvement. Some private companies are also involved in field crop improvement in Australia.

Temperate Cereals

The temperate cereals include wheat, barley, oats, triticale and cereal rye. They are adapted to areas of cool growing conditions and moderate to low rainfall in Australia and are sown in late autumn or early winter. Most of their growth is produced in the spring and they ripen in early summer.

WHEAT

Wheat is Australia's most important agricultural crop in terms of its volume of production and export. Australian wheat is predominantly bread wheat (*Triticum aestivum*), the cultivated hexaploid species ($2n = 42$), but a small amount of tetraploid ($2n = 28$) or durum wheat (*T. turgidum*) is grown for pasta and noodle manufacture.

The main problems facing the breeder for improved adaptation of wheat in Australia are increased yield, disease resistance and baking quality. Breeding for yield involves a number of different approaches. The main activity is concerned with hybridisation and selection within crosses between cultivars and parents possessing either high levels of yield or a superior expression of one or a number of its components. The incorporation of the semi-dwarf habit in Australian wheats is an important activity in breeding for yield. The semi-dwarf habit, besides ensuring more consistent yield levels by reducing lodging under conditions of high fertility, also appears to bring with it increased yield potential in the form mainly of increased grain number per spikelet. Another approach to breeding for yield is to increase resistance or tolerance to environmental stresses such as frost and drought. Sprouting resistance, resistance to boron toxicity, and tolerance of acid soils are also breeding aims of various wheat breeding programs.

The most important diseases of wheat in Australia for which breeding for resistance is undertaken are stem rust (*Puccinia graminis tritici*), stripe rust (*Puccinia striiformis*), leaf rust (*Puccinia rubigo-vera tritici*), loose smut (*Ustilago nuda*) and flag smut (*Urocystis tritici*). Breeding work is also conducted on resistance to eelworm, or cereal cyst nematode (CCN) (*Heterodera avenae*), using sources of resistance from within both hexaploid wheat and a related wild *Triticum* species. In Australia there has been renewed activity in breeding for resistance to the diseases leaf blotch (*Septori tritici*), eyespot lodging (*Cercosporella herpotrichoides*), yellow leaf spot, crown rot and barley yellow dwarf virus.

A further important wheat breeding activity in Australia is that of improving milling and baking quality. Important components of milling quality are a high percentage of flour extraction from the grain and an appropriate flour texture. Flour texture can range from hard or vitreous to soft or starchy endosperm types. Both components are strongly influenced by growing conditions but are also of moderate heritability and can thus be modified somewhat by breeding and selection.

Baking quality is influenced by flour texture and both the quantity and quality (e.g. extensibility, dough resilience) of protein in the flour. Most Australian wheat soils are low in available nitrogen and, in consequence, wheat grain produced from such soils is low in protein content. This in turn limits the baking quality of flour produced from such wheat. Breeding programs for baking quality improvement aim to produce wheats with improved protein content and quality. Protein quantity, however, is very strongly influenced by the environment and it is difficult to genetically improve its level by any substantial amount.

More recently, a number of significant new markets have been established for Australian wheat including noodle manufacture, biscuits, steam breads (for Asia) and flat breads (for the Middle East). These end uses impose different quality requirements to those for traditional bread manufacture and are now catered for in certain Australian breeding programs.

A small number of F_1 hybrid wheat cultivars have been bred by the commercial company Cargills, based at Tamworth, New South Wales.

BARLEY

Barley (*Hordeum vulgare*) is the second most important temperate cereal in Australia and is adapted to a similar range of environments as wheat. It is grown as two different forms, two-rowed and six-rowed which respectively have two and six rows of grain in the head. The two types are completely interfertile and have similar ranges of variation for most characters. The two-rowed form is grown as a malting barley for the brewing industry. If it does not achieve malting quality it is sold for stock feed. This form is also exported in large amounts to Asian countries where it is used in the manufacture of noodles and other human foods. The grain of the six-rowed form is used for stock feed.

The main breeding aims for increased adaptation of barley in Australia are for improved grain yields and malting quality. Important criteria in selecting for increased malting quality include plump grain, high enzyme activity for starch breakdown, high malt extract and low protein. Precautions have to be taken with appropriate seed treatment against the diseases covered smut (*Ustilago hordei*) and loose smut (*Ustilago nuda*). Net blotch (*Pyrenophora teres*), scald (*Rhyncosporium secalis*), leaf spot (*Drechslera verticillata*) and stem rust are problems in various parts of Australia and are receiving more attention in the quest for a disease-resistant crop. Lodging resistance and the production of winter-type barleys for both grazing and grain are additional breeding aims.

OATS

Oats (*Avena sativa*) have a very much smaller volume of production than wheat and considerably less than barley. It is adapted to a similar range of growing conditions as wheat and barley. Oat cultivars can be divided into two groups—grain oats and grazing/grain oats, the latter producing a larger amount of early vegetative growth which renders them useful for grazing in the late autumn and early winter.

The main problems of improving adaptation of oats in Australia relate to increased grain yields in both types, increased vegetative production in the grazing types, disease resistance particularly to stem rust (*Puccinia graminis avenae*), leaf or crown rust (*Puccinia coronata*) and barley yellow dwarf virus, and CCN resistance.

TRITICALE

Triticale is an artificial hybrid cereal species between wheat and cereal rye. There are two forms, hexaploid triticale, a hybrid between tetraploid wheat and cereal rye, possessing 42 chromosomes (28 wheat plus 14 rye), and octoploid triticale, a hybrid between hexaploid wheat and cereal rye, possessing 56 chromosomes (42 wheat plus 14 rye). Hexaploid triticale appears to have the much higher yield potential, and current breeding efforts both in Australia and worldwide are essentially concentrated on this form.

There are two general areas of interest in triticale breeding—grain types and grazing/grain types. In terms of grain yield the potential of triticale appears, in some situations, to be at least equal to, if not greater than, wheat. At the International Maize and Wheat Improvement Centre (CIMMYT) in Mexico, the breeding of hexaploid triticales of dwarf stature, referred to as Armadillos, has produced types with very high yield potential. Triticale breeding in Australia is limited because there is only small demand for its grain as an animal feedstuff. However, breeding is in progress for increased grain yield and the incorporation of the semi-dwarf habit for Australian conditions.

Summer Cereals

The tropical or summer cereals are adapted to growing under conditions of high temperature and humidity and high levels of moisture availability. In temperate agricultural areas of Australia these crops are grown in the summer usually under irrigation.

MAIZE

In Australia, maize (*Zea mays*) is a crop of minor economic importance compared with the winter cereals. Its grain is grown for stock feed and for the production of starch, cornflour and breakfast foods. Maize is also sometimes grown as a green fodder crop. Two other types of maize, sweet corn (*Zea mays saccharata*) and popcorn (*Zea mays everta*), are grown as a vegetable and a confection respectively.

Maize breeding in Australia is concerned with selection for yield, appropriate maturity time, drought resistance or tolerance and disease resistance. The main diseases in Australia are leaf blight, head smut (*Sphacelotheca reiliana*) and maize dwarf mosaic. The leaf blights are of two types caused by the fungus *Drechslera turuca*, common leaf blight, and *D. maydis*, maydis leaf blight.

RICE

The first cultivation of rice (*Oryza sativa*) in Australia was in 1924–5, making it a more recent crop than any of the temperate cereals. In comparison with these cereals, rice is not beset by many significant diseases which limit high yields. Breeding in Australia for the main areas of production (the Murrumbidgee and Wakool areas) is aimed at producing high-yielding types of semi-dwarf plant stature, with medium to long grain, early maturity and brown leaf hopper resistance. In the Wakool district there is a need for a very early maturing variety with a high degree of cold-hardiness. There is increased effort to produce long-grain and fragrant types to compete with imported rice from Thailand and Pakistan.

SORGHUM

The main problems in Australia in breeding for increased adaptation in grain sorghum (*Sorghum bicolor*) are greater resistance to lodging, sorghum midge resistance, appropriate maturity (which is very dependant on temperature of the environment) and disease resistance. The main diseases of sorghum in Australia are the two fungal diseases, head smut (*Sphacelotheca sorghi*) and rust (*Puccinia purpurea*), and sugar cane mosaic virus (Johnson grass strain). Breeding for increased yield is also undertaken using both selection of open-pollinated populations and commercial (F_1) hybrids.

Grain Legumes

Grain legumes include those members of the Family *Fabaceae* which are grown as field crops for grain which is used for animal and human consumption. They are also referred to as pulses. The protein content of the grain of these crops ranges from 25 to 50 per cent and as a result the grain is an important protein source in human and animal diets. The dietary value of the protein produced is determined by its amino acid composition, which in many cases is deficient in certain of the essential amino acids for non-ruminant nutrition, particularly methionine and cysteine.

TEMPERATE SPECIES

The temperate grain legumes are grown mostly in the cereal areas of southern Australia, usually in rotation with wheat. Their capacity as legumes for nitrogen fixation is useful in enriching the soil with nitrogen, from whence higher yields and grain protein content can usually be obtained in subsequent cereal crops. Their use in rotation with wheat also provides scope for reducing the build-up of soil pathogens which cause yield losses in wheat, and allows more flexibility in the management of weeds and pests. Grain legume crops also help to diversify agriculture, both in the established cereal areas of Australia and also in the pastoral areas where alternative agricultural enterprises are being sought.

Lupins Lupins have a whole-seed protein content of 35 to 40 per cent and are therefore a potentially valuable source of animal feedstuff. They were introduced to Western Australia soon after agricultural development and became naturalised over wide agricultural areas of the state. However, they possessed a number of undesirable features as agricultural plants. These included a high alkaloid content in the foliage giving bitterness and being of low palatability to livestock, poor nodulation, seed shedding, and lupinosis, a disease of livestock caused by a fungus (*Phomopsis leptostromiformis*) growing on lupin stubble under high temperature and humidity.

The main species in Australia are the sand plain lupin (*Lupinus cosentinii*), narrow-leaved lupin (*L. angustifolius*), yellow lupin (*L. luteus*) and white lupin (*L. albus*). The National Lupin Breeding Program (based in the Western Australian Department of Agriculture) has resulted in a number of cultivars which have sweet foliage and reduced pod shattering. They belong to the species *L. angustifolius* and represent a significant achievement in breeding for closer adaptation of an introduced crop to Australian climatic and agricultural requirements. The cultivar Ultra in Western Australia is disease-resistant, with non-shattering pods, soft seeds and sweet or low alkaloid foliage. It belongs to the species *L. albus*.

Field Peas   Field peas (*Pisum sativum*) are an agricultural crop with good export potential but have not, as yet, been intensively bred and selected for closer adaptation to Australian agricultural environments. Consequently yields are low, ranging from 0.2 tonnes per hectare in dry environments to 1.2 tonnes per hectare in wet environments, compared with one to 2.5 tonnes per hectare for wheat over the same environmental range. The grain, which is approximately 25 per cent protein, is used for both human and animal consumption.

A number of problems appear to limit closer adaptation of field peas to Australian environments. One is an inappropriate flowering time, there being a general need for earlier flowering types. There also appears to be a need for extensive overseas introduction of variability in the species to obtain parents of high yield potential. A need also exists for breeding and selection for resistance to the common pests of the crop, particularly the fungous disease, black spot (*Ascochyta pisi*) and the insects native budworm (*Heliothis armigera*) and pea weevil (*Bruchus pisorum*).

Chickpeas   Attention has been given to selecting for increased adaptation, within a wide range of chickpea (*Cicer arietinum*) types, to the growing conditions of the southern Australian wheat belt. This species would occupy a similar place to field peas in that the grain would be used as a high-protein (25 to 30 per cent) animal and human foodstuff.

Oilseeds

There is considerable activity in Australia in breeding oilseed crops for increased yield, adaptability, disease resistance and oil quality. Oilseeds provide an important human foodstuff and in particular fatty acids from which such foods as margarine and cooking oils are manufactured. The oils of certain oilseed crops can also be used in the manufacture of paints, varnishes and resins. The seed residue, after the extraction of the oil, is useful as an animal feedstuff which is high in protein content.

Oilseed crops in Australia can be divided into two groups: winter-growing (canola, linseed and safflower) and summer-growing (sunflower and soybean).

WINTER-GROWING OILSEEDS

Canola   Commercial canola cultivars grown in Australia belong to *Brassica napus*. Canola breeding in Australia is centred at the Agricultural Research Institute, Wagga Wagga, NSW, and the Victorian Institute of Dryland Agriculture, Horsham.

The main breeding aims in canola are for increased yield, reduced pod (siliqua) shattering, disease resistance, especially to the fungal disease blackleg (*Leptosphaeria maculans*), and increased oil quality. A desirable type of oil quality is low to zero levels of erucic acid, low levels of linolenic and high levels of linoleic acid. In addition, low to zero levels of glucosinolates are desirables as these can cause goitre in animals fed canola meal.

Indian mustard (*Brassica juncea*) is also bred in Australia for its high oil quality.

Linseed Linseed (*Linum usitatissimum*), which is a different form of the same species as flax, was once grown much more widely in Australia than at present, the oil being used in the manufacture of paints, varnishes and resins. Its decline in production is due to the competition from petroleum-derived oils for the manufacture of these products.

Linseed breeding objectives have been to increase seed and oil yield and to incorporate resistance to rust (*Melampsora lini*) and to the disease pasmo. Research by CSIRO Canberra has produced a linseed, known as linola, with an edible oil, low in erucic acid, and this could lead to a resurgence of interest in this crop.

Safflower The seed oil of safflower (*Carthamus tinctorius*) can be used in paints, varnishes and resins. The oil is also of high quality for cooking and can be used in the manufacture of margarine and mayonnaise. Its main areas of production are Queensland and New South Wales, with smaller areas in the southern states. Breeding objectives are concerned with increased adaptability and yield and resistance to three main fungal diseases — head rot (*Botrytis cinerea*), root rot (*Phytophthora drechsleri*) and leaf rust (*Puccinia carthami*).

SUMMER-GROWING OILSEEDS

Sunflower The most important oilseed crop grown in Australia is sunflower (*Helianthus annuus*), most of which is crushed for the extraction of the edible oil, which has a high content of polyunsaturated fatty acids. The seed residue contains approximately 30–35 per cent protein and is used in animal feedstuffs.

A large proportion of the sunflower crop in Australia is now sown as F_1 hybrids, with a declining percentage sown to open-pollinated cultivars. The F_1 hybrids are generally more uniform in height and maturity than the open-pollinated types.

Breeding objectives are for the production of types with increased seed and oil yield, better adaptation to high-temperature and moisture stress, and resistance to the diseases rust (*Puccinia helianthi*), *Alternaria*, sclerotinia stem rot and wilt *(Sclerotinia sclerotiorum)*, and charcoal rot (*Macrophomina phaseoli*).

Soybeans Most of the production of soybean (*Glycine max*) comes from Queensland and northern New South Wales. The grain is high in protein and oil and thus can be used directly as a human foodstuff or crushed for the extraction of edible oils, the residue constituting a valuable animal feedstuff.

The main breeding aims for soybean improvement in Australia are for cultivars for full-season culture in the areas of its production with lodging resistance and resistance to the diseases bacterial blight and bacterial pustule, soybean rust (*Phakopsora pachyrhizi*), *Phytophthora* root rot, and stem rot. Both dryland and irrigated cultivars are bred. Coastal cultivars require resistance to manganese toxicity.

Tropical and Subtropical Crops

Cotton Since the 1970s, cotton (*Gossypium hirsutum*) has become a major crop in northern New South Wales and central and southern Queensland. A large and highly successful breeding effort at Narrabri is conducted by CSIRO in conjunction with NSW Agriculture. It has concentrated on increasing: yield, insect resistance (by the exploitation of morphological traits such as the dissected okra-type leaf), quality (fibre strength, length and fineness), adaptation to marginal environments (cooler, shorter seasons), resistance to bacterial blight (*Xanthomonas campestris* var. *malvacearum*), and increased tolerance to wilt (*Verticillium dahliae*).

Although most of the breeding effort is directed towards irrigated cultivars, dryland cotton is often grown on a large area when seasonal conditions permit. The best irrigated cultivars are usually also superior in a dryland situation.

Sugar Cane Sugar cane is grown along the coastal strip from northern New South Wales to far north Queensland. The breeding effort has concentrated on increasing yield and ratooning ability. Sugar cane is clonally propagated, consequently some effort has been directed at enhancing the germplasm base and improving crossing methods so as to provide a greater range of material for selection.

Other Crops

The following relatively minor field crops are the subject of breeding programs in Australia:

Legumes

Navy beans (*Phaseolus vulgaris*)　　Cowpeas (*Vigna unguiculata*)
Field peas (*Pisum sativum*)　　Peanut (*Arachis hypogaea*)
Faba beans (*Vicia faba*)

Pasture species

Various grasses　　Subterranean clover
　(native and exotic)　　　(*Trifolium subterraneum*)
Stylosanthes species　　Lucerne (*Medicago sativa*)
Browse shrub　　Annual medics (*Medicago* species)
　(*Leucaena leucocephala*)

Miscellaneous

Potatoes (*Solanum tuberosum*)　　Rye (*Secale cereale*)
Tobacco (*Nicotiana tabacum*)　　Millet *(Pennisetum americanum)*

Further Reading

Allard, R. W. (1960), *Principles of Plant Breeding*, Wiley, New York.
Austin, R. B., Flavell, R. B., Henson, I. E. and Lowe, H. J. B. (1986), *Molecular*

Biology and Crop Improvement. A Case Study of Wheat, Oilseed Rape and
Faba Beans, Cambridge University Press, Cambridge, U.K.

Duvick, D. N. (1990), 'The Romance of Plant Breeding and Other Myths' in
J. P. Gustafson (ed.) *Gene Manipulation and Plant Improvement II*, Plenum
Press, New York.
Fehr, W. R. (1987), *Principles of Cultivar Development* (Volume 1, *Theory and
Technique*. Volume 2, *Crop Species*), Macmillan, New York.
Frey, J. K. (1981), *Plant Breeding II*, Iowa State University Press.
Russell, G. E. (1978), *Plant Breeding for Pest and Disease Resistance*, Butter-
worths, London.
Simmonds, N. W. (1979), *Principles of Crop Improvement*, Longman, London.

References

Beckman, J. S. and Soller, M. (1986), 'RFLPs and Genetic Improvement of
Agricultural Species', *Euphytica*, **35**, 111–124.
Briggs, F. N. (1938), 'The Use of the Backcross in Plant Improvement', *American
Naturalist*, **72**, 285–292.
Buck, K. (1989), 'Brave New Botany', *New Scientist*, 3 June 1989, 32–35.
Courtice, G. (1987), 'Satellite Defences for Plants', *Nature*, **328**, 758–759.
Cousins, Y. L., Lyon, B. R. and Llewllyn, D. J. (1991), 'Transformation of an
Australian Cotton Cultivar: Prospects for Cotton Improvement Through
Genetic Engineering', *Australian Journal of Plant Physiology*, **18**, 481–494.
Day, S. (1989), 'Switching Off Genes with Antisense', *New Scientist*, 28 October
1989, 36–39.
De Candolle, A. (1909), *Origin of Cultivated Plants*, Kegan, Paul, Trench,
Trubner and Co, London.
Downes, R. W. (1990), 'Australian Plant Breeding Requirements for Current
and Future Crops in Relation to Market Forces', *Bureau of Rural Resources
Bulletin No. 4*, Australian Government Printing Service, Canberra.
Duvick, D. N. (1965), 'Cytoplasmic Pollen Sterility in Corn', *Advances in
Genetics*, **13**, 1–56.
East, E. M. (1916), 'Studies on Size Inheritance in *Nicotiana*', *Genetics*, **1**,
164–76.
East, E. M. and Mangelsdorf, A. J. (1925), 'A New Interpretation of the
Hereditary Behaviour of Self-sterile Plants', *Proceedings of National Academy
of Science*, **11**, 116–83.
Evans, D. and Sharp, W. R. (1986), Applications of Somaclonal Variation,
Bio/Technology, **4**, 528–532.
Falconer, D. S. (1981), *Introduction to Quantitative Genetics*, 2nd Edition,
Longman, London.
Finlay, K. W. and Wilkinson, G. N. (1963), 'The Analysis of Adaptation in a
Plant Breeding Programme', *Australian Journal of Agricultural Research*, **14**,
742–54.
Fisher, R. A. (1918), 'The Correlations Between Relatives on the Supposition of
Mendelian Inheritance', *Transactions of the Royal Society of Edinburgh*, **52**,
399–433.
Frankel, O. H. and Bennett, E. (1970), *Genetic Resources in Plants: Their
Exploration and Conservation*, IBP Handbook No. 11, Blackwell Scientific
Publications, London.

198   *Principles of Field Crop Production*

Frankel, O. H. and Hawkes, J. G. (1975), *Crop Genetic Resources for Today and Tomorrow*, Cambridge University Press, Cambridge, U.K.

Gasser, C. S. and Fraley, R. T. (1989), 'Genetically Engineered Plants for Crop Improvement', *Science*, **244**, 1293–1299.

Gerstel, D. U. (1950), 'Self Incompatability Studies in Gauyule: II. Inheritance', *Genetics*, **35**, 482–86.

Harlan, J. R. (1976), *Crops and Man*, American Society of Agronomy, Foundation of Modern Crops Science Series.

Harlan, J. R., De Wet, J. M. J. and Price, E. G. (1973), 'Comparative Evolution of Cereals,' *Evolution*, **27**, 311–25.

Helback, H. (1954), 'Prehistoric Food Plants and Weeds in Denmark', *Danmark Geologiske Undersgelse*, II R, 80.

Hughes, M. B. and Babcock, E. B. (1950), 'Self Incompatability in *Crepis foetida* (L.) subsp. *rheodifolia* (Bieb.)', *Genetics*, **35**, 570–88.

Jinks, J. L. (1954), 'The Analysis of Continuous Variation in a Diallel Cross of *Nicotiana rustica* Varieties', *Genetics*, **39**, 768–88.

Johannsen, W. L. (1903), *Ueber Erblishkeit in Populationen and in Reinen Leinen*, Gustav Fischer, Jena.

Johannsen, W. L. (1926), *Elemente der Exacten Erblichkeitslehne*, Gustav Fischer, Jena.

Jones, D. F. (1918), *The Effects of Inbreeding and Crossbreeding on Development*, Connecticut Agriculture Experiment Station Bulletin 207, 1–100.

Jones, H. A. and Davis, G. N. (1944), *Inbreeding and Heterosis and their Relation to the Development of New Varieties of Onions*, USDA Technical Bulletin 874, 1–28.

Knott, D. R. (1963), 'The Inheritance of Stem Rust Resistance in Wheat', *Proceedings 2nd International Wheat Genetics Symposium*, Sweden, 156–66.

Lazenby, A. (1986), *Australia's Plant Breeding Needs,* Australian Government Publishing Service, Canberra.

Lonnquist, J. H. and McGill, D. P. (1956), 'Performance of Corn Synthetics in Advanced Generations of Synthesis and after two Cycles of Recurrent Selection', *Agronomy Journal*, **48**, 249–53.

Lundquist, A. (1956), 'Self-incompatability in Rye: I. Genetic Controls in the Diploid', *Hereditas*, **42**, 292–348.

Maddox, J. (1989), 'The Great Gene Shears Story', *Nature*, **342**, 609–613.

Mangelsdorf, P. C. (1965), 'The Evolution of Maize' in J. B. Hutchinson (ed.) *Essays on Crop Plant Evolution*, Cambridge University Press, Cambridge, 23–49.

Mather, K. W. (1972), *Statistical Analysis in Biology*, 5th edn, Chapman and Hall, London.

Mather, K. and Jinks, J. L. (1971), *Biometrical Genetics: The Study of Continuous Variation*, Chapman and Hall, London.

Nijkamp, H. J. J., Van Der Plas, L. H. W. and Van Aartrijk, J. (1990), *Progress in Plant Cellular and Molecular Biology*, Kluwer Academic Publishers, Dordrecht, Netherlands.

Peacock, J. (1990), 'Ways to Pollen Sterility', *Nature*, **347**, 714–715.

Perani, L., Radke, S., Wilkie-Douglas, M. and Bossent, M. (1986), 'Gene Transfer Methods for Crop Improvement: Introduction of Foreign DNA into Plants', *Physiologia Plantarum*, **68**, 566–570.

Potrykus, I. (1991), 'Gene Transfer to Plants: Assessment of Published Approaches and Results', *Annual Review of Plant Physiology and Plant Molecular Biology*, **42**, 205–225.

Pugsley, A. T. (1965), 'Inheritance of a Correlated Day-length Response in Spring Wheat', *Nature*, **207**, 108.

Pugsley, A. T. (1971), 'A Genetic Analysis of the Spring-Winter Habit of Growth in Wheat', *Australian Journal of Agricultural Research*, **22**, 21–31.

Reddy, D. V. R., Nambiar, P. T. C., Rajeswari, R., Mehan, V. K., Anjaiah, V. and McDonald, D. (1988), 'Potential of Enzyme-linked Immunosorbent Assay for Detecting Viruses, Fungi, Bacteria, Mycoplasma-like Organisms, and Hormones' in *Biotechnology in Tropical Crop Improvement. Proceedings of International Biotechnology Workshop, ICRISAT*, India, January 1987, pp. 43–49.

Roberts, E. H. (1973), 'Predicting the Storage Life of Seeds', *Seed Science and Technology*, **1**, 515–27.

Rogers, J. S. and Edwardson, J. R. (1952), 'The Utilisation of Cytoplasmic Male-sterile Inbreds in the Production of Corn Hybrids', *Agronomy Journal*, **44**, 8–13.

Shull, G. H. (1909), 'A Pure Line Method of Corn Breeding', *American Breeders Association Report*, 51–9.

Simmonds, N. W. (1976), *The Evolution of Crop Plants*, Longman, London.

Smith, M. K. and Drew, R. A. (1990), 'Current Applications of Tissue Culture in Plant Propagation and Improvement', *Australian Journal of Plant Physiology*, **17**, 267–289.

Tanksley, S. D. (1983), 'Molecular Markers in Plant Breeding', *Plant Molecular Biology Reporter*, **1**, 3–38.

Vavilov, N. I. (1926), 'Studies on the Origin of Cultivated Plants', *Bulletin of Applied Botany Genetics and Plant Breeding*, **16**, 1–248.

Wilkie-Douglas, M., Perani, L., Radke, S. and Bossent, M. (1986), 'The Application of Recombinant DNA Technology toward Crop Improvement', *Physiologia Plantarum*, **68**, 560–565.

CHAPTER 4

SOILS FOR CROP PRODUCTION

R. R. Storrier and J. W. McGarity

The influence of soil properties on productivity is one of the most important and least understood of the many factors which determine the yield of crops. The biological, chemical and physical attributes of soil interact and influence plant growth in many diverse ways and it is not always known which soil management techniques need to be adopted to optimise the interaction of these attributes with plants.

Australian soils, with few exceptions, have been developed from much weathered and highly leached parent materials exposed either on ancient land surfaces or on transported and deposited erosion products derived from these landscapes. As a consequence, the soils are usually deficient in some of the major and minor nutrients required for plant growth. In addition, many of the soils are low in organic matter due to the nature of the natural vegetation, and often have poor hardsetting and crusty structures in the topsoil which may adversely affect seedling establishment, water infiltration and the stability of the soil against erosion.

Allowing for the fact that desert and semi-arid zone soils account for 63 per cent of Australia's total area, and skeletal soils on mountainous country for 14 per cent, there is only a small percentage of inherently fertile soil in areas of good rainfall (Anon., 1975).

Except in the vicinity of towns and settlements, early agriculture was confined to those districts having an adequate and reliable rainfall and to those soils with properties favourable for the production of particular crops. Soils which were difficult to cultivate or which were low in natural fertility remained underdeveloped until technology and science enabled them to be economically utilised.

In Australian agriculture, new management techniques, the result of continuing agronomic and soil research, have helped to maintain and in some cases increase crop yields on the longest farmed soils, while new and successful crop systems have been introduced in areas previously regarded as having unsuitable soils. A good example of this applied research is the current national program aimed at improving the productivity of sodic soils which are widespread in sub-humid and semi-arid farming areas.

Advances in new technology, however, have not been without disadvantages: soil deterioration in many cropping areas is now increasingly evident and widespread. It may be attributed to the use of machinery and management techniques ill-adapted to continuing optimisation of the potential productivity of the soil, and introduced without valid evaluation.

Thus, there is increasing awareness of the need to stem and reverse the drift in fertility of the older farming soils. Fertiliser and ameliorant inputs, reduced tillage, soil and water conservation, and innovative crop and pasture rotations are but a few of the most common measures advocated to achieve this end. Success will depend on finding practices tailored for particular soils and associated climate, and for particular crops; the measures must be site-specific and directed towards stabilising, supplementing and improving those soil properties which limit crop production and which differ significantly between different soils.

SOIL CHARACTERISTICS AND PLANT GROWTH

Although some crops have specific soil requirements, it is possible to produce satisfactory yields of most commercial crops on a variety of soils providing that both the climatic requirements of the crop and the soil management practices for the particular soil and crop are met. The interrelationship between the physical, chemical and biological soil properties and plant growth needs to be established therefore in each cropping situation.

In farming practice the adjustment of cultivation and cropping techniques to soil differences has relied as much on the accumulation of experience, as on any basic understanding of the underlying principles governing agronomic practices. The expansion of agriculture was greatly dependent on the former, and particular 'country' with a unique association of native vegetation and particular soils, such as the red brown earths which occupy large geographic areas, was brought into production in the early stages of development because experience had shown that such land could be readily ploughed and prepared, and that crop yields could be maintained.

The recognition that certain broad areas of soils are suitable for particular farming systems can be made much more relevant to individual farms by specifically identifying the soils and the characteristics which affect plant growth. It is necessary, then, to describe and classify soils in terms of recognisable field and some laboratory properties and to relate these characteristics to actual and potential crop growth.

Before describing the major soils used for the production of various field crops in Australia and how the physical, chemical and biological properties of these soils affect plant growth it is essential to understand both the terminology used in describing soils and the agronomic importance of some soil characteristics (see Butler and Hubble, 1977).

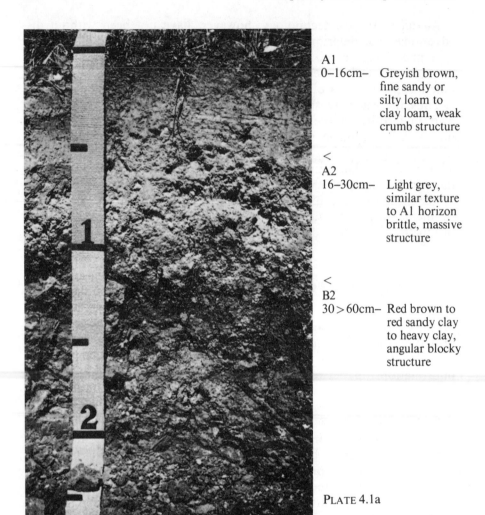

A1
0–16cm–   Greyish brown,
           fine sandy or
           silty loam to
           clay loam, weak
           crumb structure

<
A2
16–30cm–   Light grey,
           similar texture
           to A1 horizon
           brittle, massive
           structure

<
B2
30>60cm–   Red brown to
           red sandy clay
           to heavy clay,
           angular blocky
           structure

PLATE 4.1a

PLATE 4.1   Typical soil profiles illustrating horizon characteristics (after Stace *et al.*, 1968; permission of Rellim Publications, South Australia)
(a) Red podsolic soil developed on shale, near Gympie, Queensland (tape marked in feet)
(b) Grey clay, Barkly Tableland, Northern Territory

The solum   Soil develops from the breakdown of either unweathered or partly weathered rock or unconsolidated and sometimes partly altered alluvial/colluvial material deposited on the landscape. Soil is not, however, just a loose and unorganised weathered material covering the earth's surface nor is it merely a medium for plant growth. Rather the soil is seen as a dynamic, naturally occurring three-dimensional body in which chemical, physical and biological processes are interacting.

A1.1
0–3cm– Dark grey, clay,
granular to self-
mulching struc-
ture

<

A1.2
3–30cm– Grey, clay, fine
to coarse blocky
structure, vertical
cracking

<

A1.3
30–80cm– Grey, clay,
coarse blocky
to prismatic
structure, vertical
and horizontal
cracking

<

AC
> 85cm– Grey and
brown, clay,
massive, slicken-
sided structure

PLATE 4.1b

New products such as clay and humus are formed and redistributed continuously within this soil mass from the alteration and resynthesis of both the minerals derived from the parent material and the fresh organic matter additions. It is this depth of reorganised material which is referred to as the *solum*.

Horizons   As a consequence of weathering and relocation of products during soil formation, changes in colour and texture and many other properties develop in the vertical section from the surface into the underlying parent material. This vertical section is called the *soil profile*, in which are formed distinct and recognisable layers, referred to as *soil horizons*.

The relationships and differences in properties within and between soil horizons are not only the basis for the description and classification of soils but may also vitally influence plant growth and productivity.

The surface or A horizon[1] is the zone of organic matter accumulation. It may be divided into an A1 horizon of maximum biological activity which is generally darker and a lighter, pale A2 or subsurface horizon which at times may be almost white and then may be referred to as a bleached horizon. The A horizon is sometimes described as the *eluvial* horizon, because colloidal clay and iron oxide minerals, and certain other soil constituents, have been washed out, with a consequent lower content of these constituents, compared with the underlying horizons. The A2 is sometimes referred to as an E horizon (McDonald *et al.*, 1984).

The subsoil horizon is referred to as the B or *illuvial* horizon. This is a zone of accumulation of inorganic or mineral soil constituents and the texture, colour, structure and consistency of the soil are more strongly expressed than in the horizons above or below.

The C horizon is a zone of partially weathered parent material with the general appearance of the unaltered fresh rock or sediment. It shows few indications of biological activity and lacks expression of morphological properties indicative of pedological (soil-forming) development. It may be nevertheless a source of nutrients and water available to plants.

In some situations the B horizon of the solum can be underlain by a D horizon which is composed of material unrelated to the overlying soil, but possibly indirectly influencing some of its characteristics and agronomic potential.

Photographs of two soil profiles showing different arrangements of horizons are reproduced in Plate 4.1. It should be noted that all horizon combinations are not necessarily present in any one soil.

Colour   The recognition of soil horizons is most frequently made on the basis of colour, which is strongly associated with changes in other soil properties, such as texture, structure and organic matter status.

Colour also is indicative of the drainage and aeration conditions of the soil. Bright whole-coloured red and yellow subsoils are generally indicative of well-drained profiles, whereas drab, mottled bluish-grey subsoils are associated with water-logged conditions, which are unsuited to the majority of deep-rooting crops.

The correlative association of soil colour changes with other soil properties is of the utmost value in the identification and classification of soils. While colour in itself is not a property which intrinsically affects plant growth it frequently correlates with changes in edaphic properties which do influence the plant. Consequently colour can be used in a limited way to interpret soil suitability for agricultural production.

---

[1] Horizon designations conform to those used by Northcote (1971), and generally to those in the Australian Soil and Land Survey Field Handbook (McDonald *et al.*, 1984).

TABLE 4.1   Soil texture grades arranged in groups based on clay content (after Northcote, 1971) and associated available water-storage capacity (AWSC) (Williams, 1983)

Texture group	Texture grades	Clay range (%)	AWSC mm dm^{-1}
Sands	sand, loamy sand, clayey sand	0–10	13.6
Sandy loams	sandy loam, fine sandy loam, light sandy clay loam	10–20	15.5
Loams	loam, fine sandy loam, silt loam, sandy clay loam	20–30	15.8
Clay loams	clay loam, silty clay loam, fine sandy clay loam	30–35	16.9
Light clays	sandy clay, silty clay, light clay, light medium clay	34–45	13.8
Medium-heavy clays	medium clay, heavy clay	45+	11.5

Texture   The texture of the soil is a measure of the relative proportions of sand, silt and clay particles. A range of texture grades exists between sandy and clayey soils. Table 4.1 lists the texture groupings of Northcote (1971) with an indication of the range of content of clay (particle diameter <0.002 mm) present.

The total depth of the soil profile and the texture and thickness of the horizons, as well as the nature of the boundary between horizons, are important agronomic characteristics. Plants obtain the majority of their nutrients from the topsoils while the subsoil is the storage region for water. The clay-sized material is the most important part of the mineral soil in acting as a source of plant nutrients, as a reservoir to hold water, and as a bonding agent assisting in the development of soil aggregates.

Coarser, light textured soils (e.g. sandy loams) hold only small quantities of water compared with the clay soils, but most of the water in the sandy soils is generally available to plants. On finer, heavy textured soils, a high proportion of the soil water is held more firmly and is not readily available to plants. Thus, when the profiles are dry, light falls of rain are used by plants less effectively on heavy textured soils compared with sandy soils.

Ideally a combination of a moderate depth of sand overlying clay is a highly effective horizon combination for water storage. Stones and gravel, as well as shallow depth, reduce the amount of water stored.

The range of available water for the Australian texture groups is given in Table 4.1. These are generalised values which may be greatly modified by the nature of the soil structure which determines the pore space and the capacity of the soil to store water. Self-mulching clays, with stable fine structure, may have an available water range in excess of 20 mm dm^{-1}.

In general, well-structured loams and clay loams have optimum pore space, water regimes and aeration.

Structure   The arrangement of the soil particles into naturally occurring aggregates or crumbs gives the soil its structure characteristics. These natural structural units or *peds* contrast with a soil *clod* which is developed as a result of the application of some external force, such as ploughing, to the soil.

Soil aggregation is promoted by the presence of clay, iron and aluminium compounds, fine lime, organic matter, plant roots and soil organisms (bacteria, fungi, earthworms). As a result, clay loam surface soils containing a moderate amount of organic matter are likely to be better structured, whereas sandy textured soils are weakly structured, liable to form hard setting surfaces and require careful management.

It is usual to describe structure in terms of the size of the peds and their shape (crumb, granular, polyhedral, lenticular, blocky, sub-angular blocky, prismatic, columnar, platy) and the degree of aggregation, which will vary from structureless or apedal through weak and moderate to a strong grade of structure.

Peds are often characterised by being smooth-faced, which is associated with varying degrees of shininess as a result of the deposition of clay or sometimes iron or manganese on the surface. Such deposits of clay are referred to as clay-skins or clay *cutans* and impart a high degree of pedality to the soil. In contrast, peds which have a scurfy, porous surface and a general floc condition throughout, are said to have an earthy *fabric*.[2] Dense, clayey peds in which this condition dominates are referred to as rough-faced peds.

Some of the parameters used either singly or in combination to describe the morphology of soil structures can be related to plant growth either as direct measures of soil structure effects or as indirect measures of the influence of other soil properties.

Surface soil structures may vary widely. Thus some clay soils have a desirable 'self-mulching' property in which the loose, friable, aggregated surface may slake on wetting but the larger structural aggregates are reformed naturally on drying and shrinking (Plate 4.2a). This contrasts with many loam soils which have hard-setting surfaces. Such hard, compact and apparently apedal surfaces may form when structural units which are developed as a result of cultivation are dispersed on wetting but are not reformed on drying (Plate 4.2b). Crusts and seals of this kind can be broken by further cultivation, but return to the original compact state in the next cycle of wetting and drying. Surface structure can thus greatly influence crop production through its effects on water infiltration, soil aeration and seedling emergence.

---

[2] Fabric refers to the general appearance of the sand, silt and clay particles on a ped surface.

McIntyre (1955), in a study of cultivated red brown earths from South Australia, found that soil aggregates were unstable and dispersed on wetting by rain. The dispersed soil was found to have a higher density and a lower porosity and in the intensively cultivated soil, a crust up to 20 mm in thickness was developed following heavy post-seeding rains. There was a consequent 28 per cent decrease in wheat seed establishment and a poorer yield (Table 4.2). The pulverizing action of implements and dispersion by heavy rains over the years was considered responsible for the aggregate instability.

Soils which have an inherently unstable, easily slaked and dispersed surface structure or in which the structure has deteriorated due to cultivation, are liable to water erosion. Sandy textured soils which lack cohesion due to their low organic matter require protection from wind erosion.

Within the solum, compacted and sometimes cemented horizons of high density may be found. These have developed as a result of normal soil forming processes or as a result of cultivation practices, and are often referred to as 'pans'. Such horizons have reduced air and water permeability which influences the overall infiltration characteristics of the soil and hence the storage of water at depth. They can be significant barriers to root penetration (Plate 4.2c) and act as restraints to successful crop production.

Consistence Consistence comprises those attributes of the soil material which refer to the degree and kind of cohesion and adhesion, or resistance to deformation or rupture of the soil ped. It can be described in specific terms characterising strength, stickiness and plasticity (Butler, 1955; McDonald and Isbell, 1984), or in general terms such as brittle, crumbly, plastic, friable and snuffy (Soil Survey Staff, 1951). The qualities described can be useful indicators as to how the soil peds may be broken down under cultivation to form finer seedbeds, or whether the

TABLE 4.2   Physical characteristics of the surface soil (0–60 mm) of a red brown earth as affected by crop rotation and their influence on wheat yields, 1951 season (after McIntyre, 1955)

Rotation	Bulk density (mg/m³)	Total porosity (% total soil)	Unfilled porosity at field capacity (% total soil)[b]	Grain yield (kg/ha)
Fallow wheat	1.46	45	13	612
Fallow-wheat-peas-pasture	1.36	49	19	1547
Crust[a]	1.64	38	1	

[a]Sample taken in 1953 to depth of 10 mm on two-course rotation.
[b]Volume of air space when soil has been fully wetted and allowed to drain.

PLATE 4.2(a)

PLATE 4.2(b)

PLATE 4.2(c)

PLATE 4.2(di)

Plate 4.2(dii)

Plate 4.2   Soil structure and its effect on plant growth
(a) Self-mulching structure in the surface horizon of a black earth soil near Warwick, Queensland (A. A. McNeil, Wagga)
(b) Surface crust developed on a transitional red brown earth soil near Griffith, preventing the penetration of the prop roots of maize into the soil (W. A. Muirhead, Griffith)
(c) Wheat root penetration impeded by a clay B horizon in a solonised brown soil near Renmark, South Australia
(di) Subsurface (plough pan) layer developed in a well structured black earth near Emerald, Queensland
(dii) The sunflower crop failed to develop an adequate root system, resulting in moisture stress and consequent reduced crop yields

soil is at the correct status for cultivation. Farmers frequently use the term *tilth* to describe this quality of the soil.

Chemical properties   A number of chemical properties are referred to in the field descriptions of soils, and these can also have agronomic significance.

The reaction or pH is often measured for each soil horizon, and the pattern of change established for the soil profile. Extremely acid or alkaline soils are frequently found to have either deficiencies or excess quantities of the various major and minor elements (Fig. 4.1).

Aluminium occurs in soils in excessive and toxic amounts when the pH

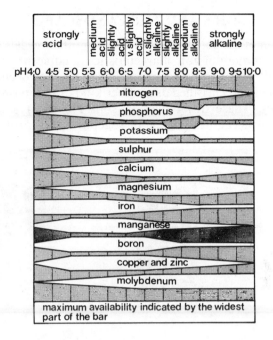

FIG. 4.1 A schematic illustration of the relation between plant nutrient availability and soil reaction (after Truog, 1946)

falls below about 4.5 as measured in a soil:water paste (1:1) or suspension (1:5). The pH indicator test used in the field is based on the 1:5 water suspension, which is the value reported in this text. Laboratory determinations are also made on soil:electrolyte (KCl, CaCl$_2$) suspensions and usually give readings which are 0.5 to 1.0 pH units lower. For this reason the method of determination should always be stated.

Some soils are characterised by the presence of calcium carbonate as fine material or concretions. Such alkaline soils can exhibit either nutrient deficiencies or toxicities, and if the precipitated carbonate forms a hardened subsoil horizon (calcrete), water penetration and root growth may be severely hindered.

Soluble salts, usually sodium chloride but sometimes sodium carbonate, accumulate in subsoils and less frequently in the surface as efflorescences. These saline properties may not only affect plant growth directly, but may lead to sodicity and to the dispersion of the clay and the subsequent development of a poor structure (see Northcote and Skene, 1972).

Frequently accumulations of iron, manganese and aluminium oxides (sesquioxides) occur in subsoil horizons. These may be present as soft spots, hard concretions or massive layers. Chemically, the sesquioxides may affect the availability of phosphorus and molybdenum by fixing these nutrient ions into insoluble compounds.

SOILS AND THE LANDSCAPE

An examination of the soil profiles in the field often reveals a continuous but slight variation in horizon characteristics in adjacent profiles. Where significantly different combinations of characteristics occur, the profile may be separated and classified as another kind of soil, termed a *soil type* or a *soil series*. These differences in soils are often associated with changes in the physical, lithological or vegetative soil-forming factors.

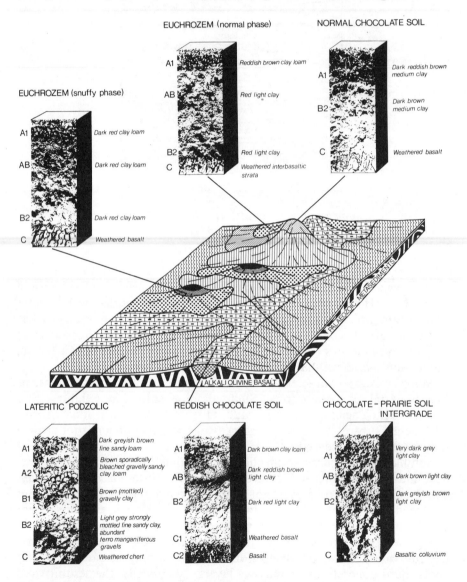

FIG. 4.2   Basaltic and metasedimentary soilscape at Arding, near Armidale, New South Wales (after Schafer and McGarity, 1980)

Thus the array of physical features found through a particular landscape component such as slope, rocky hillock, stream and floodplain will influence soil development and the resultant soil profile characteristics, and so determine the distribution of different soil series. Fig. 4.2 illustrates how soil characteristics vary in the landscape.

Because of the complexity of the distribution of soils within a landscape, it is usual to delineate soils on a map as a collection or association of soils occurring within a specifically defined landscape unit. The degree of detail depends on the scale of the map and the intensity of the survey. For intensive land use, such as irrigation developments (Blackburn and Wright, 1972) and farm planning, large-scale surveys at less than 1:20 000 are necessary so that the mapping unit contains soil types that are relatively homogeneous in the properties which are used for single-purpose interpretation. Maps with scales in the range of 1:50 000 to 1:125 000 may be used for regional planning, development projects, and broad interpretation for suitable land use (Hawkins and Driver, 1978) and erosion hazard (McKenzie, 1992).

More generalised maps of agricultural capabilities of dry land (French *et al.*, 1968), continental distribution of soils as depicted in the Atlas of Australian Soils (Northcote *et al.*, 1960–8) and soils as dominant components of landscapes (Northcote *et al.*, 1975) range between scales of 1:500 000 to 1:5 million. Currently many government agencies are undertaking surveys at 1:50 000 to 1:100 000 scales as part of the National Soil Conservation Program, an example of which is the soil and land survey of the Waggamba Shire in southeast Queensland (Thwaites and Macnish, 1991).

SOIL CLASSIFICATION

Soil classification in Australia has gradually evolved from the Great Soil Group concept which was first introduced by Prescott (1931). This concept involved the grouping of soils of generally similar morphology and with common vegetation, climatic and topographic relationships into the one great soil group.

The original ten soil groups of Prescott were subsequently expanded to forty-seven by Stephens (1962).

A modification of the Prescott–Stephens groupings was introduced by Stace *et al.* (1968). The forty-three soil groups in their system have been presented in such an order that they represent overall a progressive increase in degree of profile development and degree of leaching. Within the sequence a number of categories are recognised, beginning with those in which there is little or no profile development and ending with soils dominated by the accumulation of organic matter. Thus the dark soils follow the category with minimal profile development, and are followed in turn by categories which are progressively more strongly leached and highly differentiated.

Northcote (1971) developed a key-type identification system based on observable morphological properties:

> The basis of the key is the concept of the profile form. Profile form is the term used to express the overall visual impact of the physical soil properties in their intimate association one with another, and within the framework of the solum. Thus, the profile is regarded as a physical system and those physical properties capable of observation and record which may carry along other features or properties, be they physical, chemical or biological, are used to distinguish between groups at each step of the key. The only exception is the use of Soil Reaction Trend.

The key is based initially on distinctions made with respect to the texture pattern of the profile, that is whether there is little or no texture change with depth (uniform), gradual increases in texture (gradational), sharp changes (duplex) or organic matter dominated soil profiles. This initial distinction is then further subdivided on the basis of textural classes in the uniform texture group, the calcareous nature of those soils exhibiting a gradational texture profile, and the colour of the B horizon in the duplex profile soils. The organic soils are not subdivided.

The Principal Profile Form, which is the major unit in the system, is based on morphological characteristics considered to be most useful for each of the higher categories, rather than using similar properties for all the primary groupings. The main properties used are degree of horizon differentiation, colour, the nature of the A2 horizon, compaction of the A1 and A2 horizons, structure of the B horizon, presence of pans (e.g. calcium carbonate) and soil reaction trend.

In general, The Principal Profile Forms are more rigidly and narrowly defined taxonomic units than are the soil groups as described in the *Handbook of Australian Soils* which are currently more widely recognised by the general agriculturalist.

Unfortunately the Great Soil Group classification lacks rigorous, quantitative definition, leading to much ambiguity in identification, and the wide range of values of some of the edaphic properties does not allow a ready interpretation of the Great Soil Group for land-use capability assessment without additional qualifying information. The Northcote key, while requiring a more definitive assessment of specific morphological properties for identification of the Principal Profile Forms, also fails to include properties which can be readily correlated with agricultural productivity. The need for a new approach stemming from dissatisfaction with these classifications has resulted in a third classification being proposed (Isbell, 1993).

The adoption and widespread use of the new scheme by agriculturalists and planners will depend on how easy the classification process is to use in practice and how well it lends itself to multi-purpose interpretation. A brief account of this significant development is outlined here, although at this time the scheme has not been finalised and is simply referred to as the 'Third Approximation'.

The classification is hierarchical and multi-categorical. Ony a few of the many properties which would normally be described in examining a profile are used to assign the soil to one of the fourteen classes which makes up the highest category or *order* (Table 4.3). Sub-classes, termed *sub-order, great group, sub-group* and *family*, are then separated at progressively lower categorical levels using a wide range of differing soil properties. About 300 different sub-groups can be identified. The family is the lowest category, and the greatest total amount of information is required in order to complete the classification to this level. Soils are separated at the family level according to the depth of soil, A horizon thickness, texture, and gravel content, which are attributes governing water relations, rooting depth and nutrient availability. The use of some laboratory data as differentiating criteria is a significant conceptual departure from previous classifications.

The nomenclature and definitions used follow, in a less complicated form, those devised originally for Soil Taxonomy, a classification developed in the U.S. by the Soil Conservation Service (Soil Survey Staff, 1975). The grey clay in Plate 4.1b could be classified as an *endohypersodic* (sub-group), *self-mulching* (great group), *grey* (sub-order), *vertosol* (order); very deep, very fine, non-gravelly (family). While the name appears formidable, it translates to a very deep (>1.5 m), very fine (>60% clay), non-gravelly (<10%) grey cracking clay with a surface horizon of self-mulching structure at least 10 mm thick, overlying a deeper (>0.5 m) sodium-rich (>15% exchangeable sodium) subsoil.

Even at the sub-group level there is enough morphological and chemical information represented to enable interpretation of the suitability of the soil for a number of uses and for the transfer of technology between members of the same class.

Although different criteria have been used in this Third Approximation to derive the orders and groups compared with the Great Soil Group classification, some general relationship between the units can be established (Table 4.3).

In this brief summary of the relationships between soils and crops it has been necessary to describe the soils in broad terms, and consequently the more generalised descriptions and terminology of the *Handbook* has been retained rather than the more specific profile descriptions of the Northcote key or the Second Approximation. However, reference is also made to the more important Principal Profile Forms occurring within the range of profiles characteristic of the Great Soil Group. The intention in considering both classifications in this text is to allow agronomists to become more familiar with the Northcote key as a means of recognising and classifying soils.

However, with the gradual evolution of international soil classification schemes and the consequent improved communication, national classifications may well be superseded. Soil Taxonomy (Soil Survey Staff, 1975) is now used in many countries and offers the prospect of relating and

TABLE 4.3   Differentiating criteria and relationship of orders in the Second
Approximation to Australian Great Soil Groups

Order[a]	Major differentiating features	Equivalent Great Soil Group
Tenosols	Weak texture, structure, colour, contrast, together or solely	Solonised brown soils
Vertosols	Clay soils which crack when dry and have slicksided lenticular structures at depth	Black earths Grey, brown and red clay Humic gleys Prairie soils
Hydrosols	Profile saturated for at least several months of the year, often have gley B	Humic gleys
Sodosols	Strong texture contrast (duplex), sodic B horizon, pH>5.5	Solodic soils Solodised solonetz Red brown earth (sodic)
Chromosols	Strong texture contrast, non-sodic B (ESP<6), Ph>5.5	Red brown earth Non-calcic brown soil
Kurosols	Strong texture contrast, strongly acid B, pH>5.5	Red podsolic
Calcarosols	Lacking strong texture contrast, calcareous throughout	Solonised brown soil
Ferrosols	Lacking strong texture contrast, structured B, high free iron oxide	Krasnozem Euchrozen Reddish chocolate soil
Dermosols	Lacking strong texture contrast, structured B	Chocolate soil Prairie soil
Kandosols	Lacking strong texture contrast, massive or weakly structured B	Red earth Yellow earth

[a]Anthroposols, Organosols, Rudosols and Podosols are not represented amongst Great Soil Groups described in the text (Isbell, 1993).

transferring scientific and technological developments with greater certainty to similar soils which have been identified in other countries and continents.

DESCRIPTION OF SOILS

Not all of the soil types that have been described as occurring in Australia can be successfully utilised for crop production. Many of them, because of their shallow and stony nature or their location in adverse climatic zones, are not suitable for rainfed or irrigated agriculture. They are used for forestry and extensive pastoral management or remain undeveloped.

Nix (1974) has estimated that of Australia's total land area of 769 million hectares only about 77 million hectares are suitable for agricultural development. Of this area, about one quarter is under crop or fallow, a third under improved pastures while about another third is under-developed but is considered suitable for rainfed agriculture and forestry (Hallsworth, 1977).

Descriptions of those soils which are used in the production of various field crops are given in this section. For convenience, they are grouped into three broad climatic zones as follows:

Humid zone soils	*Sub-humid zone soils*	*Semi-arid zone soils*
Humic gleys	Euchrozems	Solodised solonetz and
Red podsolics	Yellow earths	solodic soils
Krasnozems	Red earths	Grey, brown and red
	Chocolate soils	clays
	Non-calcic brown soils	Solonised brown soils
	Red brown earths	Earthy sands
	Black earths	
	Prairie soils	

The soils located within each zone can be considered to have some common general physical and chemical properties which influence their agricultural usage, although further distinctive properties may clearly characterise the specific soil group. Some of the soils are found under a wide range of climatic conditions so that this grouping represents the most usual environment for the particular soil.

Humid zone soils are usually strongly weathered and leached, which is reflected in their mildly to strongly acidic properties, differentiated profile development and a predominance of a sesquioxidic (i.e. iron and aluminium oxide dominated) clay mineral fraction. The dominant clay mineral is kaolin, which gives the soils a low cation exchange capacity[3] and usually a low nutrient status. The soils lack calcium carbonate in the profile, but have a high level of organic matter in the surface horizon, a property which enhances the exchange capacity. Under forest conditions, organic matter may extend to moderate depths.

Within the sub-humid zone are soils which are not as highly leached, and as a result the profile is slightly acid to slightly alkaline in reaction, with traces of calcium carbonate present in the deeper subsoil of some members of the group. They contain moderate levels of organic matter and nitrogen in the surface horizon. The dominant clay minerals belong to the hydrous mica group (illite–vermiculite) and the expanding lattice clay group (montmorillonites). These give the soils a reasonable exchange capacity which is dominated by the calcium and magnesium ions.

---

[3] Cation exchange capacity (CEC)—the sum total of exchangeable cations that a soil can adsorb on the surface active clay colloids and organic colloids. Expressed in milliequivalents per 100 grams of soil (Anon., 1971), or as moles of charge per unit mass, e.g. c mol_c/kg.

Further, the expanding nature of some of the clays give the soils a high shrinking and swelling characteristic which makes them difficult to manage when wet, and they crack when dry.

The semi-arid environment soils have not been subject to intense weathering or leaching under the present climate and as a consequence they are normally alkaline in reaction and have free calcium carbonate and gypsum in the profile. Because these soils have not been significantly leached and generally have clayey subsoils with low permeability, bases and soluble salts are retained wtihin the solum. The presence of significant quantities of exchangeable sodium adsorbed on the cation exchange complex enhances the swelling of the clays and leads to the development of characteristic structure profiles and gilgai[4] features in some members of the group.

Humid Zone Soils

Humic Gleys

Waterlogged swampy soils are relatively important features of the coastal plain and delta landscape along the Australian seaboard. The soils vary from coarse to fine textured depending on the source of alluvial parent materials which are often deposited under brackish conditions.

Frequent flooding usually restricts the usage of these soils to pasture production unless adequate surface drainage and flood mitigation are undertaken as a prerequisite for successful crop production. Intensive production of sugar cane (*Saccharum officinarum*) is practised in northern areas of Australia, while in the south, vegetable, horticultural and forage crop production is significant.

Unlike the greater area of agricultural soils too much water rather than too little is often the major limitation for crop growth. Drought can usually be avoided by irrigation from subsurface supplies.

The excessive wetness of these soils, before drainage, has allowed high levels of organic matter to accumulate and become incorporated with the sand or clay in the dark surface A horizons. The organic matter decreases significantly into a subsoil which is gleyed (Bg, B2g), which indicates frequent waterlogging in the lower part of the soil profile.

The gley horizon is a definitive feature of the humic gley soils, and may be recognised by the appearance of blue-grey or green-grey colours, which are frequently mottled. Anaerobic conditions promote reduction of coloured iron and manganese compounds characteristic of well aerated soils, and the reduced compounds are colourless, so revealing the true colours of the clay and sand. Reoxidation produces rusty traces and

---

[4] Gilgai—surface micro-relief consisting of hummocks and/or hollows, caused by vertical displacement of subsoils by expansion and contraction, and associated with soils that have clay subsoils.

mottles, particularly along pores and root channels. This horizon frequently has a sour odour due to sulfur derived gases.

Stace *et al.* (1968) included the humic gleys within the group of soils whose characteristics are dominated by the presence of organic matter. Of this group of soils, the humic gleys are the only ones of agricultural significance.

These soils are included within several soil classes of the Northcote *et al.* (1975) classification. Examples include uniform and fine textured profiles with coherent and dense smooth-faced peds in the B horizon which are classified as non-cracking dense pedal clays (Uf 6.42). Black massive cracking clays (Ug 5.4) which are waterlogged and acid are also included. Other examples can be found in the grey gradational earths with smooth peds (Gn 3.9, 3.0) and also in the duplex soils with friable, pedal and mottled yellow subsoils (Dy 5.1) or dominantly gley B horizons (Dg 4.1). These latter examples are intergrades to gley podsolic soils.

DISTRIBUTION

These soils occurs as small areas throughout the humid and sub-humid zones of Australia, particularly along the eastern coast, where the average annual rainfall exceeds 750 mm. Here they are most commonly associated with the wet alluvial marine plains, extending inland along valleys and terraces of the lower reaches of rivers.

Humic gleys are also found on the lower hill slopes of the coastal valleys on colluvium-alluvium derived from a variety of parent materials which include basalts, andesites and phyllites.

Significant areas of these soils are located on the coastal plains of northern New South Wales (e.g., Dg 4.1, Uf 6.42, Ug 5.4 profiles), the Bundaberg area of Queensland (e.g., Uf 6.42 profiles) and the lower reaches of the Murray River. Many undeveloped areas occur in northern Australia.

GENERAL DESCRIPTION

The surface horizons of humic gleys, which are characteristically dark grey or black with some rusty spots, vary in thickness (0.2 to 0.4 m) and organic matter content. With depth the A horizon becomes lighter in colour with rusty and ochreous flecks, spots, and lines along root channels. The surface horizons may include some undecomposed (litter) plant material, especially in the undisturbed soil.

Soil texture ranges from an organic or peaty sand, through silts, loams and clay loams. Structure, which is strongly influenced by the texture, varies from weak crumb to strong granular with increasing clay content. The consistence of the surface structural units is soft and friable when moist but becomes hard on drying.

The transition to the mottled B horizon can be gradual and may be marked by an increase in clay content. The subsoil is dominantly a neutral grey or bluish-grey colour with variable amounts of yellowish and reddish mottles and some rusty root channels. Texture may be sandy and weakly structured, or clay with a well-developed blocky to prismatic structure. The surfaces of the peds are grey, with bright mottles appearing within the ped. Shear cutans, the result of ped surfaces sliding against each other, are present in cracking clays (Ug 5.4). The subsoil usually has a plastic and often sticky consistence when wet.

With depth the mottling decreases and the soil becomes fully gleyed below the water table. The deep subsoil texture often becomes sandy and reflects the nature of the layered sediments on which the soil has developed.

The surface soil of some representatives of the soil group shows melon-hole gilgai micro-relief where cracking clay variants occur on very fine clay sediments (Hallsworth *et al.*, 1955; Nicolls and Tucker, 1956).

FERTILITY CHARACTERISTICS

Humic gleys, particularly those associated with coarser acidic alluvial deposits, are inherently infertile. Where the soils have developed from more basic parent material and are in better-drained sites both chemical and physical fertility status are improved.

The soils are acid in the surface horizons with values as low as pH 4.0. The subsoil is normally acid but may be alkaline, depending on the parent material and local drainage conditions. Exchangeable base content is low and deficiencies arise from the lower availability of phosphorus and some trace elements. The humic gley soils of the Queensland 'Wallum' area are deficient in nitrogen, phosphorus, potassium, calcium, copper, zinc, molybdenum and boron (Andrew and Bryan, 1955).

The soils are slowly permeable and have a water table in the subsoil, which in some situations can be saline. Drainage and fertiliser application augment the agricultural potential of these soils by improving aeration and nutrient availability in the root zone.

Some of the soils contain such quantities of sulfides in the profile that artificial drainage with a consequent increase in oxidative processes causes extreme conditions of acidity to develop (Stephens, 1961). Consequently, care in the installation of drainage systems is necessary when soils are developed over estuarine clays with saline ground water.

In the Macleay River floodplain, New South Wales, excessive drainage to depth may bring about sudden and large increases in salinity and acidity by aerating significant thicknesses of estuarine sediments which contain sulfides (Walker, 1963, 1972). In this situation, drains should not be deeper than the upper limit of the estuarine strata. These sulfudic materials and associated acid sulfate soils (cat-clays) are now widely recognised on the low-lying coastal plains and inter-dune areas which fringe the continent.

LAND USE

Humic gleys, which are located in favourable topographic positions permitting drainage, can be satisfactorily developed for intensive agricultural production and consequently have a significance far exceeding their limited areal distribution. The soils are extensively used for improved pastures, vegetable production (particularly potatoes), cereals (maize) and sugar cane production.

The soil is particularly favoured for the production of sugar cane, being highly productive with adequate fertilisation. Two particular areas in which these soils are used for cane production are at Grafton (New South Wales) and Bundaberg (Queensland).

Red Podsolic Soils

Red podsolic soils occur as discrete and sizeable areas, as well as intimately intermingled in a complex pattern with yellow, grey-brown and lateritic podsolic soils and also red earths. They are found principally in the discontinuous belt of podsolic soils which fringe the eastern and southern wetter margins of the continent on the seaward side of the divide and in the tableland areas. They are also found in the coastal and subcoastal regions of the southern part of Western Australia, and in Tasmania.

Although often regarded as poor soils, the red podsolics are potentially better for agricultural production than the associated soils that are similarly utilised, as they are generally well drained and moderately well structured. As they occur in areas of high to moderate rainfall, water is generally non-limiting for crop production. Deficiencies in available plant nutrients are common, but can be corrected economically in areas close to large urban populations.

Various vegetable and horticultural crops are grown on these soils, while in dairying areas high levels of production of maize, other forage crops and improved pastures are possible. Elsewhere under drier conditions horticultural and cereal crops are produced or the soils are used for grazing.

They are included amongst those soils which show a strongly differentiated profile with a bleached subsurface (A2) horizon overlying a B horizon which is rich in sesquioxides relative to the horizons above or below (Stace *et al.*, 1968). This particular soil grades into the red brown earths and solodic soils on the drier margins and is associated with red earths and soloths under wetter, more humid conditions.

Northcote describes the red podsolics as duplex soils having hard setting A horizons and strongly pedal clayey B horizons which are acid and whole coloured red (Dr 2.21) or mottled red (Dr 3.21). Also included are soils with massive or weakly pedal, red B horizons (Dr 2.61, 2.62), and also those profiles with a structured surface horizon (not hard setting) overlying a red, strongly pedal clay B horizon (Dr 4.2, 4.4).

DISTRIBUTION

Red podsolic soils are found in the majority of Australian states, being common in the coastal and subcoastal regions and on the undulating and hilly margins of the inland areas. Average annual rainfall in these areas varies between about 550 and 1300 mm, the higher values occurring in the summer rainfall zone.

A wide range of rocks, including granites, sandstones, shales, meta-morphic rocks and sandy clay colluvium and alluvium, provide a diversity of parent materials for these soils.

GENERAL DESCRIPTION

The essential features of these soils are the pronounced texture contrast between the surface horizons and the B horizon, the existence of a pale or bleached A2 horizon, and the acid reaction throughout the solum.

The A1 horizon varies from a brownish grey to grey-brown loamy sand to clay loam, which normally has a weakly developed porous granular structure. It has a friable consistence when moist, but usually sets hard on drying.

At about 0.2 m the A1 horizon changes across a reasonably distinct boundary into a paler or bleached light grey, yellowish or pinkish A2 horizon of the same texture grade. This horizon is porous and brittle when dry and may have some ferromanganiferous concretions in the lower part of the horizon.

The change to the B horizon is clearly marked, but not necessarily abrupt. The colour is typically whole-coloured red to red-brown with a sandy clay to heavy clay texture. The structure ranges from essentially massive in the coarser textures to a moderately well developed polyhedral or angular blocky structure in the clay soils.

The peds, which are usually smooth and shiny, are friable and firm when moist but sticky when wet. The B horizon becomes weakly mottled with depth and grades into the underlying C horizon material on well-drained sites. Under conditions of poorer drainage the deep subsoils exhibit gleying.

The total depth of red podsolic soils depends on topographic position and the extent of weathering. Shallow soils are common on hilltops, but on moderate or gentle slopes and terraces, deeper, stone-free profiles, with depths of 1.0 to 1.5 m are developed. Red podsolic soils are distinguished from the red solodics and soloths by the more gradual change from the A2 to the B horizon, the more friable nature of the B horizon and the lower exchangeable sodium content of the subsoil in the red podsolics.

FERTILITY CHARACTERISTICS

Red podsolic soils are usually moderately acid in the surface horizon, and with depth the pH remains unchanged or becomes more acid. Total cation exchange capacity varies depending upon the parent material (e.g. 13 to 38 meq/100 g soil, Beckman and Reeve, 1972), and base saturation[5]

---

[5] Base saturation is the proportion of the cation exchange of a soil occupied by metal or 'base' cations such as calcium (Ca^{2+}), magnesium (Mg^{2+}), sodium (Na^+) and potassium (K^+).

is usually low. Many of the soils are lower in exchangeable calcium than magnesium and sodium particularly in the B horizon, a calcium:magnesium ratio of less than 1 usually being associated with lower yields.

The soils have a low to moderate fertility level, with low organic matter and nitrogen contents. The carbon-to-nitrogen ratio of the surface horizon of the soils developed in the north under more tropical conditions tend to be higher (20 to 28) than similar soils in southern Australia (Beckman and Reeve, 1972; Spain *et al.*, 1983).

TABLE 4.4   Percentage aggregation (water-stable aggregates >0.25 mm) of certain soils as affected by cultivation (after Downes, 1949)

	Soil group			
Treatment	Red loam	Black earth	Red podsolic	Yellow podsolic
Uncultivated	78.2	61.6	67.0	58.2
Cultivated (many years)	55.6	41.0	28.4	19.5

Red podsolics are deficient in phosphorus and sulfur, and also respond to a number of the trace elements such as molybdenum, cobalt and zinc (Stephens and Donald, 1958; McArthur, 1964). The surface structure of these soils, which is liable to set hard on drying, varies both naturally and also as a consequence of deterioration under cultivation (Table 4.4).

LAND USE

Red podsolics are used extensively for grazing but have been considered inferior for cultivation due to poor fertility, a consequence of hard-setting surface structure, lack of nutrients, and shallow stony profiles. However, when adequately fertilised and sown to improved pastures, they can be highly productive under a suitable rotation. In southern Australia, where soil depth is not limiting, they are used in clover ley rotations for winter cereal production, particularly in areas where they intergrade into the more widely used red brown earths. Many vineyards in the Hunter Valley are established on these soils.

In the subtropical areas they have been used for sugar cane and tropical fruit production. Close to urban areas they are utilised for a variety of cash crops.

Related yellow, grey-brown and lateritic podsolic soils, which are mainly used for grazing, may have a similar crop utilisation to the red podsolics when the groups are intermingled in paddock-size units.

Krasnozems

The krasnozems or 'red loams' are distinctive red earthy soils which are highly regarded and intensively farmed in many parts of Australia, extending from Tasmania (Loveday and Farquhar, 1958) to northern Queensland (Isbell *et al.*, 1976). They are generally found in areas of high rainfall and are cultivated for a variety of cash, horticultural and forage crops. In large part their widespread utilisation is a consequence of the unique friable crumb structure of the iron-rich clayey topsoil which produces an excellent stable tilth on cultivation.

Krasnozems fall into the category of sesquioxide dominant soils which are clayey in texture and have discrete structural aggregates. In the Northcote classification the group is separated into two major sub-divisions: either uniform textured, red non-cracking sub-plastic clays of low coherence (Uf 5.2) or soils exhibiting gradational texture profiles red in colour, non-calcareous throughout and with either a smooth ped fabric (Gn 3) or a rought ped fabric (Gn 4) in the B horizon.

The xanthozems are a recently recognised and associated group, having yellow subsoils which can be considered as the yellow counterpart of the krasnozems (Thompson and Beckman, 1959; Stace *et al.*, 1968).

DISTRIBUTION

Although krasnozem soils are widely distributed, their total area is small. They are principally soils of humid climates ranging from winter to summer rainfall and an annual precipitation from 750 to 3750 mm.

Significant areas are found on the Atherton Tablelands, at Kingaroy and Toowoomba in Queensland, in the Richmond–Tweed region of coastal New South Wales, Gippsland in eastern Victoria and near Burnie in northern Tasmania.

The soils are essentially developed from basalt and basic igneous rocks, although sometimes under tropical conditions they may have developed on a wider range of rocks, including the intermediate igneous parent materials such as diorites and basic sedimentary and metamorphic rocks and, in some places, ash and pyroclastic materials. They are invariably associated with well-drained sites on dissected basaltic plateaux, hilly uplands and undulating to level lowland plains.

Xanthozems are occasionally found in close association with krasnozems on lower sloping sites where the yellow subsoil is intermittently and partially saturated by seepage.

GENERAL DESCRIPTION

Typical krasnozems are red, strongly structured clays, with the clay content often exceeding 60 per cent throughout the profile. The texture either remains uniform with depth or increases gradually. Profile depth usually exceeds two metres and may reach seven metres.

Krasnozems have dark brown to dark reddish-brown surface horizons and become brighter in the subsoil, with a red to red-brown B horizon. With increasing depth the colour may become browner and possibly mottled and grey as the weathered basaltic parent rock is approached.

The clay loam to light clay surface horizon has a well-developed porous granular to sub-angular blocky structure which is friable when moist, but the peds become harder when dry.

Although the surface soils have a high clay content, many of them tend to behave and feel like loams or clay loams, particularly on initial field texturing. This condition is termed sub-plastic (Butler, 1955) and is due to the highly aggregated state of the clay, brought about by iron and aluminium oxides, giving an impression when texturing that there is a lower content of clay-sized material than actually is present. This characteristic is one of several used by Northcote *et al.* (1975) in grouping some representatives of the krasnozems into the category Uf 5.2.

The surface horizon overlies a well-structured, medium to heavy, plastic clay B horizon. The structure of the subsoil is marked by dense, interlocking polyhedral peds which are friable when moist. the subsoil peds usually have shiny surfaces or cutans, but some examples of the group have porous rough peds, particularly in krasnozems which occur in the temperate humid regions in Victoria and Tasmania. Such soils have been included in the Gn 4.1 and Gn 4.3 class of the Northcote classification.

Two further variants of the krasnozem have also been described. The lateritic krasnozem, which is a relict soil developed on the remnants of an old land surface, is characterised by an iron-rich concretionary horizon of variable depth in the profile. This grades into a light grey and red mottled heavy clay, which becomes paler with depth before passing into kaolinised parent material.

The second variant, which sometimes is found on bauxitic parent materials, is referred to as a 'snuffy krasnozem' which has an A horizon with a fine crumb structure and a friable and snuff-like, almost powdery consistence and may be water-repellent when dry.

FERTILITY CHARACTERISTICS

The strong weathering and leaching which has accompanied the formation of these soils is responsible for many of the physical and chemical characteristics of krasnozems.

Although these soils have a reputation for moderate to high fertility, loss of bases from the profile as a whole has been high, reaching as much as 90 per cent in the higher rainfall areas. The soils are therefore acid, and base unsaturated. Fortunately the surface horizon tends to retain more bases and may be only mildly acid (pH 6.0). However, pH values of 5.5 are common and the acidity remains the same or increases with depth to values as low as or even lower than pH 4.5.

Under rainforest or pastures the soils are considered to be moderately fertile, but they rapidly decline in general chemical fertility under conditions of continuous cultivation which lead to a rapid breakdown of the relatively high levels of organic matter in the surface soil and decline in organic phosphorus, exchangeable cations and available manganese (Colwell, 1958). Responses to phosphorus, nitrogen, sulfur, molybdenum and potassium have been obtained on these soils (McArthur, 1964).

The high content of sesquioxides in these soils, combined with their

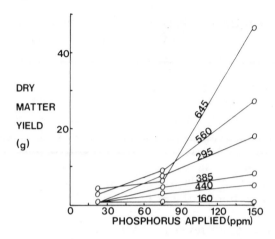

FIG. 4.3   Relationship between mean dry matter yield of tomatoes per pot and level of applied phosphorus on krasnozem soils of varying phosphorus status, 160 to 645 ppm, induced by previous superphosphate applications over a period of nine years (after Hughes and Searle, 1964)

relative acidity, results in the fixation of applied phosphorus. Consequently frequent additions of fertiliser are required.

Hughes and Searle (1964) examined the response of tomatoes to superphosphate applied to krasnozem soils which had a range of total phosphorus depending on the amount of superphosphate previously applied. The results indicated that the availability of the accumulated phosphorus was extremely low and that the increased yields on the high phosphorus soils were due to a better utilisation of the applied phosphorus as a result of reduced fixation in these soils (Fig. 4.3). This decrease in fixation was due to blocking of fixation sites by previously adsorbed phosphorus.

The well-developed structure of the surface soil is a consequence of the stabilisation of the kaolinitic clay by both sesquioxides and organic matter. The latter plays a less important role in structural stability as evidence suggests that loss of organic matter associated with cultivation does not adversely affect structural stability. However, at Redlands Bay, Queensland (S. A. Waring, pers. comm.), structure decline associated with a loss in organic matter has resulted in surface sealing and erosion on slopes, particularly when inter-row cultivation is practised. Despite the comparative stability of these soils, it would appear that under continuous cultivation elementary soil conservation measures are necessary on slopes to prevent erosion.

The available water in krasnozems (109 mm per 0.6 m) is somewhat less than for equally fine textured clay soils, such as chocolate soils (155 mm per 0.6 m) with which they are frequently intermingled. However, permeability is substantially higher and consequently water intake is efficient (McArthur, 1964). Soil erodibility is relatively low due both to this high intake and the stability of the surface aggregates, despite often prolonged and intense rainfall in the areas in which they occur.

LAND USE

The favourable physical properties and moderate fertility status of these soils make them ideal for crop production. The open, porous and stable structure allows rapid aeration and drainage following rain, excellent plant establishment, and ease of cultivation on all but the steepest slopes.

Under humid, subtropical conditions, the soils are used for sugar cane, maize and tropical fruit production. The peanut industry is largely located on these soils. In the Innisfail area, on highly weathered krasnozems, depletion of soil nutrients at one stage led to the abandonment of cane production, and these soils were only brought back into production by the application of factory filter waste, molasses, and large amounts of fertiliser (Monteith, 1966).

In the southern more temperate areas of Australia, the krasnozem soil is used for root crops, potatoes, pastures and winter cereals (Loveday and Farquhar, 1958). The lateritic and snuffy krasnozems which are usually low in fertility are not widely used for agriculture, but have been planted to exotic and indigenous softwood forests.

Sub-Humid Zone Soils

Euchrozems

The euchrozems are moderately fertile, red friable soils sporadically distributed in the sub-humid agricultural zone of eastern Australia. Major areas occur near Inverell (Hallsworth *et al.*, 1953a) and Wellington, New South Wales, and Toowoomba, Queensland (Thompson and Beckman, 1959). Other major occurrences are associated with tropical basaltic landscapes in northern Queensland (Isbell *et al.*, 1976), the Northern Territory and Western Australia, where the euchrozems are principally utilised for pastoral purposes.

Where rainfall is moderate (500 to 750 mm), wheat and other small grain crops are grown extensively, but under cool more humid conditions, horticultural crops such as apples, pears and grapes are frequently grown. Often these areas are of limited size, occurring as pockets of red soil amongst other soils in basaltic landscapes (see Fig. 4.2). Outcropping calcareous shales and limestones may produce a similar distribution of euchrozem soils and a distinctive land-use pattern.

Euchrozems are related to those soils which are dominated by sesquioxides and have red, strongly structured clays. They have morphological similarities to the krasnozems and differ mainly in their lower acidity, mild base unsaturation, and rather compact firm to friable subsoil consistence (Stace *et al.*, 1968).

They fall into several different subdivisions of principal profile forms according to Northcote. These include the non-cracking friable clays which are uniform in texture and distinctly pedal throughout the profile (Uf 6.31); soils with gradational profiles with red smooth-faced peds (Gn 3.12, 3.15); and less commonly duplex soils with friable A horizons and strongly pedal clay B horizons (Dr 4.1).

DISTRIBUTION
These soils were first described as a new soil group in New South Wales by Hallsworth *et al.* (1953a). They are developed in the sub-humid areas on strongly weathered basaltic saprolites associated with tertiary laterites or exposures of highly weathered red clays lying between successive volcanic flows. Consequently they may occur in any topographic position, but generally they are most strongly developed on the upper or mid-slope positions. As indicated, they are more prevalent in New South Wales and Queensland.

GENERAL DESCRIPTION
The surface soil is a dark reddish brown to dark brownish red clay loam to light clay with a weak to strongly developed granular to fine sub-angular blocky structure. The surface is porous and friable when moist.

The A horizon changes gradually to a brownish red to red heavy clay which has a strongly developed structure that is coarse, angular and blocky. The peds, which are smooth-faced, are coherent and the soil is compact with a firm to friable consistency when moist. Manganese stainings often form a coating on ped faces. With depth the soil grades into weathered rock or alternatively into a mottled sticky clay, but fresh rock (basalt) may occur as scattered 'floaters' or gravel in the profile.

Profile depth varies from 1 to 2 metres.

FERTILITY CHARACTERISTICS
As these soils are derived from a mixture of highly weathered and fresh rock material they vary from low to moderate fertility status. Soil reaction in the surface horizon is about pH 6.5, reaching pH 7.5 in the subsoil and may sometimes reach pH 8.0 at the base of the solum. They are deficient in available phosphorus and nitrogen.

They generally have a low capacity to store available water as a consequence of a predominance of non-expanding clays. Their surface structure, although relatively stable, does not permit a high infiltration rate and the soils are liable to sheet erosion under high intensity summer rainfall.

Crop yields are usually lower on these soils than on black earths with which they are associated.

LAND USE
These soils, which occur mainly in the summer rainfall areas, are used for cereal production (wheat and sorghum), summer and winter forage crops (eastern Darling Downs), and peanuts in some areas.

Yellow Earths

The yellow earths are scattered widely in the sub-humid and semi-arid environments under either a winter or a summer dominant rainfall. Until recently similar soils in more humid areas with structured subsoils (e.g. xanthozems) had been included in this group (Stace *et al.*, 1968). While the total area of these soils in arable regions is low, they have been utilised quite extensively in Western Australia for wheat production.

Yellow earths are soils which have a high content of sesquioxides (iron, aluminium and manganese oxides), and a yellow coloured subsoil. They are similar to the red earths in general profile characteristics, apart from colour, and sometimes a more pronounced texture gradient down the profile.

They are included in the yellow and yellow mottled massive earths (Gn 2.2, 2.6) of Northcote, and are described as soils with 'gradational texture profiles that are non-calcareous throughout and B horizons that have an earthy fabric and are whole-coloured yellow or mottled predominantly yellow'. This group includes soils with an A2 horizon which exhibits varying degrees of bleaching (Gn 2.24, 2.64, 2.7) but those soils with prominent bleached or pale A2 horizons would be seen as intergrades to the yellow podsolic soils (Stace *et al.*, 1968).

DISTRIBUTION

Yellow earths occur as a large discrete area in the north-central wheat belt of New South Wales near Coolah, and extend down into the low hilly sandstone country on the coast in the vicinity of Sydney, where they give way to podsolic soils with bleached A2 horizons. They are a significant component of the range of soils found on metamorphic rocks along the east coast north of Newcastle into Queensland.

A major area of this soil is located in southwestern Australia on landscape to which the term 'sand plain' or 'scrub plain' is applied. Other areas of the soil occur in northern and central Queensland (Coventry, 1982) and in the Northern Territory.

Yellow earths are often found in association with red earths; they develop on siliceous iron-rich parent materials on undulating plains, hill slopes and also on young colluvial-alluvial sediments derived from the erosion of older lateritic Miocene laterite surface or on the lower surfaces and pediments of lateritic colluvium and alluvium in broad shallow valleys (Bettany and Hingston, 1961).

GENERAL DESCRIPTION

The surface horizon of a yellow earth varies from a dark grey through to a yellowish-brown or yellowish-red, and from sand to clay loam, grading into an A2 or a B horizon. The pale A2 horizon, if present, is light grey-brown to light yellowish-brown or yellowish-red.

The B horizons are yellow to reddish-yellow in colour and sandy clay loam to medium clay in texture. The transition in colour and texture from the A1 through the A2 to the subsoil is gradational. The B horizon occurs at a depth of about 0.25 to 0.30 m with a total profile depth from 0.35 to 2 m.

Surface soil structure which ranges from massive hard setting to a weak granular or blocky structure, is porous and friable when moist, but tends to be compact and hard when dry. The subsoil is massive, porous and earthy with the development of a weak to moderate fine blocky structure, sometimes showing some clay skin formation on the peds.

At depth, dense clays with a blocky structure may occur, exhibiting varying degrees of mottling (yellow, light grey, red). The soils are usually acid to neutral in reaction, but calcium carbonate as nodules or soft accumulations may occur in the lower B horizons in some members of the group.

Ironstone nodules frequently occur throughout the profile, increasing in size and frequency with profile depth. The concentration of ironstone concretions may exceed 60 per cent in soils derived from lateritic material (Northcote *et al.*, 1975).

FERTILITY CHARACTERISTICS

A generally low level of fertility, with extremely low concentrations of soil phosphorus, exchangeable cations, organic matter and nitrogen (0.05 per cent), characterises these soils. They have a low cation exchange capacity and sometimes show an accumulation of exchangeable sodium in the subsoil due to accessions of cyclic salts and high water tables, particularly in the case of the Western Australian examples.

Deficiencies of trace elements (including copper, zinc, boron and molybdenum) and sulfur have been reported, particularly for those soils derived from lateritic parent materials — Table 4.5 (Bettany and Hingston, 1961; Gartrell and Glencross, 1968). Those with strongly acid subsoils have high levels of aluminium, causing reduced root growth and yield.

Moisture conservation by fallowing is successful with the finer textured members of the group, but the generally low nutrient status and low

TABLE 4.5   Wheat yield response to phosphorus, nitrogen and molybdenum on yellow earth soils (mean 1960 and 1962 trials), Gutha, Western Australia (after Doyle *et al.*, 1965)

Fertiliser application			
Superphosphate (kg/ha)	Ammonium sulfate (kg/ha)	Molybdenite (35% Mo) (kg/ha)	Grain yield (kg/ha)
168	—	—	921
168	—	0.14	987
168	126	—	1036
168	126	0.14	1271

water-holding capacity make them inherently infertile soils. However, under winter rainfall conditions, and with appropriate fertiliser applications and the incorporation of a legume-based pasture in the rotation, they have been successfully used for crop production (Callaghan and Millington, 1956).

LAND USE

The use of these soils is usually associated with larger areas of red earth and other soil types with which they are found. In Western Australia, where a minimum of 250 mm of precipitation is received during winter, these soils have been used successfully for the production of field crops, including cereals. Elsewhere, but particularly in the tropics, they are used for pastures, and cropping is confined to small areas with higher rainfall.

Red Earths

Red earths are a variable group of soils with a rather diverse range of characteristics which determine their utilisation and productivity. Significant areas in sub-humid climates are used for wheat production, while under wetter conditions the soils are utilised for tropical horticulture and forestry.

Under high rainfall conditions in Queensland coastal areas these soils have good agricultural potential. Organic matter has accumulated to a moderately high level along with nutrient ions, and structural stability has been enhanced. Such soils are used for vegetable and fruit production. In contrast, red earths found in the semi-arid 'hard red country' of New South Wales are marginal cereal growing soils susceptible to water stress, structural instability and erosion.

This diversity of properties, which is reflected in the cropping capability of the group, is a consequence of the relict nature of members of the group found in arid environments.

The red earths are included in the category of soils with predominantly sesquioxidic clay minerals and with profile characteristics which indicate that soil formation has been accompanied by some mobilisation of clay and segregation of the iron and manganese oxides.

Northcote includes them within the red massive earth group (Gn 2.1), being soils characterised by 'gradational texture profiles that are not calcareous throughout and with B horizons that have an earthy fabric and are whole coloured red'.

DISTRIBUTION

The soil group is widely distributed throughout Australia, extending in an arc from southeastern New South Wales through northern South Australia and the Northern Territory, to Geraldton in Western Australia. It also occurs on the eastern divide from northern New South Wales into northern Queensland (Coventry, 1982). Large discrete areas in the southern and central western part of New South Wales (Cobar peneplain) and also in Western Australia are used for crop production, particularly wheat.

The soils occur under a wide range of rainfall conditions (250 to 2500 mm annually), and are associated with diverse landforms, including recent or older stream levees, undulating plains, gently sloping pediments, hill slopes and remnants of old high-level surfaces. Parent material is variable, being generally siliceous and with a high iron content. Often the soils have developed on transported lateritic sediments.

GENERAL DESCRIPTION

Surface horizons of the red earth vary in texture from sandy loams through to silty clay loams, and structure gradually increases into the B horizons, which are sandy loams to medium clays.

The colour of the A1 horizon is dark grey-brown through to dark reddish brown to red, with a clear or gradual boundary to the contrasting strong red B horizon. Distinctive A2 horizons with bleaching are not characteristic of red earths, but some representatives of the group do have a light reddish brown pale A2 horizon between the darker surface horizon and the deep-coloured B horizon.

Surface horizon structure varies, with fine-textured soils tending to develop a weak crumb structure. With increasing sand content, crusted or hard-setting surfaces are more usual and if high contents of sand are present a loose structureless surface may occur.

The B horizons, which are massive, earthy and highly porous materials, are usually hard when dry but friable when moist and may have a weakly developed blocky structure. Despite the massive structure, subsoil drainage is moderate to rapid.

In the deeper subsoil there may be some development of smooth-faced peds (clay skins) with some mottling if drainage is impeded. This applies particularly to members of the group occurring on finer-textured parent materials under high rainfall.

Profiles, which are usually 1 to 2 metres in depth, may contain ironstone nodules, often in association with lesser amounts of manganese concretions at varying depths in the subsoil.

Red earths are usually non-calcareous, although isolated soft segregations, fine nodules or diffuse patches of calcium carbonate can be found in the deep subsoil. Such soils are intergrades to calcareous red earths (Stace *et al.*, 1968) and are referred to as the alkaline members of the red massive earths by Northcote *et al.* (1975).

Where the soil has developed on lateritic parent material, a horizon of ironstone nodules or, less frequently, of a cellular or reticulate sandy ironstone, may occur as a layer above a mottled kaolinised clay which grades into the parent material. Such soils are often referred to as lateritic red earths. In Western Australia, red-brown hardpan layers of cemented iron and silica may occur at a shallow depth.

The 'podsolised' red earth which is an intergrade between the red earth and the red podsolic soil has a pale A2 horizon and a B horizon showing a more marked increase in clay content.

Major areas of the red earths in sub-humid regions have surface soils which are mildly acid to neutral in reaction. The soil may become slightly more acid or alkaline at depth, depending on the extent of leaching. Under semi-arid conditions the subsoil pH may exceed 8.0.

Organic matter and soil nitrogen are low to moderate and, together with the low clay content (illite and kaolinite) in the surface horizon, give a low exchange capacity, approximately 20 meq per 100 g soil in a granite-derived soil (McGarity, 1975). Exchangeable magnesium is usually the dominant cation, with sodium rarely exceeding 10 per cent of the total exchangeable cations.

Potassium concentration is low to moderate, but available phosphorus is invariably low and as a consequence compound deficiencies in the major nutrients, nitrogen, phosphorus and to a lesser extent potassium, are common. Where the soils develop on highly weathered materials such as ironstone from laterite formation, deficiencies in trace elements such as copper, zinc, molybdenum and boron may be expected. Iron nodules also reduce water flow and soil volume, and increase wear on tillage machinery.

Where there is little organic matter, the surface horizon (although porous) may set hard during the dry season. Surface structure rapidly deteriorates under cultivation, resulting in the development of surface seals which make the soils liable to erosion. After rain, sealing may reduce water infiltration but it is generally not severe enough to prevent seedling emergence.

The porous earthy nature of the soil aids infiltration but, in those with sandier textures, water-holding capacity is low and as a result crop production depends largely upon growing-season rainfall. On the heavier textured types, the use of short summer and autumn fallow will conserve moisture for subsequent crop production.

LAND USE

Red earths in New South Wales and parts of Western Australia are used extensively for winter cereal production, in a legume-based rotation which maintains soil structure and soil nitrogen status. Superphosphate is required for successful crop production, and in Western Australia it has been found that the trace elements copper, zinc and molybdenum are normally required for successful wheat production on these soils (Gartrell and Glencross, 1968).

In Queensland, smaller areas of this soil group have been used for the production of sugar cane, peanuts near Kingaroy, tobacco near Mareeba, and general horticulture in the Gayndah—Mundabbera district.

Chocolate Soils

Chocolate soils are found on most basaltic landscapes under high or moderate rainfall, in association with either the krasnozem or black earth soil groups. In many parts of Australia the chocolate soils are not considered suitable for agriculture as they are restricted to steep slopes, knolls and plateau tops where the depth of soil development is minimal. However, under moderate rainfall in landscapes of subdued relief on the tablelands (e.g. Cooma, Orange and Guyra in New South Wales; Toowoomba in Queensland; Coleraine in Victoria) sufficient depth of soil accumulates to ensure successful crop or horticultural production.

The chocolate soils have moderate to high nutrient levels but, in the past, shallow depth and stoniness have generally precluded them from consideration and acceptance as worthwhile agricultural soils. Indeed it was only in 1952 that Hallsworth and co-workers recognised and established these soils as a distinct major soil group, thereby separating them from their previous classification as shallow lithosols or krasnozems.

Typical chocolate soils are shallow, well-drained, slightly acid, friable, brown clay loam to clay soils with weak to moderate horizon differentiation. Three subgroups (reddish, normal and grey chocolate soils) have been distinguished, chiefly on the colour of the A horizon (Hallsworth *et al.*, 1952).

Within the Northcote classification these soils are found among the classes of the shallow, uniform medium textured profiles with smooth-faced peds (Um 6.41); the non-cracking uniform textured friable clays with smooth-faced peds (Uf 6.32); and the gradational non-calcareous profiles, with a rough ped fabric and black in colour (Gn 4.4).

DISTRIBUTION

The soils, although widely distributed from Tasmania to northern Queensland, occur only on relatively scattered areas of basaltic and other basic rocks (diorite, diabase) under average annual rainfalls of 500 to 1500 mm. They are most widely distributed on the tablelands of New South Wales under moderate temperature and rainfall conditions.

Chocolate soils are usually found on the upper and mid-slope catenary positions in association with black earths on the lower slope position.

GENERAL DESCRIPTION

The normal chocolate soil is characterised by a dark brown, friable clay loam surface horizon which has a fine to medium, strongly developed crumb to sub-angular blocky structure. The A horizon grades into a reddish brown, brown or grey-brown clay loam to clay subsoil with a medium blocky structure. The subsoil is compact and firm when moist, but hard when dry. With depth, the clay becomes mottled, generally yellow-brown, due to floaters or pieces of decomposing parent rock, eventually merging into the massive weathered parent rock at from 0.5 to 1.25 m.

The reddish chocolate variant is distinguished by its colour, while the grey chocolate soil has a greyish chocolate, dark brown to almost black surface horizon overlying a yellowish chocolate light to medium clay subsoil. This latter variant may be less well drained.

Chocolate soils are normally slight to moderately acid in the surface horizon, but becomes less acid with depth. They have a high exchange capacity, associated with organic matter rather than the clay fraction which is mainly of the kaolinite-illite types.

General fertility levels are moderate, with adequate levels of phosphorus, organic matter and nitrogen, which can be as high as 1 per cent. Deficiencies in calcium, magnesium, potassium, phosphorus, sulfur and molybdenum have been reported (McArthur, 1964), although some representatives of the soil group can give satisfactory pasture production in the absence of added fertilisers.

As the dominant clay minerals are the kaolinites-illites rather than montmorillonites, there is little development of the cracking and self-mulching properties so characteristic of the black earths. Surface structure may deteriorate under increasing cultivation, causing difficulty in rewetting and in the establishment of pastures in old cultivation paddocks. Under favourable conditions of moisture and temperature and with molybdenised superphosphate, such surface structural problems have been overcome under pasture ley (McArthur, 1964).

In lower rainfall areas the available water capacity and consequent yield potential is restricted by shallow depth, but under higher rainfall the productivity may approach that of the black earths.

Chocolate soils are used mainly for grazing, but wheat, oats and forage crops are grown in some areas. Under cooler conditions, particularly in New South Wales and Tasmania, potatoes are a frequent annual crop, while excellent pome fruit orchards are often located on well-drained sites. Under tropical conditions, plantations of bananas and pineapples may be established on these soils on quite steep slopes in coastal areas.

Non-Calcic Brown Soils

Non-calcic brown soils, although of limited areal extent, are moderately widespread in the agricultural regions of Australia, particularly in wheat growing areas where they are often associated with the red brown earths which they resemble in both morphology and general land use characteristics. The essential morphological differences between the two soil groups are that the non-calcic brown soils are usually browner, shallower (0.5 to 0.75 m), and more leached soils, lacking free calcium carbonate in the solum.

They have been included in the mildly leached soils group by Stace *et al.* (1968), being distinguished within that grouping by a profile which is acid throughout or may approach neutrality in the subsoil and with a less marked texture contrast between the A and B horizons.

Northcote *et al.* (1975) include them amongst the duplex soils with hard-setting A horizons and with moderately to strongly developed pedal clay B horizons which may be red (Dr 2.11), mottled red (Dr 3.22) or brown (Db 1.11) in colour. The A2 horizon, if present, is paler than the A1, but not bleached. Isbell and Hubble (1967) classify them as neutral red duplex soils.

DISTRIBUTION

Although moderately widespread in the sub-humid regions, with a rainfall which is slightly higher than that associated with the red brown earths, the individual occurrences of the non-calcic brown soils are relatively limited. Areas of the soil occur in the southwest of Western Australia (Stephens, 1961), near Rutherglen (Victoria), central-west and northwest of New South Wales (McGarity, 1975), and parts of the subcoastal region of northeastern Queensland (Crack and Isbell, 1971).

The soils are developed from subacidic igneous, metamorphic or sedimentary rocks or the associated colluvium and occur on the mid- to upper-slope positions in rolling to hilly country of southern and eastern Australia. In Western Australia the soil is found on a variety of parent materials exposed by the dissection of the lateritic capped residuals. They are co-dominant with the lateritic soils in the area of the inner wheat belt.

GENERAL DESCRIPTION

The surface horizon, which varies in depth from 0.1 to 0.3 m, is a dark grey-brown to red-brown loamy sand to clay loam. It has a weak crumb to blocky structure, with a surface which tends to be massive and hard setting when dry. Typically there is no A2 horizon, but if present, it is paler in colour than the A1, and the transition from the A to the B horizon is clear.

A reddish brown to red, well-structured clay characterises the top of the B horizon. Clay skins are present on the peds, which are hard when dry but friable when moist. The B horizon grades into a variously mottled medium to light clay B-C horizon.

FERTILITY CHARACTERISTICS

Non-calcic brown soils are moderately fertile, with a mildly acid surface horizon remaining acid or becoming slightly alkaline in the subsoil. Soil nitrogen and organic matter levels are relatively low, although some members of the group (i.e. in Queensland) contain up to 4 per cent organic matter which gives them a darker surface horizon. They respond to phosphatic fertilisers.

The clays are dominated by the non-swelling kaolin mineral, resulting in a relative low exchange capacity and a base saturation of 50 to 80 per cent. Surface soil structure characteristics are similar to the red brown earths, and intense cultivation can lead to the breakdown of soil aggregates and to erosion problems. Some of these soils in Queensland have a low range of available water content which is considered to be their main limiting physical property (Stirk, 1957).

LAND USE

The utilisation of non-calcic brown soils is dependent on the associated soils. In southern and Western Australia they are mainly used for cereal production and their management is very similar to that of the red brown earths. In northeast Queensland they are used mainly for grazing.

Red Brown Earths

Red brown earths are important arable soils on which agriculture first became established in Australia. They have been associated extensively with wheat cultivation, and in some districts, particularly in South Australia, have been cultivated for more than 100 years.

The reason for their early and extensive utilisation was related to their ease of cultivation with horse-drawn implements, an initial fertility level which gave reasonable crop yields, and the ease of clearing the associated virgin savannah woodland.

They are distributed through the sub-humid areas generally referred to as the wheat belt, reaching their greatest extent in southern New South Wales, northern Victoria and South Australia.

While the red brown earths as a group tend to be thought of as winter rainfall, wheat growing soils, they are utilised almost as frequently in rotation for improved pastures. Fertility decline under long and continuous cultivation has necessitated ley rotations of legume pastures to restore chemical and structural fertility.

The term red brown earth has been used in the Australian classification schemes of Prescott (1931), Stephens (1962) and Stace *et al.* (1968). Initially it referred to soils with a brown, sandy-textured surface soil and a red clay subsoil with calcium carbonate in the deep subsoil, and included some soils with a strongly developed bleached A2 horizon. In the more recent proposals of Stace *et al.* (1968), soils with a bleached A2 horizon and a poorly structured clay subsoil have been separated and placed into the solodised solonetz and solodic soil groups. This is a more relevant classification for these particular soils, not only on morphological grounds but also in relation to the problems of agricultural management (McGarity, 1975).

Red brown earths are duplex soils which are characterised by a distinct texture contrast between the A and B horizons and which have a dominantly red coloured B horizon (Dr). Some soils with brown B horizons (Db) are included.

The relevant subdivisions in the Northcote classification include those soils which have hard-setting A horizons and moderately to strongly pedal clayey B horizons, exhibit an alkaline soil reaction trend down the profile and have either a pale or sporadically bleached A2 horizon or no A2 horizon (e.g. Dr 2.13, 2.23, 2.33, Db 1.13, 1.23, 1.33). Other members of the group have subsoil horizons which are mottled (Dr 3.23, Db 2.23).

DISTRIBUTION

Red brown earths are a widely distributed group in the sub-humid areas of southern and eastern Australia, extending into the tropics almost to Cape York Peninsula, and they are also found in Western Australia. They are associated with annual rainfalls of 380 to 660 mm in the temperate zone and up to 900 mm in the tropics.

The soils develop on a wide range of materials incuding alluvial and colluvial deposits, weathered sedimentary, metamorphic and subacidic to intermediate igneous rocks on plains, slopes and broad hill crests. They reach their greatest development in South Australia (north of Adelaide), the Riverina district of southern New South Wales and northern Victoria and near the Liverpool Plains in northern New South Wales.

In the Riverina, the soils are developed on a gently sloping plain formed from sediments of mixed stream and aeolian origin, resulting in a range of textures in the surface horizons of the soils.

GENERAL DESCRIPTION

Red brown earths are mildly leached soils, characterised by a grey-brown to red-brown loamy surface horizon which is weakly structured to massive, passing abruptly into a brown to red clay B horizon. The latter has a well-developed medium prismatic to blocky structure.

The surface horizon, which may be as thick as 0.5 m, varies in texture from loamy sand to clay loam. The thicker A horizons are associated with the sandy textured variants. The A1 horizon may grade into an A2 horizon which is paler in colour than the A1 or may be sporadically bleached.

The structure of the surface horizon may vary from weakly massive when the texture is sandy, through a weak blocky to a more strongly developed sub-angular blocky structure when the texture is a loam to clay loam. These structural conditions are particularly well developed under grassland, but under cultivation the structure deteriorates so that the surface becomes hard-setting and surface sealing can occur (see Table 4.2).

Red brown earths can be subdivided on the texture of the A horizon into clayey, loamy or sandy groups. This grouping has some agronomic relevance.

The transition from the surface horizons to the subsoil B horizon is usually abrupt and clearly defined. The B horizons vary in colour from a bright red to red-brown to the drabber brown colours, sometimes with yellowish brown and/or grey mottles. The subsoil has a well-developed blocky to prismatic structure with clay skins on the peds, which are hard when dry but firm to friable when moist.

With depth the colour changes into a mottled red-brown to greyish brown material with a lower clay content, moderate blocky structure and a more friable consistence when moist. The B horizon is from 0.25 to 0.5 m thick with a total profile depth from 0.75 to 1.3 m.

Typically, calcium carbonate is present as soft patches or hard concretions in the deep subsoil. When absent, the horizons are essentially base saturated, often with carbonates appearing in the C horizon.

Surface soils are typically slightly acid to neutral, grading into alkaline subsoils at depth.

Stace *et al.* (1968) indicate that several more-or-less distinct forms of red brown earth can be recognised. The soils in southern Australia are typified by a rather abrupt texture change into an illitic clay B horizon. This contrasts with northern Australia where the texture change is clear rather than abrupt, with a dominantly kaolinitic clay B horizon that is consequently friable when moist.

In both regions there are soils with distinct A2 horizons, exhibiting some bleaching, and with B horizons which contain appreciable amounts of exchangeable sodium and soluble salts. These soils intergrade with the solodic soils.

Where conditions are moister and drainage is slow, soils have darker surface horizons, a higher organic matter and better structure, overlying dull brown or grey-brown B horizons. These are considered to intergrade to the non-calcic brown and podsolic soils.

FERTILITY CHARACTERISTICS

Generally the red brown earths are considered to be inherently low to moderate in fertility status. Organic matter and nitrogen levels are low, with a concentration of 0.1 per cent nitrogen in the surface (0 to 15 cm) being common. The depletion of organic matter which takes place under cultivation on these soils has been widely reported (Clark and Marshall, 1947; Penman, 1949; Williams and Lipsett, 1961).

Intensive crop production also has a deleterious effect on soil structure, and serious losses of soil aggregation have occurred. The loss of structure, which is not necessarily associated with a change in organic matter, may result from the pulverising action of cultivation implements and the dispersion of bare soil surfaces by heavy rain (McIntyre, 1955). The associated decrease in porosity (see Table 4.2) may contribute to decreased seedling establishment (McIntyre, 1955; Millington, 1959).

The cation exchange capacity of the soil varies with the organic matter content and the amount and type of clay mineral (illite–kaolinite), but moderately high (30 meq per 100 g) values occur. Calcium and magnesium are the dominant cations, and sodium may reach values of 10 to 15 per cent of the total bases in the B horizon. The presence of excessive levels of exchangeable sodium and salt (e.g. 0.3 per cent sodium chloride) in the subsoil of some members of the group may result in a poorer structured B horizon and may also lead to salting problems if irrigation management is unsatisfactory.

The soils are deficient in phosphorus and respond well to fertiliser applications (Williams and Lipsett, 1961; Colwell, 1963). Potassium deficiencies rarely occur on these soils, but sulfur and molybdenum deficiencies, together with zinc on the more calcareous variants have been reported (McLachlan, 1952; Stephens and Donald, 1958; Northcote *et al.*, 1975).

The hard-setting surface characteristics, which develop particularly after cultivation, influence the permeability and water storage characteristics of the red brown earths. Surface sealing inhibits water infiltration and may pre-dispose the soil to serious water erosion, which is a major hazard in northern New South Wales where soils are under fallow during the period of severe summer rainstorms.

Fallowing for water conservation is practised on these soils, but is unsuccessful in the sandy textured variants if the clay B horizon is deeper than 0.3 m (French, 1963).

Soils which have a massive or strongly prismatic structure in the B horizon may suffer from temporary waterlogging, as those subsoils have a low permeability which restricts water movement and aeration.

LAND USE

These soils are extensively used for winter cereal production under a rotation with clover ley. The subterranean clover ley results in an increase in the soil nitrogen and a restoration of the soil physical conditions (Penman, 1949; French *et al.*, 1968; Taylor, 1971). Lucerne is advocated in areas with summer rainfall (Purchase *et al.*, 1949).

The resultant wheat crops outyield those obtained under a fallow wheat rotation (Table 4.6). Yield increases result from the improved mineralisation of organic nitrogen and the consequent increase in the amount of available nitrogen stored in the root zone (Barley and Naidu, 1964).

TABLE 4.6   Comparisons between wheat yield, protein content and soil nitrogen after a subterranean clover ley; each value is a mean of 3 years' observation (after Bath, 1949; Penman, 1949)

	Fallow wheat, after ley			Wheat continuously after ley		
Years after ley	Yield (tonnes/ha)	Protein (%)	Total soil N (%)	Yield (tonnes/ha)	Protein (%)	Total soil N (%)
1	—	—	—	2.33	10.1	0.135
2	2.84	11.1	0.130	3.07	10.3	0.129
3	—	—	—	2.95	9.8	0.118
4	2.56	10.3	0.120	2.54	9.7	0.120
5	—	—	—	2.53	9.7	0.112
6	2.33	9.6	0.104	2.42	9.4	0.100

Some red brown earths in the Brigalow country of Queensland are used for wheat and sorghum production under summer rainfall.

Where rainfall conditions are better or where irrigation is available, this soil is successfully used for pastures, summer fodder crops and cereals. Maize, millet, sorghum, soybeans and safflower are grown under irrigation in northern Victoria and southern New South Wales. The soils are liable to waterlogging and salinity difficulties if careful irrigation management practices are not adopted.

Under summer rainfall, adequate erosion control is essential if the soils are cropped. Mechanical measures which control run-off (e.g. banks, waterways and dams) are necessary on sloping land. Strip cropping may also reduce erosion hazards.

Black Earths

Black earths are considered to be outstanding agricultural soils, whether in the sub-humid areas of the wheat-growing zone or in more humid situations on tablelands or in coastal valleys. Moderate to high fertility due to accumulations of organic matter to depths of 0.6 m or more, combined with a capacity to trap and store moderate to high amounts of available water in the subsoil, are the principal characteristics which contribute to their desirable agricultural properties.

In general, initial crop yields are high and productivity is sustained. However, even on the renowned Darling Downs of south-eastern Queensland, yields have commenced to decline as a consequence of over 30 years of continuous crop production and the increasing influence of loss of soil by erosion.

The black earths are so named from the black colour of the wet soil to depths well beyond the plough layer. When dry they may appear dark grey or brown. They are included amongst the soils which have been mildly leached and have a dark surface horizon (Stace *et al.*, 1968).

They have been grouped with those soils which have a uniform, fine-textured profile (medium to heavy clays) and crack open periodically on drying (Ug). The profile shows pedologic organisation characterised by strong structure, smooth-faced peds, dark clay subsoils and the lack of an A2 horizon. They are usually classified as Ug 5.1 profiles.

DISTRIBUTION

The soils are associated with the more basic parent material, such as basalt, trachyte and calcareous mudstone, either forming directly on these rocks or on colluvium derived from them. These parent materials weather to produce clays and, at the same time, release large amounts of calcium and magnesium. Typically the soils are associated with undulating plains and rolling to low hilly landscapes in regions with 500 to 1000 mm annual rainfall.

The most important cropping areas of this soil group are on the Liverpool Plains and northwest slopes in New South Wales, and the

Darling Downs and Central Highlands of Queensland, which are in the summer rainfall area. There are smaller isolated areas in Victoria, South Australia and Western Australia.

Black earths occur in association with the red brown earth, grey, brown and red clays, and euchrozem soil groups in wheat-growing areas, and with chocolate soils in the higher rainfall areas.

GENERAL DESCRIPTION

Black earths are dark grey to very dark brown or black, medium to heavy clays (40 to 80 per cent clay) which vary in depth from about 0.25 to 2 m, depending upon the topographic position and extent of erosion and deposition.

The surface soil (20 to 50 mm thick) is very sticky when wet but dries to form a loose layer of granular aggregates characteristic of the self-mulching structure of this soil group. There is a clear change to a very dark brown to black clay subsurface horizon, which has a strongly developed sub-angular blocky structure. This changes to angular blocky and the structure becomes coarser and sometimes more massive with depth, generally coinciding with calcium carbonate accumulation. The profile is often designated AC because the B horizons cannot be distinguished.

Due to expansion and contraction of the predominantly montmorillonitic clays following wetting and drying, deep cracks penetrate into the subsoil. Water infiltration takes place initially down these cracks, and surface soil materials may also become mixed into subsoils by this process. The deeper structural units have polished surfaces or slickensides due to clay orientation, caused by the units expanding and moving against each other.

The shallower soils usually grade through a thin horizon of lighter brown or grey clay into the weathered, soft parent material, which frequently contains free lime. Deeper soils, which are associated with colluvial/alluvial material, grade into yellowish-brown to light grey-brown medium to heavy clay which may continue for several metres.

The surface horizon is generally neutral to alkaline, with soft patches or concretions of calcium carbonate appearing at depths of 0.4 to 0.5 m. Gypsum sometimes occurs in the deep subsoil, and in some areas the deep subsoil can be saline.

These soils may have gilgais, showing characteristic mound–depression micro-relief on deeper colluvial material.

On flat areas on the plains, circular mounds several metres in diameter occur in a repetitive pattern in association with the depressions. The mounds, which have resulted from an upheaval of the subsoil, are alkaline, and usually contain a scatter of carbonate concretions. The depressions are usually carbonate-free and neutral to slightly acid in reaction to depths of up to 0.5 m. On the gentler slopes the linear form of gilgai occurs, with depressions and mounds recurring as continuous strips downslope, separated by a wavelength of about 5 to 7 m. In many areas the gilgai phenomenon is an active and continuing process, but under continuous cultivation it tends to disappear.

FERTILITY CHARACTERISTICS

These soils are considered to be moderately fertile, with an adequate content of organic matter (average 8 per cent), particularly in the surface horizon, and a satisfactory nitrogen and phosphorus content in the virgin state. Since organic matter distribution is relatively high at depth, an excellent store of nutrients exists in the profile. However, continued cultivation has resulted in the loss of soil nitrogen and organic matter in the plough layer of the order of 50 per cent after cultivation for 30 years on the Darling Downs (Dalal, 1982), and 16 per cent after ten successive cereal crops in northern New South Wales (Hallsworth *et al.*, 1954).

Generally the soils have a high cation exchange capacity due to the presence of the clay minerals, montmorillonite and illite. The combination of a high cation exchange capacity, high clay content and low permeability reduces leaching and so maintains a satisfactory level of basic cations, particularly calcium and magnesium. In some areas of the Liverpool Plains in NSW, and southern Queensland, the magnesium greatly exceeds the calcium and an imbalance of plant uptake has resulted.

Phosphorus status varies with the parent material, but under continuous cultivation there is an increasing response by cereals to phosphate applications (Esdaile and Colwell, 1963). Other nutrient deficiencies have been associated with these soils. Zinc deficiency has occurred in linseed and legume crops and with summer and winter cereals, particularly after fallows (Duncan, 1967). In addition there have been recorded responses to sulfur and manganese (Stephens and Donald, 1958; Northcote *et al.*, 1975).

Lime-induced iron chlorosis can occur under waterlogged conditions in cotton and grain legumes in the Namoi Valley. This is attributed to excessive bicarbonate uptake rather than to any outright deficiency of iron (Hodgson, 1990).

The physical characteristics of these soils have important effects on their utilisation. The heavy texture and the stickiness of the soil when wet restricts the time during which cultivation is possible and can also affect seedling emergence during periods of high rainfall which causes temporary waterlogging.

Water-holding capacity is high, and summer fallowing is practised widely to ensure adequate soil moisture for the production of winter cereals (Fawcett *et al.*, 1974). Available soil moisture storage values of 120 to 150 mm have been measured on the Darling Downs on this soil type (Waring *et al.*, 1958a; 1958b).

Black earths are liable to water erosion, particularly where they occur on slopes with linear gilgai formations. Controlling the resultant gullies presents difficulties due to the crumbly nature of the subsoil (French *et al.*, 1968; Peasley, 1975). Extended pasture leys may be necessary to effect control and stabilisation of the situation.

LAND USE

Although these soils were readily recognised as being of high natural fertility, their utilisation for crop production was minimal because of the difficulty of cultivation. The development of modern farm machinery has been responsible for the widespread expansion of crop production on to these soils. Innovations in design have led to the development of cultivation equipment (e.g. sweep ploughs and rod weeders) which are particularly suitable for use on this soil type.

In summer rainfall areas, where summer fallow is practised, sheet and gully erosion can be severe when high intensity storms fall on soils which are shallow, lack protective cover or have been recharged with water and have a low infiltration. Stubble retention methods of farming are being increasingly accepted where hazards are high (Garland *et al.*, 1976). Some black earths have poor surface and subsoil structures which require amelioration with either gypsum or organic matter residues.

Black earths are used extensively for the production of the summer and winter cereals (sorghum and wheat), but there are also significant areas of linseed, sunflower, safflower, millet, cotton, maize and barley, the particular crop depending on climatic conditions. The use of summer fallow and phosphatic fertilisers and to a lesser extent nitrogen fertilisers is necessary for successful crop production.

Where adequate water is available, there is an increasing interest in the use of irrigation for summer grain crops, cotton and pastures.

ASSOCIATED SOILS

A soil which superficially resembles the black earths is the chernozem. The latter soil is dark coloured to depths of 0.6 m or more, neutral to alkaline, with calcium carbonate at depth but distinguished by a much lower clay content, finer and more porous structural units and being characteristically friable when dry and very friable when moist (Stace *et al.*, 1968). These soils do not exhibit a self-mulching surface or seasonal cracking. Chernozems are grouped with the friable loams with rough-faced peds (Um 6.11) and the gradational textured profiles with black, smooth-faced peds (Gn 3.4) (Northcote *et al.*, 1975).

The soils are restricted to small occurrences on basic parent materials and derived colluvium in sub-humid areas in eastern Australia. They are highly fertile and very productive, often without the use of fertilisers. They are used for cereal (maize and sorghum) production, fodder crops and vegetable production, often with irrigation.

Prairie Soils

Of the various soils grouped as mildly leached dark soils, the prairie soils are amongst those of major agricultural significance. They are related to, and often associated with the black earths and chernozems, but have formed under slightly stronger leaching conditions, generally in higher catenary positions.

Where they are associated with black earths, prairie soils are often utilised for similar crops. Since these deep fertile soils are more common in sub-humid areas, maize, lucerne, beans and other intensive cash crops are grown in summer rainfall areas, particularly with supplementary irrigation. Small grains and root crops are grown on these soils under winter rainfall.

The prairie soils are deep, neutral, dark clays to loams with the organic matter well distributed, often to depths of one metre or more. The sandy and silty members, which are generally found on more recently deposited river flats and terraces, are not as strongly structured as the darker, clayey black earth intergrades and hence have been referred to as minimal prairie soils (Walker, 1963).

Within the Northcote classification, prairie soils are included in the uniform, medium-textured, friable loams with some development of B horizons which have smooth-faced peds (Um 6.31, 6.41) and also the gradational textured profiles which are non-calcareous and have B horizons with a smooth ped fabric (Gn 3.4).

DISTRIBUTION

Prairie soils occur most commonly in coastal and subcoastal areas, particularly in eastern Australia and also under moist to sub-humid conditions on elevated tablelands and watersheds.

The soils are developed from a range of parent materials, including basalt, calcareous sandstones, shales and alluvium. They normally occur as relatively small pockets on the colluvium of foot slopes and fans in low–hilly to hilly terrain, and on alluvial terraces and levees on flood-plains in valley floors. Minimal prairie soils frequently occur on the levees which are relatively less weathered.

GENERAL DESCRIPTION

The surface horizon (A1) of a prairie soil typically exceeds 0.15 to 0.2 m in thickness and varies in colour from a dark grey-brown to brown or black. It has a moderately developed granular to angular blocky structure which is friable and crumbly when moist. The surface horizon may be subdivided into an A1 and A3, with the latter being lighter in colour due to less organic matter.

The A horizons grade into a dark yellow-grey, brown, yellow-brown or reddish brown B horizon which may show a gradual increase in texture, a moderately blocky structure and a firm to plastic consistence. In some soils, the clay B horizon has a more strongly developed blocky to prismatic structure. The B horizons are free of calcium carbonate and may grade into lighter-textured and more friable C horizon material.

Minimal prairie soils, which are usually developed on alluvium, do not exhibit the same degree of profile differentiation. Because of the original sedimentation, stratification may still be evident in the profile.

Fertility levels in the prairie soils are generally moderate to high, reflecting the basic nature of the parent material. The soils are slightly acid to neutral and generally have a satisfactory level of base saturation and nitrogen status in the natural state.

After long, intensive cultivation, responses to nitrogen, phosphorus and potassium fertilisers are likely. This may be attributed to the reservoir of nutrients contained in the organic matter which extends deep into the subsoil. Cutting and filling of these soils for irrigation, which may expose subsurface layers, often does not impair crop production, since the mineralisation of nutrients from the organic matter at depth is adequate for plant growth.

The higher contents of organic matter and expanding clays that are frequently associated in these soils provide structure characteristics which give good infiltration and permeability and allow rapid and extensive root proliferation. They are thus very suitable for irrigation.

LAND USE

Because of their general characteristics, prairie soils have been cultivated for a wide range of crops, often with irrigation when the soils are found on levees associated with major streams, in both coastal and inland valleys.

In the coastal areas, particularly in northern New South Wales and Queensland, the soils are used for maize and sugar cane production. Elsewhere they have been used for vegetable and horticultural crops (e.g. pecan nuts on the Gwydir River, New South Wales) and for forage production, particularly lucerne and oats on river flats over much of eastern Australia and Tasmania.

Prairie soils are probably the most versatile of all Australian soils, and their use is largely determined by climate and markets.

Semi-Arid Zone Soils

Solodised Solonetz and Solodic Soils

Soils of this group occur in both the sub-humid and semi-arid regions of Australia. Although generally regarded as being of somewhat poor to moderate productivity and only suitable for pastures, they are often intermingled with high-productivity cereal-growing soils. In such situations they are cropped, usually with variable and unpredictable results, as a consequence of their marked dependence on climatic conditions during the growing season.

In semi-arid, summer rainfall areas solodised solonetz soils are frequently subject to drought in the dry winter and to excessive waterlogging during the summer. Conversely, in northwest Victoria and South Australia the soils may remain waterlogged during winter due to temporary perched water tables in the solum.

The difficulty of management for crop production arises from the marked contrast in texture with depth. Sand usually overlies clay, and the interface between these horizons is generally sharp or abrupt, with the clay horizon being tough or even developing hardpan characteristics impeding water infiltration and root development. Within short periods during the growing season, the profile can swing from excessively dry to excessively wet and behave as would be expected of an arid or hydromorphic soil.

Despite the fact that these soils occupy approximately 24 per cent of the wheat belt and are considered amongst the poorest on which crop production is regularly undertaken, research on improved management techniques has been limited and inconclusive.

Although members of this group show a decrease in soil acidity with depth, thus placing them in the mildly leached soil category of Stace *et al.* (1968), they also exhibit many similar morphological and chemical characteristics to the soils within the category referred to as being strongly weathered and highly differentiated. This latter category includes the podsolic Great Soil Group.

These soils are distinguished from the podsolic soils by the strong texture differentiation across a sharp, wavy boundary between the distinctive bleached A2 horizon and a tough, dense, massive and impermeable B horizon. The A2 horizon may become almost fluid under prolonged wet conditions, so that in some districts the term 'spewy soil' is used by farmers to describe the group.

In the early classification systems (Prescott, 1931) these soils were usually grouped as variants of the red brown earths or podsolic soils. Hallsworth *et al.* (1953b) proposed their recognition as a separate soil group in view of their widespread distribution and their particular morphological and agronomic characteristics. They are closely related to soloths which are acid, leached soils of minor cropping usage found in higher rainfall areas, particularly in Queensland.

Northcote included them within the duplex soils, with hard-setting surface horizons and strongly pedal clayey B horizons which in the uppermost layer may be red (Dr 2), mottled dominantly red (Dr 3), brown (Db 1), mottled brown (Db 2), yellow (Dy 2), mottled yellow (Dy 3), black (Dd 1), mottled black (Dd 2) or mottled and dominantly grey (Dg 2). The soil reaction trend down the profile is generally alkaline and the A2 horizon is either sporadic (e.g. Dr 3.43, Dd 2.33) or conspicuously bleached (e.g. Db 1.43, Dg 2.43).

Also included in the group are the principal profile forms, which have a soft, loose and sandy A horizon overlying either a strongly pedal clay B horizon (Dy 5.43) or a massive to weakly pedal clayey B horizon (Dy 5.83).

DISTRIBUTION

The solodised solonetz and solodic soils are widely distributed over the semi-arid areas that have a winter rainfall and in the sub-humid tropics with a summer rainfall dominance. The average annual precipitation ranges from 250 to 1000 mm.

The soils, particularly the solodised solonetz, are distributed discontinuously in the belt extending from Queensland (Bundaberg–Gladstone and the western edge of the Darling Downs) through New South Wales (Warialda, Narrabri-Gunnedah, Pilliga, Peak Hill, Grenfell, the Northern Tablelands and the Hunter Valley), Victoria (Rutherglen and Horsham) and South Australia (Kybybolite and the Eyre and Yorke Peninsulas). In Western Australia, relatively large areas occur to the south of the Merredin and near Esperance. Solodic soils are widespread on the eastern side of the Great Dividing Range, in areas of lower rainfall and on geological formations of marine origin.

These soils are usually developed on fine to medium-textured colluvial and alluvial deposits associated with gently undulating riverine and coastal plains, valley floors and the mid and lower positions on slopes of low hills in a rolling to hilly topography. Although they may not dominate the soil pattern, they occupy a significant part of the soil toposequence. The quartz-rich parent materials include shales, slates, marine metasediments, felspathic sandstones, mudstones, granites and the associated sandy and clayey alluvium and colluvium.

GENERAL DESCRIPTION

The surface horizon, which varies in thickness from about 0.1 m in the finer textured soils to 0.5 m in the coarse sands, is differentiated into A1 and A2 horizons. The total depth of the solum varies from 0.75 to 1 m.

The A1 horizon, which is low in organic matter, ranges from brownish grey or light grey-brown through grey-brown to reddish brown in the more heavily textured members. The A2 is bleached, usually white or light grey, and varies from a thin capping (10 to 50 mm thick) above the clay B horizon in solodics to a horizon 0.3 to 0.4 m thick in solodised solonetz. Characteristically, small ferromanganiferous concretions are present immediately above the clay B horizon.

The structure of the surface horizon varies from apedal (single-grain or massive) to a weakly developed platy or granular structure. Hard-setting surfaces tend to develop on loam and clay loam A1 horizons when dry, although they may be firm to friable when moist.

The B horizons, which can be red, brown, black, yellow or grey, are frequently mottled, indicating temporary waterlogging. Texture varies from sandy clay to medium clay, and the soils are strongly structured. The structural units are usually very dense (bulk density values exceeding 2.0 mg m^{-3}), with a low permeability and a hard and tough consistence. The peds have smooth and shiny-faced surfaces.

In the solodised solonetz the structural units in the B horizon are arranged in coarse columns with clearly defined domes on top of the columns. The columns are typically polygonal in cross-section (0.1 to 0.25 m across) and may extend 0.3 to 0.5 m through the B horizon.

In contrast, the structure of the solodic soils is medium to coarse blocky or massive, with peds vertically orientated in a weak columnar arrangement. In both situations the surfaces of the blocky peds and faces in the columns are often stained dark brownish grey. Small black manganiferous concretions and spots may occur in the upper B horizons. In the solodised solonetz soils the bleached A2 horizon material may extend down the cracks between the columns.

The subsoil grades into a horizon of lower clay content which is usually weakly structured or massive and strongly alkaline in reaction. Soft segregations or concretions of calcium carbonate may occur in the B/C horizons which are sometimes moderately saline.

The soloth soils are a closely allied group occurring principally in the wetter climatic range. In general, they are distinguished by an acid reaction throughout the profile; otherwise their properties and agronomic behaviour are somewhat similar to the solodic and solodised solonetz soils. They may be considered intergrades between the latter and podsolic soils, which usually have a less sharply defined boundary between A and B horizons and also have a well-structured, friable B horizon.

For convenience, the soloths, which are used primarily for pastures, are included in this description of the solodised solonetz and solodic group.

FERTILITY CHARACTERISTICS

Crop production on both the solodised solonetz and solodic soils is limited by chemical infertility, waterlogging and droughtiness. The sandy solodised solonetz soils are amongst the poorest soils used for crop production in Australia. The more finely textured solodic soils are inherently superior, although by no means free of the problems which limit productivity in the group as a whole. On the Darling Downs in Queensland, the usefulness of these soils for cropping is determined primarily by the depth to the B horizon.

Some of the more heavily textured solodic soils have properties approaching those of the sandier variants of the red brown earths.

The surface horizon is low in organic matter (less than 1 per cent) and nitrogen (0.07 per cent). Soil reaction is usually acid to neutral in the surface and becomes gradually alkaline with profile depth, reaching pH 9.0 in some soils.

Cation exchange capacity, which is relatively low in the surface horizons (particularly in the coarser textured variants), increases with depth in the clay subsoil (10 to 30 meq per 100 g). The exchange sites are dominated by the base cations, with sodium and magnesium dominating in the subsoil. Exchangeable sodium may exceed 15 per cent of the exchange capacity, while sodium and magnesium combined may exceed 50 per cent.

The poor physical properties of these soils and their associated sodicity are responsible for many of their associated problems, such as seedling emergence, denitrification (McGarity and Myers, 1973), root aeration and penetration, and water intake and storage.

The high sodium content results in clay dispersion and the development of the tough, dense and impermeable B horizon. As a consequence, the impermeable B horizon creates ideal conditions for temporary saturation even though water permeability is adequate through the sandy textured A horizons of the solodised solonetz soils. Lateral drainage above the B horizon on slopes increases nutrient removal and prolongs downslope waterlogging.

A further consequence of this property is that fallowing for moisture storage is not successful on these soils, except with the more finely textured solodic soils that have shallower A horizons and more permeable B horizons.

Solodics contain less sodium, but the exchange complex is dominated by magnesium ions and has varying degrees of unsaturation. Although dispersion in the surface horizons is less, these soils nevertheless exhibit poorly structured subsoils which are difficult to improve.

The loamy-surfaced solodics and intergrades to associated groups such as red brown earths often have hard-setting surfaces when dry, due to the low organic matter, frequent wetting and drying, and low mesofaunal (earthworms, ants, termites, etc.) activity. Dryland farming exacerbates this condition (Hubble *et al.*, 1983).

These soils are liable to gully erosion in hilly landscapes. There have been many reports also of tunnel erosion, including in northeastern Victoria (Downes, 1949), the Hunter Valley (Monteith, 1954), the Riverina (Newman and Phillips, 1957) and the north coast of New South Wales (Charman, 1969, 1970). The tunnels, which subsequently collapse, form as a result of the dispersion and washing away of the sodic B and A2 horizons (Crouch *et al.*, 1986). Gully erosion often develops as a consequence (Crouch, 1976).

In some sub-humid areas, soluble salts have accumulated in the deeper weathered materials below the soil profile. Clearing and cultivation changes the water regime, and salt enters the root zone and affects crop production. Dryland salinity of this type is now widespread in many parts of eastern Australia. The soloths also exhibit nutrient deficiencies, particularly of phosphorus, sulfur, sometimes potassium and the trace elements molybdenum, zinc and copper.

LAND USE

Because of their low inherent fertility, many of these soils were not used in the developmental stages of Australian agriculture.

However, with adequate fertilisation, they are now being used more extensively for pasture and crop production under both dryland and irrigation.

The establishment of subterranean clover leys on these soils in the winter rainfall belt has improved the soil organic matter and nitrogen contents, enabling the establishment of a stable rotation involving winter cereals and pastures. A similar satisfactory and stable legume-based rotation system under a summer rainfall regime has not yet been developed.

In addition to winter cereal production in southern and western Australia, summer and winter cereals are produced successfully in central and northern New South Wales and Queensland (e.g. Callide Valley). Irrigated summer fodder crops and cereals are also produced on these soils in southern New South Wales and northern Victoria.

The use of gypsum to improve water infiltration and availability is recommended. Working gypsum into the top few centimetres of the seedbed flocculates the soil and aids seedling establishment. Gypsum applied in irrigation water acts similarly. Deeper working of the applied gypsum will improve subsoil infiltration, as does the use of the gypsum slotting technique (Jayawardane and Blackwell, 1985).

Soloths have similar uses to solodic soils where they occur in association. In better rainfall areas, pH neutral soloths are frequently used for horticulture (e.g. vineyards in the Barossa Valley, South Australia). In the coastal areas of Queensland these soils are used principally for improved pastures.

Grey, Brown and Red Clays

The grey, brown and red clay Great Soil Group characteristically occurs in the semi-arid environments of the inland, either on flood plains and terraces, or on downs-like landscapes developed on shales and mudstones. Grassland predominates over much of this soil group, but to a lesser extent scrubs of belah (*Casuarina cristata*), coolabah (*Eucalyptus coolabah*) and brigalow (*Acacia harpophylla*) are found locally in wetter sites and more humid districts.

Although they are moderately fertile, recognition of their cropping capability has had to await the development of modern technology. Crop production under both dryland agriculture and irrigation has accelerated remarkably since 1960. Much of this development has occurred in northern New South Wales and in the brigalow belt of southern and central Queensland. Under correct managements these soils are a resource as yet far from adequately exploited.

Although these soils are cracking clays (Ug), they vary considerably in their properties and yield potential by comparison with the black earths, with which group they could be confused.

The grey, brown and red clays are heavy-textured, uniform soils with a characteristic structural profile development; they include the soils previously classified as the grey and brown soils of heavy texture by Stephens (1962).

As a soil group the range of properties is so diverse that further investigation is necessary in order that new groups may be separated, possibly using properties other than colour for differentiation (Stace *et al.*, 1968). Nevertheless, no better criteria have been established, and the Third Approximation separates 6 sub-orders on the basis of colour in the upper 0.5 m of the profile (Isbell, 1993).

The soils are included in the subdivision of uniform, finely textured profiles which crack seasonally upon drying (Ug). They are strongly structured clays with dominantly smooth-faced peds throughout the profile, usually self-mulching in the surface horizon but sometimes capped with a thin (2 to 5 mm) fragile crust of laminated light clay or sandy clay. The particular soil classes included are those with a grey clay subsoil horizon (Ug 5.2), a brown or red clay subsoil horizon (Ug 5.3) or a small group with a grey clay subsoil horizon and a massive structured surface horizon (Ug 5.5).

DISTRIBUTION

These soils represent the largest area of cracking clays in Australia, extending from southeastern Australia in a broad arc to the Northern Territory, west of the Great Dividing Range. Two important areas of the soils are the Wimmera 'black' soils in Victoria and the extensive areas of grey and brown soils associated with the broad riverine plains of western New South Wales and northern Victoria. In Queensland, the soils occur in the brigalow belt which runs north of Goondiwindi and Inglewood to Roma and Claremont. One of the largest areas of these soils is associated with the Mitchell grass (*Astrebla* spp.) country in western Queensland.

Small areas of these soils are also found in South Australia along the Murray River, and in Western Australia, near Esperance.

The shallower variants of this group occur on the gently rolling uplands of central Queensland near Winton, on sedimentary rocks (mudstones, siltstones, calcareous and felspathic sandstone). The deeper types occur on the alluvial plains and are often associated with prior stream deposits. In this situation, the grey and brown members often form a complex pattern, with the grey and grey-brown members occupying the lower topographic positions and being subject to frequent flooding. The brown variants are better drained and are generally associated with the terraces of the prior streams.

Grey, brown and red clays occur in semi-arid areas with irregular rainfall (250 to 1000 mm), from the temperate winter rainfall zones to the hot monsoonal climate of the Kimberleys in Western Australia and the Victoria Downs in the Northern Territory.

GENERAL DESCRIPTION

The surface soil is usually a light to medium clay in texture overlying a dark grey, brown, red-brown to red heavy clay (45 to 80 per cent clay) B horizon which commences at a depth of 50 to 100 mm. This may grade into weathered sedimentary rock (shallower variants) or continue into deeper alluvium.

Typically, the structure of the surface horizon is strongly granular and self-mulching with sometimes a thin (50 mm) crust developing on the surface. This laminated crust, when dry, forms irregular polygonal plates 30 to 100 mm across.

The subsoil has a moderately well-developed sub-angular to angular blocky structure which becomes more coarsely angular blocky, lenticular and massive, with depth. The major structural units exhibit shear cutans, reflecting the seasonal cracking and shrinkage as a result of drying out of the clays. The cracking pattern greatly influences water recharge of the upper subsoil where massive blocky structural units bounded by large cracks may not rewet readily. Gravitational water flowing rapidly down the cracks may effectively bypass these blocky structures, as water entry would be limited by the low permeability of these units.

The surface horizon varies from acid to alkaline, and the soil usually becomes strongly alkaline with depth. However, there are variants which, although alkaline in the surface 0.3 to 0.5 m, have acid subsoils (brigalow forest areas) and others which are acid throughout the profile. Small quantities of calcium carbonate as concretions or soft aggregates, varying quantities of gypsum and moderate to high soluble salt contents are characteristic of the subsoils. In the arid areas, gypsum and salt occur even in the surface horizons, except in those soils which are flooded at irregular intervals.

Gilgais may be absent (as on large areas of the inland alluvial plain) or present, varying from small scattered circular mounds or 'puff' soils (1 to 3 m diameter) to the extreme melon-hole form which characterises the brigalow country. In the latter, the close pattern of closed circular depressions (3 to 20 m diameter) differs in vertical interval from the puff by as much as 2 metres.

A complex of grey and brown soils may occur in the gilgai formation, with the brown soil occupying the puff soil position while the grey soil occurs in the depression.

FERTILITY CHARACTERISTICS

Normally moderately fertile in the natural state, these soils reflect the nature of their parent material. They are usually lower in organic matter and nitrogen than the black earths and, as a result, the fertility status drops more quickly under a heavy cropping programme. Total nitrogen levels have decreased from 0.2 to 0.07 per cent after continuous farming in the Wimmera (Penman, 1949; Sims, 1953).

Phosphorus deficiency is widespread, particularly on soils developed on old alluvium, while responses to molybdenum, zinc, sulfur and manganese have been recorded. Zinc deficiency, in particular, occurs in situations where gilgai soils have been levelled, exposing the calcareous subsoil.

In the Wimmera, applications of zinc sulfate at 11 kg per hectare with superphosphate have increased yields by 13 per cent (Sims and Rooney, 1956). Toxic levels of boron have also been reported in this area.

Cation exchange capacity varies, depending on the clay minerals present. The exchange complex is usually calcium-saturated in the upper horizons, with sodium often exceeding 15 per cent of the exchange capacity in the subsoil, particularly in the gilgai variants.

Available water storage (<100 mm) is generally not as great as in deep black earths, due often to the presence of a higher content of non-expanding clays, a lower clay content, or poor subsoil structure which decreases infiltration. Nevertheless, with appropriate fallow systems, deep tillage and amelioration with gypsum, improved recharge of this storage capacity can greatly influence yield under dryland farming conditions.

The high sodim concentrations and the related dispersive effects influence the development of massive structures, particularly in the surface horizons where there is an absence of free lime. Such surfaces may be cracking, but massive between the cracks, a situation conducive to reduced seedling emergence.

Massive and compacted surface structure can be ameliorated by the use of gypsum in both dryland and irrigation farming. Studies in the Wimmera and more recently on the northwest plains of New South Wales have shown that the addition of gypsum at the rate of 2.5 tonnes per hectare can result in significant increases in yield of wheat, and that the effect of a single application can persist for several years (see Table 4.7, Doyle *et al.*, 1977).

Under irrigation the addition of gypsum, particularly in the irrigation water, has led to some stabilisation of the surface structure of these soils and a consequent increase in water infiltration (Davidson and Quirk, 1961). Loveday and Scotter (1966) examined the effect of gypsum applied in the irrigation water on subterranean clover seedling emergence, and noted that the gypsum treatment maintained a higher moisture content in the surface for longer periods after irrigation.

In other work (Loveday *et al.*, 1970) it was reported that gypsum applied at 22.5 tonnes per hectare improved the hydraulic conductivity and the aggregate stability of the surface horizon of these soils. There was a resultant increase in the amount of water entry into the subsoil, this

TABLE 4.7   Wheat response to an initial application of gypsum on crusty and massive grey clays (after Doyle *et al.*, 1977)

| Gypsum application (tonnes/ha) | Grain yield (tonnes/ha) | | | | | |
| | Clay loam | | | Heavy clay | | |
	1974	1975	1976	1974	1975	1976
0	0.4	1.0	0.7	0.7	1.5	a
1.25	0.5	1.6	1.0	1.4	1.5	a
2.50	0.9	2.1	1.1	1.7	2.1	a

[a]No crop due to drought.

moisture being available during peak soil moisture demands by the cotton crop.

The introduction of sufficient calcium, from gypsum, into the subsoil of some members of this soil group reduces the 'throttle' to water infiltration which develops during flooding after the surface soil is saturated. The zone of low permeability in the subsoil appears to be associated with a high clay content, low electrolyte concentration and appreciable levels of exchangeable sodium and magnesium. The added calcium increases permeability and allows water entry into the subsoil and the gradual movement of salt to the lower layers, provided there is no water table development (Loveday *et al.*, 1978). This effect is particularly important where only intermittent flooding for pastures and crops is possible, or dryland cropping is undertaken. Subsoil tillage, which breaks up compaction and massive structures, improves the effectiveness of the gypsum. Some of these soils are saline at depth and careful management is required to prevent salt accumulation in the root zone.

LAND USE

These soils were developed for dryland cereal production in the Wimmera, where fallowing was used to release soil nitrogen and to store water for satisfactory wheat production (Rooney *et al.*, 1966).

In the irrigation areas of New South Wales, they are successfully used for cereal production, particularly rice, with yields of between 6 and 8 tonnes per hectare being produced in rotations involving the use of subterranean clover leys (Anon., 1975). Phosphorus fertilisers are not necessary on these soils when used for rice production, as the waterlogged conditions apparently release adequate native soil phosphorus to satisfy the needs of the rice crop.

In the summer rainfall areas, these soils are satisfactory for the production of winter and summer cereals, usually under bare fallow or stubble mulch conditions. The northwest plains of New South Wales, an area of 1.2 million hectares extending from Narrabri to Walgett, to Boomi and to Boggabilla, is renowned for high-protein wheat production.

In central Queensland these soils have been generally avoided for dryland farming in favour of black earths. However, where the depth of the profile allows reasonable water storage and the topographic position on flatter areas is less conducive to erosion, these soils have been developed for wheat, sorghum and oilseed production in recent years.

Under irrigation, larger areas of these soils are now being utilised for the production of summer cereals, oilseeds and cotton, as at Emerald in Queensland and particularly along the Namoi and Gwydir Rivers, New South Wales. Cotton production in these areas demands intensive use of nitrogenous fertilisers.

Solonised Brown Soils

The solonised brown soils occur extensively in northern Victoria, south-western New South Wales, along the Murray Valley in South Australia and also in the wheat belt of Western Australia. The soils are found principally on those landscapes where mallee scrub vegetation (*Eucalyptus oleosa-dumosa*) is prominent. They are important agricultural soils of variable productivity, used intensively for irrigated horticulture along the Murray River, and elsewhere for dryland wheat and barley production.

Originally known as *mallisols* (Northcote, 1956) they have been more recently referred to as *calcareous earths* (Northcote *et al.*, 1975).

As initially conceived, the solonised brown soils include such a wide range of profile properties that modification and separation within the group on the basis of morphology and land-use capability is overdue. Apart from the calcareous members with a gradational texture profile, published surveys indicate that, in the past, soils which are non-calcareous and duplex (principally earthy sands and duplex red brown earths) have been included in the group.

The calcareous medium-textured soils separated by Northcote *et al.* (1975) are described as follows:

> These soils have gradational texture profiles that are calcareous throughout. In Gc 1.1, carbonates are clearly visible in the A horizons and/or the carbonate content exceeds 10 per cent by a depth of 30 cm, and the horizons of maximum carbonate concentration begin by this depth. In Gc 1.2, carbonates can only be detected in surface soils by testing with acid. The carbonate content does not exceed 10 per cent by a depth of 30 cm and the horizons of maximum carbonate concentration occur below 38 cm in depth. The Gc 2.2 soils have moderate to high pedality, whereas the others are apedal, and carbonates can be detected in surface soils only by testing with acid.

Sandy variants which are duplex (Dr) and non-calcareous have been described from the Mallee region of Victoria (Stace *et al.*, 1968).

DISTRIBUTION
This diverse group is widely distributed, occurring in Western Australia, the Yorke and Eyre Peninsulas in South Australia, the Murray Valley in New South Wales and Victoria and the central wheat belt of New South Wales around Condobolin–Rankin Springs. The soils are developed on coarse to medium-textured wind-blown sediments (*calcareous aeolianite*) that are unconsolidated, highly calcareous and usually saline. These sediments occur on undulating plains, old dunes and the lower slopes and floors of broad valleys.

GENERAL DESCRIPTION
Surface horizons vary from very dark greyish-brown through brown to red-brown, with a texture ranging from sands to clay loams. The sandy textures, which are normally structureless (apedal), loose and powdery when dry, show some tendency to develop a platy structure which can form a hard-setting surface in the more heavily textured variants.

Colour, texture and structure gradually grade into a weakly developed

B horizon. Here the colours are various shades of light brown, with a texture range from sandy loam to clay, but usually sandy. Subsoil structure is not well developed in these soils. In the more heavily textured soils, a blocky structure is found in the subsoil, which is usually earthy and porous, becoming massive as the lime content increases.

Both the surface and B horizons contain calcium carbonate, which may be present in the fine earth and as concretions. With depth, concretionary carbonate increases and may reach a maximum at about 0.3 m. The subsoil carbonate may be present as cemented concretionary nodules.

The C horizons of these soils are usually marked by a decrease in the carbonate content, compared with the B horizon, and may occur at depths in excess of 2 metres.

Where wind-blown sand has been deposited, as in dunes, deep sandy A horizons may overlie textural B horizons. Carbonate is low in the surface of such profiles, and a pale A2 may be present. These sandy members have lower pH, less organic matter and lower exchange capacity.

FERTILITY CHARACTERISTICS

The natural fertility of these soils is low to moderate, with low concentrations of organic matter and nitrogen (0.01–0.10 per cent N) characterising the surface (0 to 50 mm) horizon. The higher values are associated with the more finely textured soils.

Surface horizons are alkaline (occasionally neutral) with low soluble salt contents, but the subsoils are strongly alkaline with soluble salt contents in the range 0.2 to 0.6 per cent. Cation exchange capacity, which is moderate, particularly in loams and clay loams, is dominated by calcium, with sodium usually exceeding 15 per cent of the total exchangeable cations in the subsoil.

As a result of the salinity of these soils, production can be seriously affected when water tables lie near the surface. Salt scalds and soils with the powdery (snuffy) surfaces occur, particularly in Western Australia (Bettany and Hingston, 1961).

The soils respond to additions of superphosphate and nitrogen (see Table 4.8) and trace element deficiencies of copper, manganese, zinc and

TABLE 4.8   Response of wheat (tonnes/ha) to nitrogen and phosphorus fertiliser applications on a solonised brown soil (Gc 1.22), Walpeup, Victoria (after Colwell, 1977)

Phosphorus (kg P/ha)	Nitrogen (kg N/ha)			
	0	22.1	55.8	110.2
0	0.56	0.67	0.62	0.69
11.1	1.20	1.25	1.15	1.49
27.0	1.51	1.55	1.83	1.85
55.3	1.80	1.76	1.76	2.03

iron have been recorded in these soils, the latter particularly in citrus production on soils with a high marl content (Anon., 1962; Heard and Reuter, 1965). Potassium is in ample supply. Boron toxicity has been reported in South Australia (Cartwright *et al.*, 1991).

After clearing, the soils are easy to cultivate. Water penetration is good, even in the subsoil, where free lime prevents the development of unfavourable structural conditions due to the high sodium content. On the more finely textured soils, long-term (up to 10 months) fallowing for water storage is necessary for successful cereal production (Sims and Mann, 1957; French, 1968). Fallowing of the sandy mallee soils is not recommended due to their poor capacity to store water and the wind erosion risk. The sandy members sometimes exhibit water repellence due to waxy coatings on the sand grains which increases runoff and erosion (Hubble *et al.*, 1983).

Cultivation and long fallowing, combined with the fine sandy nature and poor structure of the surface horizons of these soils, make them liable to wind erosion. Consequently, careful management is necessary, and various legume rotations and the use of standing cereal stubble have been adopted to prevent soil drift (Guerin, 1966). Stubble retention rather than burning is necessary for continuous cereal rotations.

LAND USE

In southern Australia, where the annual rainfall exceeds 250 mm, solonised brown soils are used for the production of wheat and barley. A long fallow is necessary to conserve soil moisture but provided precautions are taken against wind erosion by the inclusion of a suitable rotation, usually based on lucerne and other medic leys, successful production can be maintained.

The soils are considered marginal for dryland crop production. Droughts and poor farming practices in the 1930s led to serious wind erosion and a decrease in soil fertility, with a consequent decline in the area under cultivation (Callaghan and Millington, 1956). With the improvement in technology and soil management this area has returned to cereal production, but is very much influenced by the market situation.

Under irrigation, the soils are usually free draining, and the sandier soils are frequently utilised for citrus. Stone fruits and grapes generally produce well on the heavier soils. Despite the high alkalinity, nutrient deficiencies are relatively easily corrected. Salinity problems can be serious due to lateral seepage at the base of slopes and where rising water tables result from uncontrolled irrigation.

Earthy Sands

The earthy sands are a relatively recently recognised soil group, the members of which were formerly included in the aeolian sands by Stephens (1962). While widely represented in the inland desert areas of

Australia, they are only of limited occurrence in rainfall areas where crops can be grown.

In Western Australia, the earthy sands are now utilised for cereal production, following the large-scale mechanised clearing of heath and sandplain in the higher rainfall coastal areas near Geraldton and on the margins of the wheat belt, east of Merredin. In general the soils have a naturally low fertility, but the widespread use of fertilisers, an assured winter rainfall and the adoption of legume ley has enabled moderately good yields and returns to be attained.

The earthy sands have been subject to minimal weathering, with a consequent lack of profile differentiation, except for an increase in organic matter in the surface horizon.

Northcote describes earthy sands (Uc 5.2) as

> uniform coarse-textured profiles that show weak pedologic development of slight colour and texture changes. They lack A2 horizons and structure development, except possibly in the immediate surface ... having coherent and porous subsoils which are characterised by earthy fabric.

DISTRIBUTION

These soils occur in a broad arc, from southwestern Australia across the Northern Territory into Queensland. The particular area of interest for crop production is in Western Australia where the soils are developed chiefly on colluvial sands. They are often found distributed on low lateritic plateaux in association with yellow earths, or in areas below and adjacent to breakaways at the edges of the plateaux.

GENERAL DESCRIPTION

These soils, which are mainly red and yellow sands to sandy loams with massive porous subsoils, occur as deep profiles on sand sheets and dunes. Shallower variants with profile depths less than 50 cm are found on siliceous rock and lateritic residuals.

The surface horizons are pale yellowish or reddish brown to dark brown sands, clayey sands or sandy loams. Structurally the surface horizons are weakly coherent or single-grained, becoming weakly massive with depth.

Subsoil horizons vary in colour from yellow through brown to red, with little change in texture. The coherence and earthy appearance of these soils is due to an even distribution throughout the profile of the clay and iron oxides which both coat the sand grains and form bridges between them.

With depth, the colour of the sand becomes paler or there is a change to the underlying rock or mottled kaolinitic or sesquioxidic horizons of the laterite formation. Soils developed on lateritic residuals frequently have ironstone gravel in the profile and occasionally at the surface.

The red earthy sands intergrade with the red earths and, where siliceous hardpans are present at shallow depths, with red and brown hardpan soils.

Earthy sands are usually acid in reaction throughout the profile and are extremely low in organic matter and nitrogen (0.03 per cent) and cation exchange capacity (less than 10 meq per 100 g). The exchange capacity is dependent on the organic matter content.

The soils are low in phosphorus, nitrogen and potassium and a range of trace elements, particularly molybdenum and copper. The highly weathered and sesquioxidic nature of these soils is responsible for outright phosphorus deficiencies, as well as unavailability due to the fixation of fertiliser phosphorus. Physically the soils are poorly structured and have low water retention properties.

Large areas of these soils are found in the desert area or provide sparse grazing but, as already indicated, areas of the soil in Western Australia are used for cereal production. Elsewhere limited use of these soils is made for a variety of crop production (e.g. horticulture) with supplementary irrigation.

Because of the nature of the soils and their low fertility the use of legumes such as subterranean clover and lupins in the rotation is widely practised, and it is necessary to apply regular fertiliser dressings. Gartrell and Glencross (1968) recommended applications of 220 kg per hectare of superphosphate containing 2.5 kg of copper sulfate, 0.75 kg of zinc oxide and 0.12 kg of molybdenum trioxide for successful cereal production.

The water infiltration characteristics of these soils are often adversely affected by the presence of water-repellent sands. As a consequence, water erosion can be a serious hazard, particularly when areas of the soil occur at the head of the watershed. Water storage capacity is also low, and consequently successful crop production depends entirely upon the growing season rain.

SOIL GROUPS AND THEIR LIMITATIONS TO CROP PRODUCTION

Soil is only one of the many interacting physical and environmental factors determining plant growth, and within the soil any one of a number of specific soil properties may set the ceiling on crop production. These edaphic properties vary in their relative importance between and within soil groups and thus make the accurate determination of potential yield virtually impossible from the field identification and limited laboratory measurement of these factors. However, in a generalised way, the properties which identify soils at the soil group level and which give an indication of the nature of the processes continually operating within the solum can be used to interpret the suitability of a soil for the production of a particular crop.

Thus loamy and fine-textured soils of the semi-arid regions frequently have hard-setting and massive structures which, if developed in the surface, lead to crusting and sealing under cultivation, with a consequent restriction in seedling emergence, or if in the subsurface, result in poor root development and penetration. The cause of the development of these structural conditions which exist in such diverse soils as the grey clays, red brown earths, and solodics, is due to restricted leaching and to consequent accumulation of higher contents of sodium in the clay fraction of the soil.

Shallowness, stoniness or pan development are properties which impede root extension and thereby effectively reduce availability of water and nutrients. These particular limitations may be present in members of the chocolate soils, earthy sands, yellow earths and podsolic groups. Acid reaction trends down the profile indicate depletion of bases and other nutrients, so that the red podsolics and krasnozems are groups with a predisposition to lowered production due to their relatively poor chemical fertility status.

Mottling and gleying are often associated with permanent or temporary water tables. Only crops adapted to conditions of periodically poor root aeration can be grown successfully on humic gleys and some grey clays which have these properties, unless the soils have been artificially drained.

Soil groups vary in their erodibility. The solonised brown soils are particularly susceptible to wind erosion, while the solodic soils and other soils with poorly structured A horizons are liable to erosion by water if there is insufficient vegetative cover. Consequently, limitations to crop production on these soils may be indirect in that the measures required to prevent soil loss restrict the intensity and kind of cropping that can be adopted.

Other attributes of the soil groups which also determine their usage and productivity include properties that influence tillage for seedbed preparation and weed control; infiltration and permeability properties which may need to be modified for irrigation and also the associated drainage; organic matter contents which affect biological and chemical fertility; and soil salinity which may be intensified under cultivation, particularly under irrigation.

The broad relationships that exist between soils identified at the soil group level and their suitability for cropping is therefore merely a first approximation. The effective management of soils requires a detailed knowledge of these identifying properties as well as others, and of the interactions of these properties when the soils are utilised for the production of particular crops and subjected to modern farming technology.

SOIL DEGRADATION AND SUSTAINABILITY

It would be remiss if this chapter did not conclude by discussing the continuing impact of cropping systems on soils and their productivity, and the long-term implications of soil degradation on the environment as a whole.

Over the past decade there has been heightened awareness of environmental issues, including accelerated soil erosion and deterioration of water resources, reduced yield and quality of farm products, fertiliser and agrichemical pollution, and harmful atmospheric emissions, all of which are, in total or part, directly related to the soil component of agricultural systems.

'Soil degradation' is the popular term which broadly depicts not only the most visually obvious scars on eroded and salinised landscapes, but all of those changes in soil properties which have led to reduced crop performance as virgin land adjusts to a new equilibrium under cultivation.

The decline in productivity in the years following the initiation of a cropping system is the predictable consequence of the change in the biological activity of the soil in response to a new set of environmental and management conditions. In general, the level of crop yields would be expected to stabilise, or show a reduced rate of decline for most soil–cropping systems well within one generation of farming. The cause of this soil fertility decline can generally be traced back to changes in the original properties of the soil, changes which vary greatly with the nature of the soil, climate, crop and management practices.

When this stabilised level of productivity for a particular soil–cropping system falls below economic expectations and viability, the system is considered to be unsustainable, at least economically, and more intensive alternative management strategies which may be even less sustainable in the long term may be put into place.

But sustainability has come to mean far more than the narrow focus of economic farm management. In the general context there is world-wide concern that if agricultural systems are not truly sustainable, the global environment may change to an extent that the quality of life will itself be seriously threatened. Sustainability for agriculture has been defined as "the successful management of resources for agriculture to satisfy changing human needs while maintaining or enhancing the natural resource base and avoiding environmental degradation" (Firebaugh, 1990). 'Soil' may be substituted for 'resources' in this definition thereby providing a suitable definition of soil sustainability.

Australian soils, when brought into cultivation, have a potential for rapid deterioration of many of the properties which govern fertility. The most obvious are those associated with the decline in organic matter, which is naturally low in many of the soils in climatically favourable agricultural regions. Nutrient levels (particularly nitrogen), cation exchange capacity, soil structure, drainage and available water, and biological activity are but a few of the properties which show a corresponding decline as organic matter is mineralised and lost from the system.

Depending on the Great Soil Group, the intensity of cultivation and the climate, crop yields may be unaffected over the first 10 to 30 years of utilisation, and economic response to fertilisers may take even longer. However, structure deterioration may be mechanically induced in a much shorter period unrelated to organic matter decline, again depending on management, soil and crop. As an example, intensively working grey clay soils for irrigated crops may lead to serious soil compaction within a few years, while the same soils under dryland farming may show little effect in the same period if minimum tillage and stubble conservation are practised.

Soil structure deterioration is of particular concern in Australia because it is associated with both wind and water erosion, which are serious causes of nutrient loss and environmental pollution.

It should be emphasised that the extent and nature of the changes in soil properties under a specified cropping system are controlled by the properties which characterise the Great Soil Groups described here. Thus black earths may respond quite differently to zero tillage compared with solodic soils. In addition, the response of soil to any cropping system will also depend on local factors such as topography and climate, and on the past and present management strategies. It is thus difficult to predict the rate and extent to which the productivity of soils will decline under a specific cropping strategy, and it is even more difficult to predict how this may be arrested and stabilised.

Where the removal of nutrients in harvested crops can be established as the major cause of reduction in yield, sustainability can be achieved by balancing these *outputs* of the system by *inputs* of fertilisers or organic residues and manures from an external source. An alternative strategy is to rely on fallow, green manure or pasture crop systems that remove nutrients from the total soil nutrient pool at a rate which matches the release by weathering of those nutrients from soil minerals and their acquisition from atmospheric sources. This low-input, low-productivity system is unlikely to be economically viable on highly weathered Australian soils once the accumulated store of original nutrients in surface soil which sustains the initial cropping is depleted.

It seems inevitable that, even in the most efficient low-input sustainable systems, there will be a need to rely on import of certain macro and micro nutrients to balance losses and natural deficiencies. In general, should the local economic factor prove favourable for low-productivity systems, the total area of cropped soil would need to expand many times to produce the same quantities of agricultural produce: an unrealistic scenario for countries dependent on export income, and for a world already short of agricultural land and with an ever-increasing population.

High fertility input – high productivity systems, even under good management, risk losing excess nutrients into the ground water and surface water, or into the atmosphere as damaging gaseous byproducts. Systems which are balanced in the amount of nutrients applied and removed, while at the same time maintaining the status quo of other edaphic properties, represent the ideal model. The difficulties of achieving such a system arise from the inherent unpredictability of the Australian climate, which affects crops and management.

Cropping systems which incorporate a pasture phase with animals, imported animal products (manures), or green manure rotations as sources of enrichment of organic matter, are not necessarily self-sustaining. Such systems require minerals and energy inputs other than nitrogen when the rate of export exceeds the inputs from mineral weathering, nitrogen fixation and added organic sources. For the present, fertiliser use will continue to be the basis of commercially sustainable cropping systems but cropping efficiency and control over potential environment hazards will depend on a better understanding of mineral cycling through research in *all* major agricultural soils. For the foreseeable future in Australia, as elsewhere, low input sustainable systems offer no threat to the manufacturers of fertilisers.

One of the most obvious signs of lack of sustainability is soil erosion. Reduction in yields at the immediate site where soil has been removed is but part of a process which eventually may extend and reduce yields elsewhere, particularly if infertile subsoil is deposited over a nearby productive soil. A discussion of measures to prevent and control erosion is beyond the scope of this chapter, but some of the soil groups (e.g. solonised brown soils) which have originated on parent materials derived from natural erosion processes, particularly wind, are those most susceptible to the ravages of accelerated erosion.

Increasing salinity and sodicity are further problems, both irrigated and dryland salting being major indicators of disequilibrium in landscapes. These areas must be brought under sustainable systems to prevent further deterioration. Dryland farming areas of Western Australia are in this category, but dryland salinity is also becoming a more significant feature of long-farmed agricultural areas in eastern Australia. The well-documented problems of irrigation salinity represent a continuing challenge of major proportions.

Other problems also exist, such as acid sulfate soils produced by lowered water tables following the drainage of wetland coastal areas, acidification produced by supposedly sustainable ley systems of farming, and the hardening of iron and silica pans due to desiccation in some sub-humid soils.

The widespread and increased manifestation of these problems in Australia has been a major factor in the heightened awareness and involvement of the community in such schemes as Landcare and Total Catchment Management. Research and extension in soils are essential prerequisites for gaining the knowledge necessary to develop sustainable systems; unless an uncompromising commitment to research and its application is maintained on a national scale, a permanent loss of productivity and deterioration of environmental quality seems inevitable.

References

Andrew, C. S. and Bryan, W. W. (1955), 'Pasture Studies on the Coastal Lowlands of Queensland: I. Introduction and Initial Plant Nutrient Studies', *Australian Journal of Agricultural Research,* **6,** 265.

Anon. (1962), *Guide Book,* Mallee Research Station Walpeup, Department of Agriculture, Victoria.

Anon. (1971), *Glossary of Soil Science Terms,* Soil Science Society of America, Wisconsin.

Anon. (1975), *Rural Industry in Australia,* Bureau of Agricultural Economics, Australian Government Publishing Service, Canberra.

Barley, K. P. and Naidu, N. A. (1964), 'The Performance of Three Australian Wheat Varieties and High Levels of Nitrogen Supply', *Australian Journal of Experimental Agriculture and Animal Husbandry,* **4,** 39.

Bath, J. G. (1949), 'The Improvement and Maintenance of Soil Fertility by Subterranean Clover Ley Rotations at Rutherglen Research Station', *Proceedings of the Specialist Conference on Agriculture,* Melbourne, HMSO, London, 448.

Beckman, G. G. and Reeve, R. (1972), 'Classification and Chemical Features of Soils of the Beenleigh-Brisbane Area, South-East Queensland', CSIRO Division of Soils Technical Paper No. 11.

Beckman, G. G., Crack, B. J. and Prebble, R. E. (1976), *Supplementary Glossary of Soil Science Terms as Used in Australia,* Australian Society of Soil Science, Publication No. 6.

Bettany, E. and Hingston, F. J. (1961), *The Soils and Land Use of the Merredin Area, Western Australia,* CSIRO Soils and Land Use Series No. 41.

Blackburn, G. and Wright, D. A. (1972), *Soil Survey of the Loxton Irrigation Area, South Australia,* CSIRO Soils and Land Use Series No. 53.

Butler, B. E. (1955), 'A System for the Description of Soil Structure and Consistence in the Field', *Journal of the Australian Institute of Agricultural Science,* **4,** 239.

Butler, B. E. and Hubble, G. D. (1977), 'Morphologic Properties' in J. S. Russell and E. L. Greacen (eds), *Soil Factors in Crop Production in a Semi-arid Environment,* University of Queensland Press, St Lucia, 9–32.

Callaghan, A. R. and Millington, A. J. (1956), *The Wheat Industry in Australia,* Angus and Robertson, Sydney.

Cartwright, B., Nable, R. *et al.* (1991), *Combating Toxic Boron in Australian Soils,* Soils Brief No. 15, CSIRO Division of Soils, 1.

Charman, P. E. V. (1969), 'The influence of Sodium Salts on Soils with Reference to Tunnel Erosion in Coastal Areas: Part I, Kempsey Area', *Journal of Soil Conservation Service, N.S.W.,* **25,** 327.

Charman, P. E. V. (1970), 'The Influence of Sodium Salts on Soils with Reference to Tunnel Erosion in Coastal Areas: Part II, Grafton Area', *Journal of Soil Conservation Service, N.S.W.,* **26,** 71.

Clarke, G. B. and Marshall, T. J. (1947), 'The Influence of Cultivation on Soil Structure and its Assessment in Soils of Variable Mechanical Composition', *Journal of the Council for Scientific and Industrial Research Australia,* **20,** 162.

Colwell, J. D. (1958), 'Observations on the Pedology and Fertility of Some Krasnozems in Northern New South Wales', *Journal of Soil Science,* 9, 46.

Colwell, J. D. (1963), 'The Effect of Fertilizers and Season on the Yield and Composition of Wheat in Southern New South Wales', *Australian Journal of Experimental Agriculture and Animal Husbandry,* 3, 51.

Colwell, J. D. (1977), *National Soil Fertility Project,* Vol. 1. *Objectives and Procedures,* CSIRO Division of Soils in collaboration with state Departments of Agriculture and the fertilizer industry.

Coventry, R. J. (1982), 'The Distribution of Red, Yellow and Grey Earths in the Torrens Creek Area, Central North Queensland', *Australian Journal of Soil Research,* 20, 1.

Crack, B. J. and Isbell, R. F. (1971), 'Studies on Some Neutral Red Duplex Soils (Dr 2.12) in North-Eastern Queensland: 1. Morphological and chemical characteristics', *Australian Journal of Experimental Agriculture and Animal Husbandry,* 11, 328.

Crouch, R. J. (1976), 'Field Tunnel Erosion—A review', *Journal of the Soil Conservation Service, N.S.W.,* 32, 98.

Crouch, R. J., McGarity, J. W. and Storrier, R. R. (1986), 'Tunnel Formation Processes in the Riverina Area of N.S.W., Australia', *Earth Surface Processes and Landforms,* 11, 157.

Dalal, R. C. (1982), 'Changes in Soil Properties under Continuous Cropping' in *Queensland Wheat Research Institute Biennial Report 1980–82,* 58.

Davidson, J. L. and Quirk, J. P. (1961), 'The Influence of Dissolved Gypsum on Pasture Establishment on Irrigated Sodic Clays', *Australian Journal of Agricultural Research,* 12, 100.

Downes, R. G. (1949), *Soil, Land Use and Erosion Survey Around Dookie, Victoria.* CSIRO Bulletin No. 243.

Doyle, R. J., Parkins, R. J., Smith, J. A. C. and Gartrell, J. W. (1965), 'Molybdenum Increases Cereal Yields on Wheat Scrubplain', *Journal of Agriculture, Western Australia,* 6, 699.

Doyle, A. D., Taylor, D. W., Yates, W. J., So, H. B. and McGarity, J. W. (1979), 'Amelioration of Structurally Unstable Grey Soils in the Northern-Western Wheat Belt of N.S.W.', *Australian Journal of Experimental Agriculture and Animal Husbandry* 19, 590.

Duncan, O. W. (1967), 'Correction of Zinc Deficiency in Wheat on the Darling Downs, Queensland', *Queensland Journal of Agriculture and Animal Science,* 24, 287.

Esdaile, R. J. and Colwell, J. D. (1963), 'Assessing Economic Rates of Phosphatic Fertilizer Application for Wheat in Northern N.S.W. by Soil Analysis', *Agricultural Gazette, N.S.W.,* 74, 282.

FAO (1974), *Guidelines for Soil Description,* Food and Agriculture Organisation, Rome.

Fawcett, R. G., Gidley, V. N. and Doyle, A. D. (1974), 'Fallow Moisture and Wheat Yields in the Northwest', *Agricultural Gazette, N.S.W.,* 87, 28.

Firebaugh, F. M. (1990), 'Sustainable Agricultural Systems: A Concluding View', in C. A. Edwards, R. Lal, I. P. Madden, R. H. Miller and G. Hause (eds), *Sustainable Agricultural Systems,* Soil and Water Conservation Society, Amkeny, Iowa, 674–676.

French, R. J. (1963), 'Facts about Fallowing', *Journal of the Department of Agriculture, South Australia,* 67, 42.

French, R. J., Matheson, W. E. and Clarke, A. L. (1968), *Soils and Agriculture of the Northern and Yorke Peninsula Regions of South Australia,* Department of Agriculture, South Australia, Special Bulletin No. 1, 68.

Garland, P. J., Wallens, P. G. and Fenton, G. (1976), 'Surface Stubble Retention Pays Dividends', *Agricultural Gazette, N.S.W.,* **87**, 17.

Gartrell, J. W. and Glencross, R. N. (1968), 'Copper, Zinc and Molybdenum Fertilizers for New Land Crops and Pastures in 1969', *Journal of Agriculture, Western Australia,* **9**, 517.

Guerin, P. D. (1966), 'Stabilizing and Managing Mallee Sandhills', *Journal of Agriculture, South Australia,* **69**, 305.

Hallsworth, E. G. (1977), 'Making the Most of Australia', *Journal of the Australian Institute of Agricultural Science,* **42**, 167.

Hallsworth, E. G., Costin, A. B., Gibbons, F. R. and Robertson, Gwen K. (1952), 'Studies in Pedogenesis in N.S.W.: II. The Chocolate Soils', *Journal of Soil Science,* **3**, 103.

Hallsworth, E. G., Colwell, J. D. and Gibbons, F. R. (1953a), 'Studies in Pedogenesis in New South Wales: V. The Euchrozems', *Australian Journal of Agricultural Research,* **4**, 305.

Hallsworth, E. G., Costin, A. B. and Gibbons, F. R. (1953b), 'Studies in Pedogenesis in New South Wales: VI. 'On the Classification of Soils Showing Features of Podzol Morphology', *Journal of Soil Science,* **4**, 241.

Hallsworth, E. G., Gibbons, F. R. and Lemerle, T. H. (1954), 'The Nutrient Status and Cultivation Practices of the Soils of the North-West Wheat Belt of New South Wales', *Australian Journal of Agricultural Research,* **5**, 422.

Hallsworth, E. G., Robertson, Gwen K. and Gibbons, F. R. (1955), 'Studies in Pedogenesis in New South Wales: VII. The Gilgai Soils', *Journal of Soil Science,* **6**, 1.

Hawkins, C. A. and Driver, R. C. (1978), *Soil Survey — Rylstone,* Department of Agriculture, N.S.W., Soil Survey Bulletin No. 5.

Heard, T. G. and Reuter, D. J. (1965), 'Manganese Spraying—Progress on Southern Yorke Peninsula', *Journal of Agriculture, South Australia,* **68**, 252.

Hodgson, A. S. (1990), Micronutrients—Are They Important Under Water-logging? *Proceedings of the Australian Cotton Conference,* Gold Coast, Australian Cottongrowers Research Association, August 1990, 165–170.

Hubble, G. D., Isbell, R. F. and Northcote, K. H. (1983), 'Features of Australian Soils', in *Soils: An Australian Viewpoint.* CSIRO–Academic Press, 17–48.

Hughes, J. D. and Searle, P. G. E. (1964), 'Observations on the Residual Value of Accumulated Phosphorus in a Red Loam', *Australian Journal of Agricultural Research,* **15**, 377.

Isbell, R. F. (1993), *A Classification System for Australian Soils (Third Approximation),* Technical Report 3/1993, CSIRO Division of Soils, pp. 1–80 (unpublished).

Isbell, R. F. and Hubble, G. D. (1967), *Soils of the Fitzroy Region, Queensland,* Resources Series, Department of National Development, Geographic Section, Canberra.

Isbell, R. F., Stephenson, P. J., Murtha, C. G. and Gillman, G. P. (1976), *Red Basaltic Soils in Queensland,* CSIRO Division of Soils Technical Paper No. 28.

Jayawardane, N. S. and Blackwell, J. (1985), 'The Effects of Gypsum-Enriched

Slots on Moisture Movement and Aeration in an Irrigated Swelling Clay',
    Australian Journal of Soil Research, **23**, 481.
Loveday, J. and Farquhar, R. N. (1958), *The Soils and Some Aspects of Land
    Use in the Burnie, Table Cape, and Surrounding Districts, North-West
    Tasmania,* CSIRO Soils and Land Use Series No. 26.
Loveday, J. and Scotter, D. R. (1966), 'Emergence Response of Subterranean
    Clover to Dissolved Gypsum in Relation to Soil Properties and Evaporative
    Conditions', *Australian Journal of Soil Research,* **4**, 55.
Loveday, J., Saunt, J. F., Fleming, P. M. and Muirhead, W. A. (1970), 'Soil and
    Cotton Responses to Tillage and Ameliorant Treatments in a Brown Clay
    Soil: I. Soil Responses and Water Use', *Australian Journal of Experimental
    Agriculture and Animal Husbandry,* **10**, 313.
Loveday, J., Watson, C. L. and McIntyre, D. S. (1978), 'Aspects of the
    Hydrology of Riverine Plain Soils', Proceedings of symposium, *Hydrogeology
    of the Riverine Plain of South-East Australia,* Griffith, N.S.W., July 1977,
    Australian Society of Soil Science, Riverina Branch, 99.
Mazloumi, H., McGarity, J. W. and Hoult, E. H. (1986), 'Effect of Gypsum and
    Tillage Treatments on the Yield and Water Use of Sunflower on a Grey Clay
    Soil', in *Proceedings of the Australian Sunflower Association,* **6**, 57.
McArthur, W. M. (1964), *Soils and Land Use in the Dorrigo-Ebor-Tyringham
    Area, New South Wales,* CSIRO Soils and Land Use Series No. 46.
McDonald, R. C. and Isbell, R. F. (1984), 'Soil Profile', in R. C. McDonald,
    R. F. Isbell, J. G. Speight, J. Walker and M. S. Hopkins (eds), *Australian
    Soil and Land Survey: Field Handbook,* Inkata, Melbourne, 83.
McGarity, J. W. (1975), 'Soils of the Australian Wheat Growing Area' in A.
    Lazenby and E. M. Matheson (eds), *Australian Field Crops I. Wheat and
    Other Temperate Cereals,* Angus and Robertson, Sydney, 227–55.
McGarity, J. W. and Myers, R. J. K. (1973), 'Seasonal Trends in the Content of
    Mineral Nitrogen in Solodized-Solonetz Soils', *Australian Journal of Experi-
    mental Agriculture and Animal Husbandry,* **13**, 423.
McIntyre, D. S. (1955), 'Effect of Soil Structure on Wheat Germination in a Red
    Brown Earth', *Australian Journal of Agricultural Research,* **6**, 797.
McKenzie, N. J. (1991), *A Strategy for Coordinating Soil Survey and Land
    Evaluation in Australia,* CSIRO Division of Soils, 1–52, Divisional Report
    No. 114.
McLachlan, K. D. (1952), 'The Occurrence of Sulphur Deficiency on a Soil of
    Adequate Phosphorus Status', *Australian Journal of Agricultural Research,*
    3, 125.
Miller, F. P. and Larson, W. E. (1990), 'Lower Input Effects on Soil Productivity
    and Nutrient Cycling', in *Sustainable Agricultural Systems,* Soil and Water
    Conservation Society, Amkeny, Iowa, 549–568.
Millington, R. J. (1959), 'Establishment of Wheat in Relation to the Apparent
    Density of the Surface Soil', *Australian Journal of Agricultural Research,* **10**,
    487.
Monteith, N. H. (1954), 'Problems of Some Hunter Valley Soils', *Journal of the
    Soil Conservation Service, N.S.W.,* **10**, 127.
Monteith, N. H. (1966), 'Preliminary Soil Mineralogy Studies on Krasnozems in
    the Innisfail District of North Queensland, Australia', *Pacific Science,* **20**,
    374.
Newman, J. C. and Phillips, J. R. H. (1957), 'Tunnel Erosion in the Riverina',
    Journal of the Soil Conservation Service of N.S.W., **13**, 159.

Nicolls, K. D. and Tucker, B. M. (1956), *Pedology and Chemistry of the Basaltic Soils of the Lismore District, N.S.W.,* CSIRO Soil Publication No. 7.

Nix, H. A. (1974), 'Land Use Planning for Commercial Agriculture' in *Proceedings of a Symposium on Land Use Planning in North Queensland,* Australian Institute of Agricultural Science, Canberra.

Northcote, K. H. (1956), 'The Solonized Brown (Mallee) Soil Group of South-Eastern Australia' in *Transactions of the VI International Congress of Soil Science,* Vol. E, 9.

Northcote, K. H. (1971), *A Factual Key for the Recognition of Australian Soils,* 3rd edn, CSIRO Division of Soils, Rellim, Adelaide.

Nortchote, K. H. *et al.* (1960−8), *Atlas of Australian Soils,* Sheets 1−10, with explanatory booklets, CSIRO Division of Soils, Melbourne University Press.

Nortchote, K. H. and Skene, J. K. M. (1972), *Australian Soils with Saline and Sodic Properties,* CSIRO Soil Publication No. 27.

Northcote, K. H., Hubble, G. D., Isbell, R. F., Thompson, C. H. and Bettany, E. (1975), *A Description of Australian Soils,* CSIRO Division of Soils, Rellim, Adelaide.

Peasley, B. A. (1975), 'Soil Conservation in the Narrabri Area', *Journal of the Soil Conservation Service of N.S.W.,* **31**, 9.

Penman, F. (1949), 'Effects on Victorian Soils of Various Crop Rotation Systems, with Particular Reference to Changes in Nitrogen and Organic Matter under Cereal Cultivation', *Proceedings of the Specialist Conference on Agriculture,* Melbourne, HMSO, London, 457.

Prescott, J. A. (1931), *The Soils of Australia in Relation to Vegetation and Climate,* CSIRO Bulletin No. 52.

Purchase, H. F., Vincent, J. M. and Ward, L. M. (1949), 'The Contribution of Legumes to Soil Nitrogen Economy in New South Wales', *Journal of the Australian Institute of Agricultural Science,* **15**, 112.

Rooney, D. R., Sims, H. J. and Tuohey, C. L. (1966), 'Cultivation Trials in the Wimmera', *Journal of Agriculture,* Victorian Department of Agriculture, **64**, 403.

Schafer, B. M. and McGarity, J. W. (1980), 'Genesis of Red and Dark Brown Soils on Basaltic Parent Materials near Armidale, N.S.W., Australia', *Geoderma,* **23**, 31.

Sims, H. J. (1953), 'Some Aspects of Soil Fertility in the Cereal Areas of Victoria', *Journal of the Australian Institute of Agricultural Science,* **19**, 89.

Sims, H. J. and Rooney, D. R. (1956), 'Wheat Fertilizer Tests in the Wimmera', *Journal of Agriculture,* Victorian Department of Agriculture, **54**, 499.

Sims, H. J. and Mann, A. P. (1957), 'Fallowing and Moisture Conservation in a Mallee Soil', Second Australian Conference on Soil Science, Melbourne, Paper No. 47.

Soil Survey Staff (1951), *Soil Survey Manual,* Handbook No. 18, U.S. Department of Agriculture, Washington D.C.

Soil Survey Staff (1975), *Soil Taxonomy: A Basic System of Soil Classification for Making and Interpreting Soil Surveys,* Handbook No. 436, U.S. Department of Agriculture, Washington D.C.

Spain, A. V., Isbell, R. F. and Probort, M. E. (1983), 'Soil Organic Matter', in *Soils: An Australian Viewpoint.* CSIRO−Academic Press, 551−564.

Stace, H. C. T., Hubble, G. D., Brewer, R., Northcote, K. H., Sleeman, J. R., Mulcahy, M. J. and Hallsworth, E. G. (1968), *A Handbook of Australian Soils,* Rellim, Adelaide.

Stephens, C. G. (1961), *The Soil Landscapes of Australia,* CSIRO Soil Publication No. 18.

Stephens, C. G. (1962), *A Manual of Australian Soils,* 3rd edn, CSIRO, Melbourne.

Stephens, C. G. and Donald, C. M. (1958), 'Australian Soils and their Responses to Fertilizers', *Advances in Agronomy,* **10,** 167.

Stirk, G. B. (1957), *Physical Properties of the Soils of the Lower Burdekin Valley, North Queensland,* CSIRO Division of Soils, Divisional Report 1/57.

Taylor, A. C. (1971), 'Wheat Protein in the South', *Agricultural Gazette N.S.W.,* **73,** 573.

Thompson, C. H. and Beckman, C. G. (1959), *Soils and Land Use in the Toowoomba Area, Darling Downs, Queensland,* CSIRO Soils and Land Use Series No. 28.

Thwaites, R. N. and Macnish, S. E. (1991), *Land Management Manual, Waggamba Shire, Parts A, B, C.* Queensland Department of Primary Industry Training Series QE 90014, Brisbane.

Truog, E. (1946), 'Soil Reaction Influence on Availability of Plant Nutrients', *Proceedings of the Soil Science Society of America,* **11,** 305.

Walker, P. H. (1963), *A Reconnaissance of Soils in the Kempsey District, N.S.W.,* CSIRO Soils and Land Use Series No. 44.

Walker, P. H. (1972), 'Seasonal and Stratigraphic Controls in Coastal Floodplain Soils', *Australian Journal of Soil Research,* **10,** 127.

Waring, S. A., Fox, W. E. and Teakle, L. J. H. (1958a), 'Fertility Investigations on the Black Earth Wheatlands of the Darling Downs, Queensland: I. Moisture Accumulation under Fallow', *Australian Journal of Agricultural Research,* **9,** 205.

Waring, S. A., Fox, W. E. and Teakle, L. J. H. (1958b), 'Fertility Investigations on the Black Earth Wheatlands of the Darling Downs, Queensland: II. Moisture and Evapotranspiration in Relation to the Wheat Crop', *Australian Journal of Agricultural Research,* **9,** 717.

Williams, C. H. and Lipsett, J. (1961), 'Fertility Changes in Soils Cultivated for Wheat in Southern New South Wales', *Australian Journal of Agricultural Research,* **12,** 612.

Williams, J. (1983), 'Physical properties and Water Relations: Soil Hydrology, in *Soils: An Australian Viewpoint,* CSIRO–Academic Press, 507–530.

CHAPTER 5

THE NUTRITION OF CROPS

O. G. Carter and A. S. Black

Mineral nutrition is just one of the factors which may limit crop yield and quality. People involved in crop production need to understand that a deficiency, at the correct stage of growth, of any one of the nutrients essential to plant growth will result in a serious reduction in the yield and quality of the crop. This will occur regardless of the adequacy of the moisture supply, weed control or seedbed preparation and of the growth potential of the variety being grown. In other words, Liebig's Law of the Minimum will apply: *'The yield of a crop is limited by the plant nutrient element present in the smallest quantity relative to the crop's requirements, all others being present in adequate amounts'.*

In many cases the correction of one nutrient deficiency leads to another nutrient becoming inadequate at the now increased level of production. Alternatively, it may result in another factor controlling growth, such as available moisture supply, becoming the limiting factor to crop growth and yield.

The aim of this chapter is to examine the principles of absorption of mineral nutrients by crop plants, the movement of nutrients in the plant, the diagnosis of nutrient deficiencies and toxicities and the use of fertilisers to overcome these conditions.

MODEL OF THE ROOT–SOIL SYSTEM

The uptake of nutrients depends on the presence of an actively growing root system in a layer of moist soil containing the nutrient in an available form. Ions enter the plant root from the soil solution, but the available supply of these ions to a crop depends very much on the rate at which the nutrients in the soil solution are replenished from the exchange surfaces, organic matter breakdown, or compounds of low solubility. The rate of supply of nutrients to the soil solution is described in detail in Wild (1988).

Fig. 5.1 is a simplified illustration of the major contributors to this complex system. The surfaces of soil organic matter, clay particles and plant roots are all negatively charged and so attract cations such as Ca^{2+},

Mg^{2+} and K^+. These exchange surfaces are important because they tend to reduce leaching and give greater nutrient retention in soils with a high exchange capacity. The exchange capacity of plant roots is also important in plant nutrition because the plants with high exchange capacities, such as legumes and herbs, tend to absorb more divalent than monovalent cations. The result is that these legume and herbage plants tend to have higher calcium levels but are also more susceptible to potassium deficiency than a grass-type crop such as wheat which has a low root cation exchange capacity.

Some plant root exchange capacities are listed in Table 5.1. Nitrogen nutrition has little effect on the root cation exchange capacity except for wheat and annual ryegrass (*Lolium rigidum*).

MAINTENANCE OF SOLUBLE NUTRIENT LEVELS IN THE SOIL SOLUTION

Mineral nutrients enter the plant root via the soil solution. Only a relatively small percentage of an available plant nutrient is in solution at any one time, and the rate of supply of nutrients to the soil solution as they are taken up by the plant root is frequently the main factor determining whether nutrients are adequate for crop growth.

Most cations are retained on the negatively charged surface of the clay, although ions can be fixed by some clay minerals (Fig. 5.1). There are

TABLE 5.1   Root cation exchange capacity of a number of plant species grown under two levels of nitrogen (Asher and Ozanne, 1961)

Plant	Root cation exchange capacity (meq/100 g dry weight of root)	
	High nitrogen	Low nitrogen
Legumes		
Field pea	32.4	33.9
Barrel medic	26.6	32.3
Subterranean clover	28.9	25.6
Lupins	16.8	19.8
Herbs		
Canola	32.7	36.3
Tomato	18.6	20.9
Capeweed	21.2	16.8
Grasses		
Annual ryegrass	22.1	12.0
Maize	12.0	8.8
Oats	7.4	9.5
Wheat	10.2	6.3
Barley	7.6	7.7

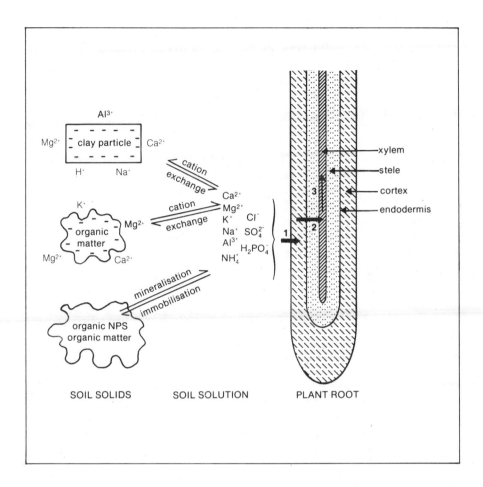

FIG. 5.1 Diagrammatic model of the soil–plant system:
(1) active uptake through the plasmalemma;
(2) secretion into the xylem vessel;
(3) translocation to the leaves via the xylem vessel.

also large amounts of other nutrients in very low solubility forms such as iron and aluminium phosphate. Soil pH can be a very important factor controlling the release of these low-solubility nutrients.

Organic matter of the kind indicated in Fig. 5.1 is a major source of nutrients such as nitrogen and sulfur. The release of nutrients following biological breakdown of soil organic matter is very sensitive to soil temperature and availability of oxygen. Waterlogging and low soil temperatures are frequently associated with temporary nitrogen and sulfur deficiency.

ION MOVEMENT INTO AND WITHIN THE PLANT ROOT

The diagrammatic plant root in Fig. 5.1 shows three stages which are critical to movement of ions into the plant root. The first stage is the entry into the cortex which may be either a passive process occurring with the bulk flow of water into the 'free space' of the cell walls or an active process into the cytoplasm through the plasmalemma membrane. The cell walls consist of an open network of cellulose which contains free negative charges; this network is the main location of the cation exchange capacity of plant roots. Movement into the cell walls is relatively insensitive to temperature and does not involve respiratory energy, but this passive movement ceases at the endodermis.

Active uptake across the plasmalemma into the cytoplasm involves the use of respiratory energy and allows the accumulation of ions against a concentration gradient. This process is very sensitive to oxygen supply and temperature because of the involvement of respiratory energy. Once in the cytoplasm, ions move freely throughout the cortex of the root, passing through the cell walls in the plasmadesmata and finally reaching a barrier at the endodermis.

A second active process (process 2 in Fig. 5.1) allows further ion concentration and transfer to the xylem.

Transport in the xylem (process 3 in Fig. 5.1) appears to be largely mass flow in the transpiration stream of the xylem and allows rapid movement into the leaves of the plant. There is some lateral transfer from the xylem to the phloem, and some mineral nutrients such as K^+ and HPO_4^{2-} can be retranslocated from leaves to other parts of the plant via the phloem. Immobile nutrients such as Ca^{2+}, tend to move into the leaves and remain there and are not retranslocated to other parts of the plant.

The movement of ions across the plant root is shown in more detail in Fig. 5.2. From the crop nutrition point of view the concentration of ions across the plasmalemma is one of the most important factors. This membrane depends on calcium to maintain its structure, and it is now known that salinity damage is largely related to the replacement of Ca^{2+} by Na^+ in the plasmalemma, thereby resulting in leakage from the cytoplasm and disruption of the primary active step in ion uptake. Salinity damage is not just a sodium effect; it is very dependent on the Na^+/Ca^{2+} ratio. Higher levels of Na^+ can be tolerated when the irrigation water or the soil solution contains higher levels of Ca^{2+}.

USE OF THE MODEL IN CROP NUTRITION

The model of the soil plant system briefly described here is examined in much greater detail by Epstein (1972). The following points illustrate how this model can be helpful in understanding the nutrition of crops.

— Cations such as Ca^{2+}, K^+ and NH_4^+ will be present in larger amounts in soils with a large cation exchange capacity. These may be soils

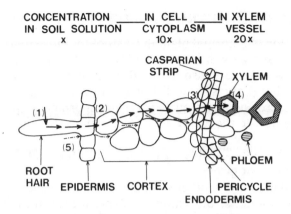

CONCENTRATION IN CELL IN XYLEM
IN SOIL SOLUTION CYTOPLASM VESSEL
x 10x 20x

FIG. 5.2   Diagrammatic outline of movement of ions through a plant root
(1) active uptake across plasmalemma membrane into the cytoplasm, with a tenfold increase in concentration;
(2) movement through the cytoplasm by cyclosis (streaming) and passing through plasmadesmata in cell walls;
(3) active uptake across the endodermis and secretion into xylem vessel with a further twofold increase in concentration;
(4) translocation to the leaves in the transpiration stream;
(5) movement in the cell wall and intercellular spaces or so-called 'apparent free space', a process occurring with the bulk flow of water.

high in clay (particularly if the dominant clay mineral is montmorillonite) or high in organic matter. Leaching will be reduced and the buffering effect of ions associated with the exchange surfaces will reduce the chances of toxicity by fertilisers. The opposite situation will prevail in coarse sandy soils low in organic matter.

— The supply of nutrients such as SO_4^{2-}, NH_4^+ and NO_3^- will depend very much on the amount and rate of breakdown of organic matter. Low temperature and waterlogging will reduce available N and S.

— Knowledge of the relative exchange capacity of different plant roots allows the prediction of relative susceptibility to potassium deficiency and relative uptake of K^+ and Ca^{2+}. Plants such as wheat are relatively low in Ca^{2+} and are more tolerant of low soil K than plants such as canola with a high cation exchange capacity on the roots.

— Recognition of the role of the plasmalemma explains the involvement of respiratory energy and the reduction in ion uptake found at lower temperature and under waterlogging conditions.

— The importance of Ca^{2+} in maintaining the plasmalemma explains the importance of high calcium in increased tolerance to Na^+ toxicity.

— Movement of ions to plant tops will be strongly influenced by transpiration because of the fact that nutrients move to leaves in the xylem.

Other examples are given as appropriate in later sections of the chapter.

ESSENTIAL NUTRIENTS FOR PLANT GROWTH

The composition of a normal plant in terms of the current list of nutrients essential for plant growth is given in Table 5.2. The levels of oxygen, carbon and hydrogen are also included, although these are not grouped as nutrients. These figures show distinctly why some mineral elements are classified as macro and micro nutrients. The importance of photosynthesis (which accounts for the C, H and O content) and nitrogen in plant nutrition is clearly illustrated.

The list of elements in Table 5.2 is complete in terms of those known to be essential for plant growth.

However, there may be other elements required by plants, presumably in very small amounts, but elaborate experimental techniques will be required before essentiality can be proven. Elements in dispute include:

Na— is essential for some halophytes such as saltbush (*Atriplex* spp.) and for those plants having the C_4 Hatch Slack pathway of photosynthesis.

Se— generally toxic, but some legumes such as *Astragalus racemosus* can accumulate selenium.

TABLE 5.2   Concentrations and yields of nutrients in crops

Element	Average elemental composition of tops	Nutrient yield/ha to grow			Ionic form adsorbed
		10 t lucerne per year	4 t/ha wheat crop		
			7 t straw	4 t grain	
	(%)	(kg)	(kg)	(kg)	
Macronutrients					
N	3–5	350	50	80	NH_4^+, NO_3^-
P	0.2–0.4	35	5	14	HPO_4^{2-}
K	2–3	200	42	17	K^+
S	0.2–0.4	35	8	10	SO_4^{2-}
Ca	0.2–3.0	100	20	8	Ca^{2+}
Mg	0.1–0.3	30	5	6	Mg^{2+}
Micronutrients (mg/kg)		(g)			
Mn	5–100	600			Mn^{2+}
Fe	45–150	1500			Fe^{2+}
Zn	10–90	450			Zn^{2+}
Cu	5–20	100			Cu^{2+}
Mo	0.1–0.5	15			MoO_4^{2-}
B	5–60	450			BO_3^{3-}
Other	(%)				
O	45				
C	45				
H	6				

Si — not regarded as essential, but very important to the physical structure of some crops such as rice. Some workers regard silicon as essential for rice. Si improves root structure in sorghum but has not been shown to be essential.

Co — not generally regarded as essential except for nitrogen fixation by *Rhizobium* but there is some evidence that cobalt is essential for subterranean clover (*Trifolium subterraneum*) and durum wheat (*Triticum durum*).

A discussion of nutrient roles is given by Epstein (1972) and Mengel and Kirkby (1987).

Table 5.2 gives an indication of the average concentrations of elements in crops, the forms taken up, and the uptake of nutrients by plant tops. Large quantities of N and K are removed by crops, especially those taken for hay (causing K deficiencies in paddocks regularly used for hay production). If the hay is used for livestock on the farm where it was made, the extent of K deficiency can be reduced by feeding the hay out on the paddocks from which it was cut. Nutrient uptake by grain crops is balanced partially by return of nutrients in the stubble. If stubbles are burnt, N and S are lost from the soil as gases.

GENOTYPE AND CROP NUTRITION

Farmers and home gardeners have for centuries claimed that particular soils suit certain crops. There are many possible reasons for this observation, but one with ample scientific backing is that crop species and even cultivars differ substantially in tolerance to deficiencies and toxicities of mineral nutrients. This is of great importance, because crops can be grown on particular soils where they are more likely to succeed, and plant breeders can breed for plant types with particular responses to plant nutrients. Much of the success in raising yields of rice, maize and wheat in recent years has been due to the breeding of new varieties with greater responses to nitrogen fertiliser. Unfortunately, because of big increases in fertiliser costs, plant breeders may be forced to breed varieties with lower nitrogen requirements and probably lower yield. The remainder of the section describes crop and cultivar differences and, where possible, explains why they are different.

A classical study by Andrew and Norris (1961) using a number of temperate and tropical pasture legumes is probably the best example of a wide range of tolerance to soil calcium and/or acid levels between different plant species (Table 5.3). The lack of tolerance to low soil calcium and/or acidity by barrel medic and the high tolerance of most tropical legumes were thought to be due to a differing ability to extract calcium from the soil, and therefore could have been predicted from the soil calcium levels where the species occur in nature. Undoubtedly, similar ranges of responses could be demonstrated between crop plants. Nitrogen fixation in the nodule is a key factor (Andrew, 1976).

Another excellent example is the susceptibility of some soybean varieties to iron chlorosis in some soils low in available iron. This has been extensively studied and appears to be controlled by a single gene (Weiss, 1943). It has also been shown by grafting experiments that the root is the controlling organ, and that the roots of cultivars susceptible to iron chlorosis do not secrete a chemical that reduces the more insoluble Fe^{3+} to soluble Fe^{2+} compounds which can be transported across the roots (Brown *et al.*, 1958; 1967). All potential new soybean varieties in the U.S. state of Iowa are screened for tolerance to iron chlorosis; the first variety with this cited as its major attribute was released for the 1977 growing season.

Studies on manganese toxicity in soybeans have shown similar wide differences in tolerance between varieties (Carter *et al.*, 1975). Grafting experiments have shown that the root is not the site of tolerance, and environmental studies have shown a strong temperature effect, with susceptibility being greatly increased at low temperatures (see Plate 5.1). In these studies the Lee variety showed considerable tolerance to manganese toxicity and the Bragg variety showed a high susceptibility. This difference appears to be controlled largely by a single gene (Heenan *et al.*, 1981).

Another example already cited is the tolerance of grass crops such as wheat to low soil potassium and the susceptibility of legumes and herbs to the same condition. Other crops have been classified as indicator plants; cauliflower, for example, is very sensitive to low molybdenum supply.

Salt tolerance is common in deciduous fruit trees such as peach, whereas evergreens such as citrus are very susceptible to salt toxicity. More examples of these genotypic deficiencies are being found each year, and many are listed by Epstein (1972).

TABLE 5.3   Relative yield of a number of pasture legumes grown in a low humic gley soil that was very low in calcium (after Andrew Norris, 1961)

Plant	Yield as % of that obtained after heavy liming
Temperate legume	
Barrel medic (*Medicago truncatula*)	0.2
Lucerne (*Medicago sativa*)	1.5
Strawberry clover (*Trifolium fragiferum*)	2.0
White clover (*Trifolium repens*)	6.5
Tropical legume	
Desmodium	25
Indigofera	40
Centrosema	52
Stylosanthes	64
Phaseolus	68

There is obviously a need to test potential new crop varieties at a number of sites in an attempt to identify genotypic differences in tolerance to high and low levels of mineral nutrients. Mineral nutrition is likely to become an increasingly important aspect of plant breeding, but those testing fertiliser responses must consider differences between varieties as well as between crop species.

ENVIRONMENT AND CROP NUTRITION

The availability of nutrients in the soil and the rate of uptake by the plant are influenced strongly by environmental conditions. The majority of field experiments which study crop nutrition have meteorological instruments to measure the aerial environment, and soil moisture is measured at the beginning of the experiment. This is very useful for interpreting the results and helps to reduce the need for experiments to be

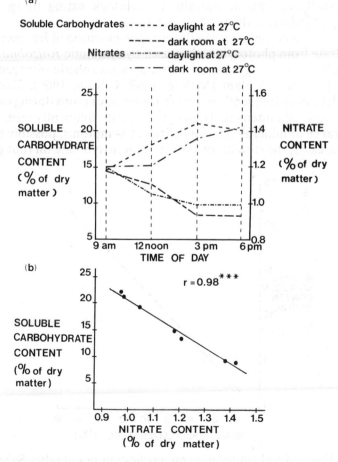

FIG. 5.3   Effect of shading on the levels of soluble carbohydrates and nitrates in oats grown under phytotron conditions (after Fernando, 1968)

repeated over a large number of seasons. The three most important environmental factors—light, temperature and moisture—are now considered.

Light

Light has a marked influence on the mineral nutrition and mineral composition of plants because of its effect on photosynthesis. Carbohydrate produced during photosynthesis is needed as a substrate for respiration to produce energy used in the active ion uptake process, the energy being required to accumulate ions from the soil solution through the plasmalemma into the cytoplasm against a concentration gradient. Low light intensity, as occurs under heavily overcast, rainy conditions, can cause a rapid drop in soluble carbohydrates in the plant and a build up in nitrate (Fig. 5.3a, b). These changes can seriously alter the flavour of fresh food products such as fruit because of lower sugar levels, and high nitrate levels can be harmful to livestock eating forage or for humans, particularly if the product is canned.

Light is also very important in the nitrogen nutrition of legumes, where carbohydrate from photosynthesis is used by symbiotic *Rhizobium* in the root nodule as an energy source for fixing atmospheric nitrogen into a form available to the plant (McKee, 1962; Gibson, 1966). Temporary nitrogen deficiency frequently occurs in leguminous crops during extended periods of low light intensity. High light intensity frequently results in the disappearance of mineral nutrient deficiency symptoms, possibly because of, in some cases, the stimulatory effect of photosynthate on root growth.

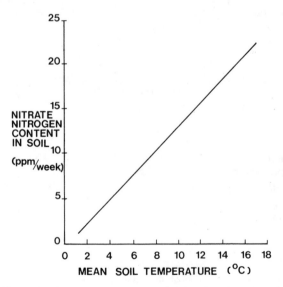

Fig. 5.4   Effect of soil temperature on nitrification in soil (after Sabey *et al.*, 1956)

LEE 33° BRAGG   LEE 21° BRAGG

PLATE 5.1   The effect of temperature on cultivar tolerance to manganese toxicity (15 ppm Mn) in soybeans; Lee variety on the left in each photo, Bragg variety on the right

Temperature

The uptake of all plant nutrients is strongly affected by temperature. Most active ion uptake processes have a Q_{10} of 2 to 2.5 (i.e. for every 10°C rise in temperature the rate of ion uptake increases 2 to 2.5 times). This is exactly what would be expected for a process which involves the use of respiratory energy. A detailed study of the effect of temperature on phosphate uptake is given by Carter and Lathwell (1967).

Temperature also has a very strong influence on the rate of organic matter breakdown and release of nutrients such as N and S which are found in the organic matter in the soil. The relationship between temperature and the rate of nitrification in the soil is shown in Fig. 5.4. Responses to nitrogen fertiliser are frequently obtained during periods of low soil temperature because of a slowdown in the rate of release of NH_4^+ and NO_3^- from organic matter. Legumes can also suffer a temporary slowdown in the rate of nitrogen fixation by *Rhizobium* in the nodule during periods of low soil temperature.

A number of deficiencies and toxicities are also influenced by temperature. A good example of this is manganese toxicity in soybeans (see Plate 5.1) which studies have found to increase markedly at low temperatures but to largely disappear at higher temperatures (Heenan and Carter, 1977).

Interactions between temperature and mineral nutrition have not been studied extensively. Undoubtedly, many more cases will be identified of the modifying effects of temperature on the development of nutrient deficiencies and toxicities through modifying the nutrient availability from the soil, increasing or decreasing the uptake by the plant, or modifying the expression of deficiency and toxicity symptoms.

Moisture

Soil moisture levels strongly influence the availability of nutrients to plant roots. Low moisture levels reduce the release of N, P and S from soil organic matter. One of the most important effects of soil moisture is on root growth, as nutrients can be absorbed only if there are actively growing plant roots in a layer of moist soil. Phosphate is frequently applied as a surface topdressing, and in periods of low moisture, when the soil surface (where the majority of P is retained) dries out, temporary P deficiency can occur as there are no actively growing roots in the zone where the P is located. Similarly, in nitrogen fertiliser experiments with soybeans in the U.S., responses were only obtained to deep placement of the fertiliser (50 cm below the soil surface) because this is where moisture and actively growing plant roots are located at the end of the growing season, when nitrogen fixation by the root nodule is unable to obtain sufficient carbohydrate to allow an adequate nitrogen supply to the plant.

Excess moisture supply can also be harmful, as some nutrients are leached from the soil profile when this occurs. This is particularly serious with NO_3^- and SO_4^{2-}, not only causing loss of plant nutrients but possibly river and lake pollution as well (Cameron and Haynes, 1988; Cullen, 1974).

The importance of movement of moisture in the soil profile is illustrated by studies at Katherine in the Northern Territory, where NO_3^- moved

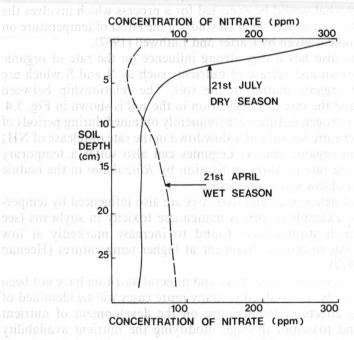

FIG. 5.5   Changes in nitrate distribution in the Tippera clay loam soil profile at Katherine, Northern Territory, in the wet and dry seasons (after Wetslaar, 1961)

down the profile in the wet season and moved back up again during the dry season (Fig. 5.5). Unfortunately, in many situations NO_3^- and other soluble nutrients may be lost from the soil profile altogether (Simpson and Freney, 1974).

Excess moisture in poorly drained soil may also result in temporary toxicity of manganese because of reduction from the insoluble manganese dioxide to the soluble Mn^{2+} ion. Boron deficiency also occurs under waterlogged conditions in some soils.

In addition, the breakdown of organic matter is slowed in waterlogged soils, so that temporary nitrogen and sulfur deficiency may result. Plant roots need oxygen for active uptake of plant nutrients, and even if these nutrients are available in the soil they may not be taken up under such low oxygen conditions.

Soil moisture level is also important in modifying fertiliser germination damage; this is discussed in a later section.

Any clear understanding of the principles of crop nutrition must take into account the modifying effect of environmental factors such as light, temperature and moisture on the availability of nutrients from the soil and their uptake by the plant.

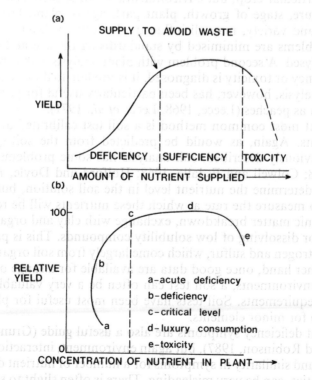

FIG. 5.6 Relationship between (a) yield and nutrient supply, and (b) relative yield and nutrient content in the plant

DIAGNOSIS OF MINERAL DEFICIENCY AND TOXICITY

Accurate diagnosis of mineral deficiencies and toxicities is vital to economic production of high-quality crops. Failure to correct nutrient deficiencies or toxicities will reduce the yield and quality of crops, while all other costs of growing the crop will still be incurred. Unnecessary use of fertiliser (particularly where fertiliser costs are high) is not only economically wasteful but can lead to serious pollution of water bodies.

The important relationships between nutrient level and yield, and between the concentration of nutrient in the plant and relative yield, are shown in Fig. 5.6. The curve in Fig. 5.6a is essentially the kind of response obtained to fertiliser in a field experiment, a technique which is generally accepted as the best method of diagnosing mineral nutrient deficiencies. It is rather costly and time consuming and is usually supplemented by other measures such as leaf analysis and soil analysis. Soil analysis and plant analysis have been reviewed by Westerman (1990), and plant analysis and visual symptoms of deficiency have been reviewed by Reuter and Robinson (1986) and Grundon (1987).

Figure 5.6b shows the relationship between the level of nutrient in the plant and the relative yield. The aim in this case is to find the critical level, c, for a particular crop, but environmental factors, such as temperature and moisture, stage of growth, plant part, age of plant, level of other nutrients and variety, may all influence the nutrient level in the plant. These problems are minimised by standardising the age and plant part being analysed. A second problem with plant analysis is that by the time the deficiency or toxicity is diagnosed, it is too late to do anything about it. Leaf analysis, however, has been particularly useful for perennial tree crops such as peaches (Leece, 1968; Leece *et al.*, 1971).

The next most common method is a soil test calibrated against field experiments. Again, as would be predicted from the soil–plant root model previously described, soil analysis has some problems (Colwell, 1967, 1968; Colwell and Esdaile, 1968; Holford and Doyle, 1992). It is simple to determine the nutrient level in the soil solution, but it is very difficult to measure the rate at which these nutrients will be replenished from organic matter breakdown, exchange with clay and organic matter surfaces, or dissolving of low solubility compounds. This is particularly true for nitrogen and sulfur, which come largely from soil organic matter. On the other hand, once good data are available for a range of soils in a range of environments, a soil test can often be a very valuable guide to nutrient requirements. Soil tests have been most useful for phosphorus but less so for minor elements.

Nutrient deficiency symptoms are also a useful guide (Grundon, 1987; Reuter and Robinson, 1987), but again environment, interaction between nutrients and similarity in symptoms for a number of nutrient deficiencies and toxicities, can be very misleading. There is often slight to severe yield reduction without any symptoms, a situation commonly referred to as 'hidden hunger'.

PLATE 5.2   The response by oats to nitrogen provided in patches of urine and dung excreted by grazing animals

Pot tests can be very helpful in the initial identification of a deficiency or toxicity problem and are often used in an initial survey. General procedures have been described by Andrew and Fergus (1976). Pot tests have been particularly useful in identifying minor element deficiencies but are usually of little use for nutrients like nitrogen and sulfur, the supplies of which change rapidly because of organic matter breakdown when the soil is placed in a pot.

Nutrient solution culture, or hydroponics, has been a very useful research tool and is usually the basis of establishing the essentiality of a nutrient. Water culture is also used to produce deficiency and toxicity symptoms and can be very helpful in studying genetic differences in mineral nutrient responses (Carter *et al.*, 1975).

The major advantages and disadvantages of various methods of diagnosing plant nutrient deficiencies are summarised in Table 5.4. All methods have some role to play but must be used and interpreted with care.

CORRECTION OF DEFICIENCIES

Most nutrient deficiencies in soil are overcome by the use of fertilisers. Australian soils are universally deficient in phosphorus, so P fertilisers are the most commonly used. Nitrogen deficiency is also a major

TABLE 5.4 Summary of advantages and disadvantages of various methods of diagnosing plant nutrient deficiencies and toxicities

Method	Advantages	Disadvantages
Soil analysis	Relatively cheap, done before crop planted so fertilisers can be applied at planting. Most useful for P, Mg, Ca and pH. N tests useful if carefully interpreted.	Very difficult to measure what the plant can use. Rate of supply to the plant is hard to estimate. Limited use for S, N or minor elements.
Plant analysis	Relatively cheap, fast and accurate. Good for perennial crops, trees, sugar cane. Avoids the impossible problem of soil sampling for trees, useful to detect toxicity.	Annual crop is growing, so often too late to correct deficiency or toxicity. Environment, plant part, plant age all strongly influence levels. Sprays—Zn, Cu and soil dust Fe, Si often cause confusion. Level of one element frequently influenced by the level of another element.
Pot tests	Moderately expensive. Measures what plant can extract from the soil. Tests can be done all year round. Very useful in initial survey particularly in establishing minor element deficiencies. Helps greatly in reducing costs and improving design of subsequent field trials.	Little use for N or S. Can be used to establish deficiency or toxicity but little use in establishing fertiliser rates, for example P. Drainage, temperature conditions changed completely.
Field plots	Always needed to calibrate other methods, must be done in the end and usually has to be done over several years. Very good for extension purposes.	Very expensive. Often lost because of low rain, flood, grass-hoppers etc. May lose whole year's work. Very susceptible to environmental factors, moisture, temperature. Strictly, results only apply to that season, that variety and site.
Symptoms	Useful to trained person who knows what to expect on that crop in that area.	Often symptoms common to a number of elements and strongly influenced by environment and levels of other elements. Too late to take any action usually. Most crops are reduced substantially in yield before any symptoms are seen.
Solution culture	Good research tool for establishing symptoms, essentiality of a nutrient for a particular plant, screening varieties of different nutrient requirements. Very useful for toxicity studies.	Totally unrelated to soil field environments. No help in diagnosis or advising farmers on fertiliser practices.

problem, but legume-based pastures grown in rotation with crops provide much of the N required in cropping areas of southern Australia. The N inputs to soil from pasture and grain legumes have been reviewed by Peoples *et al.* (1992).

The average ratio of use of N:P fertilisers on an Australia-wide basis was 1:13 until 1960. This reflected the use of single superphosphate, with legumes as the N source. Soil limitations and variable climate suggest that a combined cropping/livestock program with legume-based pastures providing N is likely to continue in the southern wheat belt areas. However between 1970 and 1985 the ratio of N:P fertilisers used in Australia dropped to 1:1 as a result of greater N fertiliser use on cotton and sugar cane and in production horticulture.

Many existing sulfur deficiencies have been overcome by the extensive use of single superphosphate. Major S deficiencies occur in the >500 mm annual rainfall belt because of S loss by SO_4^{2-} leaching (Blair and Nicholson, 1975). In coastal areas, S deficiency is less common because of S returns from the atmosphere. Potassium and Mg deficiencies are confined to sandy soils, especially if the soils are acid, or to soils where the removal of nutrients by harvesting is high.

The extent of micronutrient deficiencies in soils used for crop and pasture production was reviewed by Donald and Prescott (1975), as well as manganese by Graham *et al.* (1988), copper by Loneragan *et al.* (1981) and zinc by Weir (1987). Zinc deficiency is now widely recognised on alkaline clay soils in northern New South Wales and the Darling Downs in Queensland, where substantial yield responses to zinc supplements are obtained in grain sorghum, cotton and soybeans. Manganese and iron deficiencies occur on calcareous soils, while manganese toxicity occurs when the pH declines below 5.5 or the soils become waterlogged. Molybdenum deficiency is common where soils have become acid, because of the decreased availability at low pH. Wheat yield responses to molybdenum have been obtained on soils high in nitrate (Lipsett and Simpson, 1971). Correcting micronutrient deficiencies normally involves topdressing with micronutrient fertilisers mixed into superphosphate to ensure even distribution. Most micronutrients are applied at 3 to 10 year intervals.

Principles of Fertiliser Management

There are two key principles for successful fertiliser management.

1. Management attempts to maintain the balance between the nutrient uptake by crops and the supply from soil and/or fertiliser sources (Myers 1988). Implicit in this statement is the need to know the nutrient uptake pattern by crops. Fig. 5.7 shows the typical pattern of uptake in most crops. Of particular importance is the observation that the rate of uptake of N, P and K occurs before the high rate of dry matter accumulation. The association between N, P and K

uptake and dry matter yield for grain sorghum (Jacques *et al.*, 1975; Vanderlip, 1972), maize (Hanway, 1971) and soybeans (Hanway and Thompson, 1971) has been described in detail.

In addition, if nutrients are to be applied, it is necessary to appreciate factors which control the release of the soil nutrient reserve throughout the cropping period.

2. Management attempts to minimise adverse effects associated with fertiliser application. The effects of particular concern are:
   • reactions in soil which reduce the availability of applied fertiliser, and
   • adverse side effects including pH extremes and toxicities induced by fertilisers, salinity hazards associated with application of high rates of soluble fertiliser, and contamination of soil with non-essential elements such as Cd.

To illustrate these points, examples with N, P and K fertilisers are set out below.

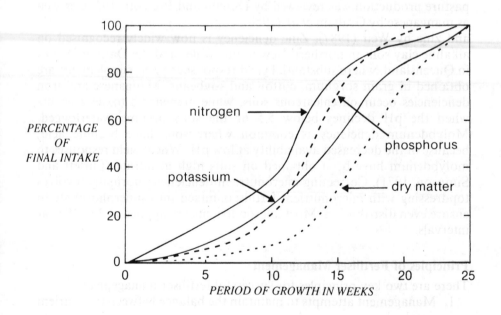

FIG. 5.7 Uptake of nutrients and production of dry matter in barley (Wild, 1988)

Nitrogen Fertiliser

The N yield in plant tops ranges from 50 to 300 kg N ha^{-1} (Table 5.2). Fig. 5.8a shows the pattern of accumulation of N by wheat grown at Harden, NSW (Stein *et al.* 1987). The period of maximum N accumulation in the tops occurred during August and early September. The principal form of N taken up by crops is NO_3^- although NH_4^+ can be utilised.

Over 95% of N present in soil occurs in organically bound forms. These forms must be mineralised by soil microorganisms to NH_4^+ before becoming available to the plant. A diverse range of organisms are responsible for mineralisation. The oxidation of NH_4^+ to NO_3^- (nitrification) is a two-stage biological oxidation carried out by a few chemoautotrophs. The processes can be written as

$$NH_4^+ + 1\frac{1}{2}O_2 \rightarrow NO_2^- + 2H^+ + H_2O \qquad (1)$$

$$NO_2^- + \frac{1}{2}O_2 \rightarrow NO_3^- \qquad (2)$$

Nitrosomonas and *Nitrosococcus* spp. are active in the first oxidation, and *Nitrobacter* spp. are responsible for the second.

Because both mineralisation and nitrification are biological processes, environmental factors such as temperature, aeration, pH and moisture content control the rate of the processes. Haynes (1986) reviewed these processes and the effect of environment on the rate of transformation. Fig. 5.8b shows an example of the close association between seasonal changes in temperature and net mineralisation under a wheat crop (Stein *et al.*, 1987). A comparison of Figs 5.8a and b shows that, in spring, the rate of N uptake and net mineralisation increase at the same time. This is a good example of a balance between N supply and plant requirement. Fig. 5.8b also shows that a proportion of soil N is mineralised in autumn, indicating a lack of balance between N supply and N uptake. Nitrate formed in autumn may be removed from the form available to plants before uptake can occur.

The coincidence of conditions reducing mineralisation in spring (for example, low soil temperatures) with the period of high plant demand can result in the N supply from soil controlling crop growth. Angus (1986) defined situations where fertiliser N responses by crops were likely, depending on conditions in the soil environment.

ADVERSE N FERTILISER REACTION

Reduced Plant Availability—Most N fertilisers are applied as:
  (i) NH_4^+, for example: sulfate of ammonia, $(NH_4)_2SO_4$
                      nitram, NH_4NO_3
                      diammonium phosphate (DAP), $(NH_4)_2HPO_4$
                      monoammonium phosphate (MAP),
                      $NH_4H_2PO_4$
  (ii) NH_4^+-producing forms, for example: urea, $(NH_2)_2CO$
                      anhydrous NH_3

An accumulation of NH_4^+ can result in loss by NH_3 volatilisation if the soil pH exceeds 8. A major review of the process was reported by Freney and Black (1988). This high pH may result from fertiliser reaction such as urea hydrolysis or occur when soils have a naturally high pH. Losses by volatilisation may be negligible but can exceed 20% if urea is broadcast onto wheat in spring (Black *et al.*, 1989). Losses are less if fertilisers are incorporated into the soil at sowing or washed into the soil by rainfall or irrigation water (Black *et al.*, 1987).

Following nitrification, further N losses can occur via dentrification or leaching. Dentrification occurs when soils become anaerobic and NO_3^- is used as the electron acceptor for soil biological oxidation processes. The products are the gases N_2O and N_2. The principal cause of low oxygen levels in soils is waterlogging. In well-drained cropping soils, losses range

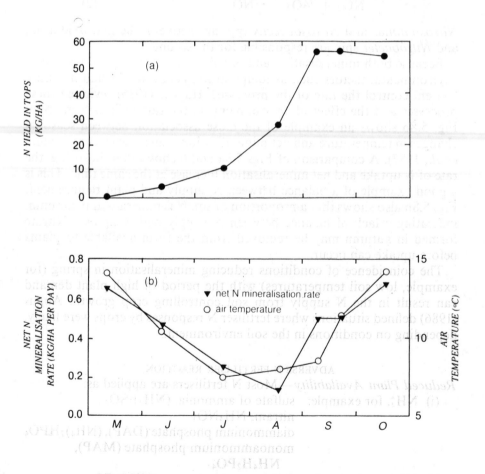

FIG. 5.8   The seasonal pattern of (a) N uptake by wheat and (b) N mineralisation rate and temperature (Stein *et al.*, 1987)

from 0 to 30% of the N fertiliser applied (Haynes and Sherlock, 1986), but where NO_3^- accumulates in soil before flooding (such as under rice) almost all NO_3^- is lost by denitrification. Simpson (unpublished data) found recovery of ^{15}N fertiliser in the soil and wheat tops of 92% in a well-drained red earth at Wagga Wagga and 46% on a poorly drained red brown earth at Whitton. Because of the impervious B horizon and associated waterlogging in the Whitton soil, the losses were attributed to denitrification.

Losses of N by denitrification can be reduced in arable crops by:

• ensuring that fertiliser application is split (the bulk of the fertiliser N being applied after the period of waterlogging),
• the use of NH_4^+ forms amended with N serve® which slows the rate of NO_3^- accumulation, and
• the use of slow release fertiliser forms.

Leaching losses of NO_3^- are rapid and most severe in high rainfall environments. Black and Waring (1976) showed that NO_3^- leached to a depth of 2 m following 2800 mm of rainfall at Redland Bay in Queensland. In contrast, Storrier (1965) found that under wheat in southern NSW in a 500 mm per annum rainfall environment, NO_3^- was leached to 0.6 m during the winter months but was later recovered by the crop during spring as the NO_3^- remained within the root zone. NO_3^- leaching represents not only a loss of N for crop growth but also results in acidification of surface soil where fertiliser forms other than NO_3^- are used and in pollution of ground waters. In the southern Australian wheat belt, the quantity of N leached can be reduced by early sowing of crops (Strong, 1992) which ensures that N mineralised during the autumn break is recovered by plant uptake before leaching commences. Leaching of NO_3^- has been reviewed by Cameron and Haynes (1986) and White (1987).

Adverse Side Effects—Injecting anhydrous ammonia or applying urea induces high pH at the point of application. This can result in NH_3 or NO_2^- (Table 5.5) toxicity, as the second stage of nitrification is inhibited by the extreme environmental conditions. Passioura *et al.* (1972) demonstrated that the accumulation of NO_2^- restricted root development in the zone of urea fertiliser placed in bands. Anhydrous and aqua ammonia should be placed at least 7 cm from the seed, while urea should not be banded with seed at rates over 40 to 60 kg N ha^{-1}.

Most solid N fertilisers are highly soluble salts. If placed near growing plants, especially germinating seedlings, dissolution of the salts will reduce the water potential (availability). The resultant damage is a salt effect similar to that which occurs in saline soils. The relative hazard from the N fertilisers is shown in Tables 5.5 and 5.6. Pelleted fertiliser is normally less harmful to seedlings as there is greater physical separation between seedlings and fertiliser. Table 5.6 summarises the complex factors which interact to cause varied levels of reduction in seedling establishment when seed is banded with fertilisers. Because of the interactions between factors such as moisture and species it is not possible to develop a set of rigid rules, but any combination of factors at the top of Table 5.6 is highly dangerous and any combination at the bottom of the table is relatively safe. Sensitive crops such as crucifers should not be banded with any fertiliser except lime super.

An application of NH_4^+ and NH_4^+-producing fertilisers can result in acidification of the soil. The protons produced during the first stage of nitrification can be neutralised, in part, by alkali excretion during NO_3^- uptake. The protons remaining in the surface soil after NO_3^- leaching are a major cause of acidification in soils (Helyar and Porter, 1989). Cregan and Helyar (1986) grouped the N fertilisers as follows:

high acidification	$(NH_4)_2SO_4$, $NH_4H_2PO_4$
medium acidification	$(NH_4)_2HPO_4$
low acidification	urea, anhydrous ammonia
alkaline fertilisers	$Ca(NO_3)_2$, $NaNO_3$

Henzell (1971) reported that, following six annual applications of $(NH_4)_2SO_4$, urea and $NaNO_3$ at 448 kg N ha^{-1} to a red podsolic soil, the change in pH to a depth of 15 cm was -1.1, -0.3 and $+1.0$ respectively.

Phosphorus Fertiliser

The P uptake by plant tops ranges from 10 to 34 kg P ha^{-1} with $H_2PO_4^-$ being the form taken up from the soil solution (Table 5.2). Between 50 and 80% of soil P is associated with organic matter (Dalal, 1976) and

TABLE 5.5 Relative toxicity of common fertiliser materials when placed in contact with seed (after Carter, 1969)

Fertiliser	Reason for reduced germination
MOST TOXIC	
anhydrous ammonia aqua ammonia calcium cyanamid urea diammonium phosphate	all release toxic ammonia in the soil
sodium nitrate ammonium sulfate monoammonium phosphate ammonium nitrate calcium ammonium nitrate	high in soluble salts which cause salt effects
potassium chloride potassium sulfate	moderate salt effects
ordinary, double and triple superphosphate	generally no effect except on very sensitive crops such as turnips
50:50 lime superphosphate	no known effect
LEAST TOXIC	

release by mineralisation is subject to environmental parameters as with N. The remainder is associated with the inorganic fraction.

ADVERSE P FERTILISER REACTIONS

Reduced Plant Availability—Under pastures, when organic matter levels are increasing, between 40 and 100% of applied phosphatic fertiliser is incorporated into organic matter. In contrast, only 13% (McLaughlin *et*

TABLE 5.6   Factors affecting seed germination injury by fertilisers. Any combination at the top of the table gives maximum injury and at the bottom minimum injury (after Carter, 1969)

Kind of fertiliser	Fertiliser concentration near seed	Soil moisture content	Soil texture	Granule size	Species of plant
		MAXIMUM DAMAGE			
Anhydrous ammonia	High rates	0–50% avail. water	Coarse sand		*Crucifers* Turnips Chou moellier
Aqua ammonia	Wide row spacing				Canola
Urea				Fine	*Legumes*
Diammonium phosphate					Garden peas White clover Cowpeas Vetches
$NaNO_3$ $(NH_4)H_2PO_4$ $(NH_4)_2SO_4$ NH_4NO_3					Subclover
Cal. ammonium nitrate		Loam		Medium	*Grasses* Phalaris Ryegrass
KCl					Sweet Sudan
K_2SO_4					
					Cereals Grain sorghum
Triple super Double super Super SF 50					Wheat Barley Oats
Gypsum	Low rates	Very dry			
Lime super Lime	Narrow row spacing	Field capacity	Fine clay	Coarse	
		MINIMUM DAMAGE			

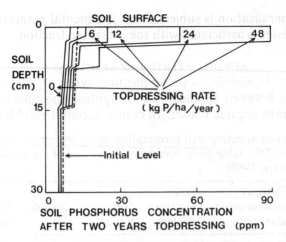

FIG. 5.9   Effect of annual topdressings with phosphorus fertiliser on vertical distribution of acid-soluble phosphate in the soil after two years (after Doll *et al.*, 1959)

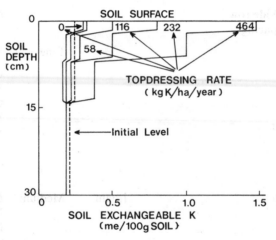

FIG. 5.10   Effect of annual topdressings with KCl fertiliser on exchangeable K in the soil profile after two years (after Doll *et al.*, 1959)

al., 1988) and 11% (Friesen and Blair, 1988) of applied fertiliser P entered the organic fraction under cropping. McLaughlin showed that 71% of applied P entered the inorganic fraction.

The mechanisms by which soluble fertiliser P enter the inorganic fraction include precipitation of insoluble Ca, Fe and Al phosphates, and adsorption onto soil mineral surfaces (Fe and Al oxides, in particular). Comprehensive reviews of these processes have geen given by Sanyal and Datta (1991) and Greenland and Hayes (1981). Following the rapid reactions which occur on fertiliser addition, slow reactions continue which result in a decrease in the levels of soluble P in soils with time.

Barrow (1974) demonstrated very clearly the decline in available P as P was exposed to soil for prolonged periods, especially at higher temperatures. These reactions are sufficiently strong that, in most soils, P is not leached and accumulates at the depth of placement (Fig. 5.9). Management of these reactions with soluble P fertilisers often involves drilling near the germinating seedling to restrict the volume of soil in contact with fertiliser and the time of contact before plant uptake.

An additional major advantage of placement is that the fertiliser is placed closer to the seedling, ensuring rapid uptake. Rudd and Barrow (1973) reported a number of field trials which showed that drilling superphosphate at sowing gave superior yield responses compared to mixing it with soil at sowing or applying the fertiliser several months before sowing.

Leaching of P fertiliser does occur in soils where the reactive clay component is absent, such as the deep sandy soils in Western Australia (Bolland, 1986) and South Australia (Lewis *et al.*, 1981).

The use of less soluble forms of P, especially finely ground reactive phosphate rock, has been suggested to reduce effects of soil reaction and leaching. Rock phosphates are variable in composition and low in solubility. Because of their low solubility, they must be mixed through soils with a pH of less than 5 (0.01 M $CaCl_2$) to achieve agronomic response in crops (Bolland *et al.*, 1988).

Adverse Side Effects—Except for the ammonium-based P fertilisers (DAP and MAP), there are generally no salt hazards associated with these fertilisers (Table 5.6). The residual acid in superphosphate can adversely influence *Rhizobium* survival if the seed is mixed with the fertiliser.

The principal adverse effect associated with P fertilisers is the addition of the heavy metals Zn, Cu, Mn and especially Cd to the soil (Tiller, 1989). These originate from the rock phosphate used in superphosphate manufacture.

Potassium Fertilisers

After N, K is the element taken up by plants in greatest quantities (Table 5.2). The pattern of K uptake by crops (Fig. 5.7) indicates that K must be present in adequate supply early in the growth period.

ADVERSE K FERTILISER REACTIONS

Reduced Plant Availability—Most available K^+ is retained on the cation exchange sites in the clay fraction so that the loss of K by leaching is not a serious problem when soils have a high cation exchange capacity (CEC).* Fig. 5.10 shows that K is retained in the surface soil layers. However,

---

* Cation exchange capacity is a measure of the total amount of exchangeable cations that can be held by the soil adsorbed to clay particles or organic matter.

when soil has a low CEC, leaching of K is a serious cause of loss (Phillips *et al.*, 1988). In soils high in the clay mineral vermiculite and to a less extent smectite, K can be fixed, reducing fertiliser availability. The fixed K is only slowly released back into the soil solution.

Adverse Side Effects—The two most common K fertilisers are KCl and K_2SO_4. Both are highly soluble and can cause salt effects if drilled at high rates with seed (Table 5.6). Although both fertilisers provide K in equally available forms, the quality of tobacco and some horticultural crops such as potatoes is impaired if Cl levels are high. For these crops the more expensive K_2SO_4 form must be used.

High rates of K fertiliser can induce Ca or Mg deficiencies, especially if the supply of these in the soil is marginal.

SUMMARY

The principles of crop nutrition examined in this chapter are summarised as follows:

— a soil–plant root model stresses the importance of the soil solution, and its rate of replenishment from exchange surfaces of clay and organic matter is the key to nutrient supply to plants;

— the entry of ions into the root involves cation exchange, diffusion and metabolic uptake across the plasmalemma: because of the involvement of respiratory energy, ion uptake is strongly affected by temperature and oxygen supply. Ca is vital to the function of the plasmalemma;

— translocation to the plant top is in the transpiration stream;

— the availability of nutrients to the plant depends to a considerable extent on location of active plant roots in close proximity to these nutrients: this is espeically true for immobile nutrients such as P;

— the cation exchange capacity of the root can influence the Ca/K in the plant top and the relative susceptibility to K deficiency;

— crop species and cultivar can strongly modify susceptibility to nutrient deficiency and toxicity;

— many methods are available for diagnosing nutrient deficiencies and toxicities: all of these methods have some advantages and disadvantages and must finally be calibrated against expensive field trials;

— the environment (temperature, moisture and light) can strongly influence nutrient uptake and plant composition;

— fertilisers can be used to overcome nutrient deficiencies: P is immobile in soils, K relatively immobile and N highly mobile and subject to leaching: therefore P and K should be banded with or near the seed while N can be applied as a topdressing provided urea is not being used;

— care must be taken to avoid germination damage which can be caused by some fertilisers under some conditions;

—patterns of nutrient uptake help in understanding the need to apply P, K and most other nutrients at planting while nitrogen can often be applied better as split dressings; and adequate nitrogen supply as the crop approaches maturity can strongly influence both yield and product quality.

References

Andrew, C. S. (1976), 'Effect of Calcium, pH and Nitrogen on the Growth and Chemical Composition of some Tropical and Temperate Pasture Legumes; I. Nodulation and Growth', *Australian Journal of Agricultural Research,* **27**, 611–23.

Andrew, C. S. and Fergus, I. F. (1976), 'Plant Nutrition and Soil Fertility' in Shaw, N. H. and Bryan, W. W. (eds) *Tropical Pasture Research,* pp. 101–133.

Andrew, C. S. and Norris, D. O. (1961), 'Comparative Responses to Calcium of Five Tropical and Four Temperate Pasture Legumes Species', *Australian Journal of Agricultural Research,* **12**, 40–55.

Angus, J. F. (1988), 'Strategies and Tactics for Fertiliser Nitrogen Usage' in *Proceedings of the 17th Riverina Outlook Conference,* pp. 73–79, Riverina-Murray Institute of Higher Education, Wagga Wagga.

Asher, C. J. and Ozanne, P. G. (1961), 'The Cation Exchange Capacity of Plant Roots and its Relationship to the Uptake of Insoluble Nutrients', *Australian Journal of Agricultural Research,* **12**, 755–66.

Barrow, N. J. (1974), 'The Slow Reactions between Soil and Anions: 1. Effects of Time, Temperature, and Water Content of a Soil on Phosphate for Plant Growth', *Soil Science,* **118**, 380–6.

Black, A. S. and Waring, S. A. (1976), 'Nitrate Leaching and Adsorption in a Krasnozem from Redland Bay, Qld. I. Leaching of Banded Ammonium Nitrate in a Horticultural Rotation', *Australian Journal of Soil Research,* **14**, 171–180.

Black, A. S., Sherlock, R. R. and Smith, N. P. (1987), 'Effect of Timing of Simulated Rainfall on Ammonia Volatilisation from Urea, applied to Soil of Varying Moisture Content', *Journal of Soil Science,* **38**, 679–87.

Black, A. S., Sherlock, R. R., Smith, N. P. and Cameron, K. C. (1989), 'Ammonia Volatilisation from Urea Broadcast in Spring on to Autumn-sown Wheat', *New Zealand Journal of Crop and Horticultural Science,* **17**, 175–182.

Blair, G. J. and Nicholson, A. J. (1975), 'The Occurrence of Sulphur Deficiency in Temperate Australia', In McLachlan, K. D. (ed.), *Sulphur in Australasian Agriculture,* pp. 137–144, Sydney University Press, Sydney.

Boland, M. D. A. (1986), 'Residual Value of Phosphorus from Superphosphate for Wheat Grown on Soils of Contrasting Texture near Esperance, Western Australia', *Australian Journal of Experimental Agriculture,* **26**, 209–15.

Bolland, M. D. A., Gilkes, R. J. and D'Antuona, M. F. (1988), 'The Effectiveness of Rock Phosphate Fertilisers in Australian Agriculture', *Australian Journal of Experimental Agriculture,* **27**, 655–68.

Brown, J. C., Holmes, R. S. and Tiffin, L. O. (1985), 'Iron Chlorosis in Soybeans as Related to the Genotype of Rootstalk', *Soil Science,* **86**, 75–82.

Brown, J. C., Weber, C. R. and Caldwell, B. E. (1967), 'Efficient and Inefficient Use of Iron by two Soybean Genotypes and their Isolines', *Agronomy Journal,* **59**, 459–62.

Cameron, K. C. and Haynes, R. J. (1986), 'Retention and Movement of Nitrogen in Soils', in Haynes, R. J. (ed.), *Mineral Nitrogen in the Plant–Soil System* 166–241, Academic Press, New York.

Carter, O. G. (1967), 'The Effect of Fertilisers on Seedling Establishment', *Australian Journal of Experimental Agriculture and Animal Husbandry,* 7, 174–80.

Carter, O. G. and Lathwell, D. J. (1967), 'Effects of Temperature on Orthophosphate Absorption by Excised Corn Roots', *Plant Physiology,* 42, 1407–12.

Carter, O. G., Rose, I. A. and Reading, P. F. (1975), 'Variation in Susceptibility to Manganese Toxicity in 30 Soybean Genotypes', *Crop Science,* 15, 730–2.

Colwell, J. D. (1967), 'Calibration and Assessment of Soil Tests for Estimating Fertilizer Requirements: I. Statistical Models and Tests of Significance', *Australian Journal of Soil Research,* 5, 275–93.

Colwell, J. D. (1968), 'Calibration and Assessment of Soil Tests for Estimating Fertilizer Response: II. Fertilizer Requirements and an Evaluation of Soil Testing', *Australian Journal of Soil Research,* 6, 93–103.

Colwell, J. D. and Esdaile, R. J. (1968), 'The Calibration, Interpretation and Evaluation of Tests for the Phosphorus Fertilizer Requirements of Wheat in Northern New South Wales', *Australian Journal of Soil Research,* 6, 105–20.

Cullen, P. W. (1974), 'Fertilizer Phosphate in Streams and Lakes,' in D. R. Leece (ed.), *Fertilizers and the Environment,* Australian Institute of Agricultural Science Symposium, Sydney.

Cregan, P. D. and Helyar, K. R. (1986), 'Non-acidifying Farming Systems', in *Proceedings of the 15th Riverina Outlook Conference,* pp. 49–62. Riverina-Murray Institute of Higher Education, Wagga Wagga.

Dalal, R. C. (1976), 'The Supply of Phosphorus from Organic Sources in Soil and Possible Manipulations' in G. J. Blair (ed.) *Prospects for Improving Efficiency of Phosphorus Utilization,* pp. 47–51, The University of New England, Armidale.

Donald, C. M. and Prescott, J. A. (1975), 'Trace Elements in Australian Crop and Pasture Production', in Nicholas, D. J. D. and Egan, A. R. (eds), *Trace Elements in Soil–Plant–Animal Systems,* pp. 7–37, Academic Press, New York.

Doll, E. C., Hatfield, A. L. and Todd, J. R. (1959), 'Vertical Distribution of Top-dressed Fertilizer Phosphorus and Potassium in Relation to Yield and Composition of Pasture Herbage', *Agronomy Journal,* 51, 645–8.

Epstein, E. (1972), *Mineral Nutrition of Plants: Principles and Perspectives,* John Wiley and Sons, New York.

Fernando, G. W. E. (1968), *The Influence of Temperature, Light and Fertilizer Nitrogen on the Yield and Chemical Composition of Oats (Avena sativa L.) and its Relation to Animal Production,* PhD thesis, University of Sydney.

Friesen, D. F. and Blair, G. J. (1988), 'A Dual Radio Trace Study of Transformations of Organic, Inorganic and Plant Residue Phosphorus in Soil in the Presence and Absence of Plants', *Australian Journal of Soil Research,* 26, 355–66.

Freeney, J. R. and Black, A. S. (1988), 'Importance of Ammonia Volatilizatin as a Loss Process', in J. R. Wilson (ed.) *Advances in Nitrogen Cycling in Agricultural Ecosystems,* pp. 156–173, C. A. B., Wallingford, U. K.

Gibson, A. H. (1966), 'The Carbohydrate Requirements for Symbiotic Nitrogen

Fixation: A "Whole-plant" Growth Analysis Approach', *Australian Journal of Biological Sciences,* **19**, 499–515.

Graham, R. D., Hannam, R. J. and Uren, N. C. (1988), *Manganese in Soils and Plants,* Klewer Academic Publishers, London.

Greenland, D. J. and Hayes, M. H. B. (1981), *Chemistry of Soil Processes,* John Wiley and Sons, Chichester.

Grundon, N. J. (1987), *Hungry Crops: a Guide to Nutrient Deficiencies in Field Crops,* Queensland Department of Primary Industries, Brisbane.

Hanway, J. J. (1971), *How a Corn Plant Develops,* Special Report No. 48, Iowa State University.

Hanway, J. J. (1977), 'Foliar Fertilizing Soybeans', *Crops and Soils,* **29**, 9–10.

Hanway, J. J. and Thompson, H. E. (1971), *How a Soybean Plant Develops,* Iowa State University Special Report No. 53.

Hanway, J. J. and Weber, C. R. (1971), 'Accumulation of N. P and K by Soybean Plants', *Agronomy Journal,* **63**, 406–8.

Haynes, R. J. and Sherlock, R. R. (1982), 'Gaseous Losses of Nitrogen' in Haynes, R. J. (ed.), *Mineral Nitrogen in the Plant–Soil system,* Academic Press, New York.

Heenan, D. P. and Carter, O. G. (1977), 'Temperature and Manganese Toxicity in Soybeans', *Plant and Soil,* **47**, 219.

Heenan, D. P. and Carter, O. G. (1976), 'Tolerance of Soybean Cultivars to Manganese Toxicity', *Crop Science,* **16**, 389–91.

Heenan, D. P., Campbell, L. C. and Carter, O. G. (1981), 'Inheritance of Tolerance to High Manganese Supply in Soybeans,' *Crop Science,* **21**, 625–7.

Helyar, K. R. and Porter, W. M. (1989), 'Soil Acidification, its Measurement and the Processes Involved', in Robson, A. D. (ed.), *Soil Acidity and Plant Growth,* pp. 61–109, Academic Press, Sydney.

Henzell, E. F. (1971), 'Recovery of Nitrogen from Four Fertilizers applied to Rhodes Grass in Small Plots', *Australian Journal of Experimental Agriculture and Animal Husbandry,* **11**, 420–30.

Holford, I. C. R. and Doyle, A. D. (1992), 'Influence of Intensity/Quantity Characteristics of Soil Phosphorus Tests on their Relationships to Phosphorus Responsiveness of Wheat under Field Conditions', *Australian Journal of Soil Research,* **30**, 343–56.

Jacques, G. L., Vanderlip, R. L. and Whitney, D. A. (1975), 'Growth and Nutrient Accumulation in Grain Sorghum: I. Dry Matter Production and Ca and Mg Uptake and Distribution. II. Zn, Cu, Fe and Mn Uptake and Distribution', *Agronomy Journal,* **67**, 607–16.

Jakobsen, P. (1974), 'Natural and Fertilizer Nitrogen in Streams and Lakes' in D. R. Leece (ed.), *Fertilizers and the Environment.* Australian Institute of Agricultural Science Symposium, Sydney.

Leece, D. R. (1968), 'The Concept of Leaf Analysis for Fruit Trees', *Journal of the Australian Institute of Agricultural Science,* **34**, 146–52.

Leece, D. R., Craddock, F. W. and Carter, O. G. (1971), 'Development of Leaf Nutrient Concentration Standards for Peach Trees in NSW', *Journal of Horticultural Science,* **46**, 163.

Lewis, D. C., Clarke, A. L. and Hall, W. B. (1981), 'Factors Affecting the Retention of Phosphorus applied as Superphosphate to the Sandy Soils in South-eastern South Australia', *Australian Journal of Soil Research,* **19**, 167–74.

Lipsett, J. and Simpson, J. R. (1971), 'Wheat Responses to Molybdenum in Southern NSW', *Journal of the Australian Institute of Agricultural Science,* **37**, 348.

Loneragan, J. F., Robson, A. D. and Graham, R. D. (1981), *Copper in Soils and Plants,* Academic Press, Sydney.

McKee, G. W. (1962), *Effects of Shading and Plant Competition on Seedling Growth and Nodulation in Birdsfoot Trefoil,* College of Agriculture Bulletin 689, Pennsylvania State University.

McLaughlin, M. J., Alston, A. M. and Martin, J. K. (1988), 'Phosphorus Cycling in Wheat Pasture Rotations. III. Organic Phosphorus Turnover and Phosphorus Cycling', *Australian Journal of Soil Research,* **26**, 343–53.

Mengel, K. and Kirkby, E. A. (1987), *Principles of Plant Nutrition,* International Potash Institute, Berne.

Myers, R. J. K. (1988), 'Nitrogen Management of Upland Crops: From Cereals to Food Legumes to Sugarcane' in Wilson, J. R. (ed.), *Advances in Nitrogen Cycling in Agricultural Ecosystems,* pp. 257–273, C. A. B., Wallingford, U.K.

Passioura, J. B. and Wetselaar, R. (1972), 'Consequences of Banding Nitrogen Fertilizers in Soil. II. Effects on the Growth of Wheat Roots', *Plant and Soil,* **36**, 461–73.

Peoples, M. B., Brockwell, J. and Bergersen, F. J. (1992), 'Factors Affecting Nitrogen Fixation: Scope for Improvements' in Angus, J. F. (ed.), *Transfer of Biologically Fixed Nitrogen to Wheat,* Grains Research and Development Corporation, Canberra.

Phillips, I. R., Black, A. S. and Cameron, K. C. (1988), 'Effect of Cation Exchange on the Distribution and Movement of Cations in Soils with Variable Charge. II. Effect of Lime or Phosphate on Potassium and Magnesium Leaching', *Fertiliser Research,* **17**, 31–46.

Reuter, D. J. and Robinson, J. B. (1986), *Plant Analysis: an Interpretation Manual,* Inkata Press, Melbourne.

Rudd, C. L. and Barrow, N. J. (1973), 'The Effectiveness of Several Methods of Applying Superphosphate on Yield Responses by Wheat', *Australian Journal of Experimental Agriculture and Animal Husbandry,* **13**, 430–3.

Sabey, B. R., Bathholomew, W. V., Shaw, R. and Pesek, J. (1956), 'Influence of temperature on nitrification in soils', *Proceedings, Soil Science Society of America,* **20**, 357–60.

Sanyal, S. K. and De Datta, S. K. (1991), 'Chemistry of Phosphorus Transformations in Soils', *Advances in Soil Science,* **16**, 1–120.

Simpson, J. R. and Freney, J. R. (1974), 'The Fate of Fertilizer Nitrogen Under Different Cropping Systems' in D. R. Leece (ed.), *Fertilizers and the Environment,* Australian Institute of Agricultural Science Symposium, Sydney.

Stein, J. A., Sageman, A. R., Fischer, R. A. and Angus, J. F. (1987), 'Soil Nitrogen Supply of Wheat in Relation to Method of Cultivation', *Soil and Tillage Research,* **10**, 23–58.

Storrier, R. R. (1965), 'Leaching of Nitrogen and its Uptake by Wheat in a Soil from Southern NSW', *Australian Journal of Experimental Agriculture and Animal Husbandry,* **5**, 323–8.

Strong, D. T. (1992), 'Early Plant Establishment Reduces Nitrate Leaching' in Helyar, K. R. (ed.) *Predicting Lime Response and Managing Acid Soil Infertility,* pp. 342–73. Final Report for the Wool Research and Development Council.

Tiller, K. G. (1989), 'Heavy Metals in Soils and their Environmental Significance', *Advances in Soil Science,* **9**, 113–142.

Vanderlip, R. L. (1972), *How a Sorghum Plant Develops,* Co-operative Extension Service, Circular No. 447, Kansas State University.

Weiss, M. G. (1943), 'Inheritance and Physiology of Efficiency in Iron Utilization in Soybeans', *Genetics,* **28**, 253–68.

Westerman, K. L. (1990), *Soil Testing and Plant Analysis,* (3rd edn), Soil Science Society of America, Madison, Wisconsin, USA.

Wetslaar, R. (1961), 'Nitrate Distribution in Tropical Soils', *Plant and Soil,* **15**, 110–33.

Wild, A. (1988), *Russell's Soil Conditions and Plant Growth,* (11th edn), Longman, Harlow.

White, R. E. (1987), 'Leaching' in Wilson, J. R. (ed.), *Advances in Nitrogen Cycling in Agricultural Ecosystems,* pp. 193–211, C. A. B., Wellingford, U.K.

Weir, R. G., Holland, J. F. and Doyle, A. D. (1987), *Zinc Deficiency in Field Crops,* Department of Agriculture N.S.W., Sydney.

CHAPTER 6

CULTURAL PRACTICES

J. E. Pratley and E. J. Corbin

The history of crop production in Australia includes a series of disasters in terms of soil degradation (Callaghan and Millington, 1956; Jenkin, 1986; Pratley and Rowell, 1987). Following the clearing of trees, European and then North American farming methods were used on Australian soils until the 1930s when severe soil erosion occurred over large tracts of southern Australian cropping lands. These events, as well as contributing to the formation of soil conservation authorities, led to the development of the ley-farming system of agriculture, particularly in southern Australia, in which crops are grown in rotation with pasture legumes. In the summer rainfall areas of the north, continuous cropping systems involve summer and winter species.

Farming systems have continued to evolve, with greater emphasis being placed on the environmental impact of agricultural practices. Eutrophication of waterways by fertilisers and pollution by pesticides, together with increasing soil acidification and soil salinisation, are issues with which farmers need to contend. At the same time, productivity needs to increase in order for financial viability to be realised.

This chapter discusses the cultural practices necessary to address these issues of sustainability, particularly as they relate to dryland crop production.

FARMING SYSTEMS

Farming systems in Australia differ according to the climatic (Chapter 2) and soil (Chapter 4) constraints of the environment. In the winter rainfall areas of southern Australia, crop production is confined to the cool season, coinciding with the incidence of rainfall. The soils are, to a large extent, structurally fragile and inherently low in nutrients. The production of crops is integrated with livestock production, thus providing through diversification a buffer against economic fluctuations. The combination of pastures and crops in rotation is called ley farming.

In the summer rainfall areas of northeastern Australia, continuous cropping is more common. Soils are usually better structured and more fertile than in the south. Further, the capacity to grow both rain-fed summer crops and stored-moisture winter crops can allow the biological and economic diversity required for a viable farming system.

Ley Farming

This farming system was introduced into southern Australia as a means by which the destructive effects of intensive cultivation on soil physical properties and soil fertility during the cropping phase could be counteracted by a period of undisturbed pasture. The length of time required for a pasture phase to restore soil fertility and physical conditions depends on pasture composition and vigour and the degree of soil degradation. In the restricted time span of the pasture phase (Figure 6.1), shallow-rooted annual legumes are limited in their ability to fully restore soil structure degraded by the excessive tillage of traditional farming practices (Stoneman, 1973; White *et al.*, 1978; Pratley, 1987). The length of the respective phases varies across the cropping zone. In the west and south the phases are short, a crop:pasture ratio of 1:1 or 2:2 being common. In the eastern portion longer rotations are the norm, being 3 to 5 years of crop followed by a comparable period of pasture.

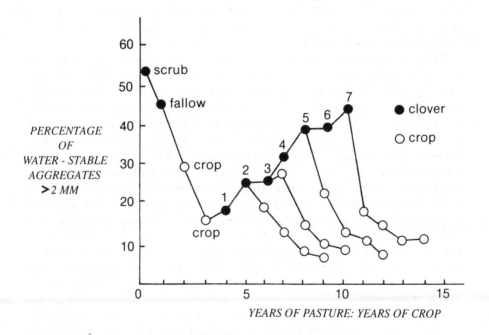

FIG. 6.1 Changes in water-stable aggregation of soil (0–7.5 centimetres) under differing rotations since clearing on Wongan loamy sand (adapted from Stoneman 1973)

Pasture Phase

It follows that if the pasture phase is a recuperative phase for soil structure and fertility, then close attention to the successful establishment of highly productive, well-nodulated pasture legumes is essential. In practice however, farmers often overlook this component.

Traditionally, subterranean clover (*Trifolium subterraneum*) has been the basis of leys in the better rainfall areas of the south, whilst medic pastures (*Medicago* species) have been used in the lower rainfall areas. Increasingly, however, lucerne (*Medicago sativa*) is being used because it extends the grazing period each year and its deep taproot provides an opportunity to utilise deep soil moisture. Greater options are now available to farmers with the development of improved cultivars, particularly of burr medic (*M. polymorpha*), murex medic (*M. murex*) and serradella (*Ornithopus compressus*) (Gillespie, 1989).

In subtropical and tropical northern Australia, pasture phases based on tropical legumes are feasible (McCown *et al.*, 1985) but ley-farming systems are rarely practised. The main contributions of the pasture phase are an enhancement of soil physical properties and an increase in soil nitrogen levels. The physical changes include an overall increase in porosity, an increase in aggregate and microaggregate stability, a decrease in bulk cohesive strength and an increase in water retention and consequently in available soil water. Improvements in water entry and permeability and decreases in runoff and erosion arise as a result of these changes (Greenland, 1971).

The rate of soil nitrogen accumulation varies with soil type, initial soil nitrogen level, pasture legume species and the environmental and management conditions prevailing during the ley phase. The average annual rate of accumulation in soils under subterranean clover pastures in medium rainfall areas of southern Australia appears to be in the order of 40 to 80 kg N per hectare (Donald and Williams, 1954; Russel, 1960; Watson, 1969; Williams, 1970; Simpson *et al.*, 1974; Carter *et al.*, 1982). This accession is in the form of organic nitrogen, which requires mineralisation to the inorganic ammonia or nitrate forms before being available to plants. Nitrogen accumulation is enhanced by increased use of phosphatic fertilisers under conditions of low soil phosphorus (Watson, 1969). Such is not the case, however, where a soil has had a history of regular superphosphate applications which have raised the soil phosphorus to adequate levels (Kohn *et al.*, 1977). It must be emphasised that the required role of the pasture phase depends upon good establishment and management.

In Western Australia and South Australia the short rotations provide for the natural regeneration of the annual legumes. This depends on high seed set at the end of the previous pasture year with sufficient hard-seededness to carry through the cropping year. Factors identified by Thorn (1989) which interfere with this practice include:

- using herbicides which prevent seed production by the pasture legume;
- extending the cropping phase, which has the effect of reducing seed reserves, particularly in subterranean clover;
- poor pasture years or high-pressure grazing not allowing adequate seed production (the effect is greater on subterranean clover than on medic pastures);
- hard-setting surface soils brought about by excessive tillage;
- tillage practices which bury the seed too deep for emergence; and
- high burdens of crop straw carryover.

Carter *et al.* (1982) reported that only 20% of stubble paddocks surveyed in South Australia had satisfactory seed reserves.

In south-eastern Australia the pastures are resown, usually under the last crop. This is a compromise between a loss in crop yield and successful establishment of the undersown pasture, since competition between them is inevitable. Dear (1989) demonstrated the effect of undersowing subterranean clover to wheat at Wagga Wagga (Figure 6.2),

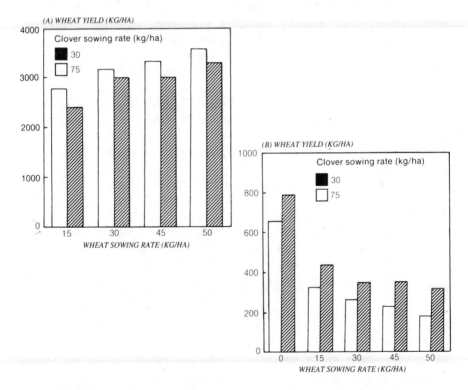

FIG. 6.2 The relationship between a wheat cover crop and undersown subterranean clover at Wagga Wagga (Dear, 1989); (a) The effect of wheat and clover sowing rate on wheat grain yield, 1987; (b) The effect of wheat and clover sowing rate on clover seed yield, 1987

where a reduction in wheat sowing rate reduced wheat yield, but it is necessary if the required seed production of about 400 kg/ha of the clover is to be obtained for productive regeneration and maximum winter yield of the pasture in the following year.

The effect on pasture establishment can be expected to be greater as rainfall declines due to competition for moisture. The impact is also greater on perennial pasture species such as lucerne which have to survive the first summer.

Poor pasture establishment results in a poor pasture phase, reducing the ability of this phase to assist the recuperation of soil fertility and soil structure and to provide high-quality feed for livestock. Farmers must therefore be prepared for lower yields in the final crop if the productivity of the whole rotation is to be enhanced.

The pasture phase also makes an important contribution to weed and disease control. Subterranean clover and lucerne leys are largely responsible for the suppression of skeleton weed (*Chondrilla juncea*) to manageable levels for crop production (Cuthbertson, 1967; Wells, 1969, 1970; Groves and Williams, 1975). This effect is shown in Table 6.1. In subterranean clover pasture, however, phosphate application can contribute to an early invasion of annual grasses (Kohn, 1974, 1975; Ayers *et al.*, 1977a) such as annual ryegrass (*Lolium ridigum*), barley grass (*Hordeum leporinum*) and silvergrass (*Vulpia* spp.), which are important weeds of cereal crops and hosts for a number of cereal diseases (Butler, 1961; Rovira, 1987). Managing such swards in the final year or two of the phase is critical to minimise the carryover of these diseases (Kidd *et al.*, 1992; McNish, 1989).

Management options for this purpose (as reviewed by Leys, 1990) include:

- spraytopping of annual grass species at early seed set with paraquat, paraquat/diquat mix or glyphosate to render the weed seed unviable (Pearce, 1973; Jones *et al.*, 1984; Davidson, 1992);
- pasture cleaning with simazine, paraquat or a simazine/paraquat mix during early winter (Leys and Plater, 1993);
- fodder conservation and heavy grazing at seedset;
- burning of grass residues, a process limited by seasonal fire regulations (Pearce and Holmes, 1976; Davidson, 1992).

The earlier the removal of grasses, however, the greater will be the control of the diseases, since sufficient times and proper conditions must be allowed for microbial breakdown of the infested plant residues. The hot dry summers of the southern cropping areas are not conducive to microbial breakdown.

The use of low rates of phenoxy herbicides on young broadleaf weeds of rosette habit, such as Paterson's curse (*Echium plantagineum*) and capeweed (*Arctotheca calendula*), minimises damage to the pasture legume, but the chemical mobilises the weed plant sugars, making it sweeter, and distorts growth, making the weed more accessible to the grazing animal (Pearce, 1973).

TABLE 6.1 The effect of lucerne ley on the skeleton weed population, soil nitrate nitrogen, wheat yield and protein content, on solonised brown soils of the Victorian Mallee (adapted from Wells, 1970)

Site	Skeleton weed population (plants/m^2)	Soil nitrate nitrogen 0–15 cm (ppm)	Wheat yield (t/ha)	Wheat protein (%)
Daalko				
Volunteer	479	6.1	1.03	6.4
Lucerne	17	11.0	1.57	13.7
Boulka				
Volunteer	263	2.2	0.10	10.5
Lucerne	71	11.0	0.72	12.4
Chillingollah				
Volunteer	372	5.1	0.35	12.4
Lucerne	19	8.3	1.32	12.7
Patchewollock				
Volunteer	372	1.7	0.17	9.0
Lucerne	46	5.2	1.25	12.1

The pasture phase has also become an important component of the strategy to manage herbicide resistance. Where herbicide options are reduced in the cropping phase due to this problem, grazing, haymaking, spray-topping and winter cleaning all become mechanisms by which the seed production of resistant plants is minimised (Powles, 1987; Powles and Holtum, 1990).

Legume dominance in pastures can also be maintained by grazing management. FitzGerald (1976) indicated that high stocking rates under a set-stocked system achieved this objective on subterranean clover pastures in southern New South Wales.

Pasture management practices have been reviewed elsewhere (Pearson and Ison, 1987; Pratley and Godyn, 1991). However, it is important to recognise the nutritional needs of the pasture in respect of its contribution to the subsequent cropping phase. In the ley farming system, fertiliser is usually only applied with each crop. Unless provision is made with each crop application for the subsequent pasture phase, the phosphorus status of the soil will decline, with implications for nitrogen fixation and also for the phosphorus nutrition of the next cropping phase. These effects are shown in Figure 6.3. Also of importance in this context is that farmers prefer to use the more concentrated forms of fertiliser such as double super, monoammonium phosphate (MAP) and diammonium phosphate (DAP) because handling is reduced and cost per fertiliser unit is often lower. Depending on the fertiliser used, sulfur may become limiting in the pasture phase because legumes have a relatively high sulfur requirement.

A wise precaution, therefore, is to use single superphosphate or sulfur-fortified double superphosphate with the last crop in the cropping phase.

If the soil is acid, the application of molybdenum is also important at this time to ensure effective nodulation of legumes for nitrogen fixation.

Cropping Phase

At the end of the pasture phase it can be assumed, if proper management has occurred, that the fields to be cropped would:
- be relatively free of weeds, particularly of annual grass weeds;
- have surface soils somewhat compacted by grazing livestock; and
- have a reasonable nitrogen fertility status, most of the nitrogen being in the organic form.

Thus decisions need to be made about the sequence of crops to be grown and the tillage and crop stubble practices to be performed, and the issues of good crop husbandry need to be addressed.

CROP SEQUENCE

The principles involved in deciding on the crop sequence are relatively simple. To maximise yields and minimise the buildup in disease levels, successive crops should not be the same or similar species. Thus, cereals should be interspersed with non-grass crops. In the winter rainfall areas, oilseed crops and grain legumes provide the necessary biodiversity, although in practice it is common for farmers to grow successive cereal crops despite the disadvantages.

With nitrogen fertility being at its highest level following pasture, crops having a greater N requirement, such as canola and hard wheats, are grown in the initial years of the cropping phase. As the fertility declines,

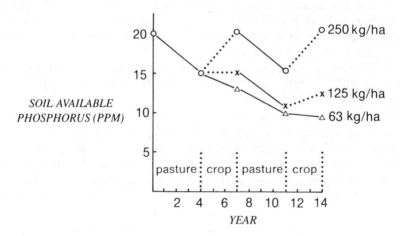

Fɪɢ. 6.3   The calculated effects of different levels of single superphosphate, applied during the cropping phase, on soil available phosphate in the top 10 cm of a red earth soil at Wagga Wagga, New South Wales (G. J. Osborne and G. D. Batten, 1976, personal communication)

grain legumes, such as lupins and field peas, provide the disease break for cereals and contribute to nitrogen fertility through nitrogen fixation. Soft wheat varieties and malting barley are grown later in the cropping phase, as high grain N levels are less important for the former and undesirable for the latter.

The contribution to N fertility by grain legumes results from their inefficiency in using soil N and from the return of organic N in the residue material remaining in the field (Evans *et al.*, 1987a). Crops with a high harvest index (that is, high grain yields are harvested from the crop) contribute less to soil N status than crops with a low harvest index, since most of the nitrogen is contained within the grain of such crops. The resultant soil N status is also affected by the species of legume and by sowing time, which also interact. For example, summer growing legumes will contribute more than winter legumes due to better nitrogen fixing conditions. Lupins also contribute more than field peas because of their longer growing season (Evans *et al.*, 1987b).

The rotation of crops also provides an advantage to producers in allowing a wide range of options for chemical control of pests. In particular, the degree of selectivity required for grass control in broadleaf crops is much less than for the same weeds in cereal crops. This greater variation in options provides an opportunity to better manage the development of herbicide resistance by allowing herbicide groups to be rotated.

Tillage Practices

Up until the 1970s in Australia, seedbed preparation involved numerous passes with cultivation equipment, resulting in substantial soil physical degradation (Pratley and Rowell, 1987). Since that time much attention has been directed towards minimum tillage techniques which preserve soil structure and facilitate crop establishment by enabling the timing of operations to be optimised. Traditional tillage practices are covered in detail by Cornish and Pratley (1987; 1991), who described the effects of cultivation in breaking down organic matter, degrading soil surface structure, increasing erodibility and creating compaction layers at plough depth. A summary of the actions of various tillage implements is given in Table 6.2, and their effect on soil is shown in Figure 6.4.

Modern methods rely heavily on herbicides instead of cultivation to control weeds. This is less damaging to soil structure, consequently improving the infiltration of moisture.

Fields coming out of the pasture phase will normally be compacted to some extent on the surface due to the presence of grazing animals. Unless the soil is particularly well structured, a cultivation with a tined implement is usually necessary to facilitate the subsequent penetration by sowing equipment. Thereafter, weed growth is managed by regular grazing by livestock, and final kill is achieved through non-selective 'knockdown' herbicides such as paraquat/diquat mixes and glyphosate.

TABLE 6.2   A comparison of the actions of various tillage implements (after Orchiston, 1965)

Action	Mouldboard plough	Disc implements (ploughs and harrows)	Tined implements (chisel ploughs, scarifiers, harrows)	Rotary implements (hoes, tillers)	Rollers
Wear	Shares require replacement	Superior to all others and remain efficient with wear of discs	Tilting action of stump-jump types wears points — new designs correct this	Comparatively rapid but generally self sharpening	Very little from rolling friction
Negotiation of obstacles	Effective if stump-jumping type	With ease without spring loading	Effective with stump-jump action	Effective if spring loaded	With ease
Movement of soil	Complete inversion of furrow at normal speeds	Partial inversion of furrow	Narrow furrows no subsoil brought to surface	A cut, slice, lift action thoroughly mixes soil	Surface compacting effect
Soil resistance	Increases with speed	Increases markedly with speed	Fairly constant all speeds — best for high speeds	Tends to decrease with speed	Constant at all speeds
Grinding or pulverising soil	Very little	Tendency; varies with adjustment — can be substantial	Less than other types	Extensive	Very little
Shattering	Completely in all directions	Partial — can be substantial	Partial	Completely; with shield	Partial
Compacting at depth of cultivation	Considerable if soil is too moist	Considerable if soil is too moist	Very little	Very considerable if soil is too moist	Absent

TABLE 6.2 continued

Action	Mouldboard plough	Disc implements (ploughs and harrows)	Tined implements (chisel ploughs, scarifiers, harrows)	Rotary implements (hoes, tillers)	Rollers
Fineness produced	Very little	Considerable	Moderate	Very considerable	Considerable
Surface trash	Well buried at normal speeds	Buries a large proportion	Left on surface	Buries in large proportion	Left on surface
Killing of old crop	Effective	Effective and ideal for scrub clearing	Partial	Effective	No effect
Seedbed preparation	Requires secondary tillage	Useful for general ploughing and cultivation. Requires secondary tillage	Ideal ploughing, crop cultivation and scarifying without making soil too fine	Makes soil fine in one operation	For compaction under drier conditions
Build-up and conservation of moisture	Effective especially on contour, unless plough pan developed	Effective unless plough pan developed	Well suited to infiltration	Surface mulching	Bring moisture to surface
Control of weeds	Unsuited	Excellent	Can be excellent	Excellent	No effect
Renovation of crop or pasture	Unsuited	Too severe	Ideally suited	Unsuited	Unsuited

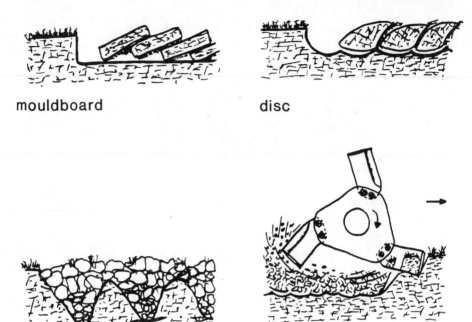

mouldboard   disc

chisel   rotary hoe

FIG. 6.4   A comparison of the effects of primary tillage implements on soil

Sowing then proceeds by drilling the seed into the minimum tilled soil. Where soils allow or seasonal conditions dictate, the pastures are sprayed and direct drilled without the cultivation process.

The timing of the tillage, where practised, also varies. Some farmers cultivate in late winter/spring to take advantage of favourable moisture conditions and to prevent weed seed set. Alternatively, the soil is cultivated at the break of season in autumn.

The need for good pasture management in the last year has been described: that is, removal of weeds through winter cleaning, spray-grazing and spray-topping. Also important is the reduction in the carryover of surface vegetation, particularly silvergrass (*Vulpia* spp.) the residues of which have strong allelopathic capabilities and can interfere significantly in the establishment of crops sown into such residue (Pratley, 1989; Pratley and Ingrey, 1990).

Stubble Management

In fields where crops have been grown in the previous season, an extra management consideration is the treatment of stubbles from the crop. The options available are grazing, incorporation, burning, or retention on the surface. A combination of grazing and another option is the usual practice. In the winter rainfall areas in particular, the high cereal yields produce a high physical stubble burden to manage. A 4 tonne/ha wheat crop produces of the order of 6 t/ha stubble, which usually goes through a dry summer without deterioration and becomes a problem for any mechanical operation. The grazing process thus becomes important in reducing the amount of stubble. Animals are usually introduced to clean up any spilt grain and weeds but modern farming requires more efficient utilisation of stubble, which has several advantages:

- a source of energy for livestock during a period of the year when feed is lacking;
- the spelling of pasture paddocks to allow species such as lucerne to produce forage for autumn; and
- a reduction in the stubble load for subsequent cropping operations such as spraying and sowing.

Research has shown that stubble utilisation is improved by:

- stocking immediately following harvest, because the stubble feed value is at its highest prior to its deterioration through rain and dew; and
- stocking with sheep initially to enable them to select out the more digestible fractions (Coombe, 1981). Cattle digest lower-quality material better than do sheep and can thus maintain their weight for longer. Mulholland *et al.* (1976) found that sheep could maintain their weight for up to 6 weeks on stubble, whereas cattle maintained their weight for up to 12 weeks on the same stubble.

Supplements such as urea, lupin grain and sulfur can improve utilisation and animal performance.

The treatment of the stubble at harvest time will also be a determinant of the success of its subsequent management. Older headers create a trash trail whereby the straw and chaff passing out the back is concentrated in a width less than a third of that of the machine. This concentration affects the physical handling later and also interferes with herbicide effectiveness.

Most modern headers can spread the straw so that it is less concentrated, but the chaff remains concentrated. The straw spreaders consist of horizontally rotating fingers. Straw choppers can assist by reducing the length of straw, and 'straw storms' reduce the straw length and blow the straw and chaff over about two-thirds of the width of the machine, but these attachments use more power and slow down the harvest process.

In a separate operation from harvesting, the standing straw can be slashed or mulched into shorter lengths, the only limitation being the risk of fire from these operations at that time of year.

Stubble Incorporation

This procedure involves the mixing of the stubble with the upper layers of soil. Turning stubbles into the soil usually causes of short-term nitrogen depletion due to the large carbon:nitrogen ratio of the material, although in the longer term soil organic matter levels benefit. Unsatisfactory decomposition can result in blockages in closely spaced tined implements that may be used in later stages of seedbed preparation.

The time required for decomposition varies with different conditions, but moisture availability is a most important factor. In lower rainfall environments, more time is usually required. Where soil temperatures are unsuitable for microbial action, the rate of decomposition will also be slower. The incorporation usually involves a disc implement, the use of which is discouraged in modern farming practice because of likely detrimental effects to soil structure.

Stubble Burning

This is the most common means of removing a physical problem during crop establishment. Traditionally, stubble was burned in autumn, as soon as conditions and fire regulations allowed. However, the role of surface cover in reducing the rate of evaporation from the soil has now resulted in burning much closer to time of sowing.

Burning is a quick, inexpensive and effective way of removing the physical burden of the straw, particularly where suitable machinery from trash working is not available. Disease carryover is also reduced. The disadvantages of burning include the loss of soil organic matter and the sublimation of nitrogen and sulfur into the atmosphere.

Stubble retention

This procedure involves the maintenance of the residues from the previous crop on the soil surface with the succeeding crop being sown through the bulk material. The advantage of better soil moisture conditions is important, and the retention of organic matter and protection from raindrop impact are also most important. Because the material is not incorporated, soil nitrogen is not locked up to any significant effect.

However, under the winter rainfall conditions of southern Australian cropping areas, the quantity and toughness of the material at the beginning of the new season create blockage problems for machinery. The passage of tined implements through the bulk of residue is described by Brown *et al.* (1986) as being affected by:

- the quantity of stubble, blockages occurring more readily with the presence of more stubble;
- the moisture content, high moisture raising the mass density of the material, the strength and toughness of the straw, and the tendency for soil and stubbles to intermix;
- the anchorage of stubble providing easier passage;
- the length of detached straw, shorter lengths allowing easier passage because they are less likely to bend ('hairpin') around the tines of the implement;
- the trash clearance of the implement, including more underframe clearance; greater horizontal distance between adjacent tines; greater

longitudinal distance between rows of tines; and adequate clearances adjacent to the wheels.

Most trash clearance implements have disc coulters preceding the tine to aid trash flow.

Continuous cropping in summer rainfall areas

In the summer rainfall areas of northern Australia, the opportunity exists for the production of rain-fed summer crops and the production of winter crops on fallow moisture. Combinations of winter and summer crops are common. Compared with southern Australia, pastures and livestock play only a minor role in the farming systems, and continuous crop and fallow cycles are practised (Holland *et al.*, 1987).

A characteristic of these areas is the high erosiveness of the summer storms, particularly where the soil is in fallow (Marston, 1978). The need for conservation farming practices of no till and stubble retention has long been realised. The effect of such practices is demonstrated in Table 6.3.

TABLE 6.3   The effect of soil management practices, soil type and slope on soil loss in south-east Queensland (Cummins, 1973)

	Soil loss (t/ha/year)		
Soil type	Bare fallow	Residue incorporation (average management)	Residue mulch (good management)
Alluvial, fertile, self-mulching clay (1–2% slope)	60	24	12
Colluvial red-brown clay (1–3% slope)	76	31	12 (with contour banks)
Colluvial red-brown clay (5–8% slope)	270	110	10 (with contour banks and crop rotation)

Stubble Retention

As in southern Australia, the retention of the stubble from the previous crop is promoted as an important conservation practice. However, it has been more widely adopted in the north for much longer because the soil erosion problem is much greater. The stubble protects the soil surface from raindrop impact and reduces the rate of flow across the surface, both of which aid water infiltration and reduce the extent to which the soil erodes (Shaw, 1971a; Garland *et al.*, 1976; Marston and Doyle, 1978; Rosewell and Marston, 1978; Lovett *et al.*, 1982).

Unlike the south, the crop residues are present during the wet season and therefore undergo considerable breakdown and cause fewer problems for machinery. However, specialised machinery has been developed for these areas to keep the residues on the soil surface during the fallow.

For primary tillage broad sweep-type implements called blade or sweep blade ploughs are pulled through the soil leaving the surface residues largely undisturbed. The two sweeps of the blade are at angles of about 105° and sweep lengths vary from 60 to 250 cm. Chisel ploughs with improved trash clearance are increasingly used because of their versatility (Marston and Garland, 1978; Sweeting and Colless, 1985).

Secondary tillage can be carried out by rod weeders, consisting of a revolving bar or rod pulled through the soil just below the surface. Weeds are pulled out by the twisting action of the rods. On heavy soils or on stony ground, however, the sweep plough and rod weeder are not suitable and tined implements are used (Garland *et al.*, 1976). Rod weeders are often attached to chisel ploughs, and return some of the trash to the surface after burial by the chisel plough (Sweeting and Colless, 1985).

Increasingly, however, no-till farming is practised in these areas, using herbicides to control weed growth.

Strip Cropping

Many of the soils in the region are black earths, which are fine-textured cracking clays with self-mulching characteristics, often occurring on flat or gently sloping areas. Infiltration rates on these soils during summer thunderstorms can be low due to sealing of the surface, and rainfall runoff rates and hence soil erosion are high (Cummins and Esdaile, 1972; Rosewell and Marston, 1978).

The technique of strip cropping has therefore become increasingly important in these areas as a means of controlling soil erosion. Crops are grown in a systematic arrangement across the slope so as to act as barriers to the flow of water. The water is spread over larger areas, slowing the rate of flow and as a consequence increasing the infiltration of the water into the soil (Hoogvliet, 1966; Kelsey, 1966; Shaw, 1971b). At any one time, therefore, there are strips of crop, stubble or fallow retarding the movement of water, thereby reducing the flow rate over the fallowed strips. The width of the strips is limited to about 100 metres. A typical system has strips of sorghum, wheat stubble and fallow. An example of this system is shown in Plate 6.1.

Strip cropping is usually limited to those areas with a landslope of 1 per cent (1 metre per 100 metres), although it can be used on landslopes up to 2 per cent with some earthworks to support it (Crawford, 1977). Cultivation approximating the contour in a strip cropping scheme also aids in reducing runoff and increasing infiltration.

One problem that is created by a strip cropping program is the location of paddock fences and farm boundaries. In this system it is desirable to have the strips as long as possible, preferably as long as the landslope and

PLATE 6.1   The layout of a strip cropping system at Quirindi, New South Wales. The dark strips represent strips prepared for wheat, the light strips are sorghum stubble. The narrow dark lines represent erosion control earthworks. (Photograph reproduced by permission of Soil Conservation Service of New South Wales)

other conditions will allow. However, property boundaries are often poorly placed in relation to natural drainage lines and many internal fencelines need to be removed and resited. This will often require co-operation between neighbouring farmers. The removal of internal fences means, however, that the area cannot be stocked unless systems of electric fencing are installed to control the livestock (Crawford, 1977).

Fallowing

Fallowing, the process of maintaining the soil weed-free for an extended period before crop sowing, has been practised in Australia since the early 1900s. Its main purpose is to store moisture from the previous rainfall season for use by the current crop, although nitrogen benefits also accrue through mineralisation. The amount of water stored by the fallow depends on rainfall during the fallow period, the length of the fallow, infiltration capacity of the soil, water holding capacity of the soil and the rate of moisture from the soil.

Fallowing has been an integral part of farming in the lower-rainfall areas of South Australia, Victoria and southern and central New South Wales (Sims, 1977; Ridge, 1986; Poole, 1987) but the increased use of reduce tillage systems, the retention of crop residues and the widespread adoption of annual legume pasture leys have made fallows less important.

French (1963) indicates that in South Australia three conditions need to be fulfilled before benefits in moisture conservation from fallowing are likely to be obtained. The first condition is that the rainfall during the growing season is significantly limiting crop production. In this context, French suggests that rainfall in the April to October growing period for wheat production should be less than approximately 460 mm before any benefits are gained. Research in other areas of Australia has been consistent with this finding (Kohn *et al.*, 1966; Tuohey *et al.*, 1972; French, 1978b).

The second condition is that the rainfall during the fallow period must be sufficient to wet the soil to a depth of at least 15 to 20 cm to avoid significant losses due to evaporation from the soil surface. This condition indicates how long the fallow needs to be in order to coincide with such rainfall. There is also the need to maintain the area free from actively growing plants during the fallow to avoid this moisture being lost via transpiration prior to crop establishment.

The third condition is that the physical features of the soil are conducive to water storage. In particular, it is necessary that the subsoil be relatively fine-textured as light soils have low water-holding capacity with the moisture being lost by percolation. Surface soil characteristics have to be conducive to ready infiltration. Ridge (1986) indicates that a clay content greater than 25 per cent is needed for moisture retention in north-western Victoria.

What is also clear from research is that a summer fallow *per se* is not efficient for moisture conservation. Any benefits in moisture content from fallow during that period result largely from minimising moisture loss by restricting transpiration, rather than by gains through rainfall which is largely evaporated in the hot conditions (Wells, 1970, 1971). In southern Australia, the winter–spring and autumn rains are most important, the fallow retaining more of the rain of the previous winter than grassland (Schultz, 1971, 1972). In Queensland, Waring *et al.*, (1958) indicated that the autumn rains were more significant than those of the summer. Nevertheless, in the northern cropping areas a summer fallow is important for the successful production of winter crops (Fawcett and Carter, 1973).

In traditional farming systems the fallow is cultivated as required to destroy weed growth and improve surface soil infiltration characteristics. The reliance on cultivation for fallow weed control has diminished as concerns for soil erosion and soil structural degradation are increased and as herbicide options have increased resulting in the development of no-till fallow techniques. Herbicides seldom completely replace cultivation, particularly because of cost, but the number of tillage operations is substantially reduced, with the positive effect of lessening the extent of soil structure degradation and erosion. The development of equipment that can discriminate green vegetation from soil or litter (Felton *et al.*, 1987) enables a substantial reduction in the amount of herbicide used and therefore the cost involved. The herbicides chosen should, wherever possible, have a different mode of action from those used for in-crop weed control, to reduce the likelihood of the build-up of herbicide resistance.

Residue retention during fallow periods can also improve infiltration (via channelling) but the main effect is to reduce the rate of evaporation from the soil surface (Lovett *et al.*, 1982; Pratley and Cornish, 1985) and to protect the soil surface from raindrop impact.

The levels of moisture conserved vary depending on the environment. In a South Australian study extending over five years, an average of 25 mm of additional water was stored as a result of fallowing in the preceding September (French, 1963). At Wagga Wagga, in southern New South Wales, additional moisture was stored in only one out of four years (Kohn *et al.*, 1961). In central western New South Wales, September fallow stored an addition 57 mm of moisture (Fettell, 1977) in one instance. The benefits of a fallow should be appraised in the light of loss of alternative income while the land is in this non-productive condition.

Fallow Efficiency

Because water is the ultimate determinant of yield in dryland farming in Australia, particularly in regions where fallowing is practised, it becomes critical to productivity to maximise the water available to plants. Thus the percentage of rainfall eventually stored in the soil for crop use (the fallow efficiency) is an important criterion for raising yields. Fallow efficiencies are low relative to water use under rain-fed conditions, being of the order of 15–30 per cent (Cornish and Pratley, 1991). The efficiency is usually reduced by cultivation but improved by residue retention due to a slower rate of evaporation.

SOWING

Time of Sowing

While each crop has an optimum sowing time, the date of actual sowing will ultimately be determined by the moisture conditions of the soil.

Delays in sowing after the optimum date may be caused by either insufficient or excessive soil moisture. Soil moisture levels in the period preceding sowing can have an indirect effect on the time the crop is sown. Insufficient or excessive moisture levels may interfere with the timing of tillage and sowing may be delayed until the required number of operations have been effected. Where direct drilling is used, optimum sowing times are more likely to be achieved because the bearing strength of the soil has not been reduced by previous cultivation.

Irrespective of the presence of a satisfactory seedbed at the optimum sowing date, sowing may be either brought forward or delayed due to particular circumstances. If a seedbed already has a high moisture content and the prospects of further, delaying rains are high, the decision may be made to sow a crop earlier than the optimum time. This will obviate an excessive delay should further rain make the seedbed too wet for the satisfactory operation of sowing implements.

Excessive soil moisture at sowing can cause too fast a rate of absorption and seed swelling, resulting in 'bursting' of the seed. In addition, water-logging of the seedbed during germination results in anaerobic conditions. These can interfere with the physiological processes in the seed and may also favour the development of fungal or bacterial pathogens on the seed.

Sowing time may be delayed beyond the optimum if insufficient moisture is available for germination. For each crop there is a minimum threshold level of available moisture below which rapid and even germination will not occur. This threshold level is influenced by seed size and the absorption capacity of the seeds. High-protein seeds such as the grain legumes require more available moisture than lower protein-containing seeds such as grasses.

Practical considerations may force a farmer to sow a crop even though seedbed moisture is less than ideal. This practice has considerable risk if enough moisture is present to initiate but not complete the germination processes. Rotting of the seed will inevitably occur unless rain falls within a few days of sowing. The risk is not present if the seedbed is very dry in the surface 8 to 10 cm. The difficulty in sowing under these very dry conditions lies in deciding at which depth to sow the seed. A shallow sowing may lead to initiation of germination by a relatively light fall of rain which may be insufficient to carry the seed through to emergence and survival until the next rain. Deeper placement may lead to difficulties in emergence, particularly where rainfall is sufficient to crust or slake the surface soil. More energy is required for seedlings to reach the surface from deeper in the soil.

A major determinant of optimum sowing time for summer crops is soil temperature. For these crops a certain minimum soil temperature is required before germination will take place, so sowing earlier in colder soils again increases the likelihood of deterioration of the seed in the soil. A soil temperature of 18°C, for example, is required for satisfactory germination of maize.

The optimum sowing time is also influenced by the occurrence of frost in relation to the timing of flowering of the crop. For summer crops, the sowing time chosen should ensure that flowering, seed set and maturation are completed before the expected first frost.

For winter-growing crops, the desirable sowing time is one that will result in the crop flowering after the last expected date of a killing frost in spring, but before the onset of the dry period in the summer (Single, 1971, 1975). The period from pre-anthesis, through flowering, into grain filling is one of increasing water demand. At this stage, rainfall is decreasing in the Australian winter cereal areas and evaporation is increasing (Nix, 1975), thus leading to soil water deficits which restrict grain yield. For each area, successful winter crop production depends on a sufficiently long period between these two major constraints for flowering and grain maturation to proceed without significant interference. Within an area, the availability of a range of crop cultivars with inherently different development patterns allows the farmer to choose from a number of optimum sowing dates depending on the availability of moisture at sowing time.

The significance of frost injury is indicated by the impact of two September frosts in 1969 when wheat yields were reduced in New South Wales and Queensland by an estimated 1.36 million tonnes (Single, 1975). The coincidence of frost and flowering impairs the reproductive function and its significance is much greater with determinate than indeterminate crops. In the latter case, where flowering is spread over a period of time, as in lupins, frost may only affect a proportion of the flowers. In determinate crops such as wheat where the whole crop flowers simultaneously, frost at that time virtually eliminates the possibility of a profitable yield.

Delayed sowings generally can be expected to yield less than sowings at the optimum time. Such sowings often experience water deficits from just prior to anthesis through the critical flowering and post-flowering period (Morgan, 1971). The decline in yields from delayed sowings has been shown for wheat by Kohn and Storrier (1970), who found a 3.7 per cent decrease in the yield of Heron wheat for each week's delay after the end of April in the southern slopes of New South Wales. Constable (1977) found an 8 per cent decline for each week's delay after mid-December for soybeans in the Namoi Valley. Similar results have been shown for delayed sowings of sorghum (Stern, 1968a), lupins (Farrington, 1974; Walton, 1976), peanuts (Stern, 1968b) and other field crops.

Under some circumstances, sowing a particular crop may be postponed to a stage where a change in cultivar cannot compensate for the delay in sowing time. In this situation, a change to a shorter-season crop is warranted. Occasionally, late breaks to the season in the southern Australian wheat belt are associated with an increase in the area sown to barley and a decline in the area sown to wheat. Barley is able to achieve more profitable yield levels in a shorter period of time than wheat.

Rate of Sowing

The rate of sowing for any crop will be determined by the plant populations required to achieve optimum yield in a particular environment. A consideration of the relationship between crop density and yield is therefore necessary.

The potential yields of any crop can only be obtained when the competition for the growth factors such as water, nutrients and light are non-limiting (Donald, 1963) and temperature and soil conditions are favourable. In practice, however, the supply of water and nutrients frequently limits production. This supply of moisture and nutrients and the general conditions of growth will determine the optimum density of a crop in a given situation.

Optimum populations and hence sowing rates are greater under less restricted rainfall conditions, or with irrigation on high fertility soils. Under low-rainfall dryland conditions, in situations where soil fertility is reduced, and in soils with low water-holding capacity, the optimum crop density will be lower. The populations of some field crops grown under dryland and irrigated conditions are compared in Table 6.4.

TABLE 6.4   The plant populations for some agricultural field crops grown under dryland and irrigation conditions in Australia

Drop	Plant population (plants/ha)	
	Dryland	Irrigation
Maize	35 000– 50 000	50 000– 65 000
Sorghum	60 000–100 000	210 000–250 000
Soybeans	170 000–200 000	200 000–350 000
Sunflowers	30 000– 40 000	60 000–100 000

Within a defined environment, the yield of grain tends to rise to a maximum as plant density increases and then falls with further increases in density. This relationship for maize is illustrated in Fig. 6.5 (Downey, 1971). Therefore, optimum densities for each crop and each environment should be determined by local research.

In addition to effects on crop yield, density changes exert significant influences on individual plant morphology. As density increases, plant height increases and stalk strength decreases. This therefore increases the risk of lodging in high populations. In maize, where the ears are higher on the stalk at high densities (Rutger and Crowder, 1967), the likelihood of lodging is further enhanced because the plant is 'top heavy'. This can also be used to advantage however because in maize, soybeans (*Glycine max*) and cotton, for example, the height of the lowest ear, pod or boll is also raised, thereby facilitating harvest.

As density increases, tillering or branching of plants declines. In winter cereals, plants can compensate to a degree for lower than optimum plant

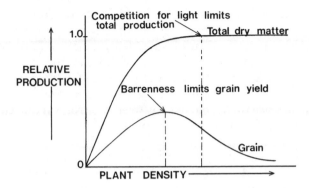

FIG. 6.5   The relationship between plant density and grain yield and dry matter production in maize (Downey, 1971)

populations caused by poor germination or disease by producing tillers. In such crops the plant population is perhaps not as critical and the relationship indicated in Fig. 6.5 is modified by a flattened peak, a greater range of densities giving maximum yields. Some indeterminately flowering crops respond to lower than optimum densities by producing secondary and tertiary branches.

The size and weight of the ear or seed head is also influenced by density, again in an inverse relationships. This has practical benefits in that smaller seed heads of sunflower, for example, dry more quickly and evenly and facilitate the harvesting process. In maize, however, densities beyond the optimum can increase the incidence of barrenness.

From the previous discussion it should be apparent that at optimum densities, the performance of individual plants within the crop is substantially less than its potential in isolation. Under field crop conditions intense interplant competition is created resulting in high yielding crops being communities of suppressed plants (Donald, 1968).

In considering plant population, crop geometry or distribution is important, particularly in row crop production. Traditionally, many crops have been sown in rows more than 90 cm apart, narrower row spacings being impractical because of the lack of equipment to grow and harvest the crop grown under these conditions. This has been a carryover from the horse-drawn equipment days. Modern machinery now provides flexibility in the choice of crop orientation for production of row crops. An increase in yield results from the more uniform distribution of plants, plants 'on the square' providing minimum competition to their neighbours. Such plants make more efficient use of sunlight, fertility and moisture. In the wider rows much sunlight falls between the rows and is wasted.

Further benefits from more evenly distributed plant populations include better weed control and better moisture utilisation. The crop competes more efficiently with weeds because the canopy develops much earlier, thereby restricting light to the weeds in the inter-row spaces. As the crop shades the ground earlier, the evaporation of soil moisture is likely to be reduced, making more moisture available to the crop. Disease incidence may also be reduced because the plants within the rows are further apart and hence there is less contact between plants to facilitate disease spread.

The change of management with narrower row spacings is important. Greater dependence on herbicides for weed control is required since post emergence mechanical cultivation is likely to cause damage to the crop. Wider row spacings may be necessary in weedy paddocks than in clean paddocks to enable inter-row cultivation to be undertaken. Fertiliser dressing and pesticide spraying, where necessary during the growing period of the crop, may be more satisfactorily carried out by aeroplane, thus avoiding damage to the crop.

In decreasing the distance between the rows it is important to note that the distance between the plants within the row will need to be increased if the same plant population is to be achieved. The effect of different row spacings on the intra-row spacing of sunflowers is illustrated in Table 6.5.

TABLE 6.5   The effect of row spacing on the number of plants per metre of row for different populations of sunflowers (assuming 85 per cent germination)

Row spacing (cm)	Plant population (plants/ha)						
	50 000	67 500	75 000	87 500	100 000	112 500	125 000
35	20	26	31	36	41	46	51
53	30	38	46	53	61	68	76
72	41	51	61	71	81	91	102
85	44	54	65	76	87	98	109
90	52	65	78	91	104	118	130
100	58	73	87	102	116	131	145
105	61	76	91	107	122	137	153

Apart from consideration of plant density and plant distribution, seed quality and time of sowing need to be taken into account in choosing a sowing rate. Seed quality is of particular importance. Seed should have a high germination percentage. Differences in seed size between different batches of seed also need to be taken into consideration. The number of seeds per kilogram may vary considerably between seasons, being influenced by the growing conditions in the year of production. Where large seed is used, sowing rates need to be higher because there are fewer seeds per kilogram. The effect of seed size on the rate of sowing of sunflowers is shown in Table 6.6.

The rate of sowing is also influenced by the timing of the sowing operation. In circumstances where sowing is delayed, sowing rates are generally increased. This partially compensates for the expected reduction in yield as a result of a reduced period in which branching and tillering can take place.

TABLE 6.6 The effect of seed size on sowing rate (kg/ha) of sunflowers grown in 35 cm rows (assuming 80 per cent germination of seed) to establish required plant populations

Plant population (plants/ha)	Number of plants per 30 cm of row	Seeds/kg		
		15 000	20 000	25 000
25 000	28	2.2	1.6	1.3
37 500	42	3.3	2.4	1.9
50 000	56	4.4	3.2	2.5
67 500	70	5.5	4.0	3.1
75 000	84	6.6	4.8	3.8
87 500	98	7.7	5.6	4.4
100 000	112	8.8	6.4	5.0

TABLE 6.7 The range in sowing rates for some agricultural field crops under dryland and irrigation conditions in Australia

	Sowing rate (kg/ha)	
Crop	Dryland	Irrigation
Wheat	20– 90	100–135
Oats	20– 90	100–135
Barley	20– 90	100–135
Rice	–	110–160
Lupins	55– 90	55– 90
Field peas	60–170	120–170
Linseed	20– 30	40– 60
Canola	3– 6	5– 9
Safflower	9– 16	20– 30
Sunflower	2– 6	4– 6
Soybeans	–	55– 80
Cotton	–	15– 40
Peanuts	–	30– 50

The range of sowing rates for a number of field crops is given in Table 6.7.

Depth of Sowing

The selection of the correct depth of sowing for any crop is based on the conflicting needs to plant deeply enough to protect the seed from desiccation within the surface layer of soil, yet shallow enough to permit easy and rapid emergence of the shoot (Donald and Puckridge, 1975). The factors that influence seed desiccation and shoot emergence need to be considered.

The risk of desiccation is obviously related to availability of soil moisture. Under low levels of soil moisture, sowing should be deeper than normal in an endeavour to locate the seed adjacent to the moisture deeper in the profile. The soil moisture/seed relations are also influenced by soil type, the lighter soils having a lower moisture-holding capacity. In these soils, therefore, deeper sowing is justified when compared with heavier soils with their higher moisture-holding capacities.

The emergence of a seedling will depend on the characteristics of the seed and the soil conditions. Small-seeded species, such as canola, have small food reserves and therefore restricted elongation potential. This necessitates much shallower sowing depths of 1 to 2 cm than is the case for larger seeds with much greater food reserves, such as soybeans, which are normally sown at depths of 3 to 5 cm. The effect of depth of sowing on emergence is more critical for the smaller seeded species but sowing practices have been generally less exacting for small seeds, perhaps because of the lack of suitable machinery. Such species, including many pasture species, are sown from a pasture seed box through swinging droppers and are covered to variable depths by trailing harrows. A more precise placement of small seeds by drilling would undoubtedly result in more even emergence and establishment and thus higher crop yields. In contrast, the placement of larger seeded species is carried out more accurately. For each set of conditions, an optimum sowing depth for any given crop exists. Sowing practices should take this into account. Sowing depths for field crops are given in Table 6.8.

TABLE 6.8   The average depths of sowing for agricultural field crops

Crop	Depth of sowing (cm)
Wheat	2–4
Oats	2–4
Barley	2–4
Maize	3–5
Sorghum	3–4
Lupins	2–4
Field peas	3–4
Soybeans	2–4
Safflower	3–5
Sunflower	2–4
Canola	1–3
Linseed	1–2
Peanuts	2–7
Cotton	2–4

The ability of a crop to emerge from the soil is influenced by the type of emergence of the seed. In crops such as soybeans which exhibit epigeal emergence (in which the cotyledons are brought above ground), the depth of sowing needs to be more shallow than for crops such as peas. The latter has an equivalent seed size but exhibits hypogeal emergence (in which the cotyledons remain below ground). The cereal crops have an advantage over many other crops because of their hypogeal emergence, together with their slender coleoptile. With the advent of semi-dwarf cultivars, however, the coleoptile length is reduced, thereby restricting the depth from which these cultivars can be successfully established.

Soil characteristics also influence emergence. In lighter soils, seeds can be sown deeper than in heavy soils because less resistance is offered to the progress of the seedlings. In poorly structured soils or where soil compaction and surface crusting are likely, emergence is impaired (McIntyre, 1955; Millington, 1959; Stoneman, 1962). In such conditions, a light surface harrowing following sowing may be necessary to break up the crust.

Seed Preparation

In field crop production, the seed sample produced by commercial harvesting machinery is seldom satisfactory for sowing another crop unless prior preparation is undertaken. This preparation is most important as it affects the results to be obtained in crop establishment.

The objectives of seed preparation are:
—to produce seed with a prospect of high percentage germination by the removal of small or cracked grains;
—to remove weed seeds; and
—to apply chemical seed dressings, or 'pickles', to control seed-borne pathogens, or to protect stored seed from insect infestation.

These objectives are accomplished by the use of an array of seed grading machinery. The choice of a particular grading machine will depend on the crop species and the species of weeds contaminating the sample. The final operation in most seed graders is the application of dusts with fungicidal or insecticidal properties.

Grain to be prepared as sowing seed is usually selected from a batch with good appearance. Weather-damaged grain or grain that has obviously been damaged by incorrect threshing techniques are avoided where possible. The batch to be graded and dusted will also need to have low moisture content levels to survive the storage period.

The most common types of grading machinery used in Australia are designed to clean cereal seed. These use perforated metal screens with round or slotted holes separating the larger seed from the small (de Silva, 1977a). Indented disc separators may also be employed in series with the screens to remove seeds of different sizes. When cleaning seed of crops other than cereals, equipment which separates seeds with different resistance to airflow may be needed. This airflow may be produced by aspiration of pneumatic methods.

The gravity table, or more correctly the specific gravity separator, will divide seeds of differing weights and size and may be useful in removing noxious weed seeds from the seed sample. Equipment is also available to separate seed on the basis of surface roughness, shape and colour.

Concurrent with the grading process, many seeds are subject to treatment peculiar to their particular morphology. Barley seed requires de-awning, cotton seed is delinted and hard-seeded legume seeds are scarified to promote water imbibition.

Chemical seed dressings are applied as a dry dust. These may be used to control seed-borne or soil-borne diseases, such as 'damping off', where the seed rots in the soil. The smut diseases of wheat are usually controlled by fungicidal dusts, as most wheat cultivars are not resistant to these diseases. Some chemicals may affect the storage life of seed and this needs to be taken into account. The effect may be more severe in some cultivars (Kuiper, 1974).

For grain legume seed, peat cultures of suitable *Rhizobium* strains are mixed with water and the resultant slurry mixed thoroughly with the seed. Methyl cellulose is often used to increase adhesion to the seed and prevent desiccation. This operation is usually restricted to less than 48 hours prior to sowing to ensure a high survival of the inoculum. Alternatively, an attachment to the sowing implement can spray the inoculum into the furrows in close proximity to the seed, avoiding the need for inoculating the seed.

Correct crop agronomy dictates that the efficiency of the seed preparation processes should be assessed. This involves a laboratory test for germination. Seed prepared for sale is usually subject to minimum germination and purity standards set by government legislation under different state Agricultural Seeds Acts. A knowledge of the germination percentage is essential. Adjustment of sowing rates to ensure optimum plant populations is made on the basis of this.

Crops which may be attacked by insects in the seedling stage are sometimes treated with systematic insecticides immediately prior to sowing. The application is usually by spraying the material on the seed while it is agitated. This application will afford protection to the crop until the plant is sufficiently developed to withstand attack. An example of this is the treatment of canola with dimethoate to enable the young seedling to withstand attack by red-legged earth mite (*Halotydeus destructor*).

Stored seed should always be kept in a cool, dry place. Seed deterioration is greatest under warm, humid conditions. In cotton and soybeans, such conditions can cause significant loss of viability in two or three months. In these circumstances seed may need to be stored in moisture-proof bags, provided that the seeds are already dry enough for sealed storage (Harrington, 1972).

Sowing Methods

The method of sowing a crop depends on a number of factors including the crop to be sown, the condition of the soil and the system of production used. The methods for sowing seed fall into two categories: surface sowing, and drilling the seed into the soil.

Surface sowing involves the broadcasting of seed on the surface of the soil by ground machine or from the air. Sowing by this method is generally inferior to placement of the seed in the soil, largely because the conditions are less conducive to good germination and establishment, with seedlings at greater risk of desiccation. Theft of seeds by ants and birds is also a problem. Aerial sowing, however, is a common method of establishing rice (*Oryza sativa*). The seed is pregerminated before being dropped into water. This method has the particular effect of advancing the growth of the crop by up to ten days and therefore is advantageous in situations where sowing is delayed or in areas with a shorter growing season such as the Murray Valley in southern New South Wales. However, crop damage by bloodworms and ducks is increased (Woodlands *et al.*, 1984).

The most common method of sowing is by *drilling* the seed into the soil at a prescribed depth. The winter cereals and many small grain crops are usually sown by a *combine*, a grain drill with a fertiliser box attached, thereby resulting in the seed being placed in the soil adjacent to a band of fertiliser. From the seedbox the seed passes through a metering device, commonly a fluted wheel or a double run, into droppers which extend to a prescribed depth in the soil behind furrow openers. The fertiliser follows a similar procedure although the metering system is different, usually being a 'star' feed. The furrow openers vary, there being single disc, double disc and tine types.

Traditional tine drills have poor plant residue handling capabilities and blockages occur unless the residues have been fragmented, are dry and are in small amounts (Kamel, 1975; Brown *et al.*, 1986). This is important for stubble retention farming, necessitating a change in tine geometry to greater trash clearance both within the tine row and between rows of tynes. Tines are also less useful under heavy wearing conditions (Stonebridge and Mackie, 1969) such as sandy soils. Problems arise where disc openers are used under conditions where soil is likely to adhere to the discs (Kamel, 1975).

Many combines have a small seeds box attachment preferably in conjunction with a band seeder through which pasture seed or small seeded crops such as canola are sown at prescribed depths. Alternatively, these species can be sown by mixing them with the fertiliser, providing consideration is given to likely fertiliser germination damage.

Combines sow the crops in rows approximately 18 cm apart although this can be varied in multiples by blocking the appropriate seed openings. The metering systems in conventional combines are not particularly precise and considerable variation in sowing rate and fertiliser rate both within and between rows is frequent. In many crops, however, this is relatively unimportant, but in the case of concentrated fertiliser, small differences represent considerable variation in actual nutrient supply.

Many combines have sets of cultivating tines or discs ahead of the furrow openers to allow the farmer a final cultivation during the sowing process.

In recent years in Australia, higher horsepower tractors capable of pulling cultivating and sowing equipment much wider than the traditional combine have become available. Typical 'wideline' implements may cover a span of 8 to 14 metres in one pass. These implements often are combined with seed and fertiliser metering equipment to allow very rapid sowing. The usual arrangement consists of a large capacity wheeled hopper trailing immediately behind the tractor. The wide span machine is attached to the rear of the hopper. After metering, the seed and fertiliser are blown along flexible plastic tubes to the sowing tines. The fan providing the pneumatic propulsion is driven either from the tractor power takeoff shaft or a separate motor. To allow this system to be used to its maximum, the seed and fertiliser are handled in bulk, being augered or elevated from large truck-mounted bins. A major difficulty with wide span machinery is obtaining consistent depth of sowing on unlevel ground.

The *sodseeder* is very similar to the combine but there is no cultivation, the seed and fertiliser being introduced into an undisturbed soil. The sodseeder places the seed and fertiliser in a narrow slit in the soil which falls back over the seed immediately following the operation. The success of sodseeding depends on soil type and soil moisture at planting. This technique is used for sowing rice, but requires heavy grazing to leave the area as bare as possible (Boerema, 1969). Following germination of the rice seedlings, the area can be grazed by sheep between irrigation 'flushings' before permanent water is applied. The rice plants are less palatable than the associated species and are therefore avoided. Alternatively, the competitive herbage can be removed by non-residual contact herbicides.

Row crop *planters* are used for sowing of row crops which require wider row spacings than small grain crops, there being a separate seed hopper for each row. Row crops require more accurate seed placement within the rows than, for example, wheat, the metering device in this case commonly being a revolving plate with notches to hold a single seed of specific size. The plate revolves as the planter is drawn forward, releasing single seeds at precise intervals. Different crops can be sown by changing the plates and the row spacings.

After the seed is placed in the soil by the combine or planter, it must be covered by sufficient soil to ensure germination. Some covering may occur with soil displaced by cultivating tines, or by the soil flowing around the sowing tine itself. However, trailing harrows are usually fitted behind the sowing implement to smooth out the ridges, and so cover the seed at the bottom of the furrow. Increasingly, press wheels are used to firm soil around the seed (Sweeting and Colless, 1985). This can aid the establishment of seedlings considerably where conditions for germination are difficult, such as where the soil is drying out (Doyle and Garland, 1975; de Silva, 1977b). Press wheels are also useful in improving seed–soil contact where the seedbed contains a large amount of crop residue or is in a rather cloddy state (Kamel, 1975; Marston and Garland, 1978).

The establishment of tobacco (*Nicotiana tobacum*) and sugar cane (*Saccharum officinarum*) warrants special mention because these crops are substantially different in culture to most agricultural crops.

Tobacco seed is extremely fine and is planted in nursery seedbeds. The seedbeds have movable covers of plastic or canvas which are used for protecting seedlings from adverse weather conditions and for seedbed fumigation. In late spring or early summer the seedlings are transplanted into the field and normal cultural operations proceed.

The production of sugar cane is confined to the subtropical–tropical regions of Australia. The plant is a perennial, and up to four crops may be obtained from the one planting. The first crop, known as the 'plant' crop, takes from 8 to 24 months to be ready for harvest; the ratoon (or subsequent) crops are harvested annually thereafter.

The crop requires high soil fertility and high soil organic matter, so often a green manure crop is used just prior to planting in order to build up the organic matter. Soil preparation is very important because it is only done about once in four years. A poor preparation may result in three poor crops.

Because the crop is grown in tropical–subtropical conditions, weed growth is prolific and early management of a crop consists of intensive weed control.

The crop is propagated vegetatively with cane cuttings, or setts, being planted mechanically by a 'cutter-planter'. This machine cuts the cane into setts, plants the setts in drills and applies fertiliser, all in the one operation. If conditions are favourable and the setts used are sound, shoot emergence should be uniform and rapid. However, some gaps often occur in the plantation, caused by non-emergence of setts. Gaps in the rows greater than about 0.5 m are replanted as soon as possible, a procedure known as 'supplying'. Whilst supplies rarely catch up to the original crop, they do provide the useful function of ground cover, thereby preventing weed growth.

POST-SOWING MANAGEMENT

Post-sowing management for most agricultural field crops grown under dryland conditions is confined largely to the control of weeds and pests, much of which is discussed in Chapter 8. This is done by using herbicides and insecticides which are applied by ground boom sprays or by aeroplane. The latter avoids any physical damage to the crop, the amount of damage increasing with age of the plants. Early spraying of herbicides enables lower rates to be used more effectively on weeds, which are generally more susceptible when young. Maximum crop benefits are achieved by early spraying.

A particular management problem resulting from prolonged use of pesticides is pesticide resistance. Since the late 1980s, farmers in the winter rainfall areas have increasingly experienced herbicide failure, particularly with annual ryegrass infestations. This problem, brought

about mainly by the persistent use of a particular herbicide, becomes serious when resistance is also found to other herbicides. The management strategies that need to be implemented include:
—prevention of seed set of resistant populations, by spray-topping or cutting for hay or silage, or collect the seed at harvest for subsequent disposal;

TABLE 6.9   Herbicide groups. Agricultural herbicides grouped according to biochemical mode of action (Leys and Dellow, 1991)

Group	Chemical	Trade name
Inhibitors of lipid biosynthesis		
Aryloxyphenoxypropionates		
(fops)	Diclofop	Hoegrass®, Nugrass®
	Fluazifop	Fusilade®
	Haloxyfop	Verdict®
	Quizalofop	Targa®
	Fenoxaprop	Puma®
Cyclohexanediones (dims)	Sethoxydim	Sertin®
	Tralkoxydim	Grasp®
Inhibitors of branched-chain amino acid biosynthesis		
Sulfonylureas	Chlorsulfuron	Glean®
	Metsulfuron	Ally®
	Triasulfuron	Logran®
	Bensulfuron	Londax®
	Thifensulfuron + metsulfuron	Harmony M®
Imidazolinones	Imazethapyr	Pursuit®
Inhibitors of aromatic amino acid biosynthesis		
Glycines	Glyphosphate	Roundup®, Roundup CT®, Glyphosate®, Glyphosate 360®
Inhibitors of photosynthesis at photosystem II		
Triazines	Simazine	Various
	Atrazine	Various
	Cyanazine	Bladex®
	Terbutryn	Igran®
	Prometryn	Gesagard®
	Metribuzin	Lexone®, Sencor®
Ureas	Diuron	Various
	Linuron	Afalon®, Linuron®
	Methabenzthiazuron	Tribunil®
Nitriles	Bromoxynil	Various
	Ioxynil	Totril®
Benzothiadiazoles	Bentazone	Basagran®
Acetamides	Propanil	Ronacil®

—prevention of the introduction of resistant seed onto the farm;
—reduction of dependence on chemical options by crop/pasture
 rotations and strategic tillage;

TABLE 6.9 continued

Group	Chemical	Trade name
Inhibitors of photosynthesis at photosystem I		
Bipyridyls	Paraquat	Gramoxone®, Shirquat®
	Diquat	Reglone®
	Paraquat + diquat	Sprayseed®
Inhibitors of chlorophyll biosynthesis		
Diphenyl ethers	Acifluorfen	Blazer®
	Oxyfluorfen	Goal®
Inhibitors of carotenoid biosynthesis		
Nicotinanilides	Diflufenican	Brodal®
Disruption of plant hormone action		
Phenoxys	2,4-D	Various
	2,4-DB	Various
	MCPA	Various
Benzoic acids	Dicamba	Banvel®, Dicamba®
Pyridines	Picloram	Various
	Clopyralid	Lontrel®
	Fluoxypyr	Starane®
	Triclopyr	Garlon®
Inhibitors of tubulin formation		
Dinitroanilines	Trifluralin	Various
	Oryzalin	Surflan®
	Pendimethalin	Stomp®
	Trifluralin + oryzalin	Yield®
Herbicides with multiple sites of action		
Thiocarbamates	Triallate	Avadex BW®
	FPTC	Eptam®
		Ordram®, Molinate®
	Molinate	Saturn®
	Thiobencarb®	
Amides	Metolachlor	Dual®
	Propachlor	Ramrod®
	Propyzamide	Kerb®
	Napropamide	Devrinol®
Carbamates	Carbetamide	Carbetamex®
	Asulam	Asulam®
Amino propionates	Flamprop-methyl	Mataven®

—rotation of herbicides with different modes of action (Table 6.9).

Herbicide resistance is most serious in annual ryegrass and increasingly in wild oats. Other species that have developed herbicide resistance include barley grass and capeweed (Powles, 1987).

Fertilisers may also be applied, particularly if a nutrient deficiency becomes evident during the course of growth of the crop. The most common uses of fertilisers, however, are split applications of nitrogen fertiliser in wheat (at sowing and at completion of tillering) to raise grain protein, and an application for rice at panicle initiation for yield response.

Under some conditions, winter cereal crops may need grazing to arrest their development. This need for grazing most commonly occurs when early sown crops make rapid growth under mild winter temperature conditions. This practice ensures that flowering of the crop does not take place while frosts are likely to occur. Grain yield is usually reduced by grazing, although the reduction is far less than in severely frosted crops (Dann, 1976). In the non-cereal winter crops, grain yield reduction is particularly severe (Dann *et al.*, 1977) and the practice is not recommended. However, the winter wheats, oats and barley are particularly valuable for winter forage, and specific cultivars have been bred to give maximum forage production and grain production. If these crops are to be used to obtain high grain yields, the grazings have to be lenient and completed before ear initiation of the crop in late winter (Dann *et al.*, 1983).

In tobacco crops an additional process is required. Because the leaf is the economically important part of the plant, management aims at maintaining the crop in the vegetative phase. This necessitates topping, which removes the flowerheads, and suckering, which removes suckers that develop in the leaf axils following topping, as the plant endeavours to complete its reproductive cycle. Chemical control of suckers has replaced what was a tedious, labour-intensive and hence costly task.

HARVESTING

Maturity

The harvesting process for each crop occurs at a particular stage of maturity of the crop. It is important to note the distinction between *physiological maturity* and *harvest maturity*.

Physiological maturity refers to that stage of a crop where the seed reaches its maximum dry weight. In general, grains cease the accumulation of dry matter at approximately 40 per cent moisture. Harvesting crops prior to physiological maturity can be expected to lower the quaity of the seed and produce shrivelled grains that have low test weights.

Harvest maturity refers to the stage of development of the crop at which harvesting will produce the best combination of yield and quality. The important factor is the moisture content of the grain. The appropriate moisture content will depend in particular on the crop grown and to some extent on the machine being used.

The problems of harvesting after physiological maturity but before harvest maturity include:

—spoilage in storage if drying facilities are not available. The maximum moisture contents suitable for grain storage are given in Table 6.10;

—mechanical damage to kernels which are high in moisture and swollen;

—an inefficient harvesting process, the machinery being unable to effect total harvest because of 'green' growth.

Delaying harvesting beyond the harvest maturity period will generally cause a reduction in yield and quality. Yields can be affected by lodging and shattering, although the degree will depend to a large extent on prevailing weather conditions and on the crop to be harvested. Quality usually deteriorates under excessively wet conditions, commonly resulting in lower test weights. Sprouting and colour loss may also occur and seed crops may lose their germinability. In wheat, rain on the mature grain causes production of α-amylase enzymes which seriously damage baking quality (Simmonds, 1989).

TABLE 6.10 A summary of the moisture contents for storage of grain of some agricultural field crops in Australia

Crop	Moisture content (%)
Wheat	12
Oats	12
Barley	12
Maize	14
Sorghum	13
Rice	16
Lupins	12
Field peas	12
Soybeans	12
Safflower	9.5
Sunflower	9
Rapeseed	8
Linseed	10
Peanuts	12

Extremes of temperature can also affect grain quality. In rice, suncracking may occur if very dry grain is exposed to very low or very high temperatures under alternate wetting and drying conditions (Blakeney, 1984).

The date of harvest of some crops, notably cotton, can be advanced by using chemicals which defoliate or desiccate the plant; this may not, however, be the overriding reason for their use. Cotton grade is assessed on colour, fibre character and trash content. Much of the leaf harvested is large and leathery and is easily cleaned in the ginning process, and therefore presents little problem. The quality of the product declines if the powdered, dry leaf and bracts are present. This is called 'pin trash' or 'leaf pepper'. Sappy green leaves in a harvest sample caused significant staining of the cotton fibre (Swann, 1971). Defoliants have the advantage of removing leaves, which would otherwise block the spindles, add trash, or stain fibre. Only mature leaves are successfully removed by defoliants and therefore in rank crops and crops containing considerable second growth, desiccants must be used. Harvesting, however, must be done before the leaves begin to crumble, otherwise pin trash will be increased (Swann, 1971). Premature use of either defoliants or desiccants may also cause serious loss in yield and quality.

Grain drying overcomes the problem of storage deterioration through high moisture contents. However, several precautions have to be noted in the grain drying process. Where the grain is to be used as seed for future crops, the drying temperature should not exceed 43°C. Where the grain protein content is important in processing, such as wheat for bread manufacture, high temperatures cause a degree of denaturation of the protein and product quality consequently suffers. Under these conditions drying temperatures should not exceed 60°C. The maximum recommended temperatures for drying grain are shown in Table 6.11. A detailed account of seed drying is given by Brandenburg *et al.* (1961).

TABLE 6.11   The maximum temperatures (°C) for drying grain

Grain	Grain for seed	Grain for commercial use	Grain for stock feed
Maize	43	54	80
Grain sorghum	43	60	80
Wheat/oats	43	60	80
Barley	40	40	80
Soybeans	43	49	80
Oilseeds	—	46	—

Harvesting Methods

Two general categories of harvest operations can be described although within the categories the requirements of crops may differ considerably.

Direct Mechanical Harvesting

In this method of harvest the economic yield of the plant is gathered and processed by suitable machinery in one operation. For most seed and grain crops, the all-purpose header or harvester removes the seed heads, threshes the seed from the seed head, separates the grain from leaf and stem material, and discards the trash.

The principle of the header-harvester is now described. The stems of the standing crop are cut by a reciprocating knife, and are swept or dragged into an elevator by a revolving reel or auger. This elevator conveys the material to a threshing drum which revolves at an adjustable controlled speed, usually between 500 and 1500 revolutions per minute. A stationary concave (a curved open grate) is fitted around the lower half of the drum. The clearance between drum and concave is adjustable. Metal rasp or rubbing bars are fitted to the drum and concave. As the crop material passes between the drum and concave, most of the grain is threshed out of the seeds or pods. This grain and chaff fall through the concave onto the grain pan underneath.

Behind the drum is a revolving beater, which assists in removing straw from the drum, at the same time removing grain carried in the straw. Any grain thus removed falls through a grate, also onto the grain pan. The threshed straw passes on to the straw walkers, which have an oscillating motion that carries the straw out to the back of the harvester. The need for straw-spreading mechanisms has previously been described. Any unthreshed or partly threshed heads or pods fall through the straw walkers on to chutes which return to the grain pan.

The mixture of grain, chaff and partially threshed material on the grain pan is carried to an oscillating top riddle or sieve and travels towards the rear of the harvester. An air blast passes upwards through the riddle, blowing the chaff out of the rear of the machine. Unthreshed heads or pieces, called 'gleanings', ride over the riddles and fall off the rear into a repeat auger which conveys them back to the drum for rethreshing. The free grain falls through the top riddle onto a lower one with smaller openings, where further chaff if removed by the blast of the fan. The grain travels backward on this lower riddle, and is dropped on to an inclined plane and so conveyed to the grain auger. From this auger it is fed to the grain elevator and lifted into the grain box on top of the header. The grain box is discharged when full through a built-on auger.

Successful harvesting requires wide experience. However, the following broad principles apply. The ground speed of the header should be such that only that amount of material which will give efficient separation is allowed to enter the header. High speed in heavy crops does not provide sufficient time for separation to occur and grain losses can be considerable. The clearance between the rasp bars and concave should be adjusted to give maximum threshing with minimum grain damage. The speed of the drum can also be altered to achieve this objective. The volume of the air blast is controlled by varying the fan speed. Excessive blast will blow grain out of the machine, while insufficient will result in a chaffy sample. Correct sieve type and sizes must be chosen for each crop to ensure a clean sample.

For maize, a specialised picker-sheller is used whilst particular harvesting equipment is required for crops such as sugar cane and cotton.

The *sugar cane* harvest takes place in the cooler months of the year when the cane has a high content of sugar. Customarily, much of the trash is removed from the crop by burning shortly before harvest. The harvest is done with a 'chopper', a machine which cuts the cane into lengths and transfers these into bins which are then transported quickly to the mill. The need to process the cane as soon as possible after harvest by chopper machines is due to the increase in the number of cut ends in the cane. Rapid deterioration of cane can result (Egan, 1971) due to the entry of the *Leuconostoc* microorganism into the cut ends of the cane. Mechanical harvesting, however, has created problems of stool damage and soil compaction when harvest has taken place under wet conditions (Sturgess *et al.*, 1976).

The *cotton* crop is ready for harvest 7 to 10 days after defoliant sprays have been applied. The crop is picked by 'cotton pickers', machines which straddle the cotton rows. They have barbed spindles which pick the seed cotton from the bolls after which suction pipes blow the cotton into a storage bin. The storage bins are compressed into modules which can be left in the paddock until convenient to transport them to the gin. Growers aim to harvest 70 to 90 per cent of the crop at the first picking, which should be completed before frosts (May–June); the balance of the crop is harvested at a second pass three to four weeks later. Once the harvested material reaches the gin, the ginning process removes the fibre, or lint, from the seed; most of the seed is then transported to oil extraction plants.

Indirect Mechanical Harvesting

This method requires more than one operation before the economic portion of the crop is obtained. The most common example of this method is where the crop is cut and windrowed prior to being picked up by a header for threshing. This method is useful where crops are close to the ground as in the case with *field peas* or with crops that have lodged. It is also common in canola, where shattering losses can be high, particularly if uneven ripening occurs. In crops with a high green weed component, excessive moisture levels may be imparted to the grain and increased losses may occur through blockage of the machine with normal harvesting. Under such conditions, windrowing will be a sound proposition, as the cut weeds can be allowed to dry in the swathe before harvesting.

In crops such as *soybeans*, desiccants are sometimes used to hasten the maturation process, particularly where a bad weed problem exists. However it predisposes the crop to weather damage and is usually more expensive than grain drying (McInerney *et al.*, 1991).

The harvesting of *peanuts* requires several operations. The crop should be cut and pulled when the greatest number of nuts show darkened veins on the inside of the shell. Two-row or four-row mid-mounted cutter bars are passed under the nut clusters of the peanut plants to cut the taproot and loosen the soil around the plants. The cutters are followed by rear mounted pullers that compress the tops and raise the plants at the same time. After the plants have wilted but not dried out, it is usual to combine two to four plant rows into a windrow using a side delivery rake. Extreme care is needed in all operations to minimise losses. Once windrowed, the peanuts are left to cure for two to five days before threshing, drying and bulk handling. Threshing is carried out with pick-up peanut threshers which either deliver into bulk bins or into bags. The moisture content of the pods is usually high, thus requiring artificial drying immediately following threshing. Where such drying facilities are not available the farmer may have to leave the crop in windrows for 7 to 14 days or longer.

Handling Grain at Harvest

The handling of the wheat harvest in the early days of the industry was in bags. The opening of the first public silo at Peak Hill, New South Wales, in 1918, followed by the completion of the Sydney Terminal Elevator in 1921, marked the beginning of the era of bulk handling of wheat and subsequently other grains. The Australian Wheat Board abolished bag stacks completely in the late 1960s.

Bulk handling solves many problems associated with handling, transporting and storage of grain (Magill, 1977). These include the cost of the bags, the problems of broken bags (accentuated by rodents and insects), the delays caused by having to sew bags before transport and storage, and delays in loading and unloading for transport.

However, considerable capital outlay is involved in bulk handling of grain. At all times during harvest, all trucks with bulk bins must be ready to receive the harvested grain from the header. Delays at this time are costly and increase the risk of weather damage to the crop. Farmers often use hessian-lined welded mesh field bins or silos to hold grain temporarily until the trucks can load. An auger is used to transfer grain from the bins to the trucks. The provision of field bins provides insurance against any delays in grain transportation.

Bulk storage of grains in silos however has not been without its problems. These include grain insect infestations, moisture migration and biochemical degradation (Callaghan, 1975). Insecticides and fumigants have been used successfully for control of insects but the development of resistance to these chemicals by insects emphasised the need to control more closely such factors as moisture and temperature, which influence infestation level. The development and introduction of moisture meters for monitoring grain moisture contents prior to, and during, storage has practically eliminated moulds and crusting. Deterioration in the quality of the grain is thereby reduced and insect infestations are better controlled. Grain with more than the maximum acceptable moisture content will not be received into silo systems. Farmers are required to hold or dry the grain until the moisture content has dropped to satisfactory levels. Aeration, which equalises and controls temperatures in the silos, has also had a significant effect on controlling infestations and on moisture migration.

A major factor in the development of infestations of grain insects is poor storage hygiene. This applies equally to on-farm storage as well as to bulk storage facilities controlled by state authorities. It is essential, therefore, that the highest standards of hygiene and storage conditions are maintained on the farm (Greening, 1975).

POTENTIAL YIELD CONCEPT

Research has identified that, because water is such a limited resource in dryland farming in Australia, a close relationship exists between crop yield and the amount of water utilised by the crop during the growing season (French and Schultz, 1984; French, 1987; 1991; Cornish and Murray, 1989).

The concept of potential yield therefore is based on the premise of maximising water-use efficiency (i.e. yield per mm of crop water use). Potential yields can be estimated according to the amount of water available to plants in any one year or any one region. Water-use

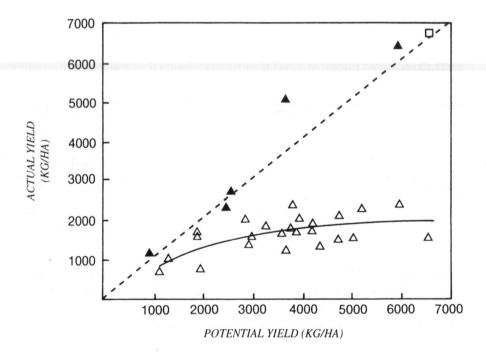

FIG. 6.6 District yields for the City of Wagga Wagga (△) compared with predicted potential yields based on rainfall and evaporation at the Agricultural Research Institute, Wagga Wagga, 1960–84 (Cornish and Murray, 1989) △ Experimental yields 1980–84; □ 1 ha plot in 1983; — — actual equals predicted potential yield

efficiencies under winter rainfall conditions for wheat range from 10 to 20 kg/mm.

The value of potential yield is that it provides a target yield for farmers to work towards and to evaluate their current performance—that is, a performance indicator. Cornish and Murray (1989) for Wagga Wagga, NSW (Fig. 6.6), French (1987; 1991) for South Australia, and Mead (1992) for Cowra, NSW, have shown a large difference between potential and actual farm yields, with the gap widening as the water-limited potential yield increases. Thus, in the best yielding years, farmers fail to capitalise on seasonal conditions because production inputs and other agronomic practices are inadequate.

Conservation farming is a system whereby soil structure, particularly at the soil surface, is enhanced to maximise the availability of water for crop use. It follows that, unless the water contributes to an increase in productivity, its alternative role is as an agent of degradation through processes such as leaching which increases the rate of acidification, and contributes to the rise and eutrophication of water tables, and may lead to dryland salinity.

Sustainability, from an environmental perspective, is a function of maximising the amount of water available to plants for productive use. Sustainability from an economic perspective is a function of maximising productivity. Both aspects are achieved by maximising crop water use. The role of the farmer-agronomist is to ensure that the principles of good agronomy are applied to ensure maximum water utilisation by crops can be achieved as far as possible.

THE NEED TO MONITOR

The need to improve productivity in order for farmers to survive financially is ever present. Increased managerial skills are required, and the need to keep records of both physical and financial aspects is a necessary component. The physical records involve regular monitoring of the crop's performance so that, between paddocks or over time, comparisons can be made. Decisions can then be made about what went wrong or what went right, so that continual adjustments can be made to each succeeding year's crop program.

This requirement is emphasised in New South Wales where monitoring programs are in place for various crops. These programs (for example, Wheatcheck, Ricecheck, Soycheck, Canolacheck) involve the close monitoring of crops by groups of farmers, so that key factors influencing yield and quality can be identified within a season by comparisons between data from different farms.

Regular testing for soil nutrient and pH status and the presence of herbicide resistance, and measuring groundwater levels, provides a firm basis for the determination of the level and type of inputs necessary to achieve a particular target yield.

References

Ayres, J. F., McFarlane, J. D., Gilmour, A. R. and McManus, W. R. (1977a), 'Superphosphate Requirements of Clover-ley Farming. I: The Effects of Topdressing on Productivity in the Ley Phase', *Australian Journal of Agricultural Research,* **28**, 269.

Ayres, J. F., McFarlane, J. D., Osborne, G. J. and Corbin, E. J. (1977b), 'Superphosphate Requirements of Clover-ley Farming. II: The Residual Effects of Topdressing', *Australian Journal of Agricultural Research,* **28**, 287.

Barley, K. P. and Graecen, E. L. (1967), 'Mechanical Resistance as a Soil Factor Influencing the Growth of Roots and Underground Shoots', *Advances in Agronomy,* **19**, 1.

Blakeney, A. (1984), 'Rice grain quality', in A. Currey (ed.) *Rice Growing in New South Wales,* Department of Agriculture NSW and Rice Research Committee.

Boerema, E. B. and McDonald, D. J. (1964), 'Draining Rice for a High Moisture Harvest', *Agricultural Gazette of N.S.W.,* **75**, 1031.

Bromfield, S. M. and Simpson, J. R. (1974), 'Effects of Management on Soil Fertility under Pasture. 2: Changes in Nutrient Availability', *Australian Journal of Experimental Agriculture and Animal Husbandry,* **14**, 479.

Brown, G. A., Quick, G. R. and Wasko, E. (1986), 'Trashflow through tined tillage implements', in *Proceedings of Conference on Recent Advances in Weed and Crop Residue Management,* Wagga Wagga, 48.

Butler, F. C. (1961), 'Root and Foot Rot Diseases of Wheat', N.S.W. Department of Agriculture, Science Bulletin No. 77.

Callaghan, A. R. (1975), 'Bulk Handling in Australia' in A. Lazenby and E. M. Matheson (eds), *Australian Field Crops. I: Wheat and Other Temperate Cereals,* Angus and Robertson, Sydney, 384.

Callaghan, A. R. and Millington, A. J. (1956), *The Wheat Industry in Australia,* Angus and Robertson, Sydney.

Carter, E. D., Wolfe, E. C. and Francis, C. M. (1982), 'Problems of Maintaining Pastures in the Cereal-Livestock Areas of Southern Australia', *Proceedings of 2nd Australian Agronomy Conference,* Wagga Wagga, 68.

Constable, G. A. (1977), 'Effect of Planting Date on Soybeans in the Namoi Valley, New South Wales', *Australian Journal of Experimental Agriculture and Animal Husbandry,* **17**, 147.

Coombe, J. B. (1981), 'Utilisation of Low Quality Residues' in F. H. W. Morley (ed.) *Grazing Animals,* Elsevier, Amsterdam, 319.

Cornish, P. S. and Murray, G. M. (1989), 'Low Rainfall Rarely Limits Wheat Yields in Southern New South Wales', *Australian Journal of Experimental Agriculture,* **29**, 77.

Cornish, P. S. and Pratley, J. E. (eds) (1987), *Tillage—New Directions in Australian Agriculture,* Inkata, Melbourne.

Cornish, P. S. and Pratley, J. E. (1991), 'Tillage Practices in Sustainable Farming Systems', in V. Squires and P. Tow (eds) *Dryland Farming—A Systems Approach,* Sydney University Press, Sydney, 76.

Crawford, K. L. (1977), 'Strip Cropping for Erosion Control in Northwest New South Wales', *Journal of Soil Conservation Service of New South Wales,* **33**, 83.

Cummins, V. C. (1973), *A Land Use Study of the Wyreema–Cambooya Area of the Eastern Darling Downs,* Division of Land Utilisation, Queensland Department of Primary Industry Technical Bulletin 10.

Cummins, E. J. and Esdaile, R. J. (1972), 'A Successful Strip Cropping Demonstration in Northwestern New South Wales', *Journal of Soil Conservation Service of New South Wales,* **28**, 137.

Cuthbertson, E. G. (1967), 'Skeleton Weed', N.S.W. Department of Agriculture Bulletin 68.

Dann, P. R. (1976), 'Using Stubbles and Forage Crops', *Proceedings of 4th Riverina Outlook Conference,* Wagga Wagga, 3.

Dann, P. R., Axelsen, A. and Edwards, C. B. H. (1977), 'The Grain Yields of Winter-grazed Crops', *Australian Journal of Experimental Agriculture and Animal Husbandry,* **17**, 452.

Dann, P. R., Axelsen, A., Dear, B. S., Williams, E. R. and Edwards, C. B. H. (1983), 'Herbage, Grain and Animal Production from Winter-grazed Cereal Crops', *Australian Journal of Experimental Agriculture and Animal Husbandry,* **23**, 154.

Davidson, R. M. (1992), 'Population dynamics of herbicide resistant annual ryegrass, *Lolium rigidum,* in pasture' in *Proceedings of the 1st International Weed Control Congress,* Melbourne, 141.

Dear, B. S. (1989), 'Establishment of Subterranean Clover Under Direct Drilled Cereal Crops' in *Proceedings of the 5th Australian Agronomy Conference,* Perth, 407.

de Silva, S. (1977), 'Eccentric Grain Cleaner', N.S.W. Department of Agriculture, South West Region, Bulletin.

Donald, C. M. (1963), 'Competition among Crop and Pasture Plants', *Advances in Agronomy,* **15**, 1.

Donald, C. M. (1968), The Breeding of Crop Ideotypes. *Euphytica,* **17**, 385.

Donald, C. M. and Puckridge, D. W. (1975), 'The Ecology of the Wheat Crop' in A. Lazenby and E. M. Matheson (eds), *Australian Field Crops. 1: Wheat and Other Temperate Cereals,* Angus and Robertson, Sydney.

Evans, J., O'Connor, G. E., Turner, G. L. and Bergersen, F. J. (1987a), 'Influence of Mineral Nitrogen on Nitrogen Fixation by Lupin (*Lupinus angustifolius*) as Assessed by [15]N Isotope Dilution Methods', *Field Crops Research,* **17**, 109.

Evans, J., Turner, G. L., O'Connor, G. E. and Bergersen, F. J. (1987b), 'Nitrogen Fixation and Accretion of Soil Nitrogen by Field-grown Lupins (*Lupinus angustifolius*), *Field Crops Research,* **16**, 309.

Felton, W. L., McCloy, K. R., Doss, A. F. and Burger, A. E. (1987), 'Evaluation of a Weed Detector, in *Proceedings of the 8th Australian Weeds Conference,* Sydney, 80.

French, R. J. (1987b), 'The Effect of Fallowing on the Yield of Wheat. II. The Effect on Grain Yield', *Australian Journal of Agricultural Research,* **29**, 653.

French, R. J. (1987), 'Future Productivity on our Farmlands' in *Proceedings of the 4th Australian Agronomy Conference,* Melbourne, 140.

French, R. J. (1991), 'Monitoring the Functioning of Dryland Farming Systems' in V. Squires and P. Tow (eds) *Dryland Farming — A Systems Approach,* Sydney University Press, Sydney, 222.

French, R. J. and Schultz, J. E. (1984), 'Water Use Efficiency of Wheat in a Mediterranean-type Environment. I. The Relationship Between Yield, Water Use and Climate', *Australian Journal of Agricultural Research,* **35**, 743.

Gillespie, D. J. (1989), 'Roles for New Pasture Legume Species in Southern Australia' in *Proceedings of the 5th Australian Agronomy Conference,* Perth, 116.

Holland, J. F., Doyle, A. D. and Morley, J. M. (1987), 'Tillage Practices for Crop Production in Summer Rainfall Areas' in P. S. Cornish and J. E. Pratley (eds), *Tillage—New Directions in Australian Agriculture,* Inkata Press, Melbourne, 48.

Jenken, J. J. (1986), 'Western Civilisation' in J. S. Russell and R. F. Isbell (eds), *Australian Soils—The Human Impact,* University of Queensland Press, Brisbane, 134.

Jones, S. M., Blowes, W. M., England, P. and Fraser, P. K. (1984), 'Pasture Topping Using Roundup Herbicide', *Australian Weeds,* 3(4), 150.

Kamel, T. R. (1975), 'Machinery Problems Raised by Farmers in South-east Queensland' in *Tillage Practices of the Wheat Crop,* CSIRO Land Resources Laboratories Discussion Paper No. 1, 172.

Kelsey, R. F. S. (1966), 'Strip Cropping Means Less Turning', *Queensland Agricultural Journal,* 92, 162.

Kidd, C. R., Leys, A. R., Pratley, J. E. and Murray, G. M. (1992), Effect of Time of Removal of Annual Grasses from Pastures on the Carryover of Take-all to Wheat' in *Proceedings of the 6th Australian Agronomy Conference,* Armidale, 555.

Kohn, G. D. (1974), 'Superphosphate Utilization in Clover Ley Farming: 1. Effects on Pasture and Sheep Production', *Australian Journal of Agricultural Research,* 25, 525.

Kohn, G. D. (1975), 'Superphosphate Utilization in Clover Ley Farming. 2. Residual Effects of Pasture Topdressing in the Cropping Phase', *Australian Journal of Agricultural Research,* 26, 93.

Kohn, G. D., Storrier, R. R. and Cuthbertson, E. G. (1966), 'Fallowing and Wheat Production in Southern New South Wales', *Australian Journal of Experimental Agriculture and Animal Husbandry,* 6, 604.

Kohn, G. D. and Storrier, R. R. (1970), 'Time of Sowing and Wheat Production in Southern New South Wales', *Australian Journal of Experimental Agriculture and Animal Husbandry,* 10, 604.

Kohn, G. D., Osborne, G. J., Batten, G. D., Smith, A. N. and Lill, W. J. (1977), 'The Effect of Topdressed Superphosphate on Changes in Nitrogen: Carbon: Sulphur: Phosphorus and pH on a Red Earth Soil During a Long Term Grazing Experiment', *Australian Journal of Soil Research,* 15, 147.

Kuiper, J. (1974), 'Suppression of Wheat Seedling Establishment by Maneb', *Australian Journal of Experimental Agriculture and Animal Husbandry,* 14, 391.

Leys, A. R. (1990), 'Control of Annual Grasses in Pastures of Southern Australian and Implications for Agriculture' in *Proceedings of the 9th Australian Weeds Conference,* Adelaide, 354.

Leys, A. R. and Dellow, J. J. (1991), 'Herbicide Resistance', Agnote DPI/43, New South Wales Department of Agriculture and Fisheries.

Leys, A. R. and Plater, B. (1993), 'Simazine Mixtures for Control of Annual Grasses in Pastures', *Australian Journal of Experimental Agriculture,* 33.

Lipsett, J. and Simpson, J. R. (1965), 'Some Effects of Subterranean Clover Pastures on the Fertility of Tableland and Wheat Soils in Southern New South Wales', *Field Station Record,* 4, 9.

Lovett, J. V., Hoult, E. H., Jessop, R. S. and Purvis, C. E. (1982), 'Implications of Stubble Retention', *Proceedings of 2nd Australian Agronomy Conference, Wagga Wagga,* 101.

MacNish, G. C. (1989), 'Root Diseases as a Major Constraint in High Rainfall Cropping Systems' in *Proceedings of the 5th Australian Agronomy Conference,* Perth, 160.

McCown, R. L., Jones, R. K. and Peake, D. C. I. (1985), 'Evaluation of a No-till, Tropical Legume Ley-farming Strategy' in R. C. Muchon (ed.), *Agro-Research for the Semi-Arid Tropics: North-West Australia,* University of Queensland Press, St Lucia, 450.

McInerney, P., McCaffery, D. and Lacy, J. (1991), *Soycheck 1991/92,* NSW Agriculture.

McIntyre, D. S. (1955), 'Effect of Soil Structure on Wheat Germination in a Red Brown Earth', *Australian Journal of Agricultural Research,* **6**, 796.

Magill, R. J. (1977), 'Transport in Agriculture', *Agricultural Gazette of N.S.W.,* **88**, 6.

Marston, D. (1987a), 'Conventional Tillage Systems as they Affect Soil Erosion in Northern N.S.W.', *Journal of Soil Conservation Service N.S.W.,* **34**, 194.

Marston, D. (1978b), 'Reduced Tillage Systems—Their Potential for Soil Conservation in Northern N.S.W.', *Journal of Soil Conservation Service N.S.W.,* **34**, 194.

Marston, D. and Doyle, A. D. (1978), 'Stubble Retention Systems as Soil Conservation Management Practices', *Journal of Soil Conservation Service N.S.W.,* **34**, 210.

Mead, J. A. (1992), 'Rotation Systems and Farming Systems: the Current Situation', in G. M. Murray and D. P. Heenan (eds), *Rotations and Farming Systems,* NSW Agriculture, 5.

Millington, R. J. (1959), 'Establishment of Wheat in Relation to Apparent Density of Surface Soil', *Australian Journal of Agricultural Research,* **10**, 487.

Morgan, J. M. (1971), 'The Death of Spikelets in Wheat Due to Water Deficit', *Australian Journal of Experimental Agriculture and Animal Husbandry,* **11**, 349.

Mulholland, J. G., Coombe, J. B., Freer, M. and McManus, W. R. (1976), 'An Evaluation of Cereal Stubble for Sheep Production', *Australian Journal of Agricultural Research,* **27**, 881.

Nix, H. A. (1975), 'The Australian Climate and its Effects on Grain Yield and Quality' in A. Lazenby and E. M. Matheson (eds), *Australian Field Crops. I: Wheat and Other Temperate Cereals,* Angus and Robertson, Sydney, 183.

Orchiston, H. D. (1965), 'The Cultivation Needs of Different Soils', *Power Farming,* **74**, 25.

Pearce, G. A. (1973), 'Faster Weed Germination with Early Cultivation', *Journal of Agriculture, Western Australia,* **14**, 134.

Pearce, G. A. and Holmes, J. E. (1976), 'The Control of Annual Ryegrass', *Journal of Agriculture, Western Australia,* **17**, 77.

Pearson, C. J. and Ison, R. L. (1987), *Agronomy of Grassland Systems,* Cambridge University Press, Sydney.

Poole, M. L. (1987), 'Tillage Practices for Crop Production in Winter Rainfall Areas' in P. S. Cornish and J. E. Pratley (eds), *Tillage—New Directions in Australian Agriculture,* Inkata Press, Melbourne, 24.

Powles, S. B. (1987), 'A Review of Weeds in Australia Resistant to Herbicides' in *Proceedings of the 8th Australian Weeds Conference,* Sydney, 103.
Powles, S. B. and Holtum, J. A. M. (1990), 'Herbicide Resistant Weeds in Australia' in *Proceedings of the 9th Australian Weeds Conference,* Adelaide, 185.
Pratley, J. E. (1987), 'Soil, Water and Weed Management—the Key to Farm Productivity in Southern Australia', *Plant Protection Quarterly,* 2(1), 21.
Pratley, J. E. (1989), 'Silvergrass Residue Effects on Wheat' in *Proceedings of the 5th Australian Agronomy Conference,* Perth, 472.
Pratley, J. E. and Cornish, P. S. (1985), 'Conservation Farming—a Crop Establishment Alternative or a Whole Farm System' in *Proceedings of 3rd Australian Agronomy Conference,* Hobart, 95.
Pratley, J. E. and Godyn, D. L. (1991), 'Pasture management' in D. J. Cottle (ed.) *Australian Sheep and Wool Handbook,* Inkata, Melbourne, 267.
Pratley, J. E. and Ingrey, J. D. (1990), 'Silvergrass Allelopathy on Crop and Pasture Species' in *Proceedings of the 9th Australian Weeds Conference,* Adelaide.
Pratley, J. E. and Rowell, D. L. (1987), 'From the First Fleet—Evolution of Australian Farming Systems' in P. S. Cornish and J. E. Pratley (eds) *Tillage—New Directions in Australian Agriculture,* Inkata Press, Melbourne, 2.
Reeves, T. G. and Smith, I. S. (1975), 'Pasture Management and Cultural Methods for the Control of Annual Ryegrass (*Lolium rigidum*) in Wheat', *Australian Journal of Experimental Agriculture and Animal Husbandry,* 15, 527.
Ridge, P. E. (1986), 'A Review of Long Fallows for Dryland Wheat Production in Southern Australia', *Journal of Australian Institute of Agricultural Science,* 52(1), 37.
Rosewell, C. J. and Marston, D. (1978), 'The Erosion Process as it Occurs Within Cropping Systems', *Journal of Soil Conservation Service N.S.W.,* 34, 186.
Rovira, A. D. (1987), 'Tillage and Soil-borne Diseases of Winter Cereals' in P. S. Cornish and J. E. Pratley (eds), *Tillage—New Directions in Australian Agriculture,* Inkata Press, Melbourne.
Russel, J. S. (1960), 'Soil Fertility Changes in the Long Term Experimental Plots at Kybybolite South Australia. 1: Changes in pH, Total Nitrogen, Organic Carbon and Bulk Density', *Australian Journal of Agriculture Research,* 11, 902.
Rutger, J. N. and Crowder, L. V. (1967), 'Effect of High Plant Density on Silage and Grain Yields of Six Corn Hybrids', *Crop Science,* 7, 182.
Schultz, J. E. (1971), 'Soil Water Changes under Fallow-crop Treatments in Relation to Soil Type, Rainfall and Yield of Wheat', *Australian Journal of Experimental Agriculture and Animal Husbandry,* 11, 236.
Schutz, J. E. (1972), 'Effect of Surface Treatments on Soil Water Storage and Yield of Wheat', *Australian Journal of Experimental Agriculture and Animal Husbandry,* 12, 299.
Shaw, J. H. (1971a), 'Soil Conservation in Western Downs—Maranoa—1', *Queensland Agricultural Journal,* 97, 277.
Shaw, J. H. (1971b), 'Soil Conservation in Western Downs—Maranoa—2', *Queensland Agricultural Journal,* 97, 316.

Simmonds, D. H. (1989), *Wheat and Wheat Quality in Australia,* CSIRO, Adelaide.

Simpson, J. R., Bromfield, S. M. and Jones, O. L. (1974), 'Effects of Management on Soil Fertility under Pasture. 3: Changes in Total Soil Nitrogen, Carbon, Phosphorus and Exchangeable Cations', *Australian Journal of Experimental Agriculture and Animal Husbandry,* **14,** 487.

Sims, H. J. (1977), 'Cultivation and Fallowing Practices' in J. S. Russel and E. L. Greacen (eds), *Soil Factors in Crop Production in a Semi-Arid Environment,* University of Queensland Press, Brisbane.

Single, W. V. (1971), 'Frost Damage in Wheat Crops', *Agricultural Gazette of N.S.W.,* **82,** 211.

Single, W. V. (1975), 'Frost Injury' in A. Lazenby and E. M. Matheson (eds), *Australian Field Crops. I: Wheat and Other Temperate Cereals,* Angus and Robertson, Sydney.

Stern, W. R. (1968a), 'The Influence of Sowing Date on the Yield of Grain Sorghum in a Short Summer Rainfall Environment', *Australian Journal of Experimental Agriculture and Animal Husbandry,* **8,** 594.

Stern, W. R. (1968b), 'The Influence of Sowing Date on the Yield of Peanuts and Sorghum in a Short Summer Rainfall Environment', *Australian Journal of Experimental Agriculture and Animal Husbandry,* **8,** 599.

Stonebridge, W. C., Fletcher, I. C. and Lefroy, D. B. (1973), '"Spray-Seed", the Western Australian Direct Sowing System', *Outlook on Agriculture,* **7,** 155.

Stoneman, T. C. (1962), 'Loss of Structure of Wheatbelt Soils', *Journal of Agriculture, Western Australia,* **3,** 493.

Stoneman, T. C. (1973), 'Soil Structure Changes under Wheatbelt Farming System', *Journal of Agriculture, Western Australia,* **14,** 209.

Storrier, R. R. (1962), 'The Availability of Mineral Nitrogen in a Wheat Soil from Southern New South Wales', *Australian Journal of Experimental Agriculture and Animal Husbandry,* **2,** 185.

Sturgess, O. W., Leverington, K. C., Hogarth, D. M. and Ridge, D. R. (1976), 'Sugar Options and Constraints' in *Limits to Growth and Options for Action,* Australian Institute of Agricultural Science National Conference, Canberra, 85.

Swann, I. F. (1971), 'A Place for Defoliation in Cotton Crops', *Queensland Agricultural Journal,* **97,** 167.

Sweeting, H. (1974), 'How Disc Tillage Tools Work', *Power Farming Magazine,* **83,** 6.

Sweeting, H. and Colless, R. H. (1985), 'Agriculture Machinery for Conservation Farming' in P. E. Charman (ed.) *Conservation Farming,* Soil Conservation Service of New South Wales, 55.

Tuohey, C. L., Robson, A. D. and Rooney, D. R. (1972), 'Moisture and Nitrate Conservation and Responses to Fallowing after Medic Ley in the Wimmera', *Australian Journal of Experimental Agriculture and Animal Husbandry,* **12,** 414.

Thorn, C. W. (1989), 'Management of Annual Legume Pastures' in *Proceedings of the 5th Australian Agronomy Conference,* Perth, 123.

Walton, G. H. (1976), 'Agronomic Studies on *Lupinus angustifolius* in Western Australia: Effect of Cultivar, Time of Sowing and Plant Density on Seed Yield', *Australian Journal of Experimental Agriculture and Animal Husbandry,* **16,** 893.

Waring, S. A., Fox, W. E. and Teakle, L. J. H. (1958), 'Fertility Investigations on the Black Earth Wheatlands of the Darling Downs, Queensland. I: Moisture Accumulation Under Fallow', *Australian Journal of Agricultural Research,* **9**, 205.

Watson, E. R. (1969), 'The Influence of Subterranean Clover Pastures on Soil Fertility. 3: The Effect of Applied Phosphorus and Sulphur', *Australian Journal of Agricultural Research,* **20**, 447.

Wells, G. J. (1969), 'Skeleton Weed (*Chondrilla juncea*) in the Victorian Malle. 1: Competition with Legumes', *Australian Journal of Experimental Agriculture and Animal Husbandry,* **9**, 521.

Wells, G. J. (1970), 'Skeleton Weed (*Chondrilla juncea*) in the Victorian Mallee. 2: Effects of Legumes on Soil Fertility, Subsequent Wheat Crops and Weed Population', *Australian Journal of Experimental Agriculture and Animal Husbandry,* **10**, 622.

Wells, G. J. (1971), 'Effects of Fallowing on Wheat Yields and Weed Populations', *Australian Journal of Experimental Agriculture and Animal Husbandry,* **11**, 313.

White, D. H., Elliott, B. R., Sharkey, M. J. and Reeves, T. G. (1978), 'Efficiency of Land-use Systems Involving Crops and Pastures', *Journal of Australian Institute of Agricultural Science,* **44**, 21.

Williams, C. H. (1970), 'Pasture Nitrogen in Australia', *Journal of the Australian Institute of Agricultural Science,* **36**, 199.

Woodlands, K., Fowler, J., Lacy, J. and Clampett, W. (1984), 'Crop Establishment and Management', in A. Currey (ed.), *Rice Growing in New South Wales,* Department of Agriculture New South Wales and Rice Research Committee.

CHAPTER 7

IRRIGATION FOR CROP PRODUCTION

J. E. Pratley and K. G. Beirne

Most of Australia lies within the arid belt between latitudes of 15°S and 35°S where rainfall is meagre and unreliable by world standards. The continent receives on average only 420 mm of rainfall per annum compared to the world average of 660 mm.

During the last century, periodic droughts have had disastrous economic effects on primary production throughout Australia. From 1864 to 1992 there were thirteen droughts of major proportions as well as numerous localised droughts of lesser national significance. In the 1895–1903 drought, the sheep population was reduced from 106 million to 53 million. It took 30 years for these numbers to build up again, only to be substantially reduced again in the devastating drought of the mid-1940s.

The lack of reliable water supplies in many parts of Australia and the consequences of drought on the national economy, and in particular of livestock and crop production, highlighted the need to initiate water conservation programs to access underground water supplies.

There was considerable public enthusiasm for irrigation development, particularly in the early 1900s. The first irrigation scheme was established in 1859 on a farm fronting the Yarra River near Melbourne. In 1886, the Irrigation Act of Victoria was enacted to rationalise all surface water resources and to enable the formation of irrigation trusts to construct and administer the various works (Anderson, 1974).

At the turn of the century, as many as ninety trusts were operating in Victoria. However, by 1905, the administration of water resources in Victoria was assumed by the newly created State Rivers and Water Supply Commission (Anderson, 1974). All trusts were abolished except for the First Mildura Irrigation Trust, which is still in operation. This scheme, as well as a similar pumping scheme at Renmark in South Australia, was established in 1889 by the Chaffey Brothers, who had developed irrigation projects in California.

The development at Renmark gave the impetus for further irrigation development along the Murray River, mainly for horticultural crops such as citrus and grapes.

The Victorian and South Australian irrigation activities prompted some action in New South Wales. The first farm irrigation scheme was started in 1890, at Curlwaa, at the junction of the Murray and Darling rivers. Two years later, the Hay Irrigation Trust on the Murrumbidgee River was established.

In 1899, Sir Samuel McCaughey bought North Yanko estate on the banks of the Murrumbidgee River and developed his ideas of irrigation which were to become the start of the Murrumbidgee Irrigation Areas. McCaughey used a system of pumps, a main canal and distribution channels to successfully irrigate forage crops and lucerne.

The Burrinjuck and Murrumbidgee Construction Act was passed in 1906, authorising the construction of Burrinjuck Dam on the Murrumbidgee River near Yass and the design and construction of an irrigation scheme centred on Yanco, N.S.W.

Water for the Murrumbidgee Irrigation Areas was first supplied on 13 July 1912 to Yanco. The scheme was initially administered by the Department of Public Works, but the N.S.W. government soon realised that water conservation and irrigation required special attention. Thus, in January 1913, the Water Conservation and Irrigation Commission was established to undertake the administration and control of all water conservation and irrigation schemes in rural areas of New South Wales. This body was superseded in April 1976 when the Water Resources Commission was formed, with additional responsibilities including underground water resources.

Irrigation development occurred much later in Queensland. The first large-scale irrigation project in tropical Australia was the Mareeba–Dimbulah Irrigation Area, which was completed in 1959 (Anderson, 1974). The Burdekin River Scheme was developed in 1986 for the sugar cane and rice growing industries. All water resources in Queensland are controlled by the Queensland Water Supply Commission, including underground waters on which a large proportion of the state's irrigation depends.

The Murray–Darling Basin continues to be the most important area of irrigation. Despite the fact that this region comprises only 15 per cent of the Australian land mass, it contains about 72 per cent of all irrigated land, including the major irrigation areas and districts of southwestern New South Wales and the northern plains of Victoria. The river systems are of vital importance to these irrigation activities, the major contributions coming from the Murray, Murrumbidgee and Goulburn rivers. In order to ensure continuity of water supplies for irrigation, emphasis has been placed on the construction of large storage dams and the development of government-operated irrigation districts and areas. In New South Wales and Victoria, government-operated irrigation areas and districts account for more than 60 per cent of the total irrigated land in those states. The extent of irrigation in Australia is shown in Fig. 7.1. A more comprehensive account of early developments is given in Smith *et al.* (1983).

AVAILABILITY OF WATER FOR IRRIGATION

It is clear from the analysis of the climatic features of the Australian environment in Chapter 2 that plant productivity is limited by a water insufficiency over the majority of the continent. This productivity could be raised by ensuring a reliable supply of water throughout the year. A survey of the water resources in Australia is therefore warranted, taking into account not only the amount of water but also its dependability and quality.

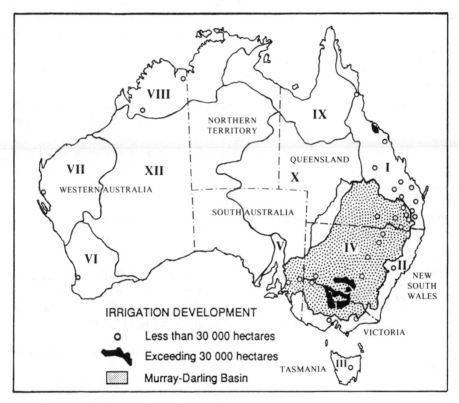

FIG. 7.1 The distribution of the principal irrigation lands in Australia in association with the drainage system of the continent (adapted from Meyer, 1992)

I	North-east Coast	VII	Indian Ocean
II	South-east Coast	VIII	Timor Sea
III	Tasmania	IX	Gulf of Carpentaria
IV	Murray–Darling Basin[1]	X	Lake Eyre
V	South Australian Gulf	XI	Bulloo–Bancannia
VI	South-west Coast	XII	Western Plateau

---

[1] The Murray–Darling contains the most extensive irrigation areas in Australia.

Topographical Features

The lack of high mountain ranges over most of the continent is a major factor affecting rainfall and drainage patterns. Only two per cent of Australia lies above 1000 metres. The most notable range is the Great Dividing Range in the east, Mt Kosciusko being the highest point at 2228 m. However, these mountains are close to the eastern seaboard, giving rise to relatively short coastal river systems which drain to the sea and account for most of the annual discharge. A significant exception is the Murray–Darling system.

The remaining 70 per cent of the continent is interior lowlands with drainage patterns characteristic of semi-arid and arid conditions (Fig. 7.1). These include disconnected ephemeral river systems and terminal salt lakes.

Runoff

The total surface runoff from the continent, as average annual river discharge, is about 440 million megalitres, which is equivalent to 57 mm per year, or 12 per cent of the rainfall (Anon., 1992). Almost half of this runoff flows to the Gulf of Carpentaria and the far northeast coast, where no significant irrigation exists. If the Western Plateau, which has no runoff, is excluded from consideration, the runoff from the remainder of the continent is less than 76 mm. This compares unfavourably with the world average runoff of 230 mm.

Seasonality

The pronounced seasonality of rainfall over the continent creates a similar pattern with respect to runoff. In northern Australia the flow is concentrated in the summer months, whilst in the south the discharge is concentrated, though to a lesser extent, in the winter months.

TABLE 7.1   Utilisation of surface water resources, Australia, 1983–84 (Anon., 1992)

State	Area (thousand km²)	Annual divertible resource (million ML)	Water made available (per cent)
Qld	1727	32.7	11.7
NSW	802	16.9	47.2
ACT	2	0.2	60.6
Vic	228	9.8	61.1
SA	984	0.4	32.3
WA	2526	11.7	24.3
NT	1346	17.7	0.3
Tas	68	10.9	9.3
Australia	7683	100.3	21.5

This pronounced fluctuation results in large rivers in the north, for example, having flow rates of 30 ML per second or more, in high flood. The same rivers may cease to flow altogether in the winter dry season.

Because of this seasonality, any regional development must necessarily provide for surface water storage to take account of the variability of flow of surface waters. The costs of this storage are high because, in addition to storage for irrigation and other purposes, there is a need to provide adequate spillway capacity to ensure safety in flood times. These costs are accentuated by the very high evaporation losses from surface water storages.

Surface Waters

In Australia, only about 21 per cent of the total divertible water resource is utilised (Anon., 1991), the majority being too remote or too expensive to develop. Of the proportion used, about 80 per cent comes from surface water.

Most of the developed resources are in south-eastern Australia, particularly the Murray–Darling Basin. In this region development of surface waters has already reached 60 per cent of the total resources (Table 7.1). This is considered to be close to the feasible limit. Table 7.2 indicates the major storages for irrigation in Australia.

Water Quality

Although some Australian rivers are naturally saline, the quality of surface water is generally good. However, most types of water pollution experienced in advanced industrial countries also occur in Australia. The main problems relate to sewage and industrial effluents in populated areas and to increasing salinity and eutrophication from agricultural activities, mainly in the irrigated areas. The presence of nutrients, particularly nitrogen and phosophorus, from these sources in the water supplies encourages the prolific growth of algae, which exhausts the water's oxygen supply and kills other aquatic life. Many of the algal blooms are highly toxic to livestock and to humans. The occurrence is greater in still water, particularly in summer months (Anon., 1992).

Salinity affects two main areas of the country. In southern Australia clearing of the natural vegetation has been followed by a rise in the water table levels (Colclough, 1973) with a consequent increase in the discharge of groundwater to river flow. As the groundwater has a naturally high salt content, both the soil and the surface water have been adversely affected. More than four million hectares of land have been rendered sterile and difficult to till due to dryland salinity resulting from poor land use (Anon., 1982).

In south-eastern Australia, salinity problems are of concern in the Murray Valley, where the salt content of the river is less than 30 mg per litre total dissolved salts in the upper reaches but increases progressively

TABLE 7.2 Major water storages in mainland Australia for irrigation (adapted from Anon., 1986)

State	Name	Location	Gross capacity (thousand (ML)	Year completed[a]
Queensland	Beardmore	Balonne River	101	1972
	Fairbairn	Nogoa River	1440	1972
	Glenlyon	Pike Creek	254	1976
	Leslie	Sandy Creek	108	1985
	Fred Haigh	Kolan River	586	1975
	Tinaroo Falls	Barron River	407	1958
	Wuruma	Nogo River	194	1968
	Boondooma	Boyne River	212	1983
	Burdekin	Burdekin River	186	1986
New South Wales	Eucumbene	Eucumbene	4807	1958
	Blowering	Tumut River	1628	1968
	Burrinjuck	Murrumbidgee River	1026	1927 (1956)
	Copeton	Gwydir River	1364	1976
	Glenbawn	Hunter River	362	1958
	Hume	Murray River	3038	1936 (1961)
	Burrendong	Macquarie River	1677	1967
	Keepit	Namoi River	423	1960
	Wyangala	Lachlan River	1218	1936 (1971)
	Menindee Lakes	Darling River	1794	1960
	Talbingo	Tumut River	921	1971
	Jindabyne	Snowy River	688	1967
	Lake Victoria	Murray River	680	1928
	Windamere	Cudgegong River	368	1984
	Glennies Creek	Hunter Valley	284	1983
	Tantangara	Murrumbidgee River	254	1960
	Lake Brewster	Lachlan River	150	1952
Victoria	Cairn Curran	Lodden River	149	1958
	Dartmouth	Mitta Mitta River	4000	1979
	Eildon	Upper Goulburn River	3392	1927 (1958)
	Thomson	Thomson River	1175	1984
	Eppaloch	Campaspe River	312	1964
	Glenmaggie	Macalister River	190	1927 (1958)
	Mokoan	Winton Swamp, Benalla	365	1971
	Waranga	Rushworth	411	1910
	Yarrawong	Murray River	117	1939
	Toolondo	Natural depression, Horsham	107	1952 (1960)
Western Australia	Ord (Lake Argyle)	Ord River	5720	1971
	Wellington	Collie River	185	1933 (1944, 1960)

[a]Dates in brackets indicate date of completion of enlargement of the storage.

downstream. At Waikerie in South Australia, for example, salinity levels in the Murray River are about 15 times those at Jingellic, NSW (Anon., 1986; 1992). This increase in salinity is caused by the additions to the groundwater from channel seepage and irrigation water, resulting in a rise in the water table and increased drainage of this saline groundwater into the river (Blackburn, 1978; McGowan, 1984).

The development of the salt problem is evidenced by the change in the annual salt balance in the Murrumbidgee Irrigation Area (Table 7.3) over two decades. This trend cannot be sustained if productivity is to be maintained or enhanced.

TABLE 7.3   Salt balance in the Murrumbidgee Irrigation Area (Evans, 1971; van der Lelij, 1991)

	Quantity of salt (t/year)	
	1968	1990
Salt added in irrigation water	42 000	95 000
Salt removed in drainage water	44 000	62 000
Net change	−2 000	+33 000

Groundwater Supplies

About 80 per cent of Australia is significantly dependent, directly or indirectly, on groundwater supplies (Anon., 1986). The estimated annual recharge is 72 million ML. Annual groundwater use is estimated at 2.5 million ML which represents 3.5 per cent of the annual recharge and 18 per cent of Australia's total water use.

There are three main sources of groundwater:
1. Shallow, unconsolidated sediments which are found in the principal river and lake systems and as coastal dunes, deltas and narrow shoreline deposits. Since 1957, use has been made of the good-quality groundwater resources of the inland drainage systems of New South Wales, such as the alluvium of the Lachlan, Macquarie and Murrumbidgee valleys. In central Australia, this good source of quality groundwater (in unconsolidated sediment) is rare because of lower rainfall and higher evaporation rates.

    In Queensland, the sugar industry has drawn on the extensive groundwater resources of the Burdekin Delta to such an extent that it has become necessary to use surface water to recharge the aquifers artificially. In 1970 for example, 33 800 ha of cane were under irrigation, requiring an estimated 3.2 ML of water. Some 350 000 ML of river water were pumped to artificially recharge the aquifers.
2. Sedimentary basins (Fig. 7.2) which contain at least one major aquifer system. In many cases the water is unsuitable for irrigation because of the high concentrations of sodium relative to the other

cations, particularly calcium and magnesium. This is especially so
in the Great Artesian Basin, which occupies about 23 per cent of
the land mass. These waters, however, are an important source of
water for domestic and livestock purposes.

3. Fractured rocks are an important source of water particularly in
    the highlands of the southeast mainland, and in Tasmania, parts of
    South Australia, central Australia and Western Australia. These
    aquifers usually yield small quantities of water, the quality of
    which may vary considerably over short distances. The quality is
    generally good over northern and eastern Australia but poor over
    much of South Australia, the southern portion of the Northern
    Territory and southwest part of Western Australia. In the latter
    areas this variability and poor quality of groundwater are due
    largely to the low rainfall and high evaporation, coupled with low
    permeability of the strata.

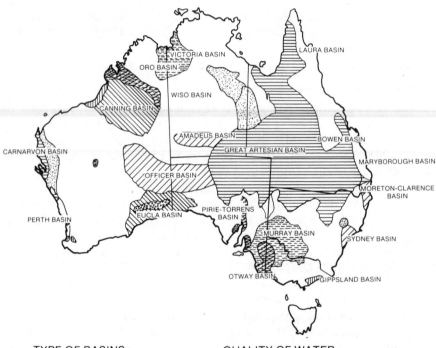

Fig. 7.2   The sedimentary basins (Anon., 1983)

Conjunctive Use of Surface and Groundwater

The utilisation of water resources in most areas of Australia has tended to concentrate on either ground or surface water. However, it has become increasingly necessary to utilise all water resources of a region in combination (Table 7.4). The relative merits of surface storage and groundwater storage are presented in Table 7.5.

TABLE 7.4   Area irrigated by various sources of water, 1989–90 (adapted from Anon., 1992)

	Area (thousand ha)					
	NSW	Vic	Qld	SA	WA + Tas	Aust (a)
Surface water						
From State schemes	399.2	418.4	89.6	16.9	16.7	940.9
From other schemes						
Direct from lakes, rivers, etc.	351.9	60.0	65.5	19.6	18.4	515.7
From farm dams	19.0	25.4	40.9	4.6	28.4	118.2
Total surface water	770.0	503.8	195.9	41.1	63.5	1574.9
Underground water	48.5	19.8	115.8	56.0	8.9	249.8
Town or country						
reticulated water	1.3	2.8	0.4	1.9	0.7	7.1
Total all water sources	819.8	526.4	312.1	99.0	73.1	1831.7

(a) Includes data for ACT and Northern Territory

TABLE 7.5   A comparison of the characteristics of surface storage and groundwater storage for irrigation

	Surface storage	Groundwater storage
Capacity	small	large
Recharge response	rapid	slow
Cost—capital	high	low
—operating	variable[a]	high[b]
Evaporation losses[c]	high	low

[a] high costs with pumping for sprinkler irrigation
[b] when pumping required
[c] losses from storage only

In Australia, the level of development of water resources has been such that combined use of these resources has not been required to any extent. As the degree of overall development proceeds and as the value of water rises, there is increasing pressure for greater amounts of water to be supplied at higher levels of reliability.

In the Namoi Valley, New South Wales, and in the Callide Valley in Queensland, irrigation from groundwater sources has lowered the water table. The conjunctive use of different sources of water is used to overcome this problem. Surface sources are used during periods of river flow, thereby also allowing recharge of groundwaters for use during the remainder of the time.

A further example of conjunctive use is where the groundwater source by itself is unsuitable for irrigation because of its high salt content. If mixed at an appropriate dilution with high-quality surface water, the resultant water is then suitable for irrigating crops and pastures.

Control of Water Resources

Australia's water resources are managed by about 800 irrigation authorities, metropolitan water boards, local government councils and private individuals. State authorities dominate the assessment and control of water resources because, under the Australian Constitution, primary responsibility for water management rests with individual state governments. The Australian government participates indirectly, through financial assistance, or directly in the co-ordination or operation of interstate projects.

In recognition of the important environment problems facing the Murray–Darling Basin, the governments of South Australia, Victoria, New South Wales and Queensland agreed in 1985 to form the Murray–Darling Basin Ministerial Council. The Council has a charter to promote and co-ordinate effective planning and management for the equitable, efficient and sustainable use of the water, land and environmental resources of the Basin (Blackmore, 1989). The Murray–Darling Basin Commission operates as the executive committee of the Council.

The proper management of water resources is essential to the maintenance of both quantity and quality of supplies and to the ecological balance of the environment in general. Since water is an agent of erosion and deposition, the consequences of its mismanagement can be very damaging.

METHODS OF IRRIGATION

The method of irrigation depends on water availability, the method of supply, the crop to be grown and the region in which it takes place. The main systems comprise surface, sprinkler, trickle and micro-irrigation, the most important for agricultural field crops being surface irrigation. Sprinkler, trickle and micro-irrigation are confined largely to horticultural crops (Table 7.6).

Surface Irrigation

Almost all irrigated field crops in Australia are surface irrigated, that is the water is applied at ground level to the surface of the soil. It is distributed over the land:

—in borders or bays, where it is confined between low banks of soil,
—in check banks, erected at regular intervals down the slope of the field, or
—in furrows.

Border Check

In this method, the water is directed down the slope of the land between a system of low parallel banks called check banks (Fig. 7.3). Under ideal conditions the rate of flooding is such that the water is uniformly absorbed by the soil as the sheet of water flows down the strip. This method is suited to land which is comparatively flat and has uniform slope. It is used on slopes ranging from 0.12 per cent to 3 per cent.

The check banks are formed by grading across the slope of the bay, i.e. *parallel* to the contours, and are then formed into the required slope and height by means of a grader or crowder. The banks should be low and

TABLE 7.6   Area of land (thousand hectares) under irrigated culture in Australia, 1983–84 (Anon., 1986)

			Method of irrigation		
Crop	Sprinkler	Surface	Trickle including microsprays	Other and multiple methods	Total area
New South Wales					
Pastures	42.6	197.5	n.a.	10.6	250.8
Pure lucerne	24.2	13.1	n.a.	1.1	38.4
Cereals	28.9	207.6	n.a.	8.1	244.7
Vegetables	8.9	3.4	0.2	0.7	13.2
Fruit	5.1	4.9	5.7	0.5	16.3
Grapevines	1.4	5.2	1.2	0.1	7.9
All other crops	4.5	84.3	0.1	1.2	90.1
TOTAL	115.7	516.3	7.2	22.3	661.5
Queensland					
Pastures	14.6	3.1	n.a.	1.6	19.6
Pure lucerne	12.4	0.1	n.a.	0.8	13.3
Grain sorghum	3.8	5.8	n.a.	0.9	10.6
Other cereals	10.6	11.6	n.a.	1.4	23.6
Sugar cane	50.7	43.5	0.5	9.4	104.0
Cotton	0.3	29.1	0.1	0.6	30.1
Soybeans	6.3	7.1	n.a.	1.0	14.4
Vegetables	14.8	3.1	0.9	1.3	20.2
Fruit (including grapevines)	3.7	0.2	4.9	0.8	9.5
All other crops	6.1	1.9	0.2	0.8	9.0
TOTAL	123.6	105.6	6.6	18.5	254.4

TABLE 7.6   continued

Crop	Method of irrigation				
	Sprinkler	Surface	Trickle including microsprays	Other and multiple methods	Total area
South Australia					
Pastures	14.5	15.4	n.a.	1.1	31.0
Pure lucerne	10.8	4.3	n.a.	0.5	15.6
Cereals	1.6	0.8	n.a.	0.1	2.5
Vegetables	5.4	0.3	0.1	0.5	6.4
Fruit	7.4	1.4	3.2	0.8	12.8
Grapevines	5.2	6.5	3.7	1.1	16.5
All other crops	1.2	0.9	>0.1	>0.1	12.1
TOTAL	46.1	29.7	7.0	4.2	87.0
Victoria					
Pastures	36.3	408.8	n.a.	12.0	457.1
Pure lucerne	4.4	7.6	n.a.	0.2	12.3
Cereals	4.8	24.5	n.a.	1.5	30.9
Tobacco	2.1	0.1	n.a.	0.2	2.3
Vegetables	12.7	4.1	0.5	2.3	19.6
Fruit	4.9	4.6	3.4	0.6	13.6
Grapevines	3.4	10.1	0.6	0.2	14.3
All other crops	1.8	3.1	>0.1	0.2	5.1
TOTAL	70.5	463.0	4.6	17.1	55.1
Tasmania					
Pastures	8.9	7.9	n.a.	0.7	17.5
Pure lucerne	0.9	0.1	n.a.	>0.1	1.0
Cereals	1.5	0.1	n.a.	0.1	1.7
Vegetables	11.0	0.1	>0.1	0.8	4.7
Fruit	1.1	0.1	0.8	0.2	2.2
All other crops	3.8	0.3	>0.1	0.4	4.4
TOTAL	27.2	8.5	0.9	3.4	39.9
Western Australia					
Pastures	1.4	10.5	n.a.	0.4	12.3
Pure lucerne	0.9	0.5	n.a.	0.1	1.6
Cereals	0.4	1.0	n.a.	>0.1	1.5
Vegetables	3.0	0.5	0.2	0.5	4.1
Fruit	1.5	0.4	1.6	0.3	3.7
Grapevines	0.2	>0.1	0.3	0.1	0.7
All other crops	0.4	1.0	0.3	0.1	1.8
TOTAL	7.9	14.0	2.4	1.4	25.7
TOTAL, AUSTRALIA	391.0	1137.1	28.7	66.9	1623.6

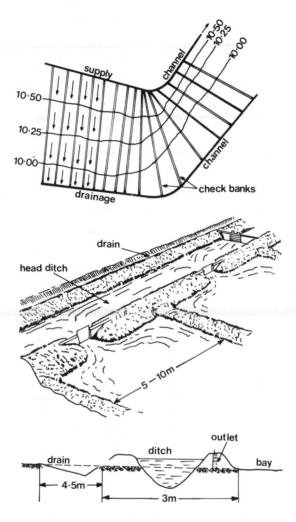

FIG. 7.3   The border check method of irrigation (Water Resources Commission of New South Wales, 1978, unpublished)

wide to permit the crossing of farm machinery and to allow for cultivation so that full use is made of the land. The banks therefore should be of the order of 15 to 20 cm high and approximately one metre wide at the base.

The dimensions of the bays depend on a number of factors, including soil type, land slope, type of crop, depth of application and rate of supply of water. Bays vary from 10 to 20 metres wide and from 100 to 200 metres in length. Narrow short bays are used on the steeper slopes and on sandy soils, in the latter case to move the water over the land quickly to avoid excessive deep-percolation losses. Where the land is flat or where the soils are heavy, wide, long bays can be used. Table 7.7 indicates the flow rates required to irrigate strips under different conditions.

TABLE 7.7 The relation of soil, slope and depth of application for border check irrigation

Soil	Slope %	Application (mm)	Width (m)	Length (m)	Quantity[a] (ML/day)
			Dimensions of strip		
Coarse	0.25	50	15	150	19.5
or sandy		100	15	240	17.0
		150	15	400	14.5
	1.00	50	12	90	6.5
		100	12	150	6.0
		150	12	270	6.0
	2.00	50	9	60	3.0
		100	9	90	2.5
		150	9	180	2.5
Medium	0.25	50	15	240	17.0
or loamy		100	15	400	14.5
		150	15	400	8.5
	1.00	50	12	150	6.0
		100	12	300	6.0
		150	12	400	6.0
	2.00	50	9	90	2.5
		100	9	180	2.5
		150	9	300	2.5
Fine	0.25	50	15	400	10.0
or clay		100	15	400	6.0
		150	15	400	3.5
	1.00	50	12	400	6.0
		100	12	400	3.0
		150	12	400	2.0
	2.00	50	9	200	2.5
		100	9	400	2.5
		150	9	400	1.5

[a]Figures are rounded to the nearest 0.5 ML.

To increase the efficiency of the operation, the bays must be graded in order to eliminate high spots, which will be underwatered, and low spots, which will be subject to ponding. To ensure uniformity of application, the steepest slope down the length of the bay should not exceed twice the flattest slope. The use of laser-controlled levelling has been a major advance. Fields can now be levelled to a near-perfect plane, allowing rapid and even application and removal of water (Barrett, 1985) and easy management.

Contour Check

On relatively flat slopes, the contour check method is used. In this system, check banks are laid out across the slope closely following the natural contours of the land (Fig. 7.4). Water is diverted from the supply channel into the highest point of each bay and held there until the whole bay is flooded. This method is used for the production of rice and also for pastures on heavy soils. It has been the principal method of irrigation in the Murrumbidgee Irrigation Areas.

The minimum distance between banks must allow for the passage of cultivating and harvesting machinery. This is usually about 20 metres. The contour intervals at which the banks are located are designed to keep the size of the bays in the range of 1.5 to 2.5 hectares for pasture irrigation and 5.5 to 6.5 hectares for rice. The most common contour interval is 7.5 cm for the slopes encountered. This may be reduced to 4 cm on flat ground, but where the ground slope is less than 0.04 per cent, a border ditch system may be preferred. The contour interval may be as high as 15 cm on steep ground, but where the slope is steeper than 0.67 per cent, a border check system should be used. It is particularly

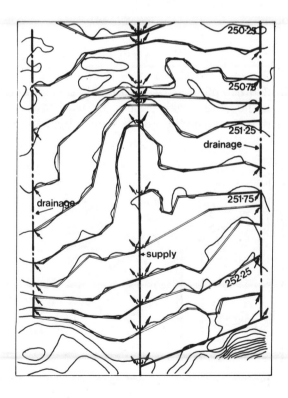

FIG. 7.4   The contour check method of irrigation (Water Resources Commission of New South Wales, 1978, unpublished)

important in the culture of rice that the slope should not be too great because of the importance of the effect of water height on the degree of stooling of the crop. Stooling is reduced if the permanent water is too deep. Under conditions of excessive slope, the need to maintain the top of the bay under permanent water results in the lower end of the bays being under excessive water.

The banks should be about 30 cm high for pastures and at least 40 cm high for rice to enable deep water to be applied at panicle initiation to counteract the effects of cool nights at that growth stage. They should be of the order of 3 metres across the base and generally no more than 800 metres long.

The advent of laser levelling, more efficient grading techniques and the need for better water control have led to a radical redesign of this method, with irregular contour bays commonly being replaced by more uniform, rectangular bays.

Whilst it takes less labour to irrigate, the contour check system will give inefficient irrigation and poor water control if used with rough grading, poor drainage, and in very large bays. It is unsuitable for permeable soils.

Border Ditch

The border ditch method is used on very flat land where drainage would be a problem if other methods were used. It is a modification of the border check method, but small ditches are used instead of the check banks. The earth from the ditches forms a small bank on each side of the ditch.

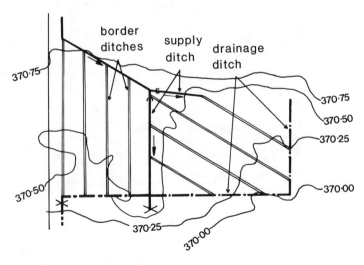

FIG. 7.5   The border ditch method of irrigation (Water Resources Commission of New South Wales, 1978, unpublished)

The ditches are run across the contours in the direction of maximum slope and may be up to 60 metres apart. Water is usually supplied at the top of the bay in the same manner as for border check systems.

The ditches are connected to the drainage system and low spots can be drained into the ditches at intervals down the bay. The bays are usually considerably larger than with border check, bay lengths of 450 metres or more sometimes being used. It is a commonly used method for irrigated pastures in the Riverina, N.S.W. An illustration of the method is shown in Fig. 7.5.

Furrow Irrigation

Field crops such as cotton, maize, sorghum and some vegetables are grown as row crops and as such are usually furrow irrigated, the irrigation water being applied in furrows between the rows (Fig. 7.6). As the water flows down the furrows, part is absorbed by the soil and penetrates to the roots of the growing plants. Cultivating the furrows controls weed growth and keeps the soil loose. This method may be used on all soil types except coarse or sandy soils. A benefit of this method is that only one fifth to one half of the total soil surface is flooded, thus reducing evaporation losses and making it possible to cultivate the soil sooner after watering.

Water is supplied from a head ditch using siphon tubes, check banks or gated piping, and distributed over the field in inter-row furrows (Fig. 7.6). Ideally, the flow rate down the furrows will be sufficient to supply the plant requirements, with a minimal loss of water through drainage at the end of the furrow.

The layout of the furrows depends on the slope of the land to be irrigated. On gently sloping land (0.05 to 1.0 per cent) straight furrows are used, but where the slope is steeper (even up to 5 per cent), contour

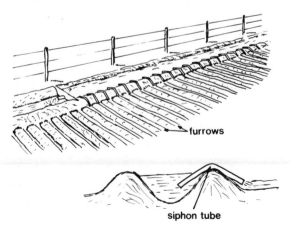

Fig. 7.6 Furrow irrigation (Water Resources Commission of New South Wales, 1978, unpublished)

furrows are used. In the latter case, the furrows are formed across the slope, sufficiently off the true contours to give the required gradient for steady flow of irrigation water down the furrows. Contour furrows are used to some extent in Queensland for the irrigation of tobacco, vegetable crops and orchards.

On sloping land, erosion of furrows can be a serious problem and requires careful control of the irrigation stream. The volume of the irrigation stream will depend largely on the slope on which furrows are to be laid. This is given by the approximate relationship

$$\text{stream flow (L per minute)} = \frac{36}{\text{slope (\%)}}$$

On flat slopes, however, erosion is not a serious problem and the carrying capacity of the furrows is probably the limiting factor. The maximum furrow length depends on the size of stream, rate of absorption, and the amount of water which is to be stored at each irrigation. The latter two depend on seasonal conditions, soil type and crop species.

The spacing of furrows is largely determined by the crops being irrigated. Normally, spacings greater than 75 cm are not desirable because the root zone is limited for much of the season, whilst it is seldom practical to construct furrows more closely spaced than 35 cm.

Furrow irrigation may be practised on slopes up to 3 per cent but should not have undulating profiles. The steepest grade on increasing slopes should not exceed twice the flattest grade, whilst on decreasing slopes the steepest should be less than 1.5 times the flattest.

The maximum allowable cross fall depends on the depth of the furrow, the relationships being explained in Table 7.8. The allowable lengths of furrows for various depths, slopes and soils are given in Table 7.9.

A modification of furrow irrigation, known as corrugation or bed irrigation, involves small furrows 100 to 150 cm apart, constructed at sowing using 25 cm scarifier points mounted behind the wheels of a combine. This method is widely used in the Moree and Trangie area of New South Wales for crops that do not require row crop treatment.

TABLE 7.8   The maximum allowable cross fall, accounting for variation in depth and slope of furrow and soil type for furrow irrigation (adapted from Water Resources Commission of NSW, 1978, unpublished)

	Maximum cross fall (%)		
Depth of furrow (cm)	Fine and medium texture	Coarse texture	Very coarse texture
7.5	0.5	0.3	—
15	3	2	1
23	8	5	2

TABLE 7.9 The allowable lengths of furrow, based on depth of application, slope of furrow and on soil type (adapted from Water Resources Commission of NSW, 1978, unpublished)

Soil texture	Depth of application (mm)	Maximum length of furrow (m)[a]							
		Furrow slope (%)							
		0.25	0.50	0.75	1.00	1.50	2.00	3.00	5.00
		Maximum allowable stream (L/minute)							
		18.0	9.0	6.0	4.5	3.0	2.5	1.5	1.00
Coarse	50	150	105	80	70	60	50	40	30
(sands)	100	220	145	115	100	80	70	45	40
	150	265	185	145	120	100	85	65	50
	200	305	205	170	145	115	100	75	60
Medium	50	250	170	135	115	95	80	65	50
(loams)	100	350	245	190	165	130	115	90	70
	150	440	295	235	200	160	135	110	80
	200	505	340	275	230	190	160	130	100
Fine	50	320	225	175	150	120	105	80	65
(clays)	100	455	310	250	230	175	145	115	90
	150	535	380	305	260	215	185	145	105
	200	650	445	350	300	245	205	170	125

[a]Figures rounded to nearest 5 m.

Sprinkler Irrigation

Sprinkler or spray irrigation requires little land preparation, such as the grading required for surface irrigation. However, the efficiency of sprinkler irrigation is improved where this has been done. This method is also useful for small area irrigation where major soil relocation is not practical or economical.

There is a high capital cost associated with sprinkler irrigation. In addition, operational costs are high because more power per unit of water applied is required than for surface irrigation due to the pressure required to operate the spray heads.

A major disadvantage with sprinkler irrigation is the distortion in the spray pattern due to wind movement. The degree to which this occurs depends on the wind speed and droplet size. Fine droplets are carried further by wind and there is considerable loss of water due to evaporation. Larger droplets are not carried to the same extent but the action of those droplets on the soil surface can result in compaction.

There are fixed and portable types of sprinkler irrigation. The fixed types are usually associated with orchards and, whilst expensive to install, minimise the labour cost. The components are shown in Figure 7.7. The portable types may be shifted manually or pulled on wheels or skids by a tractor, or may be travelling irrigators which move automatically.

The most significant development in sprinkler irrigation has been the centre-pivot travelling irrigator, which has improved uniformity and efficiency of application. The circular patterns provided place limitations on farming and land utilisation and this has led to the development of side-move systems (lateral or linear move). These can irrigate rectangular fields and can operate at lower pressures due to a constant rate of application along the lateral. Spray loss problems have been reduced by the use of droppers which release water at about ultimate crop canopy height.

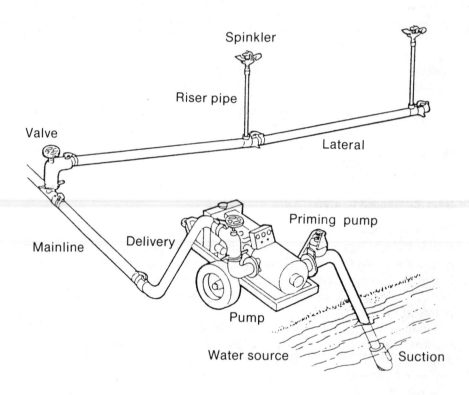

FIG. 7.7   The components of a sprinkler irrigation system (adapted from Kay, 1983)

Trickle Irrigation

The basic system of trickle or drip irrigation comprises low density polythene pipes laid along crop rows, with outlets or drippers that have orifices 0.5 to 1 mm in diameter inserted at appropriate intervals. The laterals are laid parallel to the direction of normal working and are usually left exposed, although they can be buried with only the outlets exposed. The system is designed to supply water at low rates (1 to 8 litres per hour) under low pressure directly to the soil (Murphy, 1991). Various types of trickle irrigation systems are available commercially.

Advantages of this method include labour and water savings. Irrigation efficiency is high since, if properly managed, there is no runoff, no deep percolation, minimal evaporation losses, and the inter-row spaces are not watered. Grieve (1989), working with citrus in the Sunraysia region of the Murray Valley, reported water savings of approximately 10 per cent with trickle irrigation compared with conventional sprinkler systems.

Irrigation can be carried out frequently to maintain a favourable moisture status and optimum growth. The system lends itself to auto-mation and the application of fertilisers and other chemicals through the irrigation water. Poor filtration and faulty design are the main causes of trouble and extra work. Where filtration is poor, blockages nearly always result, mainly from grit, algae and slimes, or iron oxide. Algae are the most likely source of trouble for, even when not visible, they can be present in sufficient quantity to cause frequent filter blockage. The nature of the material is such that blockage occurs over the entire screen, causing complete stoppage. Control of algae is achieved by spraying copper sulphate at the rate of one part per million over the entire water surface or, where possible, excluding light from the storage.

Water savings from this method allow small deliveries from bores and soaks to be used to irrigate areas for which the supply rate is otherwise inadequate. This method of irrigation is largely confined to horticultural crops, although it is used commercially for row crops such as tomatoes and cotton (Barrett, 1985).

Micro-irrigation

Micro-irrigation comprises both microjets and mini-sprinklers which operate at low pressures of 105 to 150 kPa (Murphy, 1991). Lateral water supply lines are usually of polythene laid along the rows of crop.

Water is applied with microjets in about 3 to 4-metre-diameter circles along the crop line at a rate of 1 to 2 litres per minute, whereas mini-sprinklers project water in circles up to about 10 metres in diameter at about 2 to 4 litres per minute.

In comparison with trickle irrigation, a greater area of soil is wetted by micro-irrigation, which thus gives a better coverage of the root zone of trees and vines, particularly on light-textured soils (Murphy, 1991). More frequent irrigation is required, however, since there is less water storage in the soil.

SOILS AND IRRIGATION

The successful production of irrigated crops depends on a range of factors, many of which are directly related to the soil. In much of the irrigation practised in Australia, crops are grown in areas where water is the limiting factor to crop production. Irrigation therefore tends to produce an artificial environment in the soil compared to its natural state. The soil water levels are higher with irrigation, and the relationship between the soil and water become more critical.

Irrigation may also affect the characteristics of the soil, depending on the nature of the soil and the quality of the water used for irrigation. Whether the soil is suitable or not for irrigation depends on a number of factors including:

—the ability of the soil to absorb water readily;
—the ability of the soil to hold adequate quantities of water to support crop growth;
—the soil's ability to drain freely;
—the susceptibility of the soil to erosion;
—the fertility status of the surface soil;
—the reaction of the soil to salts contained in the irrigation water;
—the relationship with adjoining soils or micro-relief features; and
—the depth of the root zone.

Soil Properties Affecting Irrigation Requirements

The irrigation requirement of a soil depends mainly on several properties of the soil profile. Some of these properties relate directly to the immediate profile situation at the point of water application, some relate to associations within the micro-relief while others are chemical properties which affect the physical condition of the soil. The physical properties also have considerable effect on the water-holding capacity of the soil.

Effect of Texture

Texture can be defined as the particle size distribution in soils; that is, the proportions of sand, silt and clay (Table 7.10). In the field, estimations of texture can be obtained by a feel test. In all, nineteen textural classes are recognised, ranging from sand which has less than 5 per cent clay to heavy clay which can contain 50 per cent or more of clay. In the field, coarser textured soils (i.e. sand with large particle size) have higher infiltration rates than clays and drain more freely. Furthermore, the amount of water held in soils at field capacity and permanent wilting point increases as the clay content increases. The available moisture

TABLE 7.10   International standard particle diameters for textural separates

Textural separate	Particle diameter (mm)
Gravel	Above 2.00
Coarse sand	2.00–0.20
Fine sand	0.20–0.02
Silt	0.02–0.002
Clay	Below 0.002

capacity of the soil is also influenced by texture up to the point where the soil contains about 25 per cent clay. Up to that clay content, the available water-holding capacity increases with clay content and above that level the available water-holding capacity is somewhat reduced. Permeability and vertical movement of water downwards also decreases as texture becomes finer, i.e. as the soil grades from coarse texture to fine texture.

The factors which affect the intake of water and its movement through the soil are the pore size distribution, the adhesion of water to soil particles, and gravity. Pore size distribution can be determined from the texture. The adhesion of water to the particles affects both infiltration and transmission of water in the soil. Gravity also contributes to downward vertical movement of water.

In soils of uniform texture, sands have a greater proportion of large pores (>0.02 mm diameter) and are free draining. Infiltration into these soils is rapid and easy and the water is not firmly held. In a uniform clay soil, there is a greater proportion of fine pores (<0.002 mm diameter). Infiltration is slow and the water is firmly held by adhesion, so that movement downwards is slow and lateral movement is more significant. Gravity has an effect on the drainage of sands, so furrows need to be more closely spaced than in clay soils where furrow irrigation is practised (Fig. 7.8).

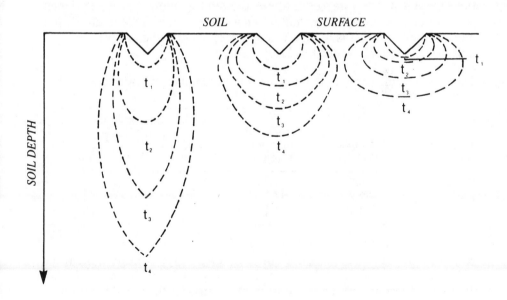

FIG. 7.8 Representation of water penetration from furrows with time (t)

Under field conditions many irrigated soils are not uniform. Texture changes from horizon to horizon and there is a substantial range of pore sizes which affects water movement through the soil profile. Hence, drainage patterns will not be exactly as shown in Fig. 7.8. Clay content tends to increase with depth down the profile and the downward movement of water slows as a result. The ability of a soil to transmit water through the profile is known as its permeability.

Water-holding characteristics also change from horizon to horizon, depending on the texture. Both field capacity and permanent wilting point change throughout the profile. Depending on this, the amount of irrigation water applied and the irrigation interval will also vary depending on the waterholding capacity of the soil. For example, crops grown on sandhill soils at Berrigan, New South Wales, require irrigation every three days in midsummer as against every seven days on loam soils nearby. The relationship between texture and water-holding capacity is shown in Table 7.11

While the amount of water held in the soil at both field capacity and permanent wilting point increases as the clay content increases, that is, as the texture becomes finer, it can be seen from Table 7.11 that the greatest volume of available water is held by the loam which can be regarded as a soil of medium texture. The amount of water held in a 30 cm depth of loam soil is equivalent to 63 mm compared to 48 mm in the same depth of clay soil and 20 mm in a sand. This means that a crop using water at the rate of 4 mm per day under peak demand would suffer water deficit stress after three days in the sand, after six days in the clay and after eight days in the loam if it is assumed that stress occurs when approximately half of the available water has been used.

TABLE 7.11   Influence of texture on soil water characteristics (Salter and Williams, 1965)

Textural class	Field capacity ($\%H_2O$)	Permanent wilting point ($\%H_2O$)	Available water-holding capacity (per 30 cm depth)	
			(%)	(mm)
Sand	6.7	1.8	4.9	20
Loam	32.2	11.8	20.4	63
Clay	39.4	22.1	17.3	48

Two physical forces, cohesion and adhesion, are responsible for the way water moves through soil in response to gravity. Cohesion, the force holding water molecules together, is important in saturated soil conditions. Adhesion, also called tension or suction, is the force holding the water film to soil particles. It is important in unsaturated soil conditions which operate after a soil has been irrigated and while it drains. Where a soil contains significant amounts of clay, adhesion becomes more important in slowing water movement down the profile.

Texture may be modified either by cementing agents such as silica, calcium carbonate, oxides of iron and aluminium, and humus, or by compaction due to cultivation, to the extent that a cemented or compacted layer is formed within the profile. The compacted layer does not transmit water as rapidly as the soil layers above it. As a result, the rate at which water moves through the soil will be determined by the transmission rate of the compacted layer. In turn, the infiltration rate of the soil will be affected. In extreme cases, the subsoil may never attain field capacity because of the very slow water movement through the restrictive upper layers of soil. Under irrigation, affected soils show shallow water penetration, thus restricting soil water reserves and root development, and inducing water stress in summer. Surface waterlogging can also be caused by excessive irrigation or by rain following the irrigation.

Effect of Structure

Soil structure is the organisation of individual particles into larger groupings called peds. The proportion of the soil mass occupied by the peds and their geometry contribute towards the permeability of the soil. Therefore if the soil is made up of a major proportion of large, regularly shaped peds which intermesh to a tight mass, the soil will not be as permeable as it would if the peds were smaller and rounder and had more pores and channels in the soil through which irrigation water could move more freely.

It is important to consider the influence of structure in soils of the same textural class. For example, a well-structured clay soil will be more permeable than one which is poorly structured. The well-structured clay soil can exhibit water infiltration and transmission characteristics similar to those of a coarser soil, i.e. a loam. For irrigation, the stability of the structure is also important. The topsoil will be required to withstand the degrading effects of both extensive land forming and seedbed preparation for irrigated crops. Where the structure is readily broken down, the surface may compact easily or crust, and water entry and movement in the soil can be reduced.

In clay soils, the nature of the clay is also another important factor affecting both infiltration and transmission of water. In those soils, where swelling and shrinking of the clay is concerned, pore space and permeability change while the soils are wetting. Infiltration and transmission are both much affected because the swelling of the clay tends to reduce effective pore size. The water content of the soil during irrigation could well modify the effect of texture and structure. Macropore flow is most important in clay soils since there is little matrix flow.

The nature of the clay mineral affects the water characteristics of the irrigated soil. In the cracking clay soils of northern New South Wales and southern Queensland, the drying out of the soils produces extensive vertical cracking together with the formation of the self-mulching surface. These vertical cracks can be quite deep. The irrigation of the soil in the dry condition is accompanied by high initial infiltration rates, and the soil is capable of taking in large amounts of water. The depth and size of the cracks can give direct access to the subsoil. As the cracks fill with water, their dimensions are reduced by expansion of the clay. This quickly reduces the absorption rate of water, so water application rates need to be adjusted. Cracking clay soils do not have to crack extensively to affect water intake. The soil can be self-mulching, with only fine cracks developing as the soil dries. Alternatively the soil may have limited shrinkage in the surface layers. With extensively cracking soils, plant roots can be broken as cracking occurs, so that plants suffer a water deficit.

Effect of Surface Condition

The initial infiltration rate of a dry soil is governed by the condition of the soil surface. The condition can be natural or can be brought about by cultivation, although in this case the effect is usually only temporary. Under normal conditions, infiltration is controlled by texture and macropore structure when surface irrigation methods are used. With sprinkler irrigation systems, the infiltration rate of the soil can be controlled by the application rate of the sprinklers, provided the application rate is less than the infiltration rate of the soil.

Infiltration of irrigation water is reduced when the surface soil sets hard. This condition may occur naturally when the soil dries or it may be induced as a crust. With a normally hard setting surface the problem can be overcome temporarily by cultivation. Alternatively, the infiltration rate gradually improves as the soil surface becomes moist. Where cultivation is used to overcome the effect of the hard setting surface, there is a mechanical degradation of surface soil structure and a progressive decrease in organic matter content. The surface soil can become more prone to slaking, where the finer particles can run together to form a crust. The crust acts as a barrier to water entry, and the thickness and strength of the crust determine the amount of resistance offered to water entry. A crust is more likely to develop in finer soils, i.e. loams to clays, than in the coarser soils. In furrow irrigations, crusts can form from sediments washed from the sides of the furrows, thus reducing intake (Brown *et al.*, 1988). Crusts can be formed also as the result of high levels of exchangeable sodium (Northcote and Skene, 1972) due to dispersion of the surface when wetted.

Major changes in soil structure need to occur before the soil's water balance is affected significantly (Cresswell *et al.*, 1992). These changes are associated with the development of surface crusts and plough pans, the most common cause of both in irrigated soils being excessive cultivation.

The self-mulching surface is ideal for the absorption of irrigation water. As it is initially very open and porous it will accept water readily. Its principal disadvantage, however, is its association with the cracking clay soils, in which the montmorillonitic clay swells on wetting. This swelling reduces the size of the non-capillary pores and also blocks the micropores. Both actions contribute to the reduction in infiltration rate and transmission of water.

A consideration of soil surface condition must also include the effect of ground cover on water entry into the soil. Generally, increased amounts of organic matter in the surface layers contribute to better structure and thus better infiltration (Table 7.12). Organic matter itself can contribute directly to increased water storage as it is able to absorb from two to six times its own weight of water. However, its contribution to more stable aggregates and improved structure can be considered as its more important role. Under sprinkler irrigation for example, a soil which contains adequate amounts of organic matter is more able to resist the battering action of the raindrops and retain its infiltration capacity over longer periods. The soil can therefore have a higher final infiltration rate.

TABLE 7.12 Effect of ground cover on infiltration (Wiesner, 1970)

Type of cover	Infiltration rate (mm/hr)
Bare ground	
—crusted	10
—cultivated	15
Weed cover	20
Permanent pasture	
—heavily grazed	30
—moderately grazed	40
—lightly grazed	50
Old pasture (with heavy mulch)	70

As well as the effect on soil structure (resulting from increased organic matter from, for example, roots and leaf litter), surface cover helps to improve water penetration by slowing the velocity of flow over the soil surface and allowing more time for soakage. This effect is most common in border check and border ditch systems of irrigation with pasture. In furrow irrigation, the presence of weeds in the furrow acts in a similar way but, in this case, the free flow of water along the furrow is interrupted. Water penetration under these conditions is uneven, resulting in uneven watering of the crop with some areas in the soil becoming overwatered and other areas becoming underwatered.

Effect of Profile Characteristics

The nature of strata underlying the root zone of an irrigated soil has an important effect on the behaviour of plants grown in that soil. Two important functions, infiltration and drainage, are, as previously described, dependent on texture and structure. The presence of a restricting layer or of coarse material lower down the profile will affect or modify the behaviour of the soil in the root zone. Also to be considered is the depth at which this layer exists.

In the case of the restricting layer, the principal effect is the development of a perched water table. The restricting layer may be rock or clay or it may be a pan. The permeability of this layer, being much lower than that of the overlying soil, tends to restrict the downward movement of water, resulting in waterlogging. If this layer occurs within 2 or 3 metres of the surface, continued irrigation can lead to surface waterlogging. Cockroft and Bakker (1966) found that, in the Goulburn Valley of Victoria, peach trees were killed by waterlogging where a restricting layer existed less than 3 metres down the profile. The waterlogging can be controlled by tile drainage below the root zone.

Subsoil waterlogging can occur where the soil overlies a porous stratum which can act as an aquifer. The situation is worsened where the porous stratum itself overlies a cemented or impeding layer. This in turn can lead to an elevation of the water table, generally over a large area. Control in this case usually involves the construction of tube wells or bores into the aquifer and relies on pumping to lower the level of the water table. In the Murrumbidgee Irrigation Area of New South Wales many of the irrigated horticultural soils overlie prior stream gravels and sand beds which are tapped for deep drainage schemes. This has proven to be an effective control measure to lower water tables over considerable areas.

Alternatively, where the strata are too permeable, it may lead to the exclusion of certain areas for particular irrigated crops such as rice (Van der Lelij and Talsma, 1978).

In the Darlington Point area, these prior stream aquifers have also been tapped to supply irrigation water to properties outside the irrigation areas.

The presence of sand beds does not automatically increase the transmission characteristics of the overlying soil. The rate of inflow and downward movement of water are determined by the conductivity of the least permeable layer. If, then, the surface soil is a clay of low permeability, water moves slowly down to the sandy layer. The slow drainage in turn can cause surface waterlogging when the application rate exceeds the infiltration rate. The drainage water will enter the coarser sand only when there is sufficient build-up of water to overcome the effect of the greater suction in the clay soil. Alternatively, when a sand lens is interposed between a loam and an impermeable clay layer, drainage can be rapid until the water reaches the interface between the sand lens and clay. The rate of downward movement of the water then decreases to a value equal to the hydraulic conductivity of the clay. Under prolonged irrigation, the soil develops a water table in the sand lens. Eventually both the root zone and the soil surface layers become waterlogged.

Effect of Salting

Salt-affected soils are widespread in Australia, representing about 33 per cent of the total area (Northcote and Skene, 1972). Many of the larger irrigation schemes have been located on salt-affected soils. Prolonged irrigation in these areas has been accompanied by waterlogging and salinity problems (Beecher, 1991; Blackburn, 1978; Mehanni, 1978; Muirhead, 1978). The problems that develop in salt-affected soils are salinity (high soluble salts) and sodicity (high exchangeable sodium levels).

SOLUBLE SALTS

The most commonly occurring soluble salts in soils are sodium chloride, calcium sulfate and the carbonates of calcium and magnesium. Bicarbonates may also be present but are frequently converted to the less soluble calcium and magnesium carbonates. The most common salt in Australian soils is sodium chloride. The concentration of soluble salts is normally determined by the electrical conductivity of a soil water extract.

The criteria for salinity levels have been described by Northcote and Skene (1972). Soils with sodium chloride contents above these criteria are not suitable for irrigation unless the salt can be leached from the root zone. The textural class of the soil and the method of irrigation are also important factors contributing to the accumulation of salt in the root zone. Finer soils, i.e. clay loams and clays, are more liable to salinisation than coarser soils, largely because the latter drain more freely and remove salt more efficiently in the drainage water.

Saline soils usually occur in areas which receive salt transported from elsewhere. Water is the principal carrier, either as saline irrigation water or as groundwater. However, the total amount of salt, its location in the profile, the salt content of the irrigation water, the method of irrigation, the ease of drainage and the depth to the water table all contribute to the development of a salting problem. Soils in which the salt occurs deeper than 2 metres may not be in much danger of becoming saline until a shallow water table is formed. The development of saline conditions is then due to upward movement of water from the water table and a concentration of the soluble salts at the soil surface as the result of evaporation. The evaporation from saline soils is no different in detail to that from non-saline soils (Russell, 1973). Evaporation from a saline soil leaves a saturated salt solution, which has a lower vapour pressure than a dilute solution and so takes longer to evaporate. Water continues to rise, carrying with it more salt solution, until eventually a salt encrustation develops. Once a salt area has been formed it tends to grow at the expense of surrounding soil, i.e. the salt tends to be concentrated in patches.

In free-draining soils, salt is leached into the lower horizons. Therefore, the control of salinity in less permeable soils is achieved by controlling water movement and preventing a shallow water table by the use of artificial drainage.

EXCHANGEABLE SODIUM

The presence of large concentrations of exchangeable sodium in the soil poses problems under irrigation. In the heavy clays of the Riverine Plain, the high sodium content causes infiltration problems and interferes with seedling emergence. Both are the result of dispersion of the clay when the soil is irrigated and the development of a surface crust when the soil dries.

Germination and establishment of annual pasture species can be improved by the addition of gypsum in the irrigation water (Davidson and Quirk, 1961). The effectiveness of the treatment depends on the amount of clay, the exchangeable sodium percentage (ESP) and evaporation conditions at sowing.

The critical value for exchangeable sodium is variable. The United States Salinity Laboratory has set the value at an ESP of 15, i.e. soils with an ESP of 15 or more show reduced permeability and poor infiltration. Under Australian conditions, however, dispersion and permeability problems occur in soils high in swelling clays at much lower ESP values. Emerson (1967) obtained dispersion of dry aggregates in water at an ESP of 7, while Northcote and Skene (1972) recommend an ESP of 6 as the lower limit. Low subsoil permeability and waterlogging in two soils (the Lemnos loam and Shepparton fine sandy loam) in the Goulburn Valley, Victoria, occurred at low ESP values (Bakker et al., 1973). However, in these cases it was suggested that these problems were caused by high exchangeable magnesium.

The importance of high levels of soluble salts and exchangeable ions in irrigated soils cannot be ignored. The soluble salts have a more direct effect on plant growth and production by making water less available to plants as a result of the increased osmotic pressure of the soil solution. The effect of the exchangeable sodium is more indirect as it changes the transmission characteristics of the topsoil or subsoil which then reduces plant production due to waterlogging or secondary salinisation.

The rate of accumulation of salts in the root zone depends on their concentration in the applied water, the soil type and the extent of natural leaching provided by rainfall. The salinity status of the soil should be monitored regularly and extra water supplied periodically to leach the salts below the root zone (Barrett, 1985).

Effect of Other Environmental Factors

Factors other than profile characteristics also affect the water regime of irrigated soils. These include the slope and the micro-relief.

Slope   The slope of the natural surface principally affects the form of irrigation used. On very gentle slopes (e.g. 0.1 per cent), border check irrigation might be used whereas on steeper slopes (e.g. 5 per cent) sprinkler irrigation would generally be used. Surface slope and texture together will determine the infiltration rate of water, and in addition affect lateral seepage and surface drainage.

Micro-relief In this context, the association of position in the micro-relief, elevation and slope all combine to affect surface drainage, lateral seepage, intake rate and the need for artificial drainage (Fig. 7.9). Where these are also combined with the association of permeable soils on the elevated parts and soils of relatively poor permeability on the lower levels, waterlogging and salinity occur. Situations akin to this exist along the Murray Valley, especially in South Australia (Northcote, 1949). It is interesting to note that soil associations frequently follow positions in the micro-relief, i.e. particular soils are nearly always in water-shedding situations while others are always in water-receiving positions.

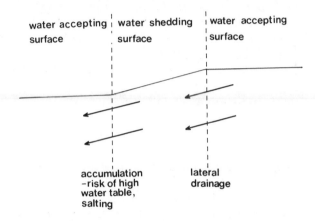

water accepting surface   water shedding surface   water accepting surface

accumulation
-risk of high
water table,
salting

lateral
drainage

FIG. 7.9 Simplified diagram of the effect of micro-relief on water movement in irrigated soils

Micro-relief also influences the local micro-climate, particularly with respect to frost. This can be particularly important with areas intended for horticultural crops.

WATER MANAGEMENT

Crop yield is strongly correlated with the level of evapotranspiration, a reduction of which results in a loss of yield. In irrigated crop management there is a need to match the supply of water to the needs of the crop.

As irrigation resources become increasingly regulated and expensive, more attention is being focused on improved layout and better water management. Barrett (1985) describes important management practices as

—timing water application for maximum yields;

—timing deficits to minimise yield reductions when water shortages are inevitable; for example, slight water deficit stress in such crops as

maize and sorghum can be imposed in the young vegetative stage without undue penalties in yield;
—avoiding overwatering;
—beginning the season with adequate water in the root zone; and
—sprinkling at night to minimise evaporation and wind drift.

Sound water management involves applying the proper amount of water to the crop at the correct time. Numerous techniques are used to determine the schedule of irrigation. These include:

(a) *visual crop symptoms*, such as wilting or colour changes. It is worth noting that plant growth is reduced when the relative turgidity of the plants declines below 90 per cent, due largely to closure of the stomates which thereby restricts carbon dioxide entry into leaves for assimilation. Growth stops when visible wilting occurs. This means therefore that considerable productivity is lost before signs of deficit stress are apparent in the crop and the efficiency of water use is reduced.

(b) *subjective sampling of soil water* down to at least 75 cm using a shovel, auger or soil sampler. Surface soil conditions are unreliable for indicating moisture availability in the root zone.

(c) *plant measurement* using pressure chambers to measure leaf water tension. This equipment requires trained operators for correct use and interpretation.

Infra-red thermometers to measure crop canopy temperatures (T_F) are also used commercially (Smith, 1986). Diagnosis of crop health is based on the deviation of the crop canopy temperature from the air temperature (T_A). In unstressed crops, the foliage is cooled by transpiration. However, as transpiration is reduced, the foliage becomes hotter and the foliage-air differential ($T_F - T_A$) is increased. A crop water stress index (CWSI) can then be determined by comparing $T_F - T_A$ of the crop with the differential of an unstressed crop (which represents the lower limit, LL) and of a completely stressed crop (the upper limit, UL). Thus,

$$\text{CWSI} = \frac{T_F - T_A - LL}{UL - LL}$$

CWSI varies from 0 (for an unstressed crop) to 1 (for a totally stressed crop). However, Stockle and Dugas (1992) have shown, using computer models, that the empirical CWSI gives a late indication of irrigation need, after some water stress has already developed. Thus the method may be unsuitable for use with water-sensitive crops.

(d) *pan evaporation and the 'crop factor'* is a common method of determining irrigation scheduling and is therefore described in more detail.

Crops vary in their water requirements (Table 7.13), and for a particular crop the irrigation needs depend on the evaporative conditions operating at particular stages of growth. A crop which completely covers the ground and is well supplied with subsoil moisture transpires at approximately the same rate as water from a pan evaporimeter (Downey, 1971). However, a crop which covers only part of the ground transpires less water than that lost by pan evaporation, depending on the wetness of the surface soil. This is particularly important for row crops such as maize, sorghum and cotton because they cover relatively little ground when young but there is almost complete ground coverage when they are mature. Therefore, young row crops generally lose less water than a standard evaporimeter but as they mature they transpire at about the same rate, particularly when the soil surface is dry (Downey, 1971).

TABLE 7.13   Average consumptive use of water by crops

	Water use (mm)
Canola	450
Peanuts	550
Sorghum	560
Soybeans	600
Maize	700
Sunflower	700
Wheat	600
Cotton	800
Sugar cane	900
Rice	1200

This variation in transpiration has the effect of changing the crop factor, (the amount of water transpired by a crop, which is estimated as some proportion of the evaporation from a standard pan evaporimeter). In some situations a constant crop factor (e.g. pastures 0.9 × pan evaporation; citrus trees 0.5 × pan evaporation) can be used. However, for row crops, a constant crop factor will produce waterlogging in young crops and drought stress with maturing crops. Consequently a useful guide is to multiply the pan evaporation by a crop factor of 0.2 at or just after sowing, 0.5 when there is 30 per cent ground cover, and 0.9 for 60 per cent ground cover to maturity. Under dry summer conditions of low humidity and strong winds, the crop factor will often reach 0.95 to 1.0 and may exceed 1.0 in crops such as lucerne and soybeans.

The amount of irrigation applied for efficient water use is dependent upon the depth of the root zone, which is defined as the depth of soil where 90 per cent of the roots occur (Anon., 1974). In some cases this is determined by the depth of soil available to the plants, but more particularly by the crops being grown. According to species, plants can be classified as deep or shallow rooted (Table 7.14). The depth of the root zone will therefore influence the amount of irrigation water applied because of the need to avoid loss of water through deep percolation. Shallow-rooted crops require more frequent irrigations than deep-rooted species growing on the same soil.

The water requirements of the plant vary with stage of growth. A critical stage is from ear initiation through to physiological maturity in grain crops. However, irrigation is an expensive operation and avoidance of water stress at any stage of the development of the crop is of paramount importance if satisfactory returns are to be expected. Where water supplies are limited and have to be rationed, slight deficit stress in crops such as maize and sorghum can be imposed in the young vegetative stage with minimal penalties in yield. The available soil water for plants has been described as that between field capacity and permanent wilting point (Fig. 7.10). Field experience has shown the desirability of irrigating the soil before it is completely dry because plants tend to suffer once half the available moisture has been used (Thomas and Moore, 1968; Wiesner, 1970; Meyer and Green, 1980; 1981).

Given the evaporation conditions (Table 7.15), the crop factor, soil type and available water (Table 7.16) and the root depth of the crop, it is

TABLE 7.14 Effective watering depths (root zones) for different crops grown under irrigation

Crop	Effective watering depth (cm)
Rice	15– 25
Peanuts	25– 35
Sugar cane	45– 75
Maize	60– 90
Sorghum	60– 90
Tobacco	60–135
Cotton	60–180
Wheat	75–100
Sunflower	80–150
Barley	90–105
Linseed	100–150
Safflower	150–250

TABLE 7.15 Mean daily evaporation rates (mm) from US Class A pans (Dale, 1984)

	October	November	December	January	February	March	April
Moree	6.7	9.1	10.5	9.3	8.5	7.0	5.1
Gunnedah	6.0	7.8	9.3	8.6	7.8	6.8	5.0
Trangie	5.5	7.4	9.5	9.5	8.2	6.4	4.6
Condobolin	5.7	8.1	10.8	10.3	9.4	6.9	4.6
Griffith	5.8	7.9	8.9	9.3	8.6	6.4	4.1

possible to calculate the irrigation requirements. Allowance must also be made for the efficiency of the irrigation process, defined as the proportion of the irrigation water involved in evapotranspiration (i.e. for crop yield) relative to the amount of irrigation water used. Losses include seepage and evaporation from channels, drains and water storages, uncollected runoff, and percolation below the root zone. Irrigation efficiency may

TABLE 7.16   The average plant available water in different soils types (adapted from Boyle, 1969)

Soil type	water available to plants (mm/m depth)
Sand	50
Fine sand	75
Sandy loam	110
Fine sandy loam	144
Loam	170
Silt loam	178
Light clay loam	178
Clay loam	170
Heavy clay loam	152
Clay	144

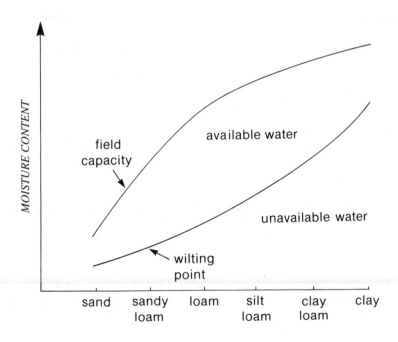

FIG. 7.10   Water-holding capacities of different textured soils (adapted from Brady, 1984)

range from 35 to 80 per cent depending on the quality of management and the irrigation facilities. These issues are reviewed in Finkel (1982).

The following example demonstrates the irrigation schedule for a flowering sunflower crop in January at Griffith, New South Wales (adapted from Dale, 1984).

Evaporation (Table 7.15)	9.3 mm/day
Crop factor	0.95
Soil capacity (Table 7.16)	170 mm/m depth
Root depth (Table 7.14)	80 cm
Available water in root zone	136 mm
Available water utilised by crop (say 50%)	68 mm

$$\text{Evapotranspiration} = 9.3 \times 0.95 = 8.8 \text{ mm/day}$$

$$\text{Days of moisture supply} = \frac{68}{8.8} = 7.7 \text{ days}$$

A crop requirement of 60 to 70 mm would be needed in approximately eight days if rain is not received on the crop during the period. After an allowance for an irrigation efficiency of, say, 70 per cent, an irrigation of 85 to 100 mm would fulfil the requirement at that time.

By recording daily rainfall and ascertaining the daily evapotranspiration throughout the growing period, the moisture balance for a given root zone can be calculated on a day-to-day basis. However, summer falls of less than 25 mm are seldom effective and can be largely ignored (Thomas and Moore, 1968).

(e) *neutron moisture meter* The use of this device requires a trained operator. It involves the installation of access tubes into the root zone following planting of the crop. Soil moisture levels are monitored regularly during the season by inserting a probe into the access tubes and taking measurements (Fig. 7.11). The probe counts the hydrogen atoms in the soil (and hence indirectly measures water changes) by emitting fast neutrons from a radio-active source. As these neutrons collide with atoms of approximately equal mass, i.e. hydrogen atoms, they are slowed down. The slow neutrons are then detected and counted.

The main advantages of the neutron moisture meter are its direct measurement and its ease of operation, recording and interpretation for the trained operator. It provides root-zone water profiles and can identify compaction layers and water tables. A detailed evaluation of the technique is given in Greacen (1981).

(f) *computer water-balance models* These have the potential to provide more reliable and consistent advice than most other methods (Browne, 1984). With computer systems it is possible to adjust the crop factor to account for increased evaporation from wet soils, to connect weather stations automatically to the computer, and to provide water-use information quickly for irrigation decision-making on a daily basis.

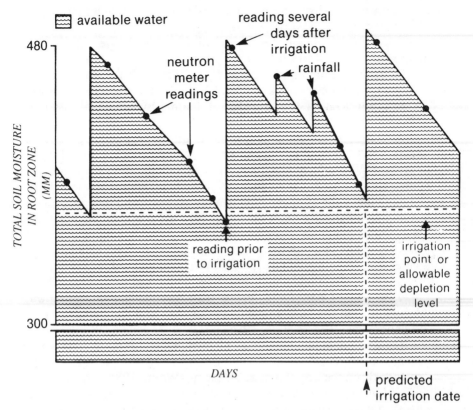

FIG. 7.11 Example of soil moisture recorded with a neutron probe. In the absence of intervening rainfall, an irrigation date can be predicted (Browne, 1984)

Drainage

Waterlogging occurs when the soil is saturated for a short period of time. It can occur as a result of overwatering, poor soil structure and poor surface and subsoil drainage. Crop yields are reduced and the risks of rising water tables and salinity are enhanced. The management options include:

(a) *irrigation scheduling* — as waterlogging frequently results from the application of water in amounts greater than required by the crop, scheduling procedures previously outlined match crop requirements with water availability. This reduces water wastage and the risk of waterlogging.

(b) *subsurface drainage* — depending on soil type, a subsurface drainage system will keep the water table at a depth which would help prevent long-term waterlogging. If water quality is reasonable the drainage water can be reused, sometimes in dilution with high-quality water to reduce the level of salinity. Where the salinity of the effluent is so high it cannot be reused, it must be disposed of to evaporation basins where the water is evaporated and the salt harvested for sale (Wood, 1984).

Subsurface drainage is effective only where soil types allow good water movement (Gilbert and Marston, 1986). The costs of installation are high and are usually only feasible for horticultural land. Most of the horticultural land in the Murrumbidgee Irrigation Area is serviced by tile drainage and some by tube wells (Wood, 1984).

(c) *surface drainage*—poor irrigation layout on a paddock or farm basis results in poor surface drainage with temporary waterlogged patches. Whole-farm landforming is therefore imperative to overcome this problem. As surface-irrigated fields frequently lose 10 to 35 per cent of the applied water as tailwater runoff, it is appropriate in any landforming exercise to provide for reuse of the drainage water. The potential benefits from on-farm recycling of drainage water include savings in water charges, improved efficiency of water use, retention of fertilisers on the farm and reduced pollution of rivers and wetlands with nutrients, salts and pesticides. There may be risk of damage to crops from herbicide contamination, depending on the susceptibility of the crop and the persistence of the chemical (Weerts *et al.*, 1986).

Regional drainage schemes already exist in many irrigation areas, but to be effective they require well laid out and maintained surface drainage on all irrigation farms (Wood, 1984).

(d) *tree planting*—where appropriate, strategic planting of trees can assist in maintaining water tables at depth by intercepting significant quantities of water which would otherwise add to the water table.

On-farm Storage

The construction of large on-farm storages is common, particularly in northern New South Wales and the Darling Downs of Queensland (Barrett, 1985). Although most storages incorporate the tailwater return system, their prime purpose is for harvesting unregulated stream flows. Storages provide the advantage of timeliness of application, particularly where there is a lag time for delivery of water. If the ordered water is not required, due to rainfall, it can be stored for subsequent use and is therefore not lost.

References

Anderson, H. (1974), *Fruit and Irrigation*, Australian Industries Series No. 12, Lothian, Melbourne.

Anon. (1963), *Review of Australia's Resources,* Department of National Development, Canberra.

Anon. (1974), *Principles of Irrigation,* State Rivers and Water Supply Commission, Melbourne.

Anon. (1978), Water Resources Commission of NSW, *unpublished.*

Anon. (1982), *Working Party on Dryland Salinity in Australia Report.*

Anon. (1983), *Rural Industry in Australia,* Australian Government Publishing Service, Canberra.

Anon. (1986), *Year Book Australia,* Australian Government Publishing Service, Canberra.

Anon. (1991), *Year Book Australia,* Australian Government Publishing Service, Canberra.

Anon. (1992), *Striking a Balance! Australia's Development and Conservation,* Australian Government Publishing Service, Canberra.

Bakker, A. C., Emerson, W. W. and Oades, J. M. (1973), 'The Comparative Effects of Exchangeable Calcium, Magnesium and Sodium on Some Physical Properties of Red-Brown Earth Subsoils: I. Exchangeable Reactions and Water Contents for Dispersion of Shepparton Soil', *Australian Journal of Soil Research,* **11**, 143–50.

Barrett, J. W. H. (1985), 'Advances in Irrigation Technology', *Journal of Australian Institute of Agricultural Science,* **51** (4), 263.

Baumnardt, R. L., Wendt, C. W. and Moore, J. (1992), 'Infiltration in Response to Water Quality, Tillage and Gypsum', *Soil Science Society of America Journal,* **56**, 261–266.

Beecher, H. G. (1991), 'Effect of Saline Water on Rice Yields and Soil Properties in the Murrumbidgee Valley', *Australian Journal of Experimental Agriculture,* **31**, 819–23.

Blackburn, G. (1978), 'Assessment of Salinity Hazards for Irrigation in the Shepparton Region of the Murray Valley' in R. R. Storrier and I. D. Kelly (eds), *The Hydrogeology of the Riverine Plain of South-East Australia,* Australian Society of Soil Science, 141–53.

Blackmore, D. (1989), 'Water—Past Developments, Future Consequences' in '*Murray–Darling Basin—A Resource to be Managed*', Australian Academy of Sciences and Engineering Invitation Symposium **13**, 41–69.

Boyle, J. W. (1969), *Irrigated Maize,* N.S.W. Department of Agriculture, Division of Plant Industry Bulletin P332.

Brady, N. C. (1984), *The Nature and Properties of Soils,* 9th Edition, Macmillan, New York, 103.

Brown, M. J., Kemper, W. D., Trout, T. J. and Humphreys, A. S. (1988), 'Sediment, Erosion and Water Intake in Furrows', *Irrigation Science,* **9**, 45–55.

Browne, R. L. (1984), *Irrigation Management of Cotton,* N.S.W. Department of Agriculture, Agfact P5.3.2.

Cockroft, B. and Bakker, A. C. (1966), 'Peach Tree Waterlogging in the Goulburn Valley Area', *Journal of Australian Institute Agricultural Science,* **32**, 292.

Colclough, J. D. (1973), 'Salt', *Tasmanian Journal of Agriculture,* **44**, 171.

Cole, K. S. (1977), *Irrigation in Western Australia,* W.A. Department of Agriculture, Bulletin 3929.

Chieu, F. A. S. and McMahon, T. A. (1992), 'An Australian Comparison of Penman's Potential Evapotranspiration Estimates and Class A Evaporation Pan Data', *Australian Journal of Soil Research,* **30** (1), 101–112.

Cresswell, H. P., Smiles, D. E. and Williams, J. (1992), 'Soil Structure and the Soil Water Balance', *Australian Journal of Soil Research,* **30**, 265–283.

Dale, A. B. (1984), *Sunflower Growing,* N.S.W. Department of Agriculture, Agfact P5.2.3.

Davidson, J. L. and Quirk, J. P. (1961), 'The Influence of Dissolved Gypsum on Pasture Establishment on Irrigated Sodic Soils', *Australian Journal of Agricultural Resarch,* **12**, 100.

Dowling, A. J., Thorburn, P. J., Ross, P. J. and Elliott, P. J. (1989), 'Estimation of Infiltration and Deep Drainage in a Furrow Irrigated Soil', *Australian Journal of Soil Research,* **27** (2), 363–75.

Downey, L. (1971), *Is Your Crop Factor Right?,* N.S.W. Department of Agriculture, Division of Plant Industry Bulletin P430.

Emerson, W. W. (1967), 'A Classification of Soil Aggregates Based on their Coherence in Water', *Australian Journal of Soil Research,* **5**, 47–57.

Evans, G. N. (1971), 'Drainage from Irrigation Areas as a Resource' in *Proceedings of the ANZAAS Congress,* Brisbane.

Finkel, H. J. (1982), *CRC Handbook of Irrigation Technology,* CRC Press, Florida.

Gilbert, G. and Marston, D. (1986), 'Waterlogging in Irrigated Lands', *Farmers Newsletter,* IREC, No. 129, 23.

Greacen, E. L. (ed.) (1981), *Soil Water Assessment by the Neutron Method,* CSIRO Melbourne.

Grieve, A. M. (1989), 'Water use Efficiency, Nutrient Uptake and Productivity of Micro-irrigated Citrus', *Australian Journal of Experimental Agriculture,* **29**, 111–118.

Kay, M. (1983), *Sprinkler Irrigation — Equipment and Practice,* Batsford Academic and Educational Ltd, London.

Langford-Smith, T. and Rutherford, J. (1966), *Water and Land,* Australian National University Press, Canberra.

Leeper, G. W. (ed.) (1970), *The Australian Environment,* 4th edn, CSIRO/ Melbourne University Press.

Mehanni, A. H. (1978), 'Changes in Water-tables with Irrigation in the Goulburn Valley' in R. R. Storrier and I. D. Kelly (eds), *The Hydrogeology of the Riverine Plain of South-East Australia,* Australian Society of Soil Science.

Meyer, W. S. (1991), 'Irrigation Research in Australia — Achievements and Challenges' in *Proceedings of Conference on Irrigation in South Africa,* Durban, South Africa.

Meyer, W. S. and Green, G. C. (1980), 'Water Use by Wheat and Plant Indicators of Available Soil Water, *Agronomy Journal,* **72**, 253–57.

Meyer, W. S. and Green, G. C. (1981), Plant Indicators of Wheat and Soybean Crop Water Stress, *Irrigation Science,* **2**, 167–76.

Muirhead, W. A. (1978), 'The Effects of Management of Irrigated Crops on Salt Distribution Above a High Water Table' in R. R. Storrier and I. D. Kelly (eds), *The Hydrogeology of The Riverine Plain of South-East Australia,* Australian Society of Soil Science.

Murphy, J. (1991), 'Irrigation systems' in *Energy Efficient Irrigation,* State Electricity Commission of Victoria, 18.

Northcote, K. H. (1949), 'The Horticultural Potential under Irrigation of Soils of the Highland Areas of the Mid Murray River Valley,' *Journal of Australian Institute of Agricultural Science,* **15**, 122.

Northcote, K. H. and Skene, J. K. M. (1972), *Australian Soils with Saline and Sodic Properties,* CSIRO Soil Publication No. 27.

Prathapur, S. A., Robbins, C. W., Meyer, W. S. and Jayawardene, N. S. (1992), 'Models for Estimating Capillary Rise in a Heavy Clay Soil with a Saline Shallow Water Table', *Irrigation Science,* **13**, (1), 1–17.

Russell, E. W. (1973), *Soil Conditions and Plant Growth,* 10th edn, Longman, London, 446.

Salter, P. J. and Williams, J. B. (1965), 'The Influence of Texture on the Moisture Characteristics of Soils', *Journal of Soil Science,* **16**, 310.

Smith, R. (1986), 'Has Your Cotton Crop Got a Temperature?', *Australian Cotton Grower,* November, 36–7.

Smith, R. C. G., Mason, W. K., Meyer, W. S. and Barrs, H. D. (1983). Irrigation in Australia: development and prospects. In *Advances in Irrigation,* Vol. 2 (ed. D. Hillel), Academic Press: London, pp. 99–152.

Stockle, C. O. and Dugas, W. A. (1992), 'Evaluating Canopy Temperature-based Indices for Irrigation Scheduling', *Irrigation Science,* **13** (1), 31–37.

Thomas, G. N. and Moore, S. D. (1968), 'Irrigating River Flats in the Upper Murray', *Journal of Agriculture, South Australia,* **72**.

Van der Lelij, A. and Talsma, T. (1978), 'Infiltration and Water Movement in Riverine Plain Soils Used for Rice Growing' in R. R. Storrier and I. D. Kelly (eds), *The Hydrogeology of the Riverine Plain of South-East Australia,* Australian Society of Soil Science, 89–98.

Weerts, P. G., Bowmer, K. H. and Korth, W. (1986), 'On-farm Recycling of Drainage Water', *Farmers Newsletter,* IREC, No. 128, 25.

Wiesner, C.J. (1970), *Climate, Irrigation and Agriculture,* Angus and Robertson, Sydney.

Wood, B. S. C. (1984), 'Water Management for Waterlogging and Soil Salinisation Control', *Farmers Newsletter,* IREC Special Edition, 13.

CHAPTER 8

CROP PROTECTION

R. J. Banyer, E. G. Cuthbertson and P. D. Slater

Agriculturally, protection implies action to prevent the undesirable activities of plant pests. In this context, however, pests are assumed to include such organisms as nematodes, insects, molluscs, bacteria, fungi, myco-plasmas and viruses, as well as some higher animals (e.g. rodents) and weeds.

The science of pest control is not of recent origin. The writings of the classical Mediterranean authors like Theophrastus (fourth century BC), Cato (234–149 BC) and Pliny (AD 23–79) contain many references to pest control. Yet, while the principles of seed treatment, fumigation, tree banding and physical or mechanical weed control were understood, true scientific investigation spans little more than 100 years. As a result of these researches, recent progress has been rapid, particularly in the field of chemical control.

PEST DAMAGE

Pests are undesirable because they reduce the economic return from the land. Just how and when the reduction occurs varies. The overall effects are reduced yield and lower quality of the marketable product allied to an increase in production costs.

Reduction in Yield

Agricultural pests reduce yield in three ways: by reducing plant populations, by diverting nutrients from the crop and by destroying the marketable product. In each of these situations the cost of control must be set against the increase in return obtained.

Reduction in Plant Numbers

Seed, particularly surface sown seed, is often removed by ants, rodents and birds (Campbell 1966; Nelson *et al.*, 1970; Campbell and Swain, 1973; Johns and Greenup, 1976). At the same time, naturally occurring germination inhibitors diffusing from plant organs or plant residues in the soil prevent the germination of other seeds present in the soil (Ballard and Grant-Lipp, 1959; Nielson *et al.*, 1960; Ferguson, 1968) although there is argument that the absorbent powers of the soil minimise this allelopathic response (King, 1952). When the reduction in plant numbers so occasioned exceeds the compensating ability of the remaining plants, yield loss occurs.

Established crops may experience density changes. They can be devastated by the Australian plague locust (*Chortoicetes terminifera*) or the armyworm (*Mythimna convecta*), while some loss of plant numbers occurs as a result of grazing by rabbits, hares and other wildlife.

Reduction in Plant Vigour

More frequent but less obvious loss occurs as a result of attack by sucking insects, pathogenic micro-organisms, or the presence of weeds. All of these pests divert plant foods from the crop plant to their own use. Consequently, crop growth-rate and yield are reduced commensurately.

Diseases like stem rust of wheat (*Puccinia graminis* var. *tritici*) and blackleg of crucifers (*Leptosphaeria maculans*) or attack by the red-legged earth-mite (*Halotydeus destructor*) can be devastating (Norris, 1948; Brown, 1975). Less catastrophic but still debilitating is the presence of diseases like Septoria tritici blotch (*Mycosphaerella graminicola*), leaf rust (*Puccinia recondita*) root-rots including 'take-all' (*Gaeumannomyces graminis* var. *tritici*) and crown rot (*Fusarium graminearum*), and cereal-cyst nematode (*Heterodera avenae*). These diseases regularly reduce cereal yields by 5 to 50 per cent (Phipps, 1938; Banyer, 1966a; Brown *et al.*, 1970; Price, 1970; Keed and White, 1971; Brown, 1975; Kuiper, 1976).

Weeds using light, water and mineral nutrients which would otherwise be available to the crop plant, reduce growth rate and yield in proportion to their competitive ability and density (Donald 1958; Aspinall, 1960; Fryer and Evans, 1968). In this regard the time of weed emergence relative to the crop is particularly important. The small canola seedling, for example, offers little interference to weeds until its leaf area is sufficiently well developed (Hubbard, 1968), whereas with fully developed leaves, the canola plant competes strongly with weeds by shading them (Weinberger, 1963).

As in the case of seed germination, allelopathic inhibition of the growing plant can also be an important factor in yield loss. Certain weeds in some way are able to prevent the uptake of nutrients by crops, thereby reducing their growth rate.

Destruction of the Marketable Product

Nowadays, grain-replacement diseases such as loose smut (*Ustilago tritici, U. nuda*) and bunt of wheat (*Tilletia* spp.) occur sporadically and rarely cause economic loss (Brown, 1975). Occasionally, however, loss may exceed 20 per cent of individual crops (Brown, 1975). Insect attack may also be directed at developing fruit, as in the case of the bollworm (*Helicoverpa* spp.), limiting cotton yield significantly (Wilson *et al.*, 1972).

Of probably greater impact in this area is the damage done to stored products by disease and insect pests. The Australian wheat industry spends tens of millions of dollars each year to protect its product from pest damage and to maintain its reputation as a pest-free trader.

Another source of yield loss, though not strictly loss of the stored product, is that occasioned by trampling of the ripened crop by kangaroos and emus.

Reduction in Quality

Blemished, discoloured, lightweight products with off-flavours, the usual result of pest attack, may not be marketable. Bunt balls in wheat grain, covered smut (*Ustilago hordei*) in barley grain, low oil content of sunflower seed and other similar problems reduce product quality and cause dockage or rejection (Brown, 1975; Anon., 1976a). Similar action occurs when too many weed seeds are present in the grain sample. The presence of Hexham scent (*Melilotus indica*) in wheat grain, for example, results in complete rejection because of its tainting characteristics (Whittet, 1968).

Increase in Production Costs

The presence of pests automatically increases labour and equipment requirements. Specialised equipment such as rod-weeders, spraying units and seed-cleaning units as well as pesticides must be purchased. Their use increases the input of labour and energy required to produce the crop. These additional costs must be set against the potential additional return obtainable in determining profitability.

Apart from their direct effect on production costs, weeds add indirectly to these costs by acting as alternative hosts to diseases and insect pests. Barley grass (*Hordeum leporinum*) is an important natural host of the take-all fungus and the barley-yellow-dwarf virus (Butler, 1961; Smith, 1964; Brooks, 1965; Price, 1970), while capeweed (*Arctotheca calendula*) acts as host to numerous insect pests including the Rutherglen bug (*Nysius vinitor*), red-legged earth mite and the blue oat mite (*Pentathleus major*). The essential control of weeds of this nature in the non-productive areas of the property adds immensely to production costs.

Reduction in Land Values

The presence of perennial weeds such as serrated tussock (*Nassella trichotoma*) and skeleton weed (*Chondrilla juncea*) often reduces land values drastically. The build-up of the unpalatable serrated tussock, for example, occurs as a result of selective grazing and seriously reduces carrying capacity (Green, 1956; Goodyear, 1964). Well-managed sown pastures, however, give permanent control of the weed (Campbell, 1960). On the other hand, the capital cost of establishing these permanent pastures is high and banking institutions may not afford the financial assistance required (Fallding, 1957; Campbell, 1965). Purchasers take such factors into consideration in the price offered. It is on record (Goodyear, 1964) that the value of one property infested with serrated tussock fell by 74 per cent in ten years.

Pests and Human Efficiency

Agricultural pests not only reduce human efficiency by causing sickness and occasional death, e.g., hay fevers, mycotoxicosis such as ergot poisoning (Culvenar, 1974), but they add directly to living costs, affecting the whole community. Pest control is a major part of the farm program, adding to the cost of production and hence to the cost of food and fibre.

Technical development over the past fifty years, however, particularly in the realm of chemical pest control, has increased farm efficiency. As a result of these advances, the physical burden of work has been reduced, fewer workers are required thereby minimising cost increases, leisure time has increased and land use and human health have been improved.

THE DEVELOPMENT OF PEST PROBLEMS

It must be remembered that organisms become agricultural pests only when their numbers reach such a level as to reduce the yield of marketable product to uneconomic levels. In nature, all organisms exist in equilibrium one with the other under the control of the environment, namely climate and biological pressures. In this context, climate is the result of the integration of radiation, temperature, rainfall and evaporation parameters, while biological pressures are inter-specific and intra-specific competition, predation and parasitism. This balance is only disturbed by calamity such as fire, flood, avalanche, earthquake, or tree fall. When this happens pioneer plants develop rapidly in the disturbed area, soon covering the scar. This additional food source leads to an increase in the organisms using the pioneer plants as food and consequently also in their parasites and predators. The vigour of the pioneer plants is reduced, allowing the entry of less vigorous plants, a process that continues until the original climax vegetation and fauna have been re-established.

Population growth has been studied in detail by Woods (1974) but can be expressed simply as

$$P_2 = P_1 + B - D + M$$

where P_1 = initial population, P_2 = final population, B = births or seed produced, D = deaths and M = migration either out of (emigration) or into (immigration) the population centre. If one ignores the seasonal changes in population due mainly to changes in the elements of climate, then P_2 is a function of changes in the biological pressures; parameters that are completely upset by the artificially maintained cropping habit.

This unique agro-ecosystem provides an ideal situation for the development of pest species. Crop production rarely fully exploits an area. Cultivation and row sowing, by creating bare areas, are essentially the same as the natural calamity and leave the community open to invasion by pioneer (weed) species. At the same time the multiplicity of single species sites provides an extremely large host population for potential pests. Moreover, crop plants selected for yield, nutritive value and digestibility are a most attractive food supply. Increases in pest populations are therefore inevitable.

Because 'big fleas have little fleas and so on *ad infinitum*', factors effecting changes in host-parasite relationships can be very important. Broad-spectrum insecticides applied to cotton for example, killed more of the beneficial insects controlling two-spotted mite than mites, a response which led to devastating outbreaks of the mites. Recognition of this type of response was implicit in early recommendations for the chemical control of the spotted alfalfa aphid following its introduction into Australia. Lower rates of chemical than are normally used were recommended in order to maintain the numbers of the ladybird beetle (*Coccinella repanda*) and other natural enemies which attack the aphid.

The situation is especially dangerous when associated with the introduction of crop or pest species into countries or districts where they have not previously existed. Indigenous pest species may find the newcomer much more attractive than the native plants and rapidly achieve pest status. In contrast, the absence of natural enemies and the production techniques employed locally, as in the case of the introduction of skeleton weed into southern New South Wales, can result in a dramatic increase in the pest population, with a commensurate crop loss (McVean, 1965).

ECONOMICS OF PEST CONTROL

The Relative Importance of Pests in Crop Loss

In Australia, critical determination of the importance of specific pests in crop production, until recently, has been neglected. This is despite the fact that in the field of plant pathology Australia can claim to be a pioneer in the methods of relating different levels of disease with actual crop loss (Cobb, 1892). Unfortunately, pest status evaluation, particularly on a national and even a regional scale, is laborious, time-consuming and, therefore, costly. In addition, from the individual's point of view, the incentive to undertake such studies is generally lacking because the researcher gains little recognition and there is little professional status associated with such projects.

Until quite recently most attention was given to the spectacular pests like stem rust of wheat, plague grasshoppers and skeleton weed: in effect, to pests whose effects can be catastrophic. The less spectacular, insidious pests including the cereal root-rots, *Septoria tritici* blotch, lucerne budmite (*Aceria medicaginis*), silverleaf nightshade and others had remained largely ignored. Current investigations, however, have shown some of these 'minor' pests to be of considerable importance. The ubiquitous *Septoria tritici* blotch of wheat, for example, which regularly causes significant yield loss (Kuiper, 1976), was once considered part of the 'norm' by farmers who, along with many field advisers, thought the premature death of diseased foliage in heavy crops of wet years was beneficial. Indeed, the deleterious effects of many soil-borne insect pests and the root-rot diseases often go unrecognised by the farmer who attributes poor crop vigour to other causes. Declining yields or yields consistently below the district average are an indication that pests may be responsible and must be investigated.

Clearly, the relative importance of pests needs to be established. In the current economic climate of limited funds and/or personnel available, such data are essential to determine research priorities and to justify applications for research grants. Moreover, accurate assessment of pest intensity and the resultant crop loss in many situations will eventually lead to accurate pest forecasting and minimise the instability in farming by budgeting for pest control as a fixed cost.

The Crop Information Service established by the Victorian Department of Agriculture is the first serious attempt in Australia at state-wide surveillance to obtain comprehensive data and report on pest activity. Its centralised computer data bank program lends itself for adoption nationwide. Such a scheme will also improve the likelihood of early detection of any new pests introduced.

Estimated annual losses of $228 million in pasture production due to redlegged earth mite (*Halotydeus destructor*), lucerne flea (*Sminthurus viridis*) and blue-green aphid (*Acyrthosiphon kondoi*) have stimulated research funding in this area (Sloan *et al.*, 1988). This has implications for field crop production after a pasture phase.

External Modifying Factors

Although expenditure on pest control programs must be justified by measured worthwhile increases in crop production, factors other than maximising crop productivity also need to be considered. For example, aspects of climate such as abundant rainfall often favour both crop growth and the activity of pests. Thus the paradox may exist where overproduction, occurring as a result of pest control, leaves the farmers with a lower unit price and even a lower total net return for their product. This kind of response is usually more of a problem with high-value crops (e.g. fruit and vegetables) supplied to small, widely fluctuating local markets, than with large-area field crops. In the latter situation, not only is the individual farmer's contribution to the total marketable product small, but price stabilisation schemes (where they exist) mask the response to market fluctuations.

Changes in demand may also change the importance of specific pests. Lupins and canola, for example, are now regularly grown as alternatives to wheat in southeastern Australia. In some areas, capeweed, which is a minor weed of cereals in most cases, is a serious weed of these crops and more especially of canola. The control of a broadleaf weed in a broadleaf crop presents completely different problems to those encountered in the cereal crop but broad-spectrum herbicides like napropamide provide useful control (Leys and Cuthbertson, 1976).

National Control Programs

Ultimately the consumer pays the cost of pest control, either directly in the price of food and fibre or indirectly in taxes. The immediate economic value of quarantine activities and biological research, a major part of national pest control, can only be estimated, but to veterinarians and graziers, the potential entry of diseases like foot and mouth disease is a frightening prospect. In contrast, the benefits of biological control programs, the other major function in national control, is more readily assessable. Successful control of some weeds and a number of insect pests has been reported. The classic example is undoubtedly the control of prickly pear (*Opuntia* spp.) by the introduced moth *Cactoblastis cactorum*, which added about 40 million hectares of land to the agricultural and grazing areas of eastern Australia. More recently Australia pioneered the use of a fungus, *Puccinia chondrillina*, for the control of skeleton weed. Within six years of its introduction, and for an outlay of little more than \$2 million, the annual wheat crop value rose by nearly \$18 million (Cullen, 1977) and the use of the herbicide 2,4-D on skeleton weed nearly ceased.

Crop Loss Assessment

The procedures used in assessing crop losses involve methods of determining various levels of pest intensity and correlating these with associated crop losses.

In all cases, correlation of pest intensity and crop loss is obtained by statistical treatment of the raw data obtained, ideally, from experiments at several sites over a number of seasons. Pest appraisal, however, must necessarily have a practical degree of accuracy. The permissible error of the estimates depends on the purpose of the exercise as well as the considerations of time and finance. Nevertheless, it is important to develop standards that apply from one worker to another and from one season to another. It is therefore imperative that standardised keys and conversion factors be used to translate pest intensity into anticipated crop loss, and to define stage of crop development as well as organs or tissues affected at the time of sampling. The ability of neighbouring crop plants to compensate for weakened or dead plants as a result of pests is also an important consideration in these studies. In practice, probably more advances have been made in relating pest intensity to anticipated crop loss from insect pests and weeds than from diseases. This is because of the greater difficulty in assessing the amount of disease and measuring the density of infective propagules.

Estimate Pest Intensity

Pest intensity is used here to describe the population density of the pest in a specific area, on the crop or in the soil at a given time. It is not necessarily directly correlated with the eventual damage to the crop and the loss of yield. It is usually measured by one of the following methods:
   (a) Visual rating of pest numbers, damaged plants or plant parts;
   (b) Actual counts of pest numbers, damaged crop plants or affected plant parts or, in the case of weeds, determination of sample dry weight or percentage ground cover.

VISUAL RATING

This method is useful in determining pest incidence and distribution over large areas. Such surveys require little labour and are often an essential preliminary to more intensive sampling but may nevertheless yield worthwhile information (Banyer, 1966b). Surveys provide useful and accurate data if the various intensities are put into defined classes and the pest or damaged crop plant is visually prominent.

In the case of foliar pathogens the amount of disease is usually estimated as the percentage area infected. Several standardised area diagrams and keys have been devised to aid this type of visual assessment (James, 1971, 1974).

The method as a whole loses accuracy where the pests are of limited extent or where the habitat shields individuals. In addition there is always a tendency to bias on the part of single assessors, so independent estimates of several workers should be averaged when possible.

REMOTE SENSING

Currently, more sophisticated techniques such as remote sensing are being employed in many large-area pest surveys. These techniques, which are broadly defined by Shepherd and Totterdell (1974) as 'the acquiring of information about ground conditions from an aerial platform by using techniques which record energy from within the electromagnetic spectrum', provide a permanent record of events less subject to the bias errors in ground assessment data of individual workers. Several variants of this concept as it applies to pest incidence, including the use of satellites (Landsat) orbiting the earth (Clare, 1974; Duggan *et al.*, 1978), have been adopted with varying degrees of success.

The techniques used are based on the ability of modern photographic film to detect, with the aid of appropriate filters, different bands of the electromagnetic spectrum. Thus vegetation stressed by disease, insect pest or weed infestation can be recorded on film. Wallen and Jackson (1971) have used this method in association with data extraction computers to assess the yield loss of field beans affected by bacterial blight.

The use of aerial photography for the detection of pest damage in crops, including the methods used, is succinctly reviewed by Chiang and Wallen (1977), who conclude 'detection of insect and disease outbreaks and their assessment by aerial photography offers several advantages over conventional methods . . . [and] should provide a useful tool for crop loss assessment'.

COUNTS OF INDIVIDUALS

Where individuals are to be counted, defined areas of known size must be sampled. These quadrats may be square, circular or rectangular, although the rectangular sample usually provides a better estimate of variability per unit area (Donald, 1946). While cost of labour makes it necessary to minimise the sample area, data variability increases as sample size diminishes. Because variability increases as pest numbers fall, sample size should increase as pest density decreases. With weed populations, counts of more than thirty individuals per quadrat are usually reliable (Cottram *et al.*, 1953).

Insect populations may be estimated using direct counts per unit of plant or soil. Alternatively, estimates can be obtained using sweep nets or various types of traps (Southwood, 1978). For crop loss assessment, priority should be given to methods which allow conversion to an absolute population density (Walker, 1988).

For example, cotton insect pest thresholds are estimated as insect numbers per metre of row (for *Helicoverpa* eggs and larvae mirid bugs) or percentage of plant parts infested (for aphids and thrips) (Shaw, 1992).

Quantitative estimation of infective propagules of plant pathogens is generally difficult. Bioassay methods may need to be used to determine levels of soil-borne inoculum, whereas with air-borne pathogens various spore-trapping techniques can be adopted for estimating spore loads in the atmosphere.

Random counts of diseased plants or plant parts are appropriate where these are directly related to yield loss, i.e., counting infected heads in the case of loose and covered smuts of cereals and number of infected plants in the case of many virus diseases.

GROUND COVER

In the case of weed populations, counts give an accurate assessment of plant density but do not necessarily give an accurate assessment of their competitive ability. Percentage ground cover, which is defined as the area of linear distance occupied by the vertical projection of the above-ground plant parts onto the ground surface (Klingman, 1971), to some extent compounds both plant size and plant numbers and gives a better picture of potential competition. Visual estimation methods can be used effectively but they are less accurate than are line intercepts, point quadrats and step-point sampling (Brown, 1954).

WEIGHT

With weeds, the weight of the aerial parts of the plant provides the best estimate of competitive ability fully compounding size and plant numbers. For comparative purposes dry weight is better than fresh weight. Cutting, drying and weighing gives the most accurate results, but estimation of weight, either directly or by correlation with measurement of leaf size and number (Evans *et al.*, 1961), can also be effective.

Estimating Crop Losses

Crop losses can be determined either from survey estimates or controlled experiments. The survey approach may consist of a general appraisal of representative sample fields or a more detailed examination of actual crop samples from a designated area. The accuracy of both methods depends upon the skill of the survey personnel, the factors considered and the timing of the operation.

Reliable estimates can be obtained by a detailed observation of samples from a series of individual survey sites. These observations must be reproducible, and exact counts are preferable to visual rating unless the time factor is limiting. In such circumstances the accuracy of sampling becomes very important, and the number and size of the samples, as well as the method of sampling, must be considered. The reader is referred to any of the standard texts on survey methods (Hansen *et al.*, 1953; Yates, 1960; Church, 1971) for further details. A word of warning, however, is not out of place. One should never attempt to estimate yield loss by comparing crop yield from fields which contain the pest and other fields without the pest. Variation in soil type, crop variety used and farm practices employed, as well as the presence of other pests, modify yield potential.

CONTROLLED EXPERIMENTATION

The objective of crop-loss experiments is a direct comparison of pest-affected and pest-free crops. This comparison is obtained in paired treatment or multiple treatment experiments depending on the requirements set out. The various levels of pest intensity desired are obtained in several ways. These are:

(a) the introduction of the pest into controlled environment conditions either in the glasshouse or in field plots. Weed species, for example, can be planted at different densities to provide varying levels of competition to the crop. Using this technique McNamara (1976) showed that wild oats (*Avena* spp.) reduced wheat yield linearly by up to 1.03 g per m^2 per day for the duration of the period of competition. Similar responses can be achieved in other pest areas, e.g. by the production of disease following artificial inoculation of healthy plants, or, with insects, by the introduction of known numbers of the pests into cages set up on field plots.

(b) the control of a natural pest attack by the use of selective pesticides, manual roguing of weeds or insect pests or any other appropriate method that will provide variation in pest densities. Variation in the time and rate of treatment may also provide different intensities of attack; retarding the progress of a disease epidemic at different times with appropriate pesticides, for example, provides different levels of disease.

(c) comparing yields from agronomically similar cultivars which differ in their resistance to certain diseases or insect pests. Isogenic lines of barley differing only in their resistance to cereal-cyst nematode attack have been used to determine yield losses due to this pest (Cotton, 1970).

Each of the above methods has its place according to a particular pest–crop situation, but all have their limitations and caution must be exercised in the interpretation of results. In some cases, as in the use of chemicals for the control of a particular pest, crop growth may respond to nutrient materials contained in the pesticide. Yield response may also be related to the control of pests other than the designated target species. Experimental design therefore must account for these problems as far as is practicable.

Detailed information on the methodology of crop loss assessment is available in the FAO manual *Crop Loss Assessment Methods*. Teng (1987) provides another useful reference in this area.

Systems Analysis

A more recent approach to the determination of the role of pests in crop loss is a mathematical systems approach. The models used recognise that crop yield is the result of a complex interaction between pests, components of the total environment, including other pests, and the crop itself. They attempt to take these interactions into account and to evaluate the contribution each component has on final crop yield.

Models have been used to account for the variability in crop yield between years and regions, but most have little relevance to the local situation where the causes of variability are less clearly defined. Stynes and Wallace (1974) used such a model on a regional basis in South Australia, collating a wide range of data on soils, moisture, crop plants, pathogens and cultural practices. Multiple regression techniques were then used to determine the relative importance of each factor in accounting for between-farm as well as within-farm yield variation. Wallace (1974) later pointed out problems and limitations which are associated with relating epidemiological measurements to yield loss, and outlined procedures which he considers answer the questions:

(i) which environmental factors are chiefly responsible for the observed variations in crop yield, and

(ii) what contribution does a particular pathogen, insect pest or low water potential make to the observed variation in yield?

He further states 'the usefulness of the systems approach in epidemiology is threefold: it provides objective diagnosis, it indicates where practical control measures can be most effectively applied and it suggests directions in which research should go'.

Studies such as these are of a complex nature and their value and applicability have yet to be fully developed.

Cost/Benefit Ratios and Control

In deciding on pest control measures the producer usually considers three main factors: the objectives, the nature of the pest problem and the cost of the control measures available.

Producer objectives, however, are rarely wholly economical; most contain a sociological factor. This sociological aspect may not only be a desire to maintain a 'clean' crop but, in the case of many subsistence farmers, there may be a critical need to minimise the risk of yield loss. In both situations there is usually a somewhat greater expenditure on control measures than the pest problem actually warrants.

The nature of the pest problem basically is the relationship between pest intensity and crop yield and product quality, i.e., crop income. Control costs incorporate direct and indirect costs, including such items as the shelf price of chemical pesticides and certified seed, the labour costs of cultural control, a proportional charge for farm hygiene operations, and potential revenue loss arising from the use of lower-yielding resistant cultivars. Totalling all control costs is most important when considering integrated pest management systems.

Threshold Level Determination

Threshold levels are defined as those population densities between which crop protection measures are economically sound. They will vary to some extent relative to the farmer's objectives but their determination involves a comparison of pest intensity–crop income and control cost–pest intensity relationships.

Budgeting crop income requires some knowledge of potential product price, the probable extent of pest damage and its associated yield loss, the effect of various pest intensities on produce quality and the cost and efficiency of the available control measures. On the assumption that the damage relationship is a linear function, Norton (1976a) shows the economic threshold of chemical control can be obtained from the equation

$$e = c/pdk$$

where c = cost of control per unit area (ha)

p = estimated return per tonne of product

d = the damage coefficient expressed as tonnes per ha lost per pest unit

k = the efficiency of control, i.e. the proportional reduction in the level of pest attack due to control.

Reeves (1976) shows that in late-sown wheat, 900 annual ryegrass plants per m² reduced yield by 55 per cent. Thus, on an average yield of 2.36 tonnes per ha, assuming cost of control (c) as $6.30 per ha, estimated return per tonne (p) as $100, the damage coefficient as 0.001 tonnes per ha per plant and the herbicide efficiency (k) as 80 per cent, then

$$e = \frac{6.30}{236 \times 0.001 \times 0.8} \text{ or 33 plants per m}^2$$

With weeds particularly and in most pest attacks, damage is rarely linear as shown by other data of Reeves (1976). In such situations the generalised pest intensity–crop income relationship is as shown by the solid line in Fig. 8.1, net crop income reducing at an increasing rate beyond an initial pest tolerance level (N).

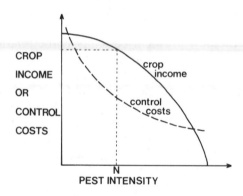

FIG. 8.1   Relationship between crop income and control cost – pest intensity levels (adapted from Carlson, 1977)

Since control measures are rarely, if ever, perfect—their efficiency is affected by the prevailing environmental conditions as well as the thoroughness with which they are applied—the cost of control rises steeply as the desired residual pest population approaches zero, i.e. as one attempts to leave fewer and fewer pests (the dashed line in Fig. 8.1). Comparing these lines provides the economic threshold for overall control costs. This is the number of pests where the slopes of the control-cost line and the net crop income line are the same. At this population the rate of increase in control costs is the same as the increase in marginal revenue (Carlson, 1977). At residual populations above this point, marginal revenue will exceed the marginal increase in control cost.

Decision Theory

Control measures can be conveniently classified as preventive or curative. The decision to use preventive control, i.e. pre-planting or pre-emergence herbicides, seed dressings, resistant varieties, or protective insecticides and fungicides, usually needs to be taken before the pest is evident. One cannot be sure that the attack will develop or just how intense it will be. Curative methods, which are applied after the attack develops, can take note of the level of pest attack and adjust marginal control costs accordingly. However, a recent innovation in the study of the economics of pest control, the use of mathematical models based on Bayesian decision theory (Officer, 1975; Norton, 1976b; Wheeler, 1976), may take much of the guesswork out of preventive pest control.

The approach provides a framework for making decisions in which costs and benefits of applying a particular control method to a pest are weighted against the probabilities of the pest occurring at various intensities or the control method failing. The model takes into account the distribution of different intensities of pest attack on the basis of past records. Thus the probability of the pest occurring at different levels can be determined and the overall expected net monetary gain or loss calculated on the basis of undertaking specific control measures. The model can be developed further by incorporating information such as the relationship between certain climatic factors and the probable pest intensity.

As an illustration, consider a crop which, in the absence of weeds, yields 2.53 t per ha, but which loses yield rapidly as weed numbers build up. The farmer has the choice of control or no control. Assuming the weed control method available gives 80 per cent control at all weed populations, then the result of each action on yield at different weed intensities is shown in Table 8.1. Where the value of the crop is $83.00 per tonne, the cost of weed control ($20.00 per ha) and other variable costs ($100.00 per ha), the yield matrix transforms to the cash return matrix given in Table 8.1.

The response shows that weed control is justified at the lower intensities. Obviously a more effective method of control is required at the higher

TABLE 8.1   Effect of weed infestation and control on crop income

Weed intensity	(a) Yield matrix yield (t/ha)		(b) Cash return matrix gross margin ($/ha)	
	No control	Control	No control	Control
Nil	2.53	2.53	110	90
Low	2.08	2.43	73	82
Moderate	1.10	2.01	−9	47
Severe	0.52	1.42	−57	−2
Average			29	54

population levels. Since it is unlikely that the different weed intensities will occur with equal probability, there is no guarantee that regular use of this method of control will be profitable over a period of several years. However, if the probability of occurrence of the different weed intensities can be estimated, then Norton (1976b) shows that the expected gross margins (EGM) can be obtained from the equation:

$$\text{EGM (action } i) = \sum_{j=1}^{2} [\text{Probability } (j) \times \text{cash value (outcome } ij)]$$

If, for example, past records suggest three light, two moderate, four severe infestations and one year without weeds in a ten-year period, then the overall EGM without control becomes:

$$\$[(110)(0.1) + (73)(0.3) + (-9)(0.2) + (-57)(0.4)]$$

or \$8.3, a small overall profit. At the same time the regular use of the control program gives an EGM of

$$\$[(90)(0.1) + (82)(0.3) + (47)(0.2) + (-2)(0.4)]$$

or \$42.2, a much more favourable return. On this basis the application of fixed costs to the return from the no control option would probably make the whole project unprofitable.

These risk-minimising strategies often conflict with profit-maximising ones, a factor that can only be overcome by expertise in forecasting the likely pest intensity.

Crop Surveillance and Pest Prediction

The timing of control methods is often critical when a disease epidemic, insect plague or serious weed infestation is to be thwarted. The ideal in any pest management system is the prediction of the probable level of pest intensity and, thus, the anticipated loss. Such predictions make possible the decisions as to whether, or which type of, control measures are warranted long before the pest reaches damaging proportions. This need is the major reason for insect and spore-monitoring techniques used in pest and disease forecasting services. These activities are associated with the surveillance concept, which, taking the broader view, can include monitoring weeds and vertebrate pests. Surveillance may be taken as the monitoring of the biotic and abiotic components of the crop ecosystem in order to assess or predict outbreaks, and is the basic element of all pest management systems. Implicit in this concept is the principle of the economic threshold or action level (ETL). This is the point at which pest suppression activity is initiated. The use of precise monitoring techniques coupled with accurate ETLs provides the most effective and efficient use of pesticides. Such an approach is essential to minimise costs, maintain stability in the agro-ecosystem and reduce the amounts of pesticides released into the environment.

Insect pest and disease forecasting services are mainly associated with high-value crops where pesticides play an important role (Richens *et al.*, 1975). Nevertheless, the application of these principles to low and medium-value crops has not been altogether neglected. In New Zealand, research has shown that chemical control of the aphid vectors responsible for infection with, and spread of, the barley-yellow-dwarf virus, increases the yield of autumn-sown wheat. A forecasting service based on crop surveys and aphid trapping in autumn has been developed. The results of these studies are used to predict the potential severity of a likely disease outbreak. Warnings on aphid flights onto the wheat crop are then issued in the main cropping areas (Smith *et al.*, 1963).

In many situations the predictions are based on an empirical approach, particularly where pest outbreaks are related to weather patterns. On the other hand, the use of regression analyses facilitates quantitative predictions of pest intensity in relation to variations in the value of readily measured environmental parameters. In addition, simulator programs, based on laboratory experiments involving the construction of a simulated pest, have been devised. Quantitative information on factors affecting the major activities of the fictitious pest are fed into the computer program and then weather data from the field are presented to the computer which predicts the course of the outbreak.

Crop surveillance, however, must be practical. It should not need to rely on sophisticated equipment or knowledge for its implementation. Currently, forecasting or prediction by sampling seems to be the more reliable method. The sample may simply be a report of the first sighting of a pest, or it may be an actual count of, say, immature insects as a means of estimating the potential number of adults.

While the most sophisticated systems mentioned are in the early stages of development at present, they emphasise the need for an interdisciplinary approach to the understanding of pest behaviour (Woods, 1974). Research into predictive models with longer-term aims is necessary but less sophisticated programs with a more immediate practical application must not be neglected.

PEST CONTROL

Identification

Recognition that a pest problem exists and the correct identification of the pest or pest complex are the first steps in formulating the most suitable strategy for control. More particularly, farmers and their field advisers should be able to recognise the early signs of attack. This is important as early detection is often essential to the application of timely control programs.

Signs and symptoms of pest activity are not always obvious, and positive identification may need confirmation by an expert with appropriate taxonomic knowledge. The diagnostician needs to be familiar with both the normal growth habits of the crop and pest characteristics, and their variations; information about the type and extent of crop damage, as well as the intensity and distribution of the pest, also often assists in the identification.

The importance of correct identification is related to the variation within species and varieties. Different strains of the pest often show differential responses to the environment, specific host crops and management practices. Many also react differently to the application of the various pesticides. The red-flowering strain of fumitory (*Fumaria* spp.), for example, tends to be resistant to 2,4-D, whereas the white-flowering strain is susceptible. Both strains are susceptible to herbicides like MCPA and bromoxynil.

Biological and Socio-economic Determinants in Control

In the absence or ignorance of scientifically based methods, farmers often achieve success in controlling certain pests by following practices developed over years of experience in a particular area. This can be considered as the art of pest control or, simply, as good crop husbandry. Farmers throughout the eastern mallee districts of South Australia, for example, discovered more than fifty years ago that the inclusion of oats in the cereal rotation reduced the incidence of a wheat disease known locally as 'hay-die', without knowing why. It is now known that, in that area, the condition is commonly caused by the soil-borne pathogen *Gaeumannomyces graminis* f.sp. *tritici* ('take-all') to which oats are resistant. Thus the oat crop very likely acted as an effective cleaning crop, reducing the carryover of infective diseased plant residues.

Pests form an integral part of the agro-ecosystem, influencing crop growth to a greater or lesser extent every season. In consequence, the recognition of pests and their contribution to crop loss, as well as an understanding of the role of their natural enemies and other regulating environmental factors, are necessary preliminaries to formulating the best strategy of control. The main aim is to manipulate these factors to the advantage of the crop but to the detriment of the pest, with least harmful effects to the overall environment, and so maintain economic crop production.

Evolutionary changes in crop production methods bring their own problems. For example, the popular practice of stubble retention as a soil conservation method in northern New South Wales and southern Queensland led to an increase in the severity of wheat yellow leaf spot disease (*Pyrenophora tritici-repentis*) (Wong, 1977). Changes in crop production also result in shifts in the genetic constitution of pest populations following processes such as mutation, sexual hybridisation and natural selection pressure. In the case of diseases, this may lead to a breakdown of host resistance at frequent intervals. The continual development of new races of black stem rust of wheat in response to the introduction of resistant cultivars is a typical example. In a like manner, a pathogen (Kuiper, 1965; Penrose, 1977), insect and weed (Holliday *et al.*, 1976; Radosevich, 1977; Nalewaja, 1978) may develop resistance to specific pesticides. Here the concentration of pesticide used is sufficient to kill all but the most resistant biotypes. With the competition removed, the less competitive but resistant biotype dominates. In Europe and North America, several weeds, including *Chenopodium album* and *Solanum nigrum*, have developed triazine-resistant biotypes. In Australia, the development of pesticide-resistant strains of the sheep blowfly, *Lucilia cuprina*, has occurred. The development of resistance in *Lolium rigidum* to dichlofop-methyl is widespread in southern Australia (Heap and Knight, 1982). Of particular interest, however, is that the dichlofop-methyl resistant biotypes of the grass are also resistant to other herbicides, including alloxydim, chlorsulfuron, fluazifop-butyl, isoproturon, propyzamide and trifluralin (Heap, 1986; Heap and Knight, 1986). Such changes impose the need for a constant reappraisal of the control measures employed in order to maintain the necessary flexibility in the control strategies available.

Established pests in field crops are almost impossible to eradicate. In these circumstances the cost of control is determined by their economically tolerable population levels in terms of acceptable crop losses. As a generalisation, suppressing pests to economically acceptable levels is better than attempting complete eradication. There is, however, at least one important exception to this rule, this being the very small pest populations which, occurring in previously 'clean' areas, act as a primary source from which further spread occurs. In this situation eradication is often advisable provided it is feasible. Such is the case in Western Australia, where government and farmers are co-operating effectively in an attempt to eradicate a number of primary infestations of skeleton weed. Left unchecked, these outbreaks would have very serious effects on the West Australian wheat industry. This type of approach has a good chance of success because pests which are concentrated in small areas or on specific weed hosts are much more susceptible to management and chemical control techniques than when they are spread over large areas or are affecting valuable crops.

The cost of eradication programs as well as the cost of reducing pest populations to tolerable levels invariably exceeds the cost of exclusion. Thus it is far better to prevent entry of pests into clean areas by suitable quarantine techniques at the local farm level as well as on the national scale.

Ideally, each crop–pest situation should be evaluated locally, taking into consideration the individual farmer's socio-economic position. The recommended means of controlling a particular pest (crop rotation, for example) may be impractical or incompatible with the farmer's preferences, farming system or current financial position. The subsistence farmer may need to use pesticides to ensure maximum productivity whereas the large-scale farmer can profitably ignore small-scale pest infestations. As already suggested, pesticide usage in the latter situation is often a matter of peer reaction.

Principles of Control

There are two major areas of pest control, preventive and curative control. Preventive control methods attempt to reduce the number of contacts or the effect of those contacts, between pest and host. Curative control methods attempt the destruction of some or all of the pests present with the host. Both methods involve the application of the principles of exclusion, avoidance, protection and eradication, applied either as regulatory, physical, cultural, biological or chemical methods of control.

Exclusion

The intention here is to keep pests out of previously 'clean' areas by the erection of artificial barriers. These barriers may be of a regulatory or of a physical character. Regulatory barriers include quarantine and inspection measures like seed certification; physical barriers include vermin-proof barriers such as the dog-proof fence separating Queensland and New South Wales, flyscreens, or chemical repellents and attractants used with chemical killing agents.

Avoidance

Cultural strategies such as time of sowing can be used either to keep the crop away from the main period of pest attack or to minimise its effects by avoiding conditions favourable for their activity. General crop management practices should aim at increasing crop vigour while, at the same time, making the environment unfavourable for the development of the pest species.

Protection

All management practices which, in the presence of the pest, favour crop development are implied in the principle of protection. Biological, physical, chemical and cultural control methods all have a role to play. General farm hygiene is essential. Where practicable, all refuse and diseased material must be destroyed. Roads, fence-lines and storage areas must be kept free of weeds; machinery and storage areas must be kept free of insect pests that attack grain.

Eradication

The destruction of the pest in the presence of the crop is the function of eradication techniques. In this context the term is not confined to the complete removal of an established pest, but refers to any significant reduction in its population.

Methods of Control

Quarantine and Regulatory Control

Quarantine refers to the isolation of one area from another by physical and commercial barriers to prevent the introduction and spread of pests. Experience has shown that introduced pests are frequently more damaging and wide-ranging in their activity in a new environment. This is partly due to their escape from the biological constraints present in the area of origin and partly because the new host populations are genetically susceptible as they have not previously been exposed to the pest.

Excluding pests is an important step in pest control programs. Freedom from pests not only lightens the economic burden of producers but opens additional markets from which infested produce may be excluded.

The huge volume of modern traffic in people and commodities prevents quarantine from being 100 per cent effective but it does slow down the entry and spread of pests.

In implementing quarantine, the community, through the action of the legislating government, recognises the need to protect established and developing agricultural industries. It is a function that requires legal sanction to be firmly applied. Failure to enforce quarantine restriction by inspections and legal sanctions leads to breakdowns in the systems. This is because quarantine reaches into all human activities and restricts the normal rights and privileges of a citizen. Many people find this irksome or economically damaging. Consequently, adequate procedures must be developed to allow safe movement of articles from infested areas. Fumigation, dipping or other commodity treatments can be applied where appropriate. Plant quarantine requirements have the effect of slowing plant introduction and thereby increasing the cost of breeding programs. With due cognisance of this disadvantage, however, Australian quarantine allowed the entry of several maize, sunflower and wheat introductions. Since 1968 these methods have allowed the development at Tamworth of the Australian wheat collection, with 14 000 introductions now available to all Australian wheat breeders. Such genetic collections not only preserve material but in fact reduce quarantine risks by rationalising plant introduction thus alleviating costly and frustrating delays in plant improvement programs.

Quarantine was not enforced in Australia until early in the twentieth century, and many of our important pests were introduced in the first hundred years of European settlement. Morschel (1973) has shown that 85 per cent of insect pests introduced into New South Wales became established prior to federal quarantine in 1909. International quarantine in Australia is administered by the various state Departments of Agriculture, usually through the chief horticulturist with advice given by practising entomologists, weed scientists, agronomists and plant pathologists. Effective liaison is maintained between the federal and state bodies, and any significant changes in policy by the federal plant quarantine bodies are introduced after joint discussions and consultation. In Australia, post-entry quarantine facilities have been established in all states and territories. Federal quarantine authorities permit entry (with appropriate safeguards) of small parcels of seed of crops prohibited or with restricted entry. This helps avoid the development of commercial crop varieties with uniform genetic bases and consequently susceptible to devastating outbreaks of diseases and pests. Release from post-entry quarantine is only permitted after the authorities are satisfied than no pest risk is present. Seed not on the restricted list may be impounded if on inspection it is found to contain propagules of potential pests. According to Evans (1975), material seen as having little or no risk and which passes into the country without a period of post-entry quarantine offers an even greater threat than seeds on the restricted list, particularly if the crop is of potential economic importance. This author also stresses the extreme caution that needs to be exercised when considering the introduction of known or potential plant pathogens for research purposes and biological control of other pests.

Quarantine requires the development of programs of containment, suppression and eradication if there is an outbreak. This has been done successfully on a number of occasions including the occurrence of boil smut of maize (*Ustilago maydis*) at Bathurst, New South Wales. Soil treatment with formalin, the burning of crops, and restrictions on maize being grown in the affected areas have prevented a recurrence of boil smut from 1940 until the early 1980s. Boil smut is now established on but restricted to the north coast of New South Wales and extending north along the coastal regions into Queensland. Seed distribution from these areas is restricted. Newer hybrids have reasonable tolerance to this disease.

Morschel (1973) lists five fundamental prerequisites on which the establishment of plant quarantine measures should rest:

1. First and foremost, the quarantine measure must be based on sound biological grounds. The pest(s) and disease(s) which the measure is designed to keep out of the country or an area must be of such a nature as to offer expected threat to substantial interests;

2. Quarantine must be established only for the prevention of introduction or the control of a pest or disease and not for the furtherance of trade or the attainment of some other objective;

3. Before a quarantine prohibition or restriction is recommended to the government, the subject needs to be carefully and thoroughly investigated and the advice of competent authorities sought;

4. Quarantine must derive from adequate law and authority and must operate within the provisions of such law; and

5. As conditions change, or as further facts become available, quarantine should be modified, either by inclusion of restrictions necessary to its success or by removal of requirements found not to be necessary.

The obligation to modify quarantine as conditions develop is a continuing obligation and should have continuing attention.

Evans (1975) evaluated these fundamental prerequisites and predicted a general relaxation of quarantine restrictions, especially on commodities that come from areas that offer little threat to the importation of diseases. In this context the Australian Quarantine and Inspection Service (AQIS) is responding to changing scientific, technological and commercial circumstances by adopting a more open and consultative policy in formulating quarantine measures.

Risk assessment is a pivotal step in deciding what quarantine measures are to be adopted. This biological analysis identifies which pests and diseases threaten Australia, where they might come from, how they might spread and how they might affect farm animals, crops and native flora and fauna. Strategies to counter means of entry and spread are considered with associated residual risks. If the assessed risk is sufficiently low, simple measures may be able to be implemented to ensure that it remains low. Thus AQIS endorses the principle of risk management in the provision of quarantine protection with minimal disruption to trade and the movement of people.

Other Regulatory Action

Pest control is affected by other legislative acts such as the Noxious Insect Act in New South Wales which allows for organised campaigns against prescribed pests, for example the Australian plague locust. Other legislative actions regulate the standards of pesticides offered for sale, the level of pesticide residues permitted in produce and the amount of produce blemished by a pest that can be offered for sale.

Physical or Mechanical Control

Mechanical sieving and grading of seed and separation by flotation are important means by which seed for sowing can be freed of various pest propagules. Included here are weed seeds, infective propagules of pathogens such as galls and sclerotes and survival stages of certain insect pests.

The importance of tillage in the control of weeds is well established and various cultural practices such as deep ploughing of diseased crop residue and burning of crop stubble are also important physical means used to suppress diseases and insect pests.

The physical agents most commonly used in controlling plant diseases are temperature (high or low) and various types of radiation, but these have greater application in intensive, high value horticultural and vegetable crop situations. Heat treatments are used for soil sterilisation (live steam, electric, hot water), controlling infections in propagative organs (hot water), and elimination of viruses from plants (hot water, virus-free apical growth at high temperatures). Post-harvest heat therapy was developed in Victoria to give good control of brown rot (*Monilia fructicola*); peaches were heat treated before full maturity for 24 hours and then ripened at 24°C to reach full ripeness without brown rot

development and with minimum losses from *Rhizopus* and other rot
fungi (Fish, 1970). Refrigeration is probably the most widely used
method of controlling post-harvest diseases of fleshy products by
inhibiting or retarding pathogenic activity.

Physical control of insect pests has assumed greater significance more
recently because of the development of resistance by insects to pesticides
and the need to avoid insecticide residues. Banks (1976) reviewed recent
developments in the physical control of insects and stresses its potential
application as a component of integrated control systems against field
pests, and its increasing importance in stored commodity protection.
Developments include physical exclusion, temperature control, environmental gas control, relative humidity control, light-induced phenomena,
sound responses, abrasive or inert dusts, physical shock, electric discharge,
ionising and neutron irradiation, bulk air movement, adhesives and
physical removal.

The effective control of insects which attack stored grain in silos and in
transit is essential if Australia is to be able to export wheat in a condition
which will continue to satisfy market demands. Care needs to be exercised
during harvest to avoid mechanical damage of seed which leads to
spoilage and destruction during storage by various fungal moulds and
insect pests. Weather and machinery damage of harvested grain and a
comparatively high content of unmillable material are factors which
favour pest activity and reduce the efficiency of chemical control measures.

It is essential that moisture levels of stored grain are kept below certain
prescribed maximum limits. The biotic potential of insect pests which
attack stored wheat, for example, is reduced in a drier environment. A 12
per cent moisture content limit for wheat receipt into central storage and
handling systems is applied by the industry. Grain driers (Sutherland,
1975), forcing of cool ambient air through stored grain (Anon., 1975),
and protecting stored grain against pests by heat, cold or suffocation
(Anon., 1976b) are all physical means which are either in commercial use
or under test to control grain pests.

Cultural Control

Cultural control is defined as the tactical use of regular farm operations
to reduce the activity of pests. It specifically excludes the use of introduced
biological agents and pesticides. Cultural control is important because it
generally involves minimum cost and avoids or reduces adverse environmental effects. The techniques involve practices which modify the
environment to favour crop growth but discourage or avoid conditions
that favour pest development.

One of the major ways in which human beings can use elements of the
macroenvironment to their advantage is by their choice of region for crop
production. While selection is based primarily on climatic and edaphic
features favourable for crop growth (e.g., rainfall, temperature, soil type)
the region can also be selected on the basis of conditions unfavourable to

the activities of certain pests. This approach is extended to a more localised level when a decision must be made as to which paddock to sow. The likelihood of pest problems as well as considerations such as fertility status of the soil should be taken into account. In most situations, however, the micro-environment (that in the immediate vicinity of crop plants) more directly influences pest activity. It can be modified in various ways. For example, alteration of plant densities is achieved by different sowing rates which in turn affect important pest-regulating factors such as relative humidity beneath the crop canopy, light intensity, surface soil moisture and temperature.

The soil environment, particularly the rhizosphere, which is the zone immediately around plant roots where most subsurface pest activity occurs, can be modified by soil amendments (artificial fertilisers, organic manure), irrigation, depth of sowing, different cultivation techniques and rotations.

In all instances, a detailed knowledge of the biology of the pest and the key factors that influence its main activities are required to take full advantage of these practices.

The following examples represent some of the commonly used practices aimed at cultural control.

MINERAL NUTRITION

The addition of well-balanced artificial fertilisers promotes vigorous crop growth which is then generally better able to withstand the deleterious effects of most pests. Mineral nutrition is an important factor affecting the predisposition of a plant to infection by disease, to attack by pests and to competition by weeds. Different types, levels, and balance of nutrients influence the severity of disease amongst the host/pathogen combinations. This results from an interaction between modified host susceptibility and pathogen virulence. In some cases, even different forms of the same nutrient are known to affect pathogen virulence differently.

In general, high levels of nitrogen increase the susceptibility of a wide range of hosts to pathogen attack. In the case of insect pests, aphid reproduction is facilitated by a high level of soluble nitrogen (Van Emden, 1969). Bardner and Fletcher (1974) report that the suitability of the host crop for growth, survival and reproduction of insect pests may be affected by water stress, fertilisers and growth regulants. There are variable reports however on the response by insect populations to the nutritional status of the host, and the reader is directed to Pritam Singh (1970) and Rodriguez (1960) for further references on this topic.

TIME OF SOWING

Adjustments to time of sowing often make it possible to avoid the period of main activity of pests. Germination of barley grass in southeastern Australia, for example, occurs mainly in the autumn-winter period. Thus heavy barley grass infestation of newly sown lucerne pastures can be avoided by a spring sowing. On the other hand, the ecological advantage is nearly always with the established plant, so that where a seasonal change in sowing time is not possible, very early sowing often provides the crop with an advantage over late germinating weeds.

Field observations often reveal that time of sowing has influenced the severity of attack by a particular pest. This type of response, however, is not always consistent and in these cases therefore not predictable. The inherent maturity character of field crops is regulated by the climate of the region which in turn determines optimum sowing time, thereby limiting the potential value of varying sowing time as a means of avoiding pest attack. With increased knowledge of pest behaviour and a better understanding of the genetics controlling phasic development of plants, new cultivars could be bred with sufficient flexibility to be sown at strategic times and so avoid damage by specific pests.

TIME OF HARVEST

Harvesting of various grains is usually timed to ensure that moisture levels are below certain critical limits. As discussed previously, higher levels are conducive to grain spoilage in bulk storage by the activity of saprophytic moulds and insect pests. Canola is best harvested during the cool of the day because this reduces seed-pod shattering. Avoiding contamination of the seed by Rutherglen bug by timely harvesting may also be important as tainted seed is unacceptable and thus downgraded.

Mature heads of grain crops are subject to colonisation of the seed by species of *Fusarium* (*Gibberella*) and other fungi which can cause discoloration of the grain. Such grain is unattractive for a variety of reasons and is subject to dockage or rejection by the miller, merchant or bulk handling co-operative. Harvesting at the earliest opportunity tends to minimise this problem.

Time of harvest is modified by practices such as rolling or cutting and windrowing, which again can serve a twofold purpose; namely, to prevent loss of grain by lodging or shattering, and reduce the deleterious effects of certain pests. Cutting and windrowing canola reduces damage by *Helicoverpa* spp. and aphids by rapidly drying out seed and stems. Rolling of cereal crops is reported to reduce losses caused by eye-spot lodging (*Pseudocercosporella herpotrichoides*) (Kuiper, personal communication).

SPECIAL CULTIVATION TECHNIQUES

Frequent cultivation is used to control many weeds and is currently employed as a means of silverleaf nightshade control in parts of the Victorian Mallee. With the regenerative capacity of skeleton weed shown to be minimal at stem initiation (Kefford and Caso, personal communication), Wells (1971) demonstrated that spring cultivations gave satisfactory control of the weed in the Victorian Mallee.

Clean fallow is used in Western Australia as a means of controlling webworm (*Hednota* spp.). Since the adult moths prefer grassed areas for oviposition and the newly hatched larvae depend on seedling grass for survival, the clean fallow provides useful control. Deep burial of crop residue is known to have beneficial effects by reducing the seasonal carryover of inoculum associated with diseases such as eye-spot lodging of cereals (Witchalls and Hawke, 1970) and bacterial blight of cotton (*Xanthomonas malvacearum*).

OTHER MANAGEMENT TECHNIQUES

Extensive use is made of sown pastures, particularly subterranean clover and perennial grasses, for the control of weeds by competition in southern Australia (Pearson, 1950; Moore and Robertson, 1964). Lucerne (*Medicago sativa*) and subterranean clover were shown to suppress skeleton weed. Shading by the clover, in association with grazing management (Kohn and Cuthbertson, 1975), reduced the weed population significantly in one to three years (Table 8.2). Other weeds successfully controlled by lucerne or subterranean clover-based pastures include St John's wort (*Hypericum perforatum*), serrated tussock (*Nassella trichotoma*), onion weed (*Asphodelus fistulosus*) and variegated thistle (*Silybum marianum*).

Where crops are irrigated, the amount, timing and method of watering can be altered to suppress the effects of disease. Overwatering, for example, is conducive to the activity of damping-off (*Pythium* spp.) and root rot (*Phytophthora* spp.). Spread of diseases such as *Septoria* spots and blotches and bacterial blights, which are normally spread by raindrop splash, is enhanced by overhead watering. Flooding where practicable reduces the level of disease-causing organisms such as plant-parasitic nematodes and certain soil-borne insect pests such as wire worm, by creating anaerobic conditions.

CROP ROTATION

Crop rotation can be effective against weeds, diseases and insect pests. It is one of the oldest approaches to disease control in cultivated plants and is just as important today; even more so where more intensive cropping systems have evolved. Its principal function is to eliminate pest populations or to sufficiently reduce their levels to allow worthwhile crop yields. With insect pests and diseases, this is usually achieved by replacing readily available sources of preferred or susceptible hosts with non-preferred or resistant hosts in the cropping sequence, thereby interrupting the cyclic activities of the pests. With weeds, populations can be reduced by cultivations and chemical sprays associated with different crop species or by the establishment of preferred species which provide more effective competition.

Although uncommon, it is possible for rotations to completely eliminate certain pests. Examples include certain obligate parasites which can

TABLE 8.2 Control of several weed species by lucerne and subterranean clover alone or in mixtures with some grasses

Pasture species	Weed	Yield of weed species		Source
		No sown species	Sown species	
Lucerne	Skeleton weed	198 plants/m²	54 plants/m²	Pearson (1950)
	Variegated thistle	16 018 kg/ha (approx.)	0	Michael (1968)
	Onion weed	938 seedlings/m²	150 seedlings/m²	Roark and Donald (1954)
Subterranean clover	Skeleton weed	198 plants/m²	42 plants/m²	Pearson (1950)
	Serrated tussock	215 plants/m²	80 plants/m²	Moore and Robertson (1964) Campbell (1960)
Phalaris tuberosa	Variegated thistle	16 018 kg/ha	2800 kg/ha (approx.)	Michael (1968)
	St John's wort	Ryegrass–white clover mixture 848 kg/ha (approx.)	4 kg/ha (approx.)	Moore and Cashmore (1942)

survive only on living host tissue, and pests whose survival depends on resting propagules in soil which are relatively short-lived. The chances of reinfestation or reinfection from outside sources, however, are usually high, either from wind-blown pest forms (winged insects, weed seeds, fungal spores) or from various introduced sources of contamination (hay, seed, soil on stock or implements).

Sanitation assumes particular significance when allied to the effective use of rotations. There is little point in growing non-host crop or pasture species to reduce the level of infective disease propagules in soil, for example, if weeds which act as an alternative host are allowed to grow unchecked. Similarly, weeds infesting neighbouring uncultivated land, such as fence lines and gullies, may constitute a ready source of reinfestation of either the weed itself or some pathogen or insect pest for which it is a host.

Crop rotation, however, is not particularly effective against pests that have exceptional powers of dispersal or are normally migratory, like the grasshopper.

Although rotation plays a significant role in modifying the effects of many pests which possess a soil-borne phase (Rovira, 1990), it is ineffective against organisms such as *Rhizoctonia solani* (bare-patch disease), a root-rot fungus which possesses a wide host range and a strongly competitive, saprophytic ability. The practice of rotation nevertheless is an essential feature for the commercially viable production of many field crops including peas, cotton and canola which would otherwise succumb to the build-up of pests.

Dissatisfaction with crop rotation as a control procedure results in part from a lack of knowledge of the biological changes induced in soil by different crop sequences. It has long been standard practice, for example, to recommend a break in the cropping sequence by the inclusion of non-host species after serious outbreaks of soil-borne diseases like cereal-cyst nematode and take-all of wheat, yet the period of interruption required before cropping can be safely recommenced may only be estimated in the light of variable field experience. Indeed, it is not uncommon for wheat crops immediately following such badly diseased crops to appear quite healthy and yield well. On the other hand, costly experience dictates that crops like peas and canola should only be grown in the same ground once in about four or five years to be certain of avoiding substantial yield losses through diseases carried over in soil as well as other pests like weeds.

Socio-economic limitations to the use of rotations have been briefly mentioned.

Further research and a better understanding of the farmer's situation is required to provide a more intelligent and sound biological basis for the effective use of rotations.

SANITATION

Sanitation includes all activities aimed at eliminating or reducing pest levels in association with plants, paddocks, machinery and storage facilities; in short, from any refuge acting as a source of spread to other healthy plants or plant products and into clean areas.

In many instances sanitation forms an important part of an overall pest control program, and in some cases the control of certain pests relies largely or solely on proper sanitary measures.

Practices vary in scale of operation—from decontamination of stock and farm machinery, spot spraying of weeds or roguing weeds or diseased plants, to broadacre disposal of diseased crop residue either by deep burial or by burning.

Normal farm hygiene (weed-free areas along fence lines and around outbuildings, clean grain storage areas) and the purchase of clean seed, are examples of sanitation practices not necessarily directed against any specific pest. On the other hand, many practices are so directed. These include the control of a weed known to be an alternative host harbouring some serious disease or insect pest, or the use of legislation to enforce proper sanitation. Such is the case with the Noxious Weeds Act, which is aimed at preventing the spread and build-up of specific weeds.

Suppressing alternative grass host species, particularly in the year prior to cropping to reduce the carryover of the take-all fungus, is a good example of effective sanitation (Kidd *et al.*, 1992).

Of considerable importance to the Australian grain industries, particularly wheat for export, is the development of resistance to insecticides (principally maldison—Malathion®) by some stored-grain insect pests which in the late 1960s marked the end of a highly successful era in the chemical control of insect pest complexes of stored grain. Uncleaned headers and other harvesting equipment acted as a source of insect-contaminated grain entering bulk storages, as did the survival of insects in treated wheat due to the natural degradation of maldison. This allowed the natural selection of maldison-resistant insect pest strains. Dispersal of resistant populations of grain insects from mills and bulk stores to farming properties in deliveries of infested grain, prepared stock feeds and empty sacks soon followed.

Rather than relying on grain handlers and millers to eradicate insects attracted to stored grain, the problem is now met at the important source of contamination, that of the farm. Officers of the New South Wales Department of Agriculture have shown that a program of farm hygiene combined with fumigation was required to prevent rapid infestation (Fishpool *et al.*, 1975).

Unfortunately the benefits of sanitation practices are often hidden and mostly impossible to quantify in cost–benefit terms. Hence, sanitation as one important means of control in a program of pest management is frequently overlooked or ignored by farmers.

One important aspect of sanitation which bears special mention because of its application to all producers is the use of clean or pest-free seed. Crop seed, like soil, is a natural repository for a wide spectrum of pests (insects, weeds and diseases) and hence is an ideal vehicle for the introduction and spread of new pests into 'clean' areas or for their re-establishment in already infested areas. Seed-borne diseases are especially important, as representatives of all major groups of disease-causing organisms (viruses, fungi, bacteria and nematodes) can be disseminated by seed.

Propagules of many pests are known to remain viable for about the same time as the seed itself; instances where this is not so present an opportunity for controlling the pest by storage.

In horticulture, seed may be taken to include propagative material such as cuttings, rhizomes and corms. Such mother material can be infected by a wide array of diseases which may be passed on directly to the vegetative growth of new progeny. All the individuals in a crop thus propagated may become diseased, and it is therefore particularly important in this industry to ensure that planting is carried out with pest-free stock.

In agricultural seeds, pests may be present as contaminants associated with foreign matter such as infested granules of soil or diseased fragments of crop residue. Pest propagules, like the hardened sclerotia of ergot-causing fungi (*Claviceps* spp.), galls of the plant parasitic nematode *Anguina* spp. (e.g. earcockle of wheat, *A. tritici*), weed seeds and resting stages of certain insect pests such as grain weevil eggs (*Sitophilus* spp.) may be of similar size to crop seed and physically mixed with it. On the other hand, resistant propagules may exist as external contaminants in close association with the seed coat. These include bacterial cells, fungal hyphae and resistant spores, as in the case of the *Tilletia* spp. causing bunt of wheat.

Still others are borne internally, deeply seated in otherwise normal viable seed (e.g., hyphae of loose smut of wheat, several phytopathogenic bacteria and some plant viruses), whilst others such as the bacterial blights of peas and cotton can be borne both externally and internally. The manner in which pests are seed-borne determines whether protective or therapeutic action is required in control.

Seed-borne pests are also important because many new infestations or infections begin from an initially very low level of contamination in the grain. Once established, however, build-up and spread usually follows reproduction and this can occur in quite spectacular fashion depending on the particular pest and the environment. This is particularly important in the case of multiple-cycle pathogens which possess a high reproductive potential and undergo several generations per growing season, and whose spores are disseminated by air from seed-borne primary lesions, resulting in secondary infections over a wide area.

One of the first considerations, therefore, of any farmer, is to take whatever steps possible to establish a crop with 'clean' seed. Unfortunately, it is usually impossible to detect contaminated seed by superficial examination. Farmers must therefore treat their own seed as appropriate (seed dressings) or rely on other agencies and schemes such as seed certification and regulatory measures (Agricultural Seeds Act), where they exist, or on seed producers or co-operatives to supply clean seed. Such seed has been produced from a parent crop inspected and found free of designated pests or has been tested for freedom from pests or treated chemically (seed dressings), physically (heat therapy-hot water) or mechanically (various mechanical graders and liquid separators) to exclude, destroy or remove specific seed-borne pests.

No guarantee can be given, however, that any seed sample is absolutely free of pests, as methods of seed testing and treatment are not foolproof nor 100 per cent effective. Not the least important problem is attempting to obtain a representative sample of seed for testing. This can only be overcome partially by statistical analysis of a large number of samples. Techniques for testing and treating seed are improving, however, and there is increasing government activity in providing seed-testing services like the unique program introduced by the New South Wales Department of Agriculture in 1974. This scheme checks on the varietal identity, purity and presence of seed-borne diseases of cereal seed offered for sale by registered seed growers and produce merchants.

Farmers failing to take advantage of such services and schemes do so at their own risk. It seems cheap insurance to outlay a little extra for certified or tested seed and thus help ensure protection against both short-term and long-term yield losses caused by pests otherwise introduced.

Where feasible, farmers may retain their own harvested seed for sowing the next season's crop—a common practice amongst many cereal growers. These farmers, however, run the risk of incurring losses by the build-up of certain seed-borne diseases such as loose smut of wheat, unless proper precautions are otherwise taken.

An important strategy in the provision of disease-free seed for some crops, such as peas, is the use of areas either free of certain diseases or areas and conditions that are not conducive to the development of their seed-borne phase. Wet-weather diseases such as Ascochyta and bacterial blights of peas, which are characteristically seed-borne, may be avoided in seed crops by growing them in arid regions. If required, supplementary water can be provided by furrow rather than overhead irrigation, the latter favouring the spread and infection of a number of foliar pathogens, particularly bacteria.

Breeding for Resistance

Because of its wider application, breeding for resistance to diseases is discussed here although many of the underlying principles would apply to the breeding of plants resistant to attack by insects, notable examples of which include lucerne varieties with multiple aphid resistance and rice that is resistant to stem borer (*Chilo* spp.), and leaf and plant hoppers.

THE LOSS OF NATURAL RESISTANCE

In the natural plant community, pests and their hosts evolve in close association. Such plants possess a generalised immunity to most of the diverse array of pests, the more advanced tending to be highly specialised. However, the dynamic balance of the natural ecosystem, which tends to be maintained to the benefit of pest and host, breaks down under human interference. The important consequences are:

(a) the loss of natural resistance which often accompanies plant improvement by selection and breeding;

(b) the encouragement of epidemics by the widespread culture of single crop species; and

(c) the lack of naturally developed immunity due to the extension of crops and pests into new areas.

Whereas wild species tend to possess a high degree of resistance to a wide spectrum of pathogens which generally cause little damage, crop plants selected almost exclusively for high yield exhibit increased susceptibility. To overcome this imbalance there is an ever-increasing emphasis on plant breeding to produce crop plants with adequate levels of resistance to the more important pathogens.

ADVANTAGES AND HISTORICAL DEVELOPMENT

The main advantages of using resistant cultivars are the saving in cost to the landholder, the absence of poisonous residues which pollute the environment, independence from high-energy sources based on fossil fuels, and easy integration with chemical control methods if warranted. Additionally there are no other economical or effective control measures available for some pests.

In Australia, breeding for resistance in field crops has been mainly concerned with the control of pathogenic organisms—virus, bacterium, nematode or fungus. The development of bunt-resistant wheat cultivars by W. J. Farrer after 1899 is one of the earliest examples of successful breeding for disease resistance (Macindoe and Walkden Brown, 1968). This was followed by cultivars resistant to flag smut (*Urocystis agropyri*), which, unlike the resistance to stem rust (*Puccinia graminis* f.sp. *tritici*), has proven extremely stable (Watson, 1958). Breeders in some states incorporate this resistance into all new wheat cultivars. McIntosh (1976) reviews the progress made in genetics of wheat and the wheat rusts since those early days.

Active breeding for resistance against cereal diseases such as *Septoria tritici* blotch (Ballantyne, 1984) and cereal-cyst nematode hitherto considered unimportant is also in progress. The breeding program for crown rot of wheat caused by *Fusarium graminearum* has a high priority to produce cultivars for the central and northern regions of the eastern wheatbelt. The pathogen has a wide host range in winter cereals and grasses and survives under reduced tillage management. Alternative crops like canola have also received attention, with resistance to blackleg being developed. Even in rice, which has few pathogenic problems in Australia, breeders realise they cannot be complacent, nor ignore the possible introduction of new pathogens. Resistance to stem rust, stripe rust and *Septoria tritici* blotch are major aims of wheat breeding programs. Resistance to stem and stripe rusts is required for release in New South Wales. The control of *Septoria tritici* blotch by use of resistant lines has been shown to give substantial yield increases, especially in wetter seasons when this disease is most destructive. The National Rust Control Programme (NRCP) conducted by the University of Sydney gives valuable assistance in breeding for rust resistance.

Breeding for resistance, while generally considered to be the ideal means of controlling plant disease, is not without its limitations and shortcomings. The relatively high cost of maintaining stem rust resistance in wheat cultivars, for example, has been queried and attempts to develop worthwhile resistance against troublesome soil-borne pathogens have been frustrated; the difficulties attending the discovery of suitable sources of resistance to the take-all fungus and the complexity of the genetics of *Rhizoctonia solani*, the organism causing bare-patch disease, are specific examples.

SOURCE OF RESISTANCE AND BREEDING METHODS

Different forms of physical and biochemical resistance occur between and within host-pathogen combinations, but all are under genetic control (Whitney, 1976). Although the aim of the breeder is to use the most suitable form of resistance, basic information on the mode of resistance is not essential for selection in applied programs. In the conventional sense, plant breeders rely on sexual recombination occurring for the transfer of resistance genes. In more recent times they have genetically engineered inter-specific transfer of genes and chromosome segments to confer resistance to several diseases. Irradiation and certain chemicals have been used to effect such wide crosses (McIntosh, 1976; Brock, 1977). These processes made possible the transfer of rust resistance from *Aegilops* and *Agropyron* species to cultivated wheat (Sears, 1956, 1972; Knott, 1961).

Sources of resistance are commonly obtained from regions where plants have developed a natural resistance to specific pests through evolutionary processes. The germ plasm from such wild species often provides satisfactory resistance after hybridisation with cultivated species.

Another common approach to identifying sources of resistance is to screen cultivated species and lines collected from a wide geographic area following either artificial inoculation or exposure to natural infection in the field. In some cases it is even possible to breed a resistant variety from a resistant individual.

Variability of pathogen virulence and the effect of environmental parameters on the host response to attack must be considered. The environment, for example, largely determines the expression of resistance. Apparent host resistance due to escape from infection is possible, in which event exposure to the pathogen under different cultural or environmental conditions may lead to successful attack. A further problem is that selecting for resistance to a single pathogen may result in selection for susceptibility to another. Varying degrees of specialisation exist amongst host–pathogen combinations and several physiologic races of the pathogen may be widely distributed in the field. This is an important consideration when screening for sources of resistance which ideally are effective against all such known races.

MODE OF INHERITANCE

It is generally considered that there are two main types of resistance, based on their different modes of inheritance and underlying resistance mechanisms:

1. resistance controlled qualitatively by a single gene (monogenic) or few major genes which convey resistance up to complete immunity against specific races of a pathogen (race-specific resistance, or vertical resistance); and
2. resistance controlled quantitatively by several genes (polygenic) which confer moderate resistance to all races of the pathogen (non-race-specific resistance, or horizontal resistance).

The physiological differences between the two modes are not clear and there is some query as to whether they are as distinctly genetically different as originally postulated (Rees, 1974; Ellingboe, 1975). Also, races are only determined when a variant of the pathogen which can attack a resistance gene is found. Until then the resistance may appear to be of the horizontal type.

Dominant or recessive characters may be involved but this is a function of the particular pathogen-host combination. Van der Plank (1963, 1968) draws a clear distinction between vertical resistance (VR) and horizontal resistance (HR).

VR is equated with Flor's (1956) gene for gene hypothesis which states that for every gene for resistance or susceptibility in the host, there is a matching or corresponding gene for virulence or avirulence in the pathogen. Thus, genes for VR are only effective against those races of the pathogen which do not possess the corresponding genes for virulence. Susceptibility results only when all resistance genes in the host are matched by the complementary virulence genes in the pathogen.

VR is simply inherited and is relatively easy to manipulate in breeding programs. Consequently, it has been the traditional basis for the development of resistant cultivars. However, where the pathogen possesses a high potential for genetic change, either through mutation, sexual recombination or other means of producing new virulent races, then cultivars with VR are not likely to remain resistant for long periods. Indeed, the widespread cultivation of resistant cultivars leads to the rapid selection of more virulent races of the pest which are able to overcome host resistance. In this situation the adoption of VR is rather self-defeating because there is a continual need to replace the cultivars in which resistance has been overcome—a process which leads to the development of still newer races of the pathogen. The classic example of this situation is the continuing battle by plant breeders against the shifting pattern of wheat stem rust races. Other methods aimed at achieving more stable VR (Brock, 1967; Watson 1974) include the incorporation of several genes for resistance in the one cultivar. Timgalen, Gatcher, Gamut and Mendos wheats, for example, possess several major genes conferring resistance, combining to combat different races of stem rust.

HR on the other hand is broadly based and usually provides greater stability because it is moderately effective against all races of the pathogen thereby reducing the selective increase of particular pathogen biotypes. It is involved in defence mechanisms which result in the reduction of the pathogen's capacity to reproduce and spread.

In contrast to VR, HR is more difficult to manipulate in breeding programs because of the complexities of its inheritance and difficulty in detection amongst progeny.

Wider diversity of resistance to *Septoria tritici* blotch, stem and stripe rusts aims to counter the risk from new races of the pathogens causing these dlseases.

ECOLOGICAL SIGNIFICANCE OF RESISTANCE

The significance of vertical and horizontal forms of resistance in relation to disease epidemics is discussed by Van der Plank (1963, 1968, 1975). He showed that VR delays the start of epidemics by reducing the effective inoculum of a pathogen while HR slows them down by reducing the rate of growth of the pathogen.

He also discussed the interrelationship between disease resistance and other methods of disease control. For example, fungicidal activity can be enhanced by the incorporation of HR into cultivars. Similarly, sanitation practices are most effective when the growth rate of the pathogen is low as can be achieved by using HR.

The recognition of the need for an interdisciplinary approach to understand and better define the genetic basis of epidemics is exemplified by Day (1977) which summarises current knowledge in this field.

FUTURE DEVELOPMENTS

Chromosome segments from rye can be translocated into wheat to introduce disease resistance. Initially, quality characteristics were affected, but with identification of marker genes, it is possible to target fairly accurately the appropriate chromosome segment required. It is believed that this will be a useful tool in future breeding programs. The ultimate aim is to develop more durable resistance thus minimising the work involved in plant breeding (Johnson, 1984). To achieve this, there needs to be a better understanding of the biochemical nature of resistance and virulence as well as a need to locate additional resistance in the germ plasm of wild and cultivated species. Controlled environmental facilities and computer techniques should prove useful in these areas.

Biological Control

In its wider sense, biological control implies control based on the management of some aspect of the biology of the pest species (genetic manipulation, breeding for resistance in crop plants, crop rotation and grazing management), which are considered separately here. In the more specialised sense, as emphasised in this section, it refers to the human manipulation of introduced and indigenous natural enemies of the pest in order to suppress it. Plant pathologists, in particular, adopt the more general interpretation which has been defined by Garrett (1965) as 'any condition under which, or practice whereby, survival or activity of a pathogen is reduced through the agency of any other living organism (except man)'. Significant developments in plant pathology consistent with this definition are also briefly discussed as are the specialised techniques such as sterilisation and natural attractants (autocidal) applicable to insect control.

There have been few successes in using natural enemies to control plant disease. On the other hand, natural enemies have suppressed insect pests and weeds, sometimes in spectacular fashion. Two examples of successful biological control in Australia are the classical control of prickly pear (*Opuntia inermis* and *O. stricta*) and, later, control of the green vegetable bug (*Nezara viridula*). Caterpillars of the moth *Cactoblastis cactorum* successfully controlled both prickly pear species, then covering 25 million hectares of agricultural and pastoral land in eastern Australia, within a few years of their introduction in 1925. In the 1950s the introduction of two geographic strains of the tiny wasp *Asolcus basalis* brought a dramatic decline in the numbers of the green vegetable bug, then a serious pest of the seeds and fruits of a wide range of field crops.

IMPORTANCE AND LIMITATIONS OF BIOLOGICAL CONTROL

Biological control is fundamental to the maintenance of the balance of nature in most stable ecosystems. This fact becomes obvious only when the natural ecosystem is disturbed, notably by the use of wide-spectrum insecticides which cause flare-ups of pests of otherwise minor significance, or in the introduction of new pest species without their associated natural enemies. The use of DDT, for example, raised the importance of some spider mites and scale insects because it killed the insects and mites feeding on them. More recently, the immediate devastation of lucerne in the Hunter Valley following the introduction of the spotted lucerne aphid

(*Therioaphis trifolii*) in autumn of 1977 (Anon., 1977) provides an outstanding example of the response by a susceptible plant to a new pest in the absence of its natural enemies.

When successful, the advantages of biological control can be summarised as:

—minimal cost to the landholder,
—on-going and self-perpetuating populations of the controlling agent rising and falling with pest populations,
—non-polluting and lacking residues potentially dangerous to beneficial species,
—independent of high-energy sources based on fossil fuels, and
—in some cases, being able to be integrated with chemical control methods.

The financial benefits are well summarised in the results of the recent project to control skeleton weed in southeastern Australia. Cullen (1977) showed that two organisms introduced to control the weed, namely the rust *Puccinia .chondrillina* and the gall midge *Cystiphora schmidtii*, added over $50 million to the value of the Australian wheat crop between their release in 1972 and 1976 (Table 8.3). The eventual recurrent annual savings are estimated as nearly $26 million out of a potential total annual loss of $29.5 million. This is an excellent return for an outlay of less than $3 million.

TABLE 8.3 Estimated annual and total savings in Australian wheat crops from the biological control of skeleton weed, *Chondrilla juncea* (Cullen, 1977)

	Saving ($ million)
1972 (year of introduction)	0.06
1973	11.23
1974	13.64
1975	17.86
Total (1972–75)	42.79

The limitations of biological control, like the advantages, are manifold and include

—the need for long-term research,
—the lack of immediacy of action when individual crops are threatened,
—the inability to prevent contamination of crops and stored products by low densities of the pest,
—its ineffectiveness in areas where the introduced pest is more suited to the environment than the beneficial species,
—the possibility that the introduced organisms may themselves become noxious, and
—the problem that most attempts at classical biological control have been failures or, at most, only partially successful.

THE FUNDAMENTALS OF BIOLOGICAL CONTROL

Underlying biological control projects are concepts related to the regulation of populations of living organisms and the balance of nature.

The balance of nature concept refers to the various natural agents which limit the movement of populations away from the mean density and bring it back to the normal. Most organisms, for example, have the potential to increase to enormous numbers if the whole of their progeny survives. This potential, fortunately, is rarely realised because of the natural control exerted on the organism.

Some populations fluctuate widely, others within narrow limits, others irregularly. All appear to fluctuate, however widely, around a characteristic long-term number usually referred to as the equilibrium level. This level represents the 'balance of nature', and is a function of the inherent characteristics of the species (its fecundity, genetic variability), its interactions with other species and the physical parameters of the environment.

Mortality factors preventing the realisation of the growth potential are both density-independent and density-dependent. Density-independent factors operate regardless of population size. In the main they are the physical parameters of the environment such as climate. For example, a frost may kill 90 per cent of the population regardless of the initial finite numbers. Density-dependent factors, however, increase their action proportionately to the host numbers. Ladybird beetles, for example, devour more aphids when they are abundant than when the aphids are scarce. Most natural enemies are density-dependent arbiters of abundance.

NATURAL ENEMIES

Natural enemies are organisms, including predators and parasites, which either consume or inhibit development of their hosts. Beneficial insects may be either predators or parasitoids (predators, e.g., ladybirds, destroy more than one individual in reaching maturity while a parasitoid destroys only a single host organism). The term biocidal is used to describe micro-organisms destroying disease micro-organisms. When such an organism inhibits rather than destroys it is said to have a 'biostatic' action. The parasitic group attacking weeds and insect pests also contains pathogenic organisms—bacteria, fungi and viruses —all of which have been used with success. Skeleton weed rust, *P. chondrillina*, represents the first successful use of a pathogenic fungus to control a weed, while the potato-moth virus offers an alternative control for the potato moth (Reed and Springett, 1971).

The bacterial pathogen *Bacillus thuringiensis* (Bt) has been used commercially (as Dipel®, Thuricide®) to control the larvae of various leipidopteran pests, including *Helicoverpa* spp. Genetically altered or transgenic cotton has recently been developed which incorporates the gene producing endotoxin, the toxin present in Bt. Thus in future, cotton (and other crops) may have an inbuilt protection against various moth pests.

MODIFYING THE ENVIRONMENT AND BIOLOGICAL CONTROL OF DISEASE

Enhancing existing levels of natural biological control of plant pathogens by manipulating biotic and abiotic environmental factors can be considered simply as an extension of good cultural practices, the underlying basis not necessarily being understood but being nonetheless effective. This is exemplified by the use of rotations which are effective against a number of soil-borne pathogens of cereals and other field crops.

The various natural biological forces at play in this type of control and suggestions as to how these can be directed against the pathogen are discussed in more detail elsewhere (Baker and Snyder, 1965; Baker, 1973; Baker and Cook, 1974; Fravel, 1988). They include considerations of widespread phenomena such as fungistasis and antagonism, the role of antibiotics in soil, soils suppressive or conducive to pathogens, saprophytic competitiveness, trap crops and breeding for features favourable to biological control.

Cook (1974) highlights the importance of pathogen-suppressive soils as a major contribution of biological control to agriculture. Of particular interest is the suppression of the take-all fungus following consecutive crops of either wheat or barley, commonly referred to as take-all decline. This specific suppression is thought to be caused by the antagonistic activity of one or a few organisms (Shipton *et al.*, 1973; Cook, 1974; Cook and Rovira, 1976; Ownley *et al.*, 1992). Non-specific or general suppression, thought to involve many soil organisms, has also been recognised (Cook, 1974).

SPECIALISED TECHNIQUES FOR CONTROL OF INSECTS

One successful technique involves the sterilisation of male insects chemically or by irradiation. Satisfactory control can be obtained with species in which the females only mate once and are unable to differentiate between sterile and non-sterile males. The technique, termed autocide, is most effective in insect populations confined in restricted areas.

Pheromones are naturally occurring chemicals which are released into the environment by animals and which modify the behaviour of a member of the same species. One group of pheromones appears to control responses like aggregation, dispersal, trail following and mating. Considerable attention has been given to the set of pheromones as an aid to control of insects. They have value in monitoring and mating disruption (Jutsum and Gordon, 1989).

FUTURE DEVELOPMENTS

There is no easy way of obtaining biological control. Because safeguards to prevent the entry of noxious pests in the guise of beneficials must be erected, a sophisticated program of research, introduction and evaluation is required. The reader is referred to DeBach (1964) and Walpshere (1974) for detailed information on this aspect. Moreover, while some natural enemies are self-perpetuating, others need to be reintroduced periodically and still others need to be supplied with refuges or food in times of stress. Thus, prediction of success in specific predator–host relationships is impossible. Nevertheless, more active searches for ways in which to develop biological control are probable in the immediate future, especially if they can be integrated with other control methods and overcome the typical initial slowness of natural enemies in crises.

Although not yet developed to any significant practical degree in Australian field crops, hyperparasitism, which includes the use of bacteriophages (bacteria-destroying viruses), mycoparasites and nematophagus fungi (nematode-trapping) in the control of disease, and inundative control of weeds, involving the application of high concentrations of fungal, bacterial or viral pathogens to susceptible weeds (Charudattan and Walker, 1982; Wilson, 1969), nevertheless holds promise for the future. Products of this kind, based on the fungal pathogens *Colletotrichum gloeosporioides* f.sp. *aeschynomene* (Bowers, 1986) and *Phytophthora palmivora* (Kenney, 1986) and hence termed 'mycoherbicides', are in commercial use in the United States. Because this method of biological control is peculiarly adapted to the control of annual weeds, Australian studies are concerned with the control of annual species of *Xanthium* by *Colletotrichum xanthii* (Nikandrov *et al.*, 1985) and a rust fungus, and on the control of annual grasses in crops with a seed-borne fungus (R. Medd, personal communication).

The phenomenon known as 'cross protection' may also be useful in the control of disease. It is defined as a mechanism whereby plants become resistant to pathogens following prior exposure to an avirulent strain usually of the same species. Traditionally associated with the use of mild virus strains to prevent infection by more virulent strains of the same virus, the practice has had some success in Australia (Simmonds, 1959; Stubbs, 1961). More recently, similar reactions have been recorded amongst other main groups of pathogenic micro-organisms. The crown gall disease caused by *Agrobacterium radiobacter* f.sp. *tumefasciens* can be controlled by the closely related non-pathogenic bacterium *A. radiobacter* f.sp. *radiobacter* through the production of bacteriocin by the non-pathogen and to which the pathogen is sensitive (Kerr and Htay, 1974, Clare *et al.*, 1990). Commercial advantage has been taken of this discovery by the protection of almond seedlings against soil-borne crown gall following inoculation with the non-pathogen. In addition, Wong (1973, 1974) has shown that prior colonisation of wheat roots by the non-pathogenic *Gaeumannomyces graminis* f.sp. *graminis* under experimental conditions has afforded some protection against the closely related pathogen *G. graminis* f.sp. *tritici*.

Of considerable interest was the discovery by Smith (1973) that ethylene is a critical regulator of microbial activity in soil as a cause of soil fungistasis (a non-specific form of dormancy) amongst aerobes including pathogens. He later postulated (Smith, 1974) that the successful manipulation of ethylene production in soil offers a valuable means of regulating important soil processes with the ultimate aim being to control soil-borne pathogens.

Integrated control programs of insects need major inputs in biological control. Intractable pests such as the bollworm (*Helicoverpa* spp.) may have to await further developments in pest management concepts. Other pests may respond to biological control especially where systematic attempts are made. Improved techniques in mass rearing, selection of suitable geographic strains of beneficials, better timing of releases and improvements in cultural practices to supply shelter and augment food supplies for natural enemies should assist such programs.

In many field crops, biological control may be more effective when host plants resistant to pests are developed, as in the successful combination of such methods to control the spotted lucerne aphid in the U.S.A.

Further information on biological control of weeds is provided in Delfosse (1985).

Genetic Engineering

Genetic engineering involving the location and cloning of genes for disease resistance and their transfer from plant to plant represents one of the most important developments in molecular plant pathology. Kerr (1987) discusses the role of *Agrobacterium tumefasciens* as a vehicle for the transfer of such resistance genes.

The commercial application of such developments is subject to regulation. The Genetic Manipulation Advisory Committee is responsible through the Minister of Administrative Services for the promulgation of guidelines for the regulation of all genetic engineering (McLean and Nicholls, 1991).

Chemical Control

Chemical control methods provide effective and often the only feasible means of crop protection. The monocultural techniques associated with most Australian cropping systems, for example, produce pest populations with the same ecological requirements as the crop. Such pests cannot readily be controlled without the use of pesticides. Chemical control, however, is of a repetitive nature and must be reapplied with each new outbreak of the pest. Nevertheless, significant savings of energy are possible as a result of pesticide use.

All methods of pest control return a high-energy output relative to energy input because of the very large yield increases usually obtained. This is most evident in areas of chemical control. Nalewaja (1974), comparing the energy relationships associated with different methods of weed control in maize, reported a 96:1 output–input ratio as a result of chemical control of weeds using the herbicide atrazine, as against a 57:1 ratio for control by cultivation plus hand-weeding.

Chemicals used as pesticides are classified according to the type of pest controlled: acaricides, fungicides, herbicides, insecticides, molluscicides, nematicides and rodenticides. Some, like DNOC (dinitro-orthocresylate), which is used as an acaricide, an insecticide and a herbicide, fall into several categories. These groups may be subdivided according to the way they achieve control—as repellents, fumigants, stomach poisons, contact poisons, residual poisons and systemic (translocated) poisons—their time of application and their mode of action. But no matter how they are classified, they are used only in one of two ways: to prevent the establishment of a pest population (called protection) or to remove (kill) all or a proportion of an existing population (cure, therapy), results which can be achieved directly or indirectly. The direct approach is seen in the application of organophosphates for the control of redlegged earth mite. A similar result can be obtained indirectly by removing alternate hosts like capeweed with herbicides, thus reducing the source of infestation.

TOXICITY

The toxicity of biologically active compounds is an assessment of the amount of pesticide required to kill or injure a living organism. It is usually expressed in one of three ways:
1. acute toxicity, or the dose which kills in a single application;
2. sub-acute toxicity, or the dose which, in repeated applications, over a short period kills or injures; or
3. chronic toxicity, or the small dose which in repeated applications over a long period causes death or injury.

Individuals vary in their response to pesticides, so that it is impossible to define a single killing dose with accuracy. To overcome this impasse the potency of pesticides is expressed as their LD_{50}, or the lethal dose required, on average, to kill 50 per cent of the test species. When comparing plant responses, however, the term ED_{50}, the equivalent dose required to reduce the growth rate of the test plants by 50 per cent, is often used instead.

SELECTIVITY

The essential feature of chemical control in the cropping situation is that the pesticide used must be selective: it must be less toxic to the crop than to the pest organism. At the same time its use in the field must not be harmful to humans, domestic animals and other beneficial life-forms. Such pesticides need not be completely harmless to the crop. All that is required is that the dose which kills or inhibits development of the pest does not seriously injure the crop. Obviously the requirements for selective activity in herbicides are somewhat more critical than those for other pesticides because the weed to be killed may be related botanically to the crop plant.

The margin of safety between crop and pest, usually termed the selective index (SI), is obtained from the ratio:

$$LD_{50} \text{ crop}/LD_{50} \text{ pest or } ED_{50} \text{ crop}/ED_{50} \text{ pest}$$

It is not necessary to confine the ratios to LD_{50} and ED_{50} values. Indeed, it is often more realistic to find the margin between pesticide dose rates giving a high degree of pest control and the dose rate causing maximum acceptable damage to the crop, e.g., ED_{20} crop/ED_{80} pest.

Determination of SIs is best done graphically using dose–response curves comparing the per cent kill or growth inhibition against the log of dose rate. Thus in Fig. 8.2, if Curve I represents the pest response and Curve II the crop response, then *a* represents the SI.

Some knowledge of these safety margins is essential because toxicity is not the sole factor determining the choice of pesticides. As Holly (1964) records, 'a highly active, moderately selective chemical is not necessarily superior to a moderately active, highly selective chemical'.

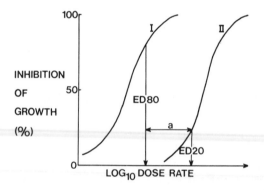

FIG. 8.2 Determination of selective indices in a crop/pest situation where Curve I represents the pest response and Curve II the crop response to pesticide application

In most species age plays an important part in determining their susceptibility to pesticides. As a general rule, young rapidly growing plants, because of their high proportion of meristematic tissue, are more susceptible to herbicides than are mature plants. Conversely, many adult insects are more susceptible to insecticides than the larval stages; in this instance usually because the insecticide cannot readily be brought into contact with the immature individual.

Age and growth rate apart, the principles used to obtain selective action are well established. They are broadly classified as physical and biochemical selectivity.

Physical selectivity   Activity in this sense is basically a lack of contact between crop and pesticide. Specific techniques are employed to bring the pesticide into more intimate contact with the pest than with the crop. These techniques include treatment of the pest before the crop is planted, special placement of the pesticide relative to the crop, using specialised equipment and the use of systemic as opposed to contact pesticides.

Retention Since the amount of chemical retained on the target species is a factor in building up a lethal dose, target morphology can be used to provide selective action. Waxy, corrugated, minutely ridged and hairy surfaces avoid injury by shedding spray droplets. This response can be modified by the addition of wetting agents or other similar adjuvants to the spray solutions. Hairy surfaces, for example, can be made to increase pesticide toxicity since, when the space between the hairs is saturated, the amount of spray retained is increased and more of the chemical is available for absorption.

Biochemical selectivity No matter how much of the pesticide is retained on the target, kill depends on a build-up of the lethal dose at the site of action. Thus selectivity may be obtained because the tolerant individual does not absorb the chemical, does not transport it to the site of action (accumulation and binding) or alters the pesticide metabolically (detoxication or activation).

Penetration Most plants and insects are covered by a non-cellular cuticle which varies in composition and sorptive capacity (Sargent, 1965; Wigglesworth, 1965; Franke, 1967). Basically the cuticle consists of a layer of cutin on and in which are embedded varying amounts of wax platelets. These waxes are hydrophobic although their water-repellent characteristics vary with their polarity (Holloway 1969). Cutins (cuticulin or chitin in insects) show both hydrophilic and lipophilic properties. As a result the cuticle reduces the rate of penetration of compounds which are soluble only in water or oil. Where applicable good pesticides should be sufficiently lipoidal to dissolve the cuticular waxes and yet have a low lipoid–water partition coefficient allowing movement into the internal aqueous phase of the organism (Hadaway, 1971).

Accumulation and binding Poisoning may also be avoided because there are barriers to the movement of the pesticide in the translocation system (O'Brien, 1967a, 1967b; Olunga *et al.*, 1977). In all cases the source–sink relationship is important to movement. Thus poisoning may be avoided because the pesticide is sidetracked into inert storage tissues (Aamisep, 1961; O'Brien, 1967a; Knuesli *et al.*, 1969); or, if translocation rate is limited by accumulation in surface cells as a result of physical or chemical conjugation with cell contents (Brian, 1960).

Enzyme systems Pesticides usually kill by inhibiting enzyme systems catalysing specific reactions (Klepper, 1974). Thus selective action is obtained when the particular enzyme system inhibited is not present, or where the critical growth factor can be synthesized by a number of alternative pathways in the crop. The fungicide Polyoxin D, for example, kills fungi by inhibiting an enzyme used in the synthesis of chitin, a component of the fungal cell wall and the insect cuticle, which is not present in higher plants and vertebrate animals (Corbett, 1974).

Degradation and activation Most toxicants are degraded in the host species by a group of enzymes known as the microsomal mixed function oxidases, which occur in mammals, insects and plants. Selective action is obtained because the rate of degradation varies from species to species; in some, degradation is rapid. In others, the change takes place slowly or not at all. In general, the process is one of detoxication but, occasionally, a relatively non-toxic compound is converted to a pesticidally active metabolite (Wain, 1953; Hodgson and Plapp, 1970; Busvine and Feroz, 1972).

Pesticide structure   There is a close relationship between chemical structure and selectivity. Each major group of pesticides, despite some overlapping in the carbamates and the substituted phenols particularly, provides a generalised type of selective action between pest types and crop. But, within each group small structural changes in the pesticide molecule have a marked effect on their activity and selectivity (Holly, 1964; Winteringham, 1969; Knowles *et al.*, 1972). This type of response is probably most clearly illustrated by referring to the symmetrical triazine herbicides. These are six-membered heterocyclic compounds with three annular nitrogen atoms at the 1, 3 and 5 positions in the ring and substituent groups appended to the alternating carbon atoms.

$$
\begin{array}{c}
R_1 \\
| \\
C \\
(2) \\
N(1) \qquad (3)N \\
\| \qquad\qquad | \\
R_3{-}C(6) \qquad (4)C{-}R_2 \\
(5) \\
N
\end{array}
$$

The most reactive group is usually designated to the 2-position (R_1) of the ring.

Gysin (1960) and McWhorter and Holstun (1961) have considered selectivity changes resulting from variation in molecular structure in some detail. Briefly the symmetrical chlorotriazines typified by simazine (2-chloro-4,6-bis-ethylamino-s-triazine) are highly selective in maize, asparagus, sugar cane, grapes and some other species. Increasing the alkylamino chain length at the 4- and 6-positions to three carbon atoms varies the selectivity. Propazine (2-chloro-4,6-bis-isopropylamino-s-triazine) for example, is also selective in sorghum, carrots and celery which are usually killed by simazine. Further, if the chlorine atom at the 2-position is replaced by an alkoxy (OCH_3) or an alkylmercapto (SCH_3) group, the selectivity in maize and grapes particularly is lost immediately. With the alkylmercapto group at the 2-position, especially with the compound 2-methylthio-4,6 bis-isopropylamino-s-triazine (prometryne), selectivity is extended to wheat and barley, species normally sensitive to the triazine herbicides.

Similar changes in selectivity occur in other pesticide groups.

Simazine

Propazine

METHODS OF APPLICATION

Formulations In order to obtain a uniform distribution of the small dose rates used, modern pesticides are applied in one of three ways:

1. as liquid drops using water or oil as a diluent (sprays);
2. as liquid drops using air as a diluent (mists); or
3. absorbed on to inert solid carriers as dusts or granules.

To make this possible they are formulated as solutions, emulsions, wettable powders, dusts or granules.

Salts of most pesticides readily form solutions, physically homogeneous mixtures of substances which cannot be separated mechanically. Some dissociation of the solute molecules always occurs in solutions. As a result they provide problems when used with 'hard' water; filters and nozzles often blocking if insoluble calcium and magnesium precipitates form.

This response is usually overcome by formulating the pesticide as an oil-soluble compound which can be used as an emulsifiable concentrate. These concentrates are applied as emulsions—one liquid dispersed in another but retaining its original identity. The emulsifying agents used, preventing direct contact between the two liquids, stop the formation of precipitates. Emulsions are of two types:

1. oil-in-water low-viscosity mixtures; and
2. water-in-oil high-viscosity mixtures which are usually called invert emulsions.

These invert emulsions, in which viscosity increases as the water-to-oil ratio decreases, require specialised application equipment. However, they are often used to minimise spray drift problems in environmentally sensitive areas.

Pesticides which cannot readily be made soluble in water or oils are very finely ground and dispersed in water as suspensions. Surfactants are included in the pack to aid dispersion and minimise 'settling'. Nevertheless, continuous agitation of the spray mixture is required to maintain the suspension when using these formulations, and nozzle wear is faster than with solutions or emulsions.

Where the dose rate of a pesticide is so high as to make spray application impractical or, as sometimes happens, low solubility makes the preparation of other formulations difficult, the pesticide is often absorbed into solid, inert materials to be applied as finely ground dusts or granules. The carriers used are mainly clays and sands, but occasionally dried plant materials are employed. Granular application has a number of advantages, including the nil requirement of water and costly spraying equipment and the lack of retention of the pesticide on crop plants. The major disadvantages of dusts and granules are their bulk, which increases freight costs, and the uneven coverage of the target area which usually occurs.

Spray application   Sprays are the commonest method of pesticide application. They give a relatively even coverage of the pesticide over a wide range of application volumes. The type of spray equipment used, however, has a marked effect on efficiency and selectivity. Spray coverage is a function of volume applied and the size of the spray droplets. For any given droplet size the cover obtained is proportional to the volume applied until the target surface is completely covered. After the limit is reached, any additional spray applied runs off the target surface. Conversely, for any given volume, cover is inversely proportional to the droplet size. Droplets with a diameter less than 200 μm, however, tend to drift in the air and lack the inertia necessary to overcome air streamlines. Spray clouds usually consist of a large spectrum of droplet sizes and the cover obtained varies with the environmental conditions pertaining at the time of application. See Metcalf and Flint (1962) for a more detailed explanation of the engineering aspects of spray production.

High, medium and low volume application   Volume requirements vary with the project. Economically one should use the lowest volume commensurate with an effective result—death of insect, fungus, weed or whatever. High-volume application at 300 to 6000 litres per ha gives a cover of coalescing droplets or a continuous film of solution over the target surface, with some 'runoff'. This is often desirable in the application of protectant materials or contact herbicides. Medium volume is when the liquid is sprayed over the surface of the target until some coalescence of droplets occurs. The normal range varies from 100 to 500 litres per hectare. Low-volume application—volumes from 5 to 50 litres per ha— gives a cover of discrete droplets with a wide range of sizes. Currently it is the most used method of application.

Ultra-low-volume (ULV) application   The economics of aerial application and the need to reduce the amount of chemical used because of residue problems resulted in the development of ULV application techniques. With application volumes ranging from 5 to 10 litres per ha the deposit on target consists of numerous discrete droplets. Coverage and distribution is improved by using application equipment that produces a narrow range of the smaller droplet sizes formed in low and high-volume methods. Drift and strip effects, which often result from aerial application, are overcome by spraying in crosswinds which, by causing turbulence, trap the finer droplets within the crop canopy.

Occasionally, when spraying close to sensitive areas, lateral drift is further reduced by using larger droplet sizes and appropriate adjuvants.

The actual volume applied and the diluent used vary with the type of pesticide. In general, effective herbicide response seems to require higher application volumes than either insecticide or fungicide.

Evaporation problems with small droplets resulted in the replacement of water by oils as the diluent. These solvents themselves, however, influence the effectiveness of the pesticide dose rate, affecting movement over the cuticle and absorption through it. For example, a persistent solvent will modify both the contact and residual toxicities of insecticides sprayed on to plant surfaces. Barlow and Hadaway (1974) indicated that there is a conflict between solvent properties that promote spreading on and absorption through the cuticle, and those that promote persistence of the pesticide. They suggested a search for additives that will alter the viscosity of the higher boiling point solvents (300°C or more) which provide persistence but limit spread and absorption.

Mists Air-blast sprayers are designed basically to inject small droplets of spray material through nozzles into an air stream from a powerful fan that carries the spray to the target area. Pesticides may be made up as dilute semiconcentrate or concentrate sprays, and so only relatively small volumes of water are required in spraying operations. Application rates of spray per hectare vary depending on the intended use but may be as little as a few millilitres when utilising ULV equipment.

Additives Substances which are added to pesticides to improve their action are often termed 'adjuvants'. They are used to assist emulsification, increase the wetting power of the solution, promote spray retention and penetration, and increase the toxicity of the pesticide. Such additives include wetting and spreading agents, emulsifiers, thickeners, stickers, penetrating agents which assist absorption by solubilising cuticle waxes and lipoids, dispersing agents and activators.

The wetting and spreading agents and many emulsifiers belong to a group of compounds which reduce interface tensions. As a result they may be referred to as surface-active agents, or surfactants. Activating agents are substances which, though not toxic themselves, improve the kill of the pesticide to which they are added.

Spraying equipment This consists essentially of a tank to hold the spray solution, a pump to force the spray through the nozzles which, in turn, form the spray droplets and direct them onto the target. Tanks must be corrosion-resistant and include some form of agitation. Hydraulic agitation is obtained by returning some pressurised solution directly back to the holding tank and is satisfactory when used with soluble and some emulsified materials. Mechanical agitation, obtained from power-driven paddles in the spray tank, is necessary when using oily emulsions and wettable powders.

There are four main types of pump in use: piston, centrifugal, gear and impeller pumps. The piston pump is probably the best and most versatile pump available, but it is expensive. Gear pumps are satisfactory in most situations and are cheap. Their major disadvantage is a limited working life when abrasive sprays like wettable powders are used.

Nozzles which transform the spray solution into droplets are of three basic types: flat-fan, cone and flooding. The flat-fan nozzle, which produces an evenly distributed spray, is most commonly used. The hollow-cone nozzle produces a cone-shaped spray with most droplets at the outer edge of the cone. Droplet distribution is not as even as with the flat-fan nozzle but is more efficient than the latter at low volume application (20 to 30 litres per ha). The flooding nozzle produces very coarse droplets which minimise drift.

Modern spray requirements, however, demand a narrow range of droplet sizes. Controlled droplet application (CDA) (Combellack, 1978) cannot be obtained with the nozzle types described, and therefore other means of droplet production like rotary atomisers (spinning discs and cages) and ultrasonic atomisers (vibrajet) are under investigation.

There is a growing awareness of the importance of technology of spray application to avoid waste and to maximise biological efficiency of pesticides. Droplet size, spatial distribution and nature of the target surface, biology of the target organism and weather conditions are all important interacting factors which influence the success of droplet deposition and retention and, in turn, how effective and reliable the pesticide is. Wind turbulence and very small droplets, less than 100 μm for example, are required for penetration of deep crops and deposition on otherwise inaccessible sites.

Dusts   These are finely ground crystalline materials mixed with inert solid carriers. They are used mainly as insecticides and fungicides as there is a serious drift problem where herbicides are concerned. Anti-caking compounds may be included in the formulation.

Seed dressings   One of the more common uses for dusts is the protection of germinating seeds from attack by fungi like *Tilletia caries* (bunt of wheat) which is present as an external seed contaminant. Seed is dusted before sowing. The older mineral dusts have been replaced by organic compounds with a much lower bulk density. As a result, effective protection may require a smaller weight of material than is usually considered necessary (J. Kuiper, personal communication).

Fumigants   Fumigants are volatile liquids which produce toxic vapours. They include compounds like methyl bromide, carbon disulfide, chloropicrin, ethylene dibromide and some other similar substances. They are of limited general use because of their volatility and they require careful preparation of enclosed space. They find considerable use, however, as sterilants in nursery beds, eliminating fungi, insects and weeds. In such situations their effectiveness depends on temperature and soil characteristics; overmoist or compacted soils, dense plough soles and hardpans restrict vapour movement and hence kill.

Granules   These are formulations containing up to 50 per cent of the pesticide in solid carriers. Materials used as carriers range from perlite, through diatomaceous earths, clay minerals and pyrophyllite to ground organic materials. They are expensive to make and transport because of their bulk. Their major advantage is their complete absence of drift, lack of foliar retention and easy application. New developments in the field of microgranules with diameters in the range of the smallest spray droplets may change this. Spherical micro-granules do not stick but disc and rod structures, particularly if partially deliquescent, are readily adherent.

PESTICIDES AND THE ENVIRONMENT

Farming activities upset the natural biological equilibrium by encouraging overpopulation of a single species—the crop. This artificially maintained, unstable habitat, in turn, fosters the development of adaptive species, which the farmer regulates by further protective measures including the use of pesticides.

Unfortunately most pesticides are not specific in their action and damage 'beneficials' as readily as pest organisms. In some instances because of a strong persistence (e.g., the organochlorine insecticides and mercurial fungicides) the compound passes from organism to organism in food with an apparent amplification of the residues, particularly in predators. Recent studies indicate that these trends are not necessarily due to efficient accumulation from the food, as suggested by Carson (1962), but can be the result of direct absorption through the skin and selective predation (Walker, 1976).

Very few chemicals, however, have such an acute toxicity that they are dangerous at the rates used in the field. On the other hand, long-term exposure may present chronic toxicity problems. The major risk of poisoning in this situation in order of decreasing danger is:

1. to the operator producing or applying the chemical;
2. to the treated animal or to the animal who eats treated fodder;
3. to the consumer who eats foods prepared from treated livestock or crops; and
4. to wildlife in general.

Because of these factors, most governments now demand long-term experiments covering the response of daily intakes of pesticides in order that acceptable residue tolerances can be established. In consequence of these strictures 'not a single case is recorded which may lead to the conclusion that residues of these substances have affected the health of any consumer or other person exposed to traces of these compounds or their breakdown products' (Anon., 1972) in nearly thirty years of intensive use in practical agricultural situations. Indeed this author goes on to say 'pesticides have contributed very considerably to the high quality of food which is enjoyed today .... and they have saved countless lives especially in tropical countries'.

Pesticides are tools devised to overcome practical farming problems. Consequently an inbuilt persistence is of immense practical value though

prolonged persistence may have serious repercussions. The major problems of persistence are potential damage to non-target organisms,
potential damage to the complex micro-organism population of the soil
necessary for the maintenance of fertility, and movement out of the target
area. Fortunately the degree of persistence of any one pesticide is not
absolute; it is a function of quantity applied, climatic variation and target
characteristics. Thus persistence depends on the rate of loss from, or
breakdown in, the target area. Such loss may occur as leaching, run-off,
evaporation, removal in resistant organisms and chemical or micro-
biological breakdown.

Overcoming the pollution problem Because of the absorptive and
degradation processes inherent in the soil complex, and assuming that
the application techniques employed minimised drift, few except the
most persistent pesticides are likely to move out of the target area. Thus
the major problem is the potential build-up of toxic residues within the
target area. Where complete degradation does not occur within the
season this is a very real danger, especially where application of the one
pesticide is made season after season. This danger can be minimised by:
1. adopting crop rotations which include wholly resistant species or
   cultivars;
2. adjusting pesticide dose rates to the average degradation rate; or
3. rotating the pesticides employed.

Under current conditions the first alternative is restrictive and somewhat
unreal, the second attainable, possibly at the expense of efficacy, and the
third wholly practicable. Rotation of pesticides is also an important
component of the strategy to minimise the build-up of pesticide resistance.

Used in accordance with the instructions given by the manufacturer,
residues are likely to be minimal at any time. Thus, as there are several
pesticides available to deal with each potential pest problem, residues can
also be minimised by changing the pesticide type applied season by
season.

Pesticides and ecological considerations Destruction of beneficials
(enemies and interspecific competitors) by broad-spectrum type pesticides
frequently allows a rapid and uninhibited reinfestation by a pest species,
and occasionally raises the status of hitherto comparatively unimportant
pests. Changing pest status can also result from ecological succession;
suppression of one pest leaves the way open for development of a
previously minor one, or a more noxious new pest to fill the vacant
ecological niche.

Continued use of specific pesticides often results in the evolutionary
selection and development of resistant individuals in a pest population, a
phenomenon well established and of considerable significance amongst
insect pests and an increasingly evident one amongst fungal pathogens
(Kuiper, 1965; Rippen and Wild, 1976; Penrose, 1977). Even weed
populations can develop resistance against certain herbicides (Nalewaja,
1978; Heap and Knight, 1982). The nature of the chemical and its specific
mode of action, dose rate, persistence and frequency of application as

well as the genetic variability of the pest population are all factors contributing to the development of resistance.

Apart from considerations of cost and pollution, there are thus strong ecological grounds for the judicious use of pesticides, particularly those that are highly persistent or of a broad-spectrum nature, or whose mode of action incites rapid build-up of resistant individuals.

Chemicals with greater selectivity and with just sufficient persistence appropriate for the effective control of the target pest species are required. However, more stringent goverment regulatory restrictions, the increasing difficulty of finding new types of compounds with high biological activity, and the development of resistant pest strains which limits the effective economic life of the chemical, all contribute to a significant increase in cost which is passed on to the farmer. Some answers may lie in the discovery, characterisation and development of naturally occurring compounds that are active against pests, as evidenced by the development and use of the synthetic pyrethroids such as bioresmethrin for insect pest control.

Protective chemicals also operate in natural plant disease resistance, being either present in disease-free plants (numerous phenolic compounds), produced in plant tissue in response to infection (phytoalexins such as the chromanocoumaran, pisatin) or produced by micro-organisms (the antibacterial antibiotic, cycloheximide, produced by *Streptomyces griseus*).

Wain (1969) discusses these and others with regard to their possible role in plant disease resistance and their synthesis as agricultural fungicides.

The reader is also directed to a comprehensive review on the modern trends in the use of natural products for controlling insect pests and diseases (Marini-Bettolo, 1978).

PESTICIDE REGULATION

The discriminant and judicious use of pesticides by good agricultural practice and common sense cannot always be relied upon, so there will always be a need to impose legal restrictions to protect people and the environment from the harmful effects of pesticides.

Various state acts have sought to regulate the introduction of new pesticides by imposing restrictions involving acceptable toxicities, persistence and efficacy data to support registration of a product before it could be marketed.

More stringent regulations were introduced in New South Wales when the old Pest Destroyers Act of 1945 was replaced by the Pesticides Act 1978, the objectives which are to:
  1. provide for the registration of pesticides, the approval of containers (for registered pesticides) and the registration of labels;
  2. control the sale, supply, use and possession of pesticides; and
  3. prevent certain foodstuffs containing prohibited pesticide residues from becoming available for consumption or export.

Education in the proper use of pesticides remains of paramount importance in the protection of the environment and human and animal health.
Significant efforts in this direction include the national pesticides users
short course supported jointly by the Agricultural and Veterinary
Chemicals Association, state Departments of Agriculture and a number
of educational institutions across Australia.

Integrated Control (IC) and Integrated Pest Management (IPM)

It was largely from ecological considerations that the concept of integrated
control was first conceived. Initially the concept combined the use of
pesticides and natural enemies in a compatible manner. The modern
approach advocates the integration of several control measures in a
unified program referred to as integrated control or integrated systems of
pest management and control. The concept is defined by the Food and
Agriculture Organization as 'a pest management system that in the
context of the associated environment and the population dynamics of
the pest species, utilises all suitable techniques and methods in as
compatible a manner as possible and maintains the population at levels
below those causing economic injury'. The determination of levels of
tolerable pest damage is thus an essential prerequisite to the development
of integrated control programs. Smith (1968) states that these threshold
levels should be determined both in terms of foreseeable crop loss and the
economics of crop production and marketing.

This modern concept embraces two terms, IC and IPM, which are
sometimes used interchangeably but between which the authors prefer to
draw a distinction in common with a number of other workers. *IC* refers
to a single target pest organism, whereas *IPM* refers to a pest complex.
This distinction is further elaborated in the following section, 'Strategies
for Solving Pest Problems'. Both approaches demand a thorough
understanding of the ecology and dynamics of pest populations.

The more complex IPM programs in particular require an understanding of the agro-ecosystem, and further research is required before
IPM practices can be elevated beyond their present empirical level.

The recognition of key pests (those against which control measures are
essential if economic production is to be maintained), however, reduces
the number of pests of immediate concern. In the meantime, IC and IPM
programs can be based on existing knowledge (Anon. 1978; Apple and
Smith, 1976).

Strategies for Solving Pest Problems

Approaches   Pest control strategies should form an integral part of farm
management decision making. Account should be taken of interactions
with the environment, economics and sociological factors. A more
integrated approach to solving pest problems is needed and much is to be

gained by the co-operation between basic and applied scientists and industry (Banyer, 1985).

Although under the influence of increasing competition for research and development funds and greater accountability, there is evidence of increasing co-operation between traditionally distant research institutions. For example, CSIRO, universities and state Departments of Agriculture have co-operative programs with industry with more emphasis on strategic mission and applied research (SIRATAC, SIRONEM, SIRAGCROP).

Approaches to solving pest problems fall into three broad categories:

  (i) the longer-term approach leading to a more permanent, environmentally sound solution, e.g. IPM programs (multiple controls of pest complexes, economic injury levels and action thresholds);

 (ii) shorter-term, more immediately achievable programs. These are still economically and environmentally based but aimed at integrated control of a single target pest (IC). Simple models are used which are more easily validated and more likely to receive better producer acceptance than IPM programs. Examples of successful programs are associated with citrus red scale, two-spotted mite in pome fruit, stripe rust of wheat and various pest forecasting services;

(iii) the short-term 'silver-bullet' approach or kneejerk response to sudden disasters. This approach epitomises the 'technological cure' mentality. It commonly relies on pesticides as the only method of control. Regrettably, experience has shown that disasters do occur and it is sometimes necessary to provide emergency short-term economic relief (e.g. use of chemicals to control stripe rust of wheat when first introduced).

There is a need for all three approaches. However, careful deliberation is required by funding authorities and policy makers so that they are not swayed by the attraction of short-term gains at the expense of longer-term research efforts. The latter are more likely to reveal strategies of improved, long-term pest management with less harm to the environment. Industry funding authorities will need to pay particular attention to this aspect.

Genetic diversity The increased genetic homogeneity of crop plants grown in monoculture systems increases the risk of devastating epidemics (Marshall, 1977).

In 1970 the widespread epidemic of the southern corn leaf blight disease rudely awakened scientists in the U.S.A. about the genetic vulnerability of their crops. Many blamed the pathogen (*Helminthosporium maydis* race T) but the real cause was the excessive homogeneity of the nation's corn crop. This prompted Ullstrup (1972) to state, 'Never again should a major cultivated species be moulded into such uniformity that it is so universally vulnerable to attack by a pathogen, insect or environmental stress. Diversity must be maintained.'

Breeders must continually strive to develop or maintain a wide genetic base in all major crops. Monitoring genetic diversity and the preservation of associated germ plasm of cultivated and wild species are national responsibilities, e.g. the national wheat seed collection, Tamworth, and the eight crop and pasture seed centres that are planned.

Production systems   Production systems should be planned to minimise the risk of epidemics. Large-scale, single-crop enterprises should be discouraged. On the other hand, the integration of cropping and animal production systems should be encouraged as good agricultural practice. This is supported by a Standing Committee of Agriculture (SCA) task force review (Reeves *et al.*, 1985).

Monitoring changing production systems for their effect on changing 'pest' status is also necessary. Conservation farming practices such as minimum tillage and stubble retention have been shown to enhance the activity of certain pathogens (Rovira and Venn, 1983), insect pests (Allen, 1982) and weed problems (Le Baron and Gressel, 1982) whilst restricting others.

Pest Control in Cotton

The following account is not intended as a comprehensive blueprint for the control of pests in cotton, nor indeed can it be because of changing strategies based on newly acquired knowledge, changing pest status, management practices and new chemicals.

Cotton production, however, illustrates both the complexities of the management involved in plant pest control and the feasibility of developing integrated control strategies. It also illustrates how an industry can be placed in jeopardy by the failure to observe basic principles of crop protection.

THE COTTON PLANT AND GENERAL AGRONOMY

Most cotton varieties grown in Australia belong to the species *Gossypium hirsutum*. However, a small industry has developed with *Gossypium barbadense* (Pima cotton). Normally a perennial shrub, cultivated cotton bred for early maturity is grown as an annual in the temperate climatic zone where the major determinants of yield are solar radiation, nitrogen and water supplies and pest and disease control. Irrigated cotton does best on moderately fertile soils of good structure over a well-drained subsoil. The main stem produces a new node every two or three days. Flower buds (squares) are produced on lateral fruiting branches, the first of which appears at the fifth to eighth mainstem node; thereafter each subsequent node subtends a branch. Node and square production continue as long as conditions are favourable. The number of bolls that the plant can carry is limited by competition for carbohydrates and nitrogen. The crop has some compensatory powers such that if young bolls are shed, due to, for example, pest damage, the production of squares is maintained at a higher rate provided other factors such as water supply and temperature are not limiting.

FACTORS PREDISPOSING COTTON TO INCREASED DAMAGE BY PESTS

Prior to the 1960s, cotton was grown in Australia on a small scale chiefly under dryland conditions. Markedly fluctuating yields and quality of the harvested lint characterised the industry because of the uneven rainfall patterns and, to some degree, spasmodic outbreaks of pests. Dryland cotton was replaced by irrigated production which generally improved yields and overall quality. With irrigation, however, units were cropped intensively. The cotton–cotton–cotton–fallow system of monoculture quickly compounded pest problems which now account for much of the variation in yield. In the Ord River region of Western Australia, insect pressures forced the over-reliance on and sometimes indiscriminate use of insecticides, particularly the broad-spectrum, highly residual types such as DDT, practices which led to the development of resistance in bollworm (*Helicoverpa armigera*). The activity of insect pests in the Ord escalated to the extent that the cost of their control was prohibitive, and cotton is now no longer considered an economic proposition in that region (Davidson, 1974).

A similar situation threatened the Namoi Valley in northern New South Wales, where in the 1972–3 season the crop was severely damaged by bollworm which showed resistance to DDT, as applications had been double that normally used.

In 1983–4, resistance to pyrethroids in *H. armigera* led to the development of an insecticide use strategy for cotton and other field crops in northern New South Wales and Queensland. The basis of the strategy is to restrict the use of pyrethroids to a maximum of three sprays against one *H. armigera* generation only. This restricts the use of pyrethroids in New South Wales from 10 January to 13 February in cotton and from 10 January to 20 February in other summer crops. The strategy, generally well accepted by users of pyrethroids, was developed jointly by industry, the New South Wales Department of Agriculture, CSIRO and Queensland Department of Primary Industries. Resistance levels are monitored by the New South Wales Department of Agriculture.

SEQUENTIAL PROGRAM OF PEST MANAGEMENT AND CONTROL

Quarantine Strict quarantine regulations aim to prevent the introduction into Australia of a number of potentially important pests. These apply particularly to those that are seed-borne, as about half the known diseases of cotton as well as some insect pests such as the pink bollworm (*Pectinophora gossipiella*) are known to be spread by infected or con-taminated seed. Cotton boll weevil (*Anthonomus grandis*) is especially feared because of its devastating effects wherever it occurs (Morschel, 1971).

Crop rotations The cyclic activities of several important pests must be broken by introducing alternative, resistant or non-preferred host crops into the rotation. Where verticillium wilt (*Verticillium dahliae*) is a problem, a cropping strategy, which combines the rotation of cotton with non-host crops such as legumes or cereals, early incorporation of crop residues after the harvest of a susceptible crop and good control of alternative weed hosts, will reduce the build-up of the disease.

The use of cultivars with resistance or tolerance to verticillium wilt would reduce losses but would not altogether obviate the need for crop rotation and sanitation. Effective rotations require good sanitation practices. Many of the weeds along fence lines and irrigation channels harbour insect pests and diseases and constitute a continuing source of infection, e.g., noogoora burr and castor oil plant. Rotation as a means of pest control, however, is limited, particularly against pests which possess a wide host range. *Helicoverpa* spp., for example, can attack summer crops such as maize, sorghum, linseed, safflower, sunflower and tobacco as well as a number of vegetable and pasture species including lucerne.

The use of winter cereals (wheat, oats and barley) as cash crops to coincide with the resting (diapause) phase of the insect has limited application because *Helicoverpa* can have a high incidence in wheat especially in the warmer months of September, October and November. If the grain ripening period of wheat is cut short by hot dry weather, the cross transmission of *Helicoverpa* moths from wheat to cotton may be very high. Irrigated wheat is more susceptible to *Helicoverpa*, especially *H. armigera*, due to higher plant density, greater succulence in plant tissue and softer ground for larval stages to burrow into and pupate.

The cotton phase in the rotation may be from three to five years, depending on soil type and the rapidity with which major pests increase to economic proportions. Where practised, rotations are one or two years of cotton followed by one year of wheat, fallow, or soybeans.

The economic practicalities of rotations could be questioned in view of the restricted areas for growing cotton under irrigation and the high capital investment in specialised equipment. Nevertheless, early mistakes of continuous monoculture are encouraging the cotton grower to adopt rotations regardless of present pest problems, rather than rotate crops only when necessary due to pest and disease incidence. Importantly there are also other benefits conferred by rotations, such as improved soil nutrient status and build-up of organic matter. Phosphorus, for example, is usually applied solely in the wheat phase of a two-year cotton, one-year wheat rotation at rates of 18 to 24 kg of P per hectare. A third benefit is that rotations with cereals help to control broad-leaved weeds such as thorn apple (*Datura* spp.), noogoora burr (*Xanthium pungens*) and deadly nightshade (*Solanum nigrum*). A further benefit of rotation is to avoid or ameliorate soil structure degradation. Unirrigated wheat allows the soil to dry out, allowing more time for land preparation for cotton.

Choice of cultivars and origin of seed   Cultivars are chosen on the basis of their agronomic suitability to particular localities and whether they possess any worthwhile resistance to pests. Cultivars as they are developed for resistance to verticillium wilt and bacterial blight, for example, should be used where these diseases are a problem. Siokra, which was bred by CSIRO, possesses good resistance to bacterial blight.

In Australia, almost all planting seed is produced and marketed by Cotton Seed Distributors, a grower-controlled organisation based in Wee Waa. Distribution of seed is arranged through the ginning organisations located in the growing areas. Pioneer Hi-bred Australia Pty Ltd and Delta Pineland Seed Company have combined to market Deltapine varieties.

Some operations involved in pure seed production either aid or are designed to reduce disease and insect pest problems. Delinting controls external contamination by bacterial blight (*Xanthomonas campestris* pv. *malvacearum*).

New transgenic cultivars may become available with inbuilt protection against *Helicoverpa* and other moth pests. In New South Wales, mother cotton seed for planting is supplied to grower-organised seed associations. Following build-up under supervision, the Namoi Cotton Seed Distributors Association carries out seed preparation in its own acid delinting plant. Once delinted, pure seed is graded, pickled, and held against orders from growers for the following season. Seed pickling with a fungicide is carried out principally to provide early protection against organisms causing damping-off, in particular *Pythium* spp. It also affords some control of bacterial blight (*Xanthomonas malvacearum*) as a contaminant on seed which has been mechanically delinted. Acid delinting controls external contaminations but not deep-seated infections. The development of resistant cultivars and a seed certification scheme to provide seed free of bacterial blight are feasible approaches to control.

Post-harvest sanitation Land preparation begins immediately after harvest when trash and crop residues are thoroughly slashed and buried by deep ploughing. This destroys the breeding grounds of the rough bollworm and reduces its survival during the off seasons. These practices also facilitate cultural operations, hasten decomposition of crop residues and help reduce primary sources of inoculum of organisms like *X. malvacearum, V. dahliae* and *Fusarium* spp. Deep cultivation also encourages the breakdown of herbicide residues.

Tillage and weed control Tillage in preparation for sowing is primarily aimed at controlling weeds and developing a suitable seedbed, but at the same time nutrients are made available from plant residue breakdown products. Weed hygiene is important throughout the year and particular attention needs to be paid to controlling weeds, including volunteer cotton, known to harbour other pests. The level of rough bollworm, for example, is directly related to the abundance of bladder ketmia (*Hibiscus trionum*) which, if allowed to grow unchecked in laneways, channel banks and headlands, results in massive reinfestations in neighbouring cotton. Cultivations usually commence with the last autumn rains. The number and timing of subsequent cultivations is aimed at weed control with the least amount of damage to soil structure.

Germination of dormant weed seed can be encouraged by irrigation or tillage and subsequently killed. Roots and rhizomes of shallow-rooted perennials may be cultivated to the surface and killed by drying or by toxic chemicals in the topsoil. Control by herbicides generally requires deep penetration into the soil or into the root system. However, trifluralin is applied pre-planting and incorporated with cultivators; paraquat or amitrole is sometimes used as a knock-down desiccant when bad weed problems occur pre-planting. Fluometuron may be applied to the surface in a separate operation immediately following the planters.

Weed control during the winter may be achieved by residual herbicides, which have the advantage of reducing the number of cultivations. They are also more efficient if perennial weeds have become established. Minimum tillage techniques now being used save on fuel costs and reduce soil structure degradation.

Land planing   Grading kills most of the weeds which remain, as the soil is worked at least 30 cm deep to facilitate hilling-up. Weeds may still establish from seeds thereafter; rhizomatous and stoloniferous weeds, if not deeply buried, may be spread by cultivation. Established weeds are difficult to control by herbicides, which are regarded as a supplement to other methods of control. Application of trifluralin at rates of 3 to 4 litres per hectare prior to planting is necessary for annual grass control.

Pre-planting irrigation   The main purposes of this practice are twofold. First, a heavy irrigation applied two to three weeks before sowing ensures adequate soil moisture for germination of cotton; second, it provides a weed-free seedbed following cultivation of germinating weeds. The damaging effects of damping-off fungi may be reduced by avoiding temporary waterlogging and poor aeration at seeding. In addition, it is believed that pre-planting flooding reduces the viability of *X. malvacearum* in infected plant debris and also helps in the control of *Helicoverpa* spp. by destroying the pupal state in soil, in addition to cultivation.

Planting   Spring sowing is delayed to avoid cold, wet conditions which are conducive to the development of verticillium wilt and encourage the damping-off disorder. Sowing too deeply delays emergence and weakens seedling vigour, contributing to poor establishment through damping-off. Stands sown too densely may increase relative humidity to levels favouring the development of fungal rots especially where insect damage has occurred.

Seeds are treated with insecticidal dusts for protection against soil-borne pests such as wireworm (*Coleoptera elateridae*) or alternatively insecticides may be incorporated in soil with the fertiliser. Compatibility of different seed dressings must be assured and correct rates applied to avoid any interference with germination which would favour the condition of damping-off. Immediately after planting, a narrow band of a pre-emergence herbicide like fluometuron is applied directly over the plant row.

Rates of application of nitrogenous fertiliser vary according to irrigation regimes and field history. Usually 95 to 130 kg of nitrogen per hectare are applied on cotton following one year of cotton while 120 to 150 kg per hectare is applied following two preceding years of cotton. For cotton following wheat with a six-month fallow period prior to planting, rates of nitrogen are as low as 50 kg per hectare. Over-application causes problems later in the season; crop development is delayed and plant metabolism is still active when leaf senescence and abscission are required. Rank growth from too much nitrogen reduces micronaire (percentage of fibre maturing) and favours a number of fungal diseases associated with boll rot.

Post-planting weed control   After planting, early weed control can be achieved by inter-row cultivation which causes only light damage to roots of cotton if carried out carefully. This practice also allows application of insecticides for early thrip and jassid damage. On the 'red' clay soils, inter-row cultivation is regarded as the rule rather than the exception for better infiltration of irrigation water.

By early December the banded pre-emergence herbicide will be broken down and its activity lost. This coincides with the stage of cotton growth too big to permit inter-row cultivation by tractor. There is still a need to control weeds through until February, and this can be achieved by hand hoeing. An alternative approach, however, is the use of 'lay-by' herbicides, which are basically pre-emergence herbicides (such as diuron, fluometuron and prometryne) which are applied at or soon after the final inter-row cultivation to provide late-season weed control. Application of herbicides by ground rig and aerial means is increasing. Fluometuron and mono-sodium methyl arsenate are only occasionally aerially applied at cotton tolerance levels.

Spot spraying may be necessary to kill localised patches of Johnson grass (*Sorghum halepense*) or couch grass (*Cynodon dactylon*) in the crop.

Irrigation   Irrigation is usually by furrow because of cost and effectiveness but has the added advantage of avoiding spread of bacterial blight by splash droplets from overhead irrigation. Excessive supplementary watering by any means is to be avoided, as it provides lush vegetative growth which further encourages activities of insect pests and diseases because pest incidence is highly related to vegetative growth.

Post-planting insect pest control   The importance of various insect pests and their effect on yield is closely related to the period of infestation, stage of plant development and seasonal factors. Cotton provides a varied and ideal source of food from planting to harvest for its many insect pests. These vary in their behavioural patterns, making control difficult. The whole insect complex must be considered because control of one pest may give little or no economic benefit because more damage may be caused by others.

Protection from devastation by insect pests is uppermost in the minds of growers because of the very high capital investment in the crop, and regular scheduling of sprays has been a common practice. This must be actively discouraged not only from a doubtful cost–benefit viewpoint but, more importantly, because of harmful effects to beneficial predators and parasites, as well as inducing insect resistance.

Resistance to synthetic pyrethroid insecticides developed in *H. armigera* in 1983 (Gunning *et al.*, 1984). The nature of resistance mechanisms has been described by Gunning *et al.* (1991) and mode of inheritance by Daly and Fisk (1992). Lower levels of resistance to the organochlorine endosulfan have also developed. To protect the useful life of the pyrethroid insecticides an 'insecticide resistance management strategy' was implemented in 1983. This strategy involves a 'window' of restricted use of pyrethroids around which other pesticide groups are used (Shaw, 1991). The strategy is a dynamic one, being continuously updated and modified in the light of resistance changes (Forrester, 1992). One of the main problems in using any insecticides is to decide when to begin and when to cease applications. Regular monitoring of insect numbers is required in an attempt to determine levels likely to go beyond the economic threshold, and to relate this to the critical period during which the crop must be protected. The decision to cease spraying is difficult because of the variability between seasons, but it usually continues until defoliation.

Attacks by thrips, aphids and jassids in the seedling and pre-squaring stages are more likely to be damaging if crop growth is relatively slow. Probably the most critical period for the crop is about twelve weeks after the first major burst of squaring, as these fruit-forms contribute most to crop yield. Regular monitoring of pest numbers conducted two to three times a week during this critical period is warranted. Various methods based on experience or on given advice may be used. It is important to recognise the different developmental stages of the major pests and apply sprays at the earliest stage possible to obtain effective control. *Helicoverpa* larvae can destroy many squares in the early instar stages but third and fourth instar larvae also do considerable damage. Usually the bigger the instar the greater their migratory ability, letting them move down the plant onto neighbouring squares. Most insecticides only have the ability to kill larvae up to the second instar and thus it is vital that they be detected and recognised before they get to this stage. Detection should be undertaken by egg counts and the appropriate spray applied on the estimated hatch-point day. If a continuous egg lay, which may stretch for many days, is found, sprays should be timed so as to be applied on the major egg-hatch day. Effective early control is also more likely to keep pest numbers at the end of the season to negligible or manageable levels. Late in January spraying is timed to control the migration of rough bollworm from its alternative weed host, bladder ketmia, to the cotton crop. Sometimes simultaneous insecticide control of two pests is required, as in the case of *Helicoverpa* and rough bollworm.

Aerial spraying facilitates more timely and fewer applications and, when combined with ultra-low-volume application to improve deposition of spray droplets on the plant, may increase the overall efficiency of insect pest control (Tunstall and Matthews, 1972).

Economics of insect pest control The development of an effective and economic insecticidal spraying program involves the careful consideration of five closely interrelated factors: the choice and dose of insecticide; time of application; method of application; crop potential; and economic value. Longworth and Rudd (1975) examine and discuss these critical questions with respect to pest management and highlight the fact that very little attention has been given to the study of the pest–parasite–predator–pathogen complex in unsprayed cotton in Australia. The University of Queensland Integrated Pest Management Unit has shown that a natural predator–parasite–pathogen complex exists in the cotton agro-ecosystem of southeast Queensland which could influence all serious cotton pests. Longworth and Rudd (1975) summarise the insect pest-controlling and pest-influencing techniques to be evaluated in Australia's cotton industry (Table 8.4) and suggest a package of policy measures for the control of the use of insecticides.

Later efforts (Sterling, 1976) have been aimed at a more accurate assessment of economic threshold levels to enable more effective management decisions with regard to chemical control. This is important in pest management because premature control decisions can result in the destruction of natural parasites and predators at a time when they may be suppressing pests to below damaging levels. Late decisions however can result in significant yield losses. Sterling (1976) gives an outline of a pest management plan for cotton insects based essentially on three sets of sequential sampling plans. The plans utilise the binomial sampling theory and treatment levels for pests during three phenological stages of cotton plant growth. The technique increases accuracy in making management decisions and frequently reduces the time taken to make them compared with other techniques.

Immediate pre-harvesting and post-harvesting control Defoliants are usually applied at 50–60 per cent open-boll maturity. After about two weeks, some 75 per cent of bolls have matured and opened, allowing harvest to commence. For the earlier-opening bolls, little can be done to prevent boll rot organisms. This practice of chemical desiccation can also reduce the late-season build-up of insect pests such as the rough bollworm and aphids.

Under irrigation, a single massive set of early fruits forms so that ripening occurs evenly throughout the crop. Consequently, even low numbers of *Helicoverpa* larvae can cause serious losses and harvesting should not be delayed.

Improved control methods and strategies have eventuated with better co-ordination and continued efforts by research institutions and organisations concerned with cotton, i.e., the agricultural chemical companies, the New South Wales Department of Agriculture, the Queensland

TABLE 8.4   Classified list of pest controlling or pest influencing techniques applicable to cotton grown in Australia (Longworth and Rudd, 1975)

Component of ecosystem manipulated	Type of control	Examples of control type
Crop plant	Regulatory	• Quarantine (e.g. prevents import of pests/diseases) • Ratoon cotton controls
	Cultural	• Variety (e.g. nectariless, high gossypol, earliness, fregobract) • Management (e.g. crop residue disposal, narrow rows, low plant density, volunteer cotton controls, planting and harvesting times, selection of crop site)
Other plants	Regulatory	• Quarantine on weeds, etc. • Eradication, etc. of weed species
	Cultural	• Elimination of unwanted plants (e.g. weed control cultivation, destruction of surrounding alternate hosts) • Management of desirable plants (e.g. as over-wintering sites for beneficial insects, as trap crops for pest species)
Insect populations	Regulatory	• Pest quarantine • Pest eradication, etc. programs • Pest population monitoring (e.g. light trapping *Helicoverpa* spp.) • Introduction of beneficial species (e.g. *Trichogramma* sp.)
	Biological	• Release of beneficial insects (e.g. predators, parasites, pathogens (e.g. *Bacillus thuringiensis,* nuclear polyhedrosis virus)
	Autocidal	• Mass release of sterile males • Genetic manipulation of pest • Pheromones
Abiotic environment	Regulatory	• Control over use of irrigation • Control over use of insecticides
	Cultural	• Water management (e.g. stressing for influencing *Helicoverpa*, flooding for control of over-wintering pupae) • Nutrient management (e.g. nitrogen stressing for influencing *Helicoverpa*)
	Chemical	• Insecticides

Note: A number of the above techniques are theoretical but those of practical value include quarantine for the exclusion of exotics, development of varieties possessing pest and disease resistance, pest population monitoring, chemicals and good crop husbandry.

Department of Primary Industries, the University of Queensland, CSIRO and the Australian Cotton Growers Research Association. The last mentioned organisation is financed by growers through a levy on each bale.

CONCLUSION

There is every incentive to devise integrated control techniques which are more effective, ecologically sound and cheaper. This must be achieved by possessing a thorough understanding of the biology and ecology of important pests and their main regulating factors both biotic and abiotic. Equally important is a knowledge of how the host crop responds to the activities of pests under different environmental conditions.

As convenient and appropriate as it sometimes is to apply a single means of control or to adopt an integrated approach against a single pest, the concept should be more broadly interpreted and apply to pest complexes of particular crops. This is embodied in the IPM concept.

The successful development of integrated control strategies ultimately depends on fostering the concept at the educational/extension level (involving farmers, and the training of specialist plant/crop protectionists), at the research level where there is a greater than ever need for interdisciplinary efforts, and at the political level to provide the necessary funds.

Longworth and Rudd (1975) pose the possibility of the creation of a new institution, namely the pest manager, for effective delivery.

References

Aamisepp, A. (1961), 'The Occurrence of 2, 4-D in Seeds from Cultivated Plants Sprayed with Chlorinated Phenoxyacetic Acids', *Annals of Royal Agricultural College of Sweden,* **27**, 445–51.

Allen, P. G. (1982), 'Changing Agricultural Systems and Insect Pests in South Australia', *Proceedings 2nd Australian Agronomy Conference,* Wagga Wagga.

Anon. (1972), 'Pesticides, Clean Food and Human Health' in *Pesticides in the Modern World,* Symposium by members of the co-operative program Agro-Allied Industries with FAO and other UN Organisations.

Anon. (1975), 'Cool Air for Insect Control in Stored Grain', *Agricultural Gazette of New South Wales,* April, 47–8.

Anon. (1976a), *45th Annual Plant Disease Survey,* N.S.W. Department of Agriculture, Rydalmere.

Anon. (1976b), 'Protecting Stored Grain: Heat, Cold or Suffocation', *Rural Research,* **91**, 18–20.

Anon. (1977), 'Action—Lucerne Aphids', *Agricultural Gazette of New South Wales,* **88**, August Supplement.

Anon. (1978), 'Managing Cotton Pests', *Rural Research,* **100**, 13–18.

Apple, J. A. and Smith, R. F. (1976), *Integrated Pest Management,* Plenum Press, New York.

Aspinall, D. (1960), 'An Analysis of Competition between Barley and White Persicaria: II. Factors Determining the Course of Competition', *Annals of Applied Biology,* **48**, 637.

Baker, K. F. (1973), *Biological Control of Soilborne Plant Pathogens,* A. W. Dimock Lectures, Dept Plant Pathology, New York State College of Agriculture and Life Sciences, Cornell University, New York.

Baker, K. F. and Cook, R. J. (1974), *Biological Control of Plant Pathogens,* W. H. Freeman and Co., San Francisco.

Baker, K. F. and Snyder, W. C. (eds) (1965), *Ecology of Soil-borne Plant Pathogens. Prelude to Biological Control,* John Murray, London.

Ballantyne, Barbara (1984), 'Resistance to Septoria tritici Blotch in Southern N.S.W.', in *Australian Wheat Breeding Today,* a treatise of the Fourth Assembly of the Wheat Breeding Society of Australia, Toowoomba, 38–42.

Ballard, L. A. T. and Grant-Lipp, A. E. (1959), 'Differential Specificity Exhibited by Two Germination Inhibitors Present in *Echium plantagineum*', *Australian Journal of Biological Science,* **12**, 343.

Banks, H. J. (1976), 'Physical Control of Insects—New Developments', *Journal of the Australian Entomology Society.*

Banyer, R. J. (1966a), *Cereal Root Diseases and their Control,* Department of Agriculture, South Australia, Leaflet No. 3848.

Banyer, R. J. (1966b) 'Factors Influencing the Field Occurrence of *Ophiobolus graminis',* National Plant Pathology Conference, **2**, 38.

Banyer, R. J. (1985), 'Agricultural Protection: Practices, Problems and Possibilities', Australian Institute of Agricultural Science, Occasional Publication No.21, *Agricultural Ecology. The Search for a Sustainable System,* 15–25.

Bardner, R. and Fletcher, K. E. (1974), 'Insect Infestation and their Effects on the Growth and Yield of Field Crops: a review', *Bulletin of Entomological Research,* **64** (1), 141–60.

Barlow, F. and Hadaway, A. B. (1974), *Some Aspects of the Use of Solvents in ULV Formulations,* British Crop Protection Council Monograph No. 11, 84–93.

Bowers, R. C. (1986), 'Commercialization of Collego™—An Industrialist's Viewpoint', *Weed Science,* **34** (Supplement 1), 24–5.

Brian, R. C. (1960), *Plant Physiology,* **35**, 773–82.

Brock, R. D. (1967), 'Disease Resistance Breeding', *Journal of the Australian Institute of Agricultural Science,* **33**, 72–6.

Brock, R. D. (1977), 'Genetic Engineering and Plant Improvement', *Journal of the Australian Institute of Agricultural Science,* **43**, 14–21.

Brooks, D. H. (1965), 'Wild and Cultivated Grasses as Carriers of the Take-all Fungus (*Ophiobolus graminis*)', *Annals of Applied Biology,* **55**, 307.

Brown, D. (1954), *Methods of Surveying and Measuring Vegetation,* Commonwealth Bureau of Pastures and Field Crops Bulletin No. 42, Farnham Royal.

Brown, J. F. (1975), 'Diseases of Wheat—Their Incidence and Control' in A. Lazenby and E. Mathesọn (eds), *Australian Field Crops,* Vol. I. *Wheat and Other Temperate Cereals,* Angus and Robertson, Sydney.

Brown, R. H., Meagher, J. W. and McSwain, N. K. (1970), 'Chemical Control of the Cereal Cyst Nematode (*Heterodera avenae*) in the Victorian Mallee', *Australian Journal of Experimental Agriculture and Animal Husbandry,* **10**, 172.

Busvine, J. R. and Feroz, M. (1972) in A. S. Tahori (ed.), *Pesticide Chemistry: Proceedings of Second International IUPAC Congress,* Vol. 2, 1–28.

Butler, F. C. (1961), *Root and Footrot Diseases of Wheat,* N.S.W. Department of Agriculture, Science Bulletin No. 77, 98.

Campbell, M. H. (1960), 'Only Well Managed Sown Pastures Provide Permanent Tussock Control', *Agricultural Gazette of New South Wales,* **71**, 9–19.

Campbell, M. H. (1965), 'It Pays to Control Serrated Tussock', *Agricultural Gazette of New South Wales,* **76**, 606.

Campbell, M. H. (1966), 'Theft by Harvesting Ants of Pasture Seed Broadcast on Unploughed Land', *Australian Journal of Experimental Agriculture and Animal Husbandry,* **6**, 344.

Campbell, M. H. and Swain, F. G. (1973), 'Factors Causing Losses During the Establishment of Surface-sown Pastures', *Journal of Range Management,* **26**, 355.

Carlson, G. A. (1977), 'Economic Aspects of Crop Loss Control at the Farm Level' in L. Chiarappa (ed.), *Crop Loss Assessment,* Supplement 2 FAO/CAB, Farnham Royal.

Carson, R. (1962), *Silent Spring,* Houghton-Mifflin, Boston, 304.

Charudattan, R. and Walker, H. L. (1982), *Biological Control of Weeds with Plant Pathogens,* Wiley, New York, 293.

Chiang, H. C. and Wallen, V. R. (1977), 'Detection and Assessment of Crop Diseases and Insect Infestations by Aerial Photography' in *Crop Loss Assessment Methods,* Supplement 2, Commonwealth Agricultural Bureau.

Chiarappa, L. (ed.) (1971), *Crop Loss Assessment Methods,* FAO Manual on the Evaluation and Prevention of Losses by Pests, Diseases and Weeds, CAB, London.

Church, B. M. (1971), 'The Place of Sample Survey in Crop Loss Estimation' in Chiarappa (1971).

Clare, B. G. (1974), 'Evaluation of Photographs from Earth Resources Technology and Sky Lab Satellites as Aids in Epidemiological Studies', *Australian Plant Pathology Society Newsletter,* **3**, 24.

Clare, B. G., Kerr, A., Jones, D. A. (1990), 'Characteristics of the Nopaline Catabolic Plasmid in Agrobacterium Strains K84 and K1026 used for Biological Control of Crown Gall Disease', *Plasmid. Duluth,* **23**(2), 126–37.

Cobb, N. A. (1892), 'Contribution to an Economic Knowledge of the Australian Rusts (*Uredineae*)', *Agricultural Gazette of New South Wales,* **3**, 60–8.

Combellack, J. H. (1978), 'The Value of Controlled Droplet Application as a Spot Spraying Technique for the Control of Noxious Weeds in Victoria', *Proceedings of 1st Conference of Council of Australian Weed Societies,* Melbourne, 15–24.

Cook, James R. (1974), 'Recent Advances in Biological Control of Plant Pathogens', *Australian Plant Pathology Society Newsletter,* **3**, 46–7.

Cook, R. J. and Rovira, A. D. (1976), 'The Role of Bacteria in the Biological Control of *Gaeumannomyces graminis* by Suppressive Soils', *Soil Biology and Biochemistry,* **8**, 269–73.

Corbett, J. R. (1974), *The Biochemical Mode of Action of Pesticides,* Academic Press, London, 330.

Cotton, J. (1970), 'Nematode Pests of Temperate Cereals', *Span* (Shell Inst. Chem. Co.), **13**, 150–2.

Cottram, G., Curtis, J. T. and Hale, B. W. (1953), 'Some Sampling Characteristics of Randomly Dispersed Individuals', *Ecology,* **34**, 741.

Cullen, J. M. (1977), 'Evaluating the Success of the Program for Biological Control of *Chondrilla juncea*' in *Proceedings IVth International Symposium for the Biological Control of Weeds,* Gainesville, Florida.

Culvenar, C. C. J. (1974), 'The Hazard from Toxic Fungi in Australia', *Australian Veterinarian Journal,* **50**, 69–78.

Daly, J. C. and Fisk, J. H. (1992), 'Inheritance of metabolic resistance to the synthetic pyrethroids in Australian *Helicoverpa armigera*', *Bulletin of Entomological Research,* **82**, 5.

Davidson, B. (1974), 'Ord River Scheme Turns Out to be a Costly Myth', *Herald,* Melbourne.

Day, P. R. (1977), *The Genetic Basis of Epidemics in Agriculture,* New York Academy of Sciences.

DeBach, P. (1964), *Biological Control of Insect Pests and Weeds,* Chapman and Hall, London.

Delfosse, E. S. (ed.) (1985), *Proceedings of the VI International Symposium on Biological Control of Weeds,* Vancouver, B.C., Canadian Government Publishing Centre, Ottowa.

Donald, C. M. (1946), *Pastures and Pasture Research,* 2nd edn, University of Sydney.

Donald, C. M. (1958), 'The Interaction of Competition for Light and Nutrients', *Australian Journal of Agricultural Research,* **9**, 421.

Duggan, M. J., Legget, E. K. and Evans, J. C. (1978), 'Applications of Satellite Remote Sensing to Agriculture in Australia', *Journal of Australian Institute of Agricultural Science,* **44**, 186–9.

Ellingboe, Albert H. (1975), 'Horizontal Resistance: an Artifact of Experimental Procedure?', *Australian Plant Pathology Society Newsletter,* **4**, 44–6.

Evans, R. A., Eckert, R. E. and Kinsinger, F. E. (1961), 'A Technique for Estimating Grass Yields in Greenhouse Experiments', *Journal of Range Management,* **14**, 41.

Evans, G. (1975), 'The Role and the Responsibility of the Plant Pathologist in Plant Quarantine', *Australian Plant Pathology Society Newsletter,* **5**, 2, 17–27.

Fallding, H. (1957), *Social Factors in Serrated Tussock Control,* Research Bulletin No. 1, Department of Agricultural Economics, University of Sydney.

Ferguson, A. W. (1968), 'Effect of Seed Extract of *Trifolium subterraneum* on Germination and Seedling Growth Rate', *Nature,* **217** (5133), 1064.

Fish, S. (1970), 'The History of Plant Pathology in Australia', *Annual Review of Phytopathology,* **8**, 13–36.

Fishpool, K. I., Garland, P. J. and Greening, H. G. (1975), 'Farm Clean-up and Fumigation to Control Grain Insects', *Agricultural Gazette of New South Wales,* **88**, 33–5.

Flor, H. H. (1956), *Advances in Genetics,* **8**, 29–54.

Forrester, N. W. (1992), 'Summer crop resistance management strategy for *Heliothis*', *NSW Agriculture & Fisheries — Cotton Irrigator,* **13**, 18.

Franke, W. (1967), 'Mechanisms of Foliar Penetration', *Annual Review of Plant Physiology,* **18**, 281–300.

Fravel, D. R. (1988), 'Role of Antibiosis in the Biocontrol of Plant Disease', *Ann. Rev. Phytopath.* **26**, 75–91.

Fryer, J. D. and Evans, S. A. (1968), *Weed Control Handbook,* Vol. I. *Principles,* 5th edn, Blackwell Scientific Publications, Oxford.

Garrett, S. D. (1965), 'Toward Biological Control of Soil-borne Plant Pathogens' in K. F. Baker and W. C. Snyder (eds), *Ecology of Soil-borne Plant Pathogens: Prelude to Biological Control,* University of California Press, Berkeley, 4–17.

Goodyear, G. D. (1964), 'Serrated Tussock — a Threat to the Grazing Lands of Victoria', *Journal of the Department of Agriculture, Victoria,* **62**.

Green, K. R. (1956), 'The Problem of Serrated Tussock in New South Wales', *Agricultural Gazette of New South Wales,* **67**, 8.

Gunning, R. V., Easton, C. S., Greenup, L. R. and Edge, V. E. (1984), 'Pyrethroid resistance in *Heliothis armigera* (Hubner) (Lepidoptera: Noctuidae) in Australia', *Journal of Economic Entomology,* **77**, 1283.

Gunning, R. V., Easton, C. S., Balfe, M. E. and Ferris, I. G. (1991), 'Pyrethroid resistance mechanisms in Australian *Helicoverpa armigera* (Hubner) (Lepidoptera: Noctuidae) in Australia', *Pesticide Science,* **33**, 473.

Gysin, H. (1960), 'The Role of Chemical Research in Developing Selective Weed Control Practices', *Weeds,* **8**, 541–55.

Hadaway, A. B. (1971), 'Some Factors Affecting Distribution and Rate of Action of Insecticides', *Bulletin,* World Health Organization, **44**, 221–4.

Hansen, M. H., Hurwitz, W. N. and Madow, W. G. (1953), *Sample Survey Methods and Theory,* Wiley, New York.

Heap, I. (1986), 'The Appearance of Herbicide Cross Resistance in a Population of Annual Ryegrass (*Lolium rigidum*)', *Working Papers — Annual Grass Weeds in Winter Crops Workshop,* Adelaide, 118.

Heap, I. and Knight, R. (1982), 'A Population of Ryegrass Tolerant to the Herbicide Dichlofop-methyl', *Journal of the Australian Institute of Agricultural Science,* **48**, 156–7.

Heap, I. and Knight, R. (1986), 'The Occurrence of Herbicide Cross-Resistance in a Population of Annual Ryegrass, *Lolium rigidum*, Resistant to Dichlofop-methyl', *Australian Journal of Agricultural Research,* **37**, 149–56.

Hodgson, E. and Plapp, F. W. Jr. (1970), 'Biochemical Characteristics of Insect Microsomes', *Journal of Agriculture Food Chemistry,* **18**, 1048–55.

Holliday, R. J., Putwain, P. D. and Dafni, A. (1976), 'The Evolution of Herbicide Resistance in Weeds and its Implications for the Farmer', *Proceedings of the 1976 British Crop Protection Conference,* 937–45.

Holloway, P. J. (1969), 'Effects of Superficial Wax on Leaf Wettability', *Annals of Applied Biology,* **63**, 145–53.

Holly, E. (1964), 'Herbicide Selectivity in Relation to Formulation and Application' in L. J. Audus (ed.), *Physiology and Biochemistry of Herbicides,* Academic Press, London, 423–64.

Howard, W. E. (1950), 'Wildlife Depredations on Broadcast Seedlings of Burned Land', *Journal of Range Management,* **3**, 291.

Htay, Khin (1974), 'Biological Control of Crown Gall', *Australian Plant Pathology Society Newsletter,* **3**, 10.

Hubbard, K. R. (1968), 'Evaluation of Herbicides for Weed Control in Oilseed Rape', *Proceedings of the 9th British Weed Control Conference,* 260–4.

James, W. Clive (1971), 'An Illustrated Series of Assessment Keys for Plant Diseases, their Preparation and Usage', *Canadian Plant Disease Survey,* **51**, 39–65.

James, W. Clive (1974), 'Assessment of Plant Diseases and Losses', *Annual Review of Plant Pathology,* **12**, 27–48.

Johns, C. G. and Greenup, L. R. (1976), 'Pasture Seed Theft by Ants in Northern New South Wales', *Australian Journal of Experimental Agriculture and Animal Husbandry,* **16**, 249.

Johnson, R. (1984), 'A Critical Analysis of Durable Resistance', *Annual Review Phytopathology,* **22**, 309–30.

Jutsum, A. R. and Gordon, R. F. S. (eds) (1989), *Insect Pheremones in Plant Protection,* John Wiley and Sons.

Keed, B. R. and White, N. H. (1971), 'Quantitative Effects of Leaf and Stem Rusts on Yield and Quality of Wheat', *Australian Journal of Experimental Agriculture and Animal Husbandry,* **11**, 550.

Kenney, D. S. (1986), 'Devine® — The Way it Was Developed — An Industrialist's Viewpoint', *Weed Science,* **34** (Supplement 1), 15–16.

Kerr, A. (1987), '*Agrobacterium:* Pathogen, Genetic Engineer and Biological Control Agent', Daniel McAlpine Memorial Lecture, **16** (3), 45–7.

Kerr, A. and Htay, Khin (1974), 'Biological Control of Crown Gall through Bacteriocin Production', *Physiological Plant Pathology,* **4**, 37–44.

Kidd, C. R., Leys, A. R., Pratley, J. B. and Murray, G. M. (1992), Effect of Time of Removal of Annual Grasses from Pastures on the Carryover of Take-all to Wheat. Proceedings 6th Australian Agronomy Conference, Armidale, 555.

King, L. J. (1952), 'Germination and Chemical Control of the Giant Foxtail Grass'. *Contributions from the Boyce Thompson Institute,* **16**, 469.

Klepper, L. (1974), 'Mode of Action of Herbicides: Inhibition of the Normal Process of Nitrite Reduction', University of Nebraska Agriculture Experiment Station, *Research Bulletin,* 259.

Klingman, D. L. (1971), 'Measuring Weed Density in Crops' in Chiarappa (1971).

Knott, D. R. (1961), 'The Inheritance of Rust Resistance: VI. The Transfer of Stem Rust Resistance from *Agropyron elongatum* to Common Wheat', *Canadian Journal of Plant Science,* **41**, 109.

Knowles, C. O., Ahmad, S. and Shrivasta, S. P. (1972), 'Insecticides' in A. S. Tahori (ed.), *Proceedings of the Second International IUPAC Congress,* Vol. 1, 77–98.

Knuesli, E., Berrer, D., Dupuis, G. and Esser, H. (1969), 'S-Triazines' in P. C. Kearney and D. D. Kaufman (eds), *Degradation of Herbicides,* Dekker, New York, 51–78.

Kohn, G. D. and Cuthbertson, E. G. (1975), 'Response of Skeleton Weed (*Chondrilla juncea*) to Applied Superphosphate and Grazing Management', *Australian Journal of Experimental Agriculture and Animal Husbandry,* **15**, 102–4.

Kuiper, J. (1965), 'Failure of Hexa Chlorobenzene to Control Common Bunt of Wheat', *Nature,* **206**, 1219–20.

Kuiper, J. (1976), 'Wheat Yield Losses Caused by *Septoria tritici*', Abstract of papers, 2nd National Plant Pathology Conference, Brisbane (Supplement to *Australian Plant Pathology Society Newsletter,* **5** (1), 41).

Le Baron, H. M. and Gressel, J. (eds) (1982), *Herbicide Resistance in Plants,* Wiley, New York, 401.

Leys, A. R. and Cuthbertson, E. G. (1975), 'Weed Control in Oilseed Rape', *Journal of the Australian Institute of Agricultural Science,* **41**, 98–105.

Longworth, John W. and Rudd, D. (1975), 'Plant Pesticide Economics with Special Reference to Cotton Insecticides', *Journal of the Australian Agricultural Economics Society*, **19**, 210–27.

Macindoe, S. L. and Walkden Brown, C. (1968), *Wheat Breeding and Varieties in Australia*, N.S.W. Department of Agriculture, Science Bulletin No. 76.

Marini-Bettolo, G. B. (1978), 'Natural Products and the Protection of Plants' in *Proceedings of a Study Week of the Pontifical Academy of Sciences 18–23 October 1976*, Elsevier, Amsterdam and New York.

Marshall, D. R. (1977), 'The Advantages and Hazards of Genetic Homogeneity', *Annals New York Academy of Science*, **287**, 1–20.

McIntosh, R. A. (1976), 'Genetics of Wheat and Wheat Rusts since Farrer' (Farrer Memorial Oration), *Journal of the Australian Institute of Agricultural Science*, **42**, 203–16.

McLean, G. D. and Nicholls, T. J. (1991), 'Australian Regulation of Biotechnology', *Australian Plant Pathology*, **20** (4), 166–8.

McNamara, D. W. (1976), 'Wild Oat Density and the Duration of Wild Oat Competition as it Influences Wheat Growth and Yield', *Australian Journal of Experimental Agriculture and Animal Husbandry*, **16**, 402–6.

McVean, D. N. (1965), 'Skeleton Weed in Australia', *New Scientist*, 764–6.

McWhorter, C. G. and Holstun, J. T. (1961), 'Phytotoxicity of S-Triazine Herbicides to Corn and Weeds as Related to Structural Differences', *Weeds*, **9**, 592–9.

Metcalf, R. L. and Flint, W. P. (1962), *Destructive and Useful Insects*, McGraw-Hill, New York.

Michael, P. W. (1968), 'Perennial and Annual Pasture Species in the Control of *Silybum marianum*', *Australian Journal of Experimental Agriculture and Animal Husbandry*, **8**, 10.

Moore, R. M. and Cashmore, A. B. (1942), *The Control of St. John's Wort (Hypericum perforatum L. var. angustifolium D. C.) by Competing Pasture Plants*, CSIR Australia, Bulletin 151.

Moore, R. M. and Robertson, J. A. (1964), 'Studies on Skeleton Weed Competition from Pasture Plants', CSIRO Australia, *Field Station Record*, **3**, 69–72.

Morschel, J. R. (1973), *An Outline of Plant Quarantine*, Australian Department of Health, Canberra.

Nalewaja, J. D. (1974), 'Herbicidal Weed Control Uses Energy Efficiently', *Weeds Today*, **6** (4), 10–13.

Nalewaja, J. D. (1978), 'Weed Control in Cereals—Now and in the Future' in *Proceedings of the First Conference of the Council of Australian Weed Science Societies*, Melbourne, 215–22.

Nelson, J. R., Wilson, A. M. and Goebel, C. J. (1970), 'Factors Influencing Broadcast Seeding in Burnt Grass Range', *Journal of Range Management*, **23**, 163.

Nielson, K. E., Cuddy, T. F. and Woods, W. B. (1960), 'The Influence of the Extracts of Some Crops and Soil Residues on Germination and Growth', *Canadian Journal of Plant Science*, **40**, 197.

Nikandrov, A., Weidemann, G. J. and Auld, B. A. (1984), 'Mycoherbicide Co-operative Project', *Proceedings of the 7th Australian Weeds Conference*, Perth, 129.

Norris, K. R. (1948), 'Seasonal Severity of the Attack of the Red-legged Earth

Mite (*Halotydeus destructor*) on Subterranean Clover', *Journal of the Council for Scienti*fic *and Industrial Research*, **21**, 7–15.

Norton, G. A. (1976a), 'Decision Making—To Spray or Not To Spray', *Shell in Agriculture*, March, Shell Printing, London.

Norton, G. A. (1976b), 'Pest Control Decision Making: an Overview', *Annals of Applied Biology*, **84**, 444–7. O'Brien, R. D. (1967a), *Insecticides: Action and Metabolism*, Academic Press, New York.

O'Brien, R. D. (1967a), *Insecticides: Action and Metabolism*, Academic Press, New York.

O'Brien, R. D. (1967b), *Federation Proceedings*, **26**, 1056–61.

Officer, R. R. (1975), 'Classical, Neo-Classical and Bayesian Decision Making: a Comparison', *Agricultural Economics Discussion Paper 4/75*, University of Queensland, St Lucia.

Olunga, B. A., Lovell, P. H. and Sagar, G. R. (1977), 'The Influence of Plant Age on the Movement of 2,4-D and Assimilates in Wheat', *Weed Research*, **17**, 213–18.

Ownley, B. H., Weller, D. M. and Thomashow, L. S. (1992), 'Influence of *in situ* and *in vitro* pH on Suppression of *Gaeumannomyces graminis* var. *tritici* by *Pseudomonas fluorescens*. *Phytopathology*, **82** (2), 178–84.

Pearson, A. (1950), 'Experiment in Control of Skeleton Weed: Effect of Long Term Pasture', *Agricultural Gazette of New South Wales*, **61**, 425–7.

Penrose, L. J. (1977), 'Fungicide Resistance Poses New Problem for Agriculture', *Agricultural Gazette of New South Wales*, **88**, 43.

Phipps, I. F. (1938), 'The Effect of Leaf Rust on Yield and Baking Quality of Wheat', *Journal of the Australian Institute of Agricultural Science*, **4**, 148.

Price, R. D. (1970), *Stunted Patches and Deadheads in Victorian Cereal Crops*, Victorian Department of Agriculture, Technical Bulletin No. 23, 165.

Pritam Singh (1970), *Host-plant Nutrition and Composition: Effects on Agricultural Pests*, Information Bulletin, Belleville Research Institute, Canadian Department of Agriculture.

Radosevich, S. R. (1977), 'The Mechanism of Atrazine Resistance in Lambsquarters and Pigweed', *Weed Science*, **25**, 316–18.

Readshaw, J. L. (1971), 'An Ecological Approach to the Control of Mites in Australian Orchards', *Journal of the Australian Institute of Agricultural Science*, **37**, 226–30.

Reed, E. M. and Springett, B. P. (1971), 'Large-scale Field Testing of a Granulosis Virus for Control of Potato Moth (*Phthorimaea opercullella* (Zell.) Lepidoptera: Gelechiidae)', *Bulletin of Entomological Research*, **61**, 223–33.

Rees, R. G. (1974), 'Epidemiological Measurement of Disease Resistance in Plants', *Australian Plant Pathology Society Newsletter*, **3**, 48.

Reeves, T. G. (1976), 'The Effect of Annual Ryegrass (*Lolium rigidum* Gaud.) on Yield of Wheat', *Weed Research*, **16**, 57–63.

Reeves, T. G., Mears, P. P. and Ockwell, A. P. (1985), *A Review of Research into Pasture-Crop-Animal Systems in Temperate and Sub-Tropical Australia*, S.C.A. Technical Report Srs.

Richens, K. F., Sproule, R. S. and Kable P. F. (1975), 'Disease Forecasting in the Murrumbidgee Irrigation Areas', *Agricultural Gazette of New South Wales*, **86**, 12–14.

Rippen, L. E. and Wild, B. L. (1976), 'Control of Benomyl Resistant Strains of Green Mould of Citrus with Sodium Ortho-Phenylphenate, 2-Aminobutane and Imazalil', *Australian Plant Pathology Society Newsletter,* **5**, 45–6.

Roark, B. and Donald, C. M. (1954), 'The Control of Onion Weed by Competing Pasture Plants', Report of Australian Weed Control Conference, Roseworthy, S.A., 1953.

Rodriguez, J. G. (1960), 'Nutrition of the Host and Reaction to Pests. Biological and Chemical Control of Plant and Animal Pests', *American Association for Advancement of Science,* **61**, 149–67.

Rovira, A. (1990), 'Ecology, Epidemiology and Control of Take-all, Rhizoctonia Bare Patch and Cereal Cyst Nematode in Wheat, 1989 Daniel McAlpine Memorial Lecture, *Australian Plant Pathology,* **19** (4), 101–11.

Rovira, A. D. and Venn, N. R. (1983), 'The Effects of Rotation and Tillage on Take-All and Rhizoctonia Root Rot in Wheat', *Proceedings Section 5 Fourth International Congress of Plant Pathology,* Melbourne.

Sargent, J. A. (1965), 'The Penetration of Growth Regulators into Leaves', *Annual Peview of Plant Physiology,* **16**, 1–12.

Sears, E. R. (1956), 'The Transfer of Leaf-rust Resistance from *Aegilops umbellulata* to Wheat', *Brookhaven Symp. Biol.,* **9**, 1.

Sears, E. R. (1972), 'Chromosome Engineering in Wheat', *Stadler Symposia,* **4**, 23, University of Missouri, Columbia.

Shaw, A. J. (1991), 'Insecticide resistance managment strategy for *Heliothis* in 1991–92, in NSW Department of Agriculture — Cotton Pesticides Guide, 1991–92.

Shaw, A. J. (1992), 'Cotton Pesticides Guide 1991–92'. N.S.W. Agriculture (Agdex 151/680).

Shepherd, C. J. and Totterdell, C. J. (1974), 'The Use of Remote Sensing for Evaluating Plant Diseases', *Australian Plant Pathology Society Newsletter,* **3**, 24.

Shipton, P. J., Cook, R. J. and Sitton, J. W. (1973), 'Occurrence and Transfer of a Biological Factor in Soil that Suppresses Take-all of Wheat in Eastern Washington', *Phytopathology,* **63**, 511–17.

Simmonds, J. H. (1959), 'Mild Strain Protection as a Means of Reducing Losses from the Queensland Woodiness Virus in the Passion Vine', *Queensland Journal of Agricultural Science,* **16**, 371–80.

Sloan, Cook and King Pty Ltd (1988), *The Economic Impact of Pasture Weeds, Pests and Diseases on the Australian Wool Industry,* Report to the Australian Wool Corporation, Melbourne.

Smith, A. M. (1973), 'Ethylene as a Cause of Soil Fungistasis', *Nature,* **246** (5431), 311–13.

Smith, A. M. (1974), 'Microbial Activity and Energy Flow in Soil Regulated by Ethylene Produced by Spore-forming Anaerobes', *Australian Plant Pathology Society Newsletter,* **3**, 15.

Smith, H. C., Lowe, A. D. and Copp, L. G. L. (1963), 'Farmer Recognition and Control of Cereal Aphid Virus', *New Zealand Journal of Agriculture,* **106**, 392–5.

Smith, H. C. (1964), 'A Survey of Barley Yellow Dwarf Virus in Australia 1963', *New Zealand Journal of Agricultural Research,* **7**, 239.

Smith, Ray F. (1968), 'Recent Developments in Integrated Control', *Pest Articles and News Summaries*, Secn. A., **14**, 201–6.

Southwood, T. R. E. (1978), *Ecological Methods: with Particular Reference to the Study of Insects*, 2nd edn, Chapman and Hall, London.

Stem, V. M., Smith, R. F., Van den Bosch, R. and Hagen, K. S. (1959), 'The Integration of Chemical and Biological Control of the Spotted Alfalfa Aphid', *Hilgardia*, **29**, 81–101.

Sterling, W. L. (1976), 'Sequential Decision Plans for the Management of Cotton Arthropods in South-east Queensland', *Australian Journal of Ecology*, **1**, 265–74.

Stubbs, L. L. (1961), 'Tristeza: Investigation of Protective Value of Mild Strains from Lisbon & Meyer Lemon and from Grapefruit', *Citrus News*, **37** (12), 139.

Stynes, B. A. and Wallace, H. R. (1974), 'A Quantitative Study of Variation in Wheat Yield', *Australian Plant Pathology Society Newsletter*, **3**, 24–5.

Sutherland, J. W. (1975), 'Good Sizes for Simple Grain Driers', *Power Farming Magazine*, **84**, 27–8.

Teng, P. S. (ed.) (1987), *Crop Loss Assessment and Pest Management*, American Phytopathological Society, Minnesota, USA.

Tunstall, J. P. and Matthews, G. A. (1972), 'Insect Pests of Cotton in the Old World and their Control' in *Cotton*, Ciba-Geigy Agrochemicals Technical Monograph No. 3, Ciba-Geigy, Basle, 46–59.

Ullstrup, A. J. (1972), 'The Impact of the Southern Corn Leaf Blight Epidemics of 1970–71', *Annual Review of Phytopathology*, **10**, 37–50.

Van der Plank, J. E. (1963), *Plant Diseases: Epidemics and Control*, Academic Press, New York and London.

Van der Plank, J. E. (1968), *Disease Resistance in Plants*, Academic Press, London.

Van der Plank, J. E. (1975), 'Hannaford Lecture: The Genetic Basis of Plant Disease Epidemics', *Australian Plant Pathology Society Newsletter*, **4**, 27–30.

Van Emden, H. F. (1969), 'Plant Resistance to Aphids Induced by Chemicals', *Journal of Science of Food and Agriculture*, **20**, 385–7.

Wain, R. L. (1953), *Plant Growth Substances*, Royal Institute of Chemistry Lecture, Monograph and Reports No. 2.

Wain, R. L. (1969), 'Naturally Occurring Fungicides', *Proceedings of a Symposium on Potentials in Crop Protection*, New York State Agricultural Experimental Station, Geneva, College of Agriculture, 26–32.

Walker, C. H. (1976), 'The Significance of Pesticide Residues in the Environment', *Outlook on Agriculture*, **9**, 16–20.

Walker, P. T. (1988), 'Measurement of Insect Populations and Injury', in P. S. Teng (ed.) (1987), *Crop Loss Assessment and Pest Management*, American Phytopathological Society, Minnesota, USA.

Wallace, H. R. (1974), 'Problems Associated with Relating Epidemiological Measurements to Loss in Yield', *Australian Plant Pathology Society Newsletter*, **3**, 47–8.

Wallen, V. R. and Jackson, H. R. (1971), 'Aerial Photography as a Survey Technique for the Assessment of Bacterial Blight of Beans', *Canadian Plant Disease Survey*, **51**, 163–9.

Walpshere, A. J. (1974), 'A Strategy for Evaluating the Safety of Organisms for Biological Weed Control', *Annals of Applied Biology,* **77**, 201–11.

Watson, I. A. (1958), 'The Present Status of Breeding Disease Resistant Wheats in Australia', Farrer Memorial Oration, *Agricultural Gazette of New South Wales,* **69**, 630–60.

Watson, I. A. (1974), 'Losses from Wheat Stem Rust in Australia—Are they Inevitable?', *Australian Plant Pathology Society Newsletter,* **3**, 64–5.

Weinberger, P. (1963), 'On Reciprocal Influences Between Barley, Rape and Various Weed Species', *Flora Jena,* 153 (cited in *Weed Abstracts,* **12**, 1729, 1963), 242.

Wells, G. J. (1971), 'Skeleton Weed (*Chondrilla juncea*) in the Victorian Mallee. 4. Effect of Fallowing on Wheat Yield', *Australian Journal of Experimental Agriculture and Animal Husbandry,* **11**, 313–19.

Wheeler, B. E. J. (1976), *Diseases in Crops,* Institute of Biology Studies in Biology No. 64, Edward Arnold, London.

Whitney, P. J. (1976), *Microbial Plant Pathology,* Hutchinson, London, 160.

Whittet, J. N. (1968), *Weeds,* 2nd edn, New South Wales Government Printer, Sydney.

Wigglesworth, V. B. (1965), *The Principles of Insect Physiology,* Methuen, London.

Wilson, A. G. L., Hughes, R. D. and Gilbert, H. (1972), 'Response of Cotton to Pest Attack', *Bulletin of Entomological Research,* **61**, 405–14.

Wilson, C. L. (1969), 'Use of Plant Pathogens in Weed Control', *Annual Review of Phytopathology,* **7**, 411–33.

Winteringham, F. P. W. (1969), 'Mechanisms of Insecticidal Action', *Annual Review of Entomology,* **14**, 409–42.

Witchalls, J. T., and Hawke, M. F. (1970), 'Eyespot in Wheat', *New Zealand Journal of Agriculture,* **120** (3), 38–45.

Woods, Arthur (1974), *Pest Control: A Survey,* McGraw-Hill, London.

Wong, P. T. W. (1973), 'The Ecology of Take-all and Related Fungi on Roots of Cereals and Grasses', *Australian Plant Pathology Society Newsletter,* **3**, 32–3.

Wong, P. T. W. (1974), 'Biological Control of Take-all by *Gaeummannomyces graminis* var. *graminis*', *Australian Plant Pathology Society Newsletter,* **3**, 15–16.

Wong, P. T. W. (1977), 'Yellow Leaf Spot Disease of Wheat', *Agricultural Gazette of New South Wales,* **88**, 10–11.

Yates, F. (1960), *Sampling Methods for Censuses and Surveys,* 3rd edn, Griffin, London.

CHAPTER 9

FARM MANAGEMENT ASPECTS OF CASH CROP PRODUCTION

D. W. Glastonbury and R. H. Wilson

THE MARKETING APPROACH

Much has been written concerning the application of the marketing concept to business operation (Boone and Kurtz, 1974) yet too little is understood about the application of these principles to farm and crop production. The farmer who makes a decision to produce a particular crop is making a marketing decision, and this decision must be made in full recognition of the particular requirements of the potential consumer in terms of such factors as the nature and quality of the product, and the time of supply.

For many years Australian farmers were secure in the knowledge that any increase in crop production would lead to higher returns and increased farm income. The present situation, however, is that the markets for many crop products are reluctant to accept increased supplies except at what might be considered to be sacrifice prices to the producer. No longer can farmers base their decision to produce a particular crop solely on the basis of the suitability of their locality for the production of that product, nor does attention to good husbandry with resulting higher yields necessarily ensure the success of a particular cropping activity.

This book has discussed the major technical aspects of crop production. While crop husbandry is important, the ultimate success of any particular crop must be determined by the level of financial return that is obtained by the grower.

This chapter aims to provide the farmer and the student with a simple practical method of measuring the financial contribution made by a crop to the whole farm income. The method involves the determination of cash crop gross margins which not only consider the yield and the price of the product but also take into account the major costs directly associated with the production of that crop.

Examples of the construction and interpretation of cash crop gross margins are given, as well as worked examples of the use of gross margins for assessing the performance of a crop and for planning the re-organisation of the structure of the farm business. These techniques are simple and could be readily adopted by farmers and those who advise farmers on aspects of crop production.

THE GROSS MARGIN SYSTEM

Whole Farm Concept

While many examples of specialised production of either crop or livestock products do exist in Australian agriculture, most farm businesses simultaneously produce one or more products which become available for market. These diversified farms may consist of a range of crop production activities, a range of livestock production activities, or some particular combination of both crop and livestock production. Therefore, in most farm situations, the whole farm unit has a discernible pattern and can be broken down into a number of separate production units.

The ultimate success of the business must be gauged by the financial performance of the whole farm, which is determined by the financial contribution of each of the separate production units. It is this whole farm income which provides the owners of the business with funds to purchase additional production inputs, to service existing debt, to make further capital investment in the development of the business and to meet family and personal living expenses. The whole farm performance represents the financial return achieved on the capital invested in the farm and the return to the farmers on their management skills.

In order to ensure that a satisfactory level of whole farm performance is achieved, the manager must firstly ensure that individual production units are operated as efficiently as possible. Consideration should be given to the technical aspects of the use of the production inputs such as seed, fertiliser and sprays for crops as well as to the financial reward of that activity.

Attention to the performance of the individual production units alone does not ensure the success of the total farming business. Managers must also be concerned that the most appropriate pattern of farm production is being followed. This means that they must consider the combination of production activities on the farm to ensure that the common resources of land, labour and capital are used most efficiently.

The success of any management considerations concerning the crop and livestock production on a farm is ultimately measured in terms of the improved whole farm performance achieved. Thus managers who are interested in improving the financial performance of the whole farm require a system of farm analysis which enables them to:

1. *evaluate the performance and efficiency of the individual production units on the farm.*
   This may lead to a change in crop and livestock husbandry practices and more efficient use of production inputs.
2. *measure the whole farm performance resulting from the combined contribution of each production unit.*

This may lead to changes in the organisation and basic structure of the farm business through expansion of some production activities and introduction of new, more profitable alternatives. The reduction of some existing farm production activities may also be required.

Understanding the Gross Margin System

Many methods have been developed for analysing the performance of the whole farm. One such system which enables the farm manager to analyse the farm business within the framework of the whole farm concept is the gross margin system of farm analysis.

The system was first described in the UK by Wallace and Burr (1963) who developed the gross margin principle into a practical and workable system. Following its development in the UK, the methods were incorporated into farm business analysis services provided by the Agricultural Development and Advisory Service (ADAS), and by Imperial Chemical Industries, UK, through its Farm Advisory Service.

An Australian example of the application of the gross margin system of farm analysis is the computerised farm analysis service provided by the University of New England[1].

The gross margin system recognises that the whole farm consists of a combination of production activities termed *enterprises.*

Enterprises

It is important to understand the enterprise concept as applied in farm management analysis. This is a way of recognising the structure of the production activities on the farm in the same way as it may be perceived by the operator of the farm business.

This enterprise or production activity structure may be different to the manner in which income is presented by the farmer's accountant as detailed in the farm's income tax return.

Enterprises on a farm may be recognised as being *crop* or *livestock* production enterprises, with each activity resulting in the production of an identifiable commodity or group of commodities. Thus a mixed farm in the New South Wales wheat belt may be considered as having *crop enterprises* such as wheat, barley, oats and lupins and *livestock enterprises* such as sheep breeding, cattle breeding and cattle fattening.

A single production enterprise may produce more than one product which is then available for sale. For example, a wheat enterprise may produce grain that is sold for manufacture, grain sold for seed or feed, and grain for storage on the farm. A sheep breeding enterprise may produce a range of products such as lambs, cull stock and wool, all of

---

[1] The Agricultural Business Research Institute at the University of New England, Armidale, provides a computerised farm analysis service for farm financial records forwarded to the Institute.

which can be sold. In each case all of these are recognised as being products of a single production enterprise.

In some cases modification of management practices or production methods is sufficient to warrant the recognition of a new farm enterprise for the purposes of farm analysis. For example, a farmer who produces both irrigation and dryland wheat may wish to consider these as separate and distinct farm enterprises. Similarly a rice farmer who has adopted different methods of sowing the crop may wish, for the purposes of analysis, to make a comparison of the performance of an aerially sown crop with a crop sown by conventional methods.

For crops, a farmer may wish to recognise different varieties of the same crop as being different production enterprises.

It is not uncommon for farmers interested in a very detailed analysis of crop performance to consider even different paddocks of the same crop to be different enterprises, particularly where these paddocks may have received different husbandry treatments or are grown in different stages of the whole farm crop and pasture rotation.

The degree of subdivision of the farm into enterprises depends on the degree of analysis which the farm manager wishes to achieve.

Enterprise Gross Margin

A measure of the financial performance of each enterprise on the farm is given by the *gross margin*, which is calculated by subtracting the sum of the variable costs associated with the production of that enterprise from the total value of output produced.

For each enterprise:

OUTPUT	minus	VARIABLE COSTS	equals	GROSS MARGIN

The *output* of an enterprise is the *value* of all production from that particular activity and thus reflects both the physical quantity and the price of the product.

Variable costs are those costs that are easily and accurately allocated to a particular enterprise and which vary in proportion to small changes in the size of that enterprise.

These terms, as well as the conventions normally adopted in the calculation of enterprise gross margins, are discussed in detail in the section of this chapter concerned specifically with cash crop gross margins.

Whole Farm Income

Once the gross margin of each enterprise has been calculated, it is possible to determine the *total farm gross margin* by summation of the gross margins of each separate farm enterprise. A measure of whole farm income is obtained by subtracting the total *fixed costs* from the *total farm gross margin*.

For the whole farm:

| TOTAL FARM GROSS MARGIN | minus | FIXED COSTS | equals | WHOLE FARM INCOME |

The *total farm gross margin* is the sum of the individual gross margins of all enterprises on the farm. In a mixed farm this necessarily includes the gross margins of each crop and livestock enterprise.

Whilst it is not the purpose of this chapter to discuss livestock enterprise gross margins, it should be noted that pasture and forage costs are usually deducted from the sum of the livestock gross margins in the determination of the total livestock gross margin (forage costs deducted) and the total farm gross margin. The reason for this is that pasture and forage costs are normally considered as inputs into the livestock enterprise and the costs of growing these must be met by the returns from livestock. This convention is normally followed even though it is recognised that certain crops may benefit from the pasture phase in the rotation. The total livestock gross margin (forage costs deducted) is added to the total cash crop gross margin to give the total farm gross margin.

Fixed costs are those costs that are common to all enterprises on the farm, which do not change appreciably with small changes in the enterprise structure of the farm or which cannot be allocated realistically or accurately to the various farm enterprises. Fixed costs may also be called *common costs* or *overhead costs*.

Costs that would normally be considered as fixed costs in the gross margin system of farm analysis are listed in Table 9.1.

The *whole farm income*, which provides a measure of the profitability of the farm business, has been shown to be the total farm gross margin less the fixed costs associated with the business. Various measures of whole farm income may be determined depending upon the costs that are included in the fixed costs section. One system of classification which has been incorporated in the Agricultural Business Research Institute computerised farm analysis service is summarised in Table 9.2.

TABLE 9.1   Fixed cost categories for the gross margin system

---
Rent and rates charges
Fuel, oil and grease
Machinery repairs and maintenance
Depreciation, machinery
Depreciation, structures and improvements
Paid permanent labour and unallocated casual labour
Miscellaneous: e.g. electricity, telephone office, insurance, registrations, accountant, travel

---

TABLE 9.2   Measures of whole farm income

TOTAL FARM GROSS MARGIN	minus	FIXED COSTS	equals	NET FARM INCOME
NET FARM INCOME	minus	OPERATOR'S ALLOWANCE	equals	OPERATING RETURN
OPERATING RETURN	minus	INTEREST AND LEASE CHARGES	equals	BUSINESS RETURN

Conventions in Applying the Gross Margin System

Concept of Output

Earlier it was indicated that the output of an enterprise is the value of all production from that particular activity and thus reflects both the physical quantity and the price of the product.

From an accounting point of view it is normal to calculate the value of output through the use of a *trading account*. The general format of a trading account, and therefore the calculation of the Total Output value, is depicted in Table 9.3.

This format would be applicable to all cash crop enterprises as well as all livestock enterprises, and is the format of the traditional trading account as used in the calculation of *gross profit* for income taxation assessment.

In those situations where the farmer does not trade (buy or sell accumulated stocks), the statement of output (trading account) may be

TABLE 9.3   Format of trading account

	$
Output	
Sales of product	
Value of product used on-farm	
Value of product transferred out	
Value of closing inventory	
Sub-total	A
less	
Cost of purchases	
Value of product transferred in	
Value of opening inventory	
Sub-total	B
Total Output	A−B

greatly simplified by deleting those items which have no monetary value to that enterprise. In such situations the statement of output may be simplified to the form as shown in Table 9.4, where *Grain sales* and *Stock on hand* would be the only items shown. In Table 9.4 it is assumed that there has been no grain used on-farm, transferred in or out, or purchased, and there was no inventory of grain on hand at the commencement of the year.

TABLE 9.4   Simplified format of trading account

Output	$
Sales of product	
Value of closing inventory	
Total Output	

Other modified versions of output as depicted in Table 9.3 would be possible depending on the use or disposal of the product of the crop. Examples of two simplified statements of output for cash crop enterprises are given in Table 9.5.

For students more familiar with accounting practices it should be appreciated that output as calculated in the gross margin system is similar to the accounting measure of *gross profit* as determined in a trading acount.

The major difference between output and gross profit is that output is usually based on a production enterprise and would therefore include, in a single statement of output, the value of all products produced by that enterprise. Gross profit is usually based on single crop or livestock product categories often produced by one or more different enterprises.

Gross profit does not recognise enterprise categories, so it would rarely show the value of a product transferred between different enterprises on the farm. In enterprise accounting and the calculation of output for separate but related enterprises, it is common for a product to be transferred between enterprises on the farm. It is therefore necessary to credit the enterprise that produced the product with its value and to charge (debit) the enterprise that received the product.

This process enables the output of separate enterprises to be determined while having no net effect on the level of whole farm profit performance.

The method of valuing inventories, transferred product or product used on the farm is also different in each case. Output in a gross margin uses values based on current market values. Gross profit is commonly based on historic values or even the lower inventory values (as may be determined by applying the average cost price method of valuation used frequently for livestock inventory valuations in income taxation calculation).

Production Year Concept

In the application of the gross margin system to analysis and budgeting, it is usual to adopt a *production year* basis in the determination of enterprise gross margins. This means that the gross margin must be based on output and variable costs which relate to the same production cycle. Income and costs associated with previous or future production years are excluded from the gross margin calculation. In order to ensure that any output is directly related to the actual costs associated with producing that output, it may be necessary for the farm manager to depart from the rigid financial year as used in traditional accounting practice and use a more flexible production (or management) year.

The further discussion of the construction and interpretation of cash crop gross margins in this chapter is based on this production year as a concept to be applied to management analysis and decision making. Enterprise gross margins drawn up on the basis of a rigid or fixed financial accounting period could be quite misleading for management purposes.

Cash and Non-Cash Items

It is also mentioned that both enterprise output and variable costs in the gross margin include the value of *all* the product obtained from that production cycle and also include *all* the costs associated with that production.

Output and variable cost components of the gross margin would include both cash and non-cash items, with cash items expected to be realised and non-cash items not realised within the time scale of the gross margin account.

Output would therefore include the value of all production, whether or not the sale proceeds have been realised. Product stored on the farm, or which has been delivered but the proceeds not yet received, are valued and included in the total output of that enterprise. Similarly, any items of input which were used in the production process but which did not incur cash costs are valued and included in the variable costs attributable to that enterprise.

Cash income and cash costs that may have been incurred but are not directly associated with the current production cycle are also deleted from the gross margin calculation for that enterprise for the current production year and would be accounted for in the production cycle to which they directly relate.

The enterprise gross margin thus gives a measure of the financial contribution made by each enterprise to the whole farm income and this gross margin would be useful for both *diagnosis* of the performance of an enterprise through comparison with other enterprises and for *planning* the reorganisation of the enterprise combination on the farm. These applications are discussed in a later section in this chapter.

Special Problems of Labour and Machinery Costs

Much of the confusion that has resulted from the practical application of the gross margin system of farm analysis has arisen because of the misunderstanding of the distinction between fixed and variable costs and, in particular, the classification of permanent labour and machinery costs.

Kerr (1968), in discussing this distinction, has indicated that much of the apparent confusion arises because of the conflict between the *budgeting* requirement to consider costs as either variable or fixed costs where changes in farm organisation are being evaluated and the *accounting* requirement to have an orderly and consistent system of cost classification for analysis.

In the budgeting application the costs that are taken as variable costs and fixed costs *depend on the situation under consideration* and for any specific change the costs that vary are considered as variable costs and the costs that remain unaltered are classified as fixed costs. Kerr (1968), however, indicated that this terminology cannot be applied precisely to accounting procedure and that there is a need for some agreement or convention as to which costs are variable costs and fixed costs under the fixed time scale of a financial account.

This chapter has adopted one convention for the classification of variable and fixed costs for use in enterprise and whole farm analysis. However, others may adopt different conventions and this may result in the gross margins prepared by other authorities not being exactly comparable.

The main argument concerns the treatment of *permanent labour* and *machinery costs*, including machinery repairs and maintenance, fuel, oil and grease. These costs are listed as *fixed costs* on the basis that they must be allocated either on an arbitrary basis or by maintaining detailed records concerning labour and machinery use. The maintenance of these records is extremely time consuming and it may be argued that the allocation of these costs does not aid significantly the analysis of the farm. In addition, permanent labour is a cost which does not alter in proportion to small changes in the size of an enterprise. One unit of labour is generally appropriate over a range of enterprise size until expansion necessitates additional labour input which is again adequate for a further range of enterprise size.

It is frequently argued that permanent labour and machinery costs are costs associated with crop production and should therefore be directly allocated to individual farm enterprises. In this regard, it must be understood that the system of gross margin analysis does not ignore these costs but for accounting purposes only considers them as fixed costs that are separately identified. In budgeting the effect of changes in the farm organisation, all cost items including labour and machinery costs must be considered. Examples of this are given in Tables 9.10 to 9.12 at the end of this chapter.

CASH CROP GROSS MARGINS

The Cash Crop Gross Margin

In the previous section of this chapter, the gross margin concept was discussed in general terms. It is now proposed to provide a specific discussion of the construction and interpretation of the cash crop gross margin.

Cash crops are those crops that are grown on the farm with the specific objective of obtaining a harvestable product which may be sold for cash. It should be noted however, that not all the product of a cash crop is necessarily sold for cash and the farm manager may make a decision to store part of the crop yield for later sale or for future use on the farm.

It has previously been shown that for each cash crop enterprise on the farm, a gross margin may be calculated. This provides a measure of the financial performance of the enterprise as well as the financial contribution that the enterprise makes to the whole farm income. The gross margin of each cash crop enterprise is determined by subtracting the variable costs associated with the production of that enterprise from the total value of the output produced.

Table 9.5 provides examples of the presentation of gross margins for two selected cash crops, wheat and canola. These gross margin statements demonstrate both the structure of cash crop enterprise gross margin statements and the principles and procedures of diagnosing and planning crop production. It is not suggested that the technical assumptions or figures used in these two crop gross margin statements would apply exactly in any given farm situation.

Similar enterprise gross margin data are made available by certain authorities, and farmers would be advised to use data more appropriate to their district and, more importantly, their farm. An example of these published data would be as presented in Walker (1993).

Structure of a Crop Gross Margin Statement

Gross margin statements for cash crop enterprises would normally be presented as in Table 9.5.

While different authorities may use different forms of structure, it is important that the cash crop enterprise statement should include the following features:

(i) The statement should specify the name of the cash crop enterprise and be sufficiently descriptive to clearly identify the particular crop being detailed. This should also include the size (scale) of the enterprise.

(ii) A gross margin statement would be presented most commonly in vertical tabular form, with Output at the top, Variable Costs below and the calculation of Gross Margin at the foot of the table.

TABLE 9.5   Example of cash crop gross margins for dryland wheat and canola
enterprises
(a) WHEAT (100 ha)

	Total	Per ha
Output		
Yield (tonnes)	250	2.5
	$	$
Grain sales: 245 tonnes @ $130/tonne net	31 850	318.50
Stocks on hand: 5 tonnes @ $130/tonne net	650	6.50
Total output	32 500	325.00
Variable Costs		
Seed: 60 kg/ha @ $0.32/kg	1920	19.20
Fertiliser: 90 kg/ha DAP @ $0.51/kg	4590	45.90
Sprays: Hoegrass 1.5 L/ha @ $20.10/L	3015	30.15
Igran 0.85L/ha @ $17.50/L	1487	14.87
MCPA 0.3 L/ha @ $6.50/L	195	1.95
Contract casual labour: spraying	250	2.50
Miscellaneous: bags, hessian, insurance	1650	16.50
Total variable costs	13 107	131.07
Gross margin	19 393	193.93

Change in wheat enterprise gross margin ($/ha) for change in crop yield and/or
grain price assumptions.

YIELD tonnes/ha	ON-FARM PRICE ($/tonne)		
	100	130	160
2.0	69	129	189
2.5	119	194	269
3.0	169	259	349

TABLE 9.5 (continued)
(b) CANOLA (100 ha)

	Total	Per ha
Output		
Yield (tonnes)	180	1.8
	$	$
Grain sales: 180 tonnes @ $250/tonne net	45 000	450.00
Total output	45 000	450.00
Variable Costs		
Seed: 4 kg/ha @ $1.75/kg	700	7.00
Fertiliser: 100 kg/ha MAP @ $0.50/kg	5000	50.00
85 kg/ha urea @ $0.42/kg	3570	35.70
Sprays: Lemat	220	2.20
Trifluralin 2.0 L/ha @ $7.75/L	1550	15.50
Supracide 0.2 L/ha @ $23/L	460	4.60
Contract casual labour: spraying	250	2.50
windrowing $30/ha	3000	30.00
Miscellaneous: levy, insurance	1500	15.00
Total variable costs	16 250	162.50
Gross margin	28 750	287.50

Change in canola enterprise gross margin ($/ha) for change in crop yield and/or grain price assumptions.

YIELD tonnes/ha	ON-FARM PRICE ($/tonne)		
	220	250	280
1.5	167	212	257
1.8	233	287	341
2.1	299	362	425

(iii) If the enterprise gross margin is to be used effectively for analysis and planning, each separate component of both Output and Variable Cost should be listed separately and as far as is possible include both the physical and financial basis of all figures in the gross margin, e.g. Grain sales: 245 tonnes @ $130/tonne net, or Spray: Hoegrass 1.5 L/ha @ $20.10/L.

(iv) All data should be presented on the basis of $ Total and $ per 'unit' figures in separate columns of the table. For cash crop enterprises the productive basis is usually considered to be land area, and the per unit figure is expressed on the basis of $ per ha (of land).

Other factors may, however, be used to assess the financial efficiency of a cash crop enterprise, and enterprise gross margins may be presented on the basis of gross margin per unit of capital invested or gross margin per unit of labour utilised.

Components of the Cash Crop Gross Margin

Output

Based on the previous discussion of enterprise output the *output* component of the cash crop gross margin must show the *total physical yield* which is produced by a single production cycle of the enterprise and the *price received for each part* of the output. The value of the output is the product of the *physical yield* and the *price received*.

The physical yield must include all the grain harvested from that crop whether the grain has actually been sold off the farm or stored on the farm for later sale or for on-farm use.

GRAIN SALES

Estimating the price received for cash crop output presents some particular difficulties. If the grain is sold off the farm, the price used is the full market price. It is frequently the case under marketing arrangements that apply to certain crops that the full proceeds of the sale are not received in one cash payment but over a series of progressive or pool payments. If the farmer has not received the full proceeds from grain sold off the farm, the price included in the enterprise gross margin must be the expected full market price, which includes all cash payments received plus estimated future payments.

STORED GRAIN AND GRAIN FOR FARM USE

Where the grain or a portion of the grain is stored on the farm, it must be valued and included in the output of the enterprise as the value of stocks on hand or the value of grain transferred out. This must be included if a realistic gross margin is to be determined for each enterprise.

These components of the crop output would be non-cash valuation items for which the monetary value has not been received within the time scale of the account.

For management purposes, the most acceptable method of valuing stored grain is to base its value on the current market price less the costs of marketing. This is the price which could be obtained if that portion of the yield was sold instead of being retained on the farm. In economic terms, this represents the opportunity cost of the decision to retain the grain rather than dispose of it on the highest price market available.

Other methods of valuing stored grain are available, including valuation based on an estimate of the variable costs of production or valuation based on set standard or average values for particular products. These methods tend to undervalue the grain in terms of the established market for the product and therefore understate the gross margin of the enterprise.

GROSS AND NET OUTPUT

It is also important in the construction of crop enterprise gross margins that some convention be established for determining *output* as either a *gross output* or *net output* figure. This consideration arises because *account sales statements*, which are the main source of product sales information, usually detail the gross proceeds of the sale as well as the selling costs and deductions. It is the net amount which is paid to the farmer or credited to his account and appears on the bank statement. Fig. 9.1 shows an Australian Wheat Board payment advice detailing the gross value, rail freight and dockage deductions and net amount payable. The

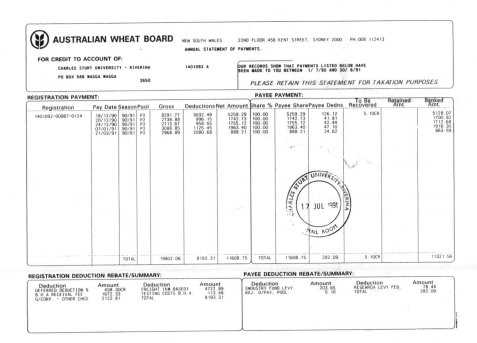

FIG. 9.1 Australian Wheat Board statement of deliveries and payment advice

statement shows a gross value of $19,802.06 and a net cash payment of $11,321.56 to the growers' account. (This amount represents only the first payment of the quantity delivered.)

The cost deductions appearing on these statements would normally be considered as part of the variable costs of production and while the use of a net sales figure in the determination of output may give a true picture of the enterprise gross margin it does not separately identify these deductions. It may be advisable to record the output as gross output and include all the deductions from sale proceeds in the variable cost section of the gross margin account.

Variable Costs

Variable costs have already been defined as those costs that can be easily and accurately allocated to a particular enterprise and that will vary in proportion to small changes in the size of an enterprise.

For accounting and analysis purposes, the costs that are normally considered as being the variable costs of crop production are listed in Table 9.5. Further subdivision of costs requires greater recording detail and may not greatly assist the analysis of the enterprise performance.

TABLE 9.6   Variable cost categories for cash crops

Seed
Fertiliser
Sprays
Contract and casual labour
Miscellaneous: e.g. crop insurance, bags

It may be satisfactory under some circumstances to recognise further variable cost categories. For example the cost of *irrigation water* may be included as a cost category when an accurate allocation of water costs can be made to each farm enterprise. Where this is not possible, irrigation water costs may be recorded as a fixed cost.

Some authorities may use different classifications of the variable costs appropriate to a cash crop enterprise. It is important, however, that the costs treated as variable costs be consistent with the definition of this cost category in the gross margin system of analysis. This is of particular importance where a comparison is being made between enterprise gross margin statements. A valid comparison can only be made when consistent conventions have been used in the classification of variable costs and the construction of the gross margin statement.

It is also important when using enterprise gross margins for budgeting purposes that only variable costs that alter in proportion to a change in the size of the enterprise are incorporated in the gross margin.

Only those costs incurred in the production of the physical yield included as the output of that production cycle may be included as the variable costs for that enterprise. Thus costs associated with previous or future production must be excluded from the enterprise gross margin for the production cycle. For example, a farmer may have purchased fertiliser on early delivery terms for use in the cropping program of the next season. The latest fertiliser purchase is excluded from the gross margin for current crops and only the cost of fertiliser actually used on this crop is included as the variable cost of fertiliser. It may be that the fertiliser used on the crop was paid for some time prior to the planning of the current cropping program.

In cash crop production, it may also be the case that the farmer did not incur a cash cost associated with a particular variable input. Where this occurs, the item must still be included as a variable cost of crop production, and the quantity of input used must be *valued* to determine a cost of the input. A common example of this situation is where a wheat farmer retains seed from a previous crop. The home-grown seed should be valued and included as the seed cost for that enterprise.

The list of variable costs of crop production includes contract and casual labour which can be directly attributed to that enterprise. It does not include an allocation of permanent labour nor any machinery costs, both of which are considered as fixed costs for reasons previously discussed. Contract and casual labour includes such costs as contract sowing, harvesting, cartage and seed cleaning and casual labour employed specifically for crop production operations.

The miscellaneous variable cost category provides an opportunity to list variable costs not covered by more specific cost categories. Examples of items which may be included in this classification are crop insurance, bags for seed wheat and chemicals for seed treatment.

PUTTING GROSS MARGINS TO WORK

The Use of Gross Margins for Diagnosis

What is Diagnosis?

In farm management it is important that the farmer evaluate the performance and efficiency of each enterprise on the farm. This ensures that correct crop production methods are used and that the enterprise makes the maximum contribution to the whole farm performance. Such an analysis of enterprise performance is frequently termed *diagnosis*.

For cash crop enterprises, diagnosis is usually aimed at identifying:
1. unsatisfactory enterprise performance and the key factors which have contributed to the level of performance achieved; and
2. opportunities for the manager to take corrective action which may lead to improved performance by that crop in future production years.

It may be recognised in cash crop production, as indeed for other farm production enterprises, that many factors contributing to the level of performance of an enterprise may be beyond the direct control of the farm manager, while other factors may be able to be influenced by the farm manager. The aim of diagnosis should be to identify those factors that may be controlled by the manager and for which managerial decisions are of importance in determining the performance of an enterprise. Diagnosis of the performance of each enterprise is of particular importance to cash crop production and is an essential aid to farm management.

Simple Diagnostic Procedures

The basis of farm management diagnosis of performance is to identify the separate factors that have contributed to the measured level of financial performance of the enterprise. The ultimate success of an enterprise's performance *must* be measured in financial terms; this is a concept that is often not appreciated fully by agricultural technologists.

It is the detailed enterprise gross margin account (for example, see Table 9.5) which is used as the basis of simple diagnostic procedures of enterprise performance. The enterprise gross margin, as well as giving a measure of the financial contribution of that enterprise to the whole

FIG. 9.2   Diagnostic procedure for examining crop enterprise gross margins

farm, also provides physical and financial details of the enterprise output and the variable costs associated with production.

The diagnostic procedure (Fig. 9.2) is firstly to consider the enterprise gross margin and determine if a satisfactory level of performance has been achieved. If the performance is assessed as being unsatisfactory, then the component parts of the gross margin must be examined. A low level of gross margin could be attributed to either a low level of output or unsatisfactory expenditure on variable production inputs.

To aid diagnosis it is usual for all components of the enterprise gross margin to be expressed on a *per unit* as well as total enterprise basis (Table 9.5). For cash crops these data are usually calculated on the basis of the unit of land being one hectare. The per unit figures are then directly comparable with appropriate performance standards. The fact that a cash crop enterprise, say wheat (Table 9.5), has given a total gross margin of $19 393 tells the farmer nothing about the relative performance of that enterprise. If this gross margin is expressed in terms of the area of land devoted to the crop, then the resulting gross margin of $194 per hectare would indicate the economic efficiency of the use of land and can be directly compared with other crops.

The same requirement applies to the diagnosis of all the components of the gross margin. Neither cash crop output nor the expenditure on the variable production inputs can be evaluated unless *each item* is expressed on a per unit basis, again usually taken as per hectare of crop sown.

Regardless of the comparison being made, it is necessary when comparing enterprise gross margins to ensure that the *same procedures and conventions* are adopted in the construction of the gross margins. It has already been discussed that different workers may adopt different conventions for the classification of the variable costs of cash crop production. This may result in the gross margins not being directly comparable and may lead to the farm manager reaching the wrong conclusion regarding the performance of the farm enterprise.

For similar reasons it is also important to ensure that the enterprises being evaluated are comparable crop enterprises and have been conducted under similar management practices and environmental conditions.

Increasing Crop Returns

It is important to appreciate that the method of diagnosis based on analysis of the enterprise gross margin does assist the farm manager to identify those aspects of crop husbandry where inefficiencies exist and which should receive attention from the manager if maximum crop returns are to be achieved.

This text has discussed crop husbandry practices in detail and it is not proposed to discuss further these technical aspects. It is known, however, that the main determinants of the level of crop gross margins are:

1. output per hectare:
   determined by crop yield and average product price.

2. variable costs per hectare:
   determined by physical quantities and price of productive inputs.

Output per hectare   Many aspects of crop husbandry influence the level of output per hectare achieved by the farmer for a particular crop. Diagnosis of this component of the gross margin involves consideration of the large number of management factors such as detailed in Table 9.7 and discussed in this text.

TABLE 9.7   Factors influencing cash crop output

Yield factors	Product price factors
stage in rotation	quality of product
crop species	(including purity)
crop variety	proximity to markets
time of sowing	timeliness of sale
rate of sowing	demand
fertiliser (type and rate)	quantity produced
weed control	continuity of supply
irrigation (if applicable)	
time of harvest	
efficiency of harvest	
weather conditions	
diseases	
pests	

Variable costs per hectare   Diagnosis of enterprise performance frequently shows that farmers may increase their cash crop returns by increasing the level of expenditure on variable inputs. Thus higher yields and higher output may be achieved by increased expenditure on such items as fertiliser, herbicides and pest control. According to economic theory, it pays the farmer to increase expenditure on any productive input up to the stage where the marginal return just equals the marginal cost (Dillon, 1977). In practice, however, many farmers adopt less than optimum levels of input usage because of limited capital and more attractive alternative opportunities for deployment of this capital.

Alternatively there may be the opportunity for cost savings resulting in higher levels of enterprise gross margin. While the cost saving is a popular concept with farmers, careful consideration must be given to the final effect of reduced expenditure, particularly when this is on inputs normally considered as variable cost items. Variable costs are, by definition, those costs directly associated with the output of an enterprise and therefore have a direct relationship to crop yield and the value of crop output. Care must be taken to ensure that any reduction in variable cost expenditure does not result in reduced crop yields through poor crop husbandry.

More frequently, increased crop returns can be achieved by savings in fixed costs, particularly labour and machinery costs through mechanisation and adoption of more efficient production methods.

Farmers must closely examine their use of inputs and evaluate the extra benefits to be achieved through increased expenditure on productive inputs or through possible cost savings.

Sensitivity Analysis

An extension of this simple informal approach to diagnosis of the enterprise performance is the technique of sensitivity analysis. This technique involves the determination of the sensitivity of an enterprise's gross margin to changes in one or more key factors identified as being critical to the financial performance of that enterprise.

Sensitivity analysis, as well as aiding diagnosis of the performance of an enterprise, also provides a range of enterprise performance levels that may be considered for incorporation in farm planning and budgeting procedures. Thus, instead of having only a single gross margin, the manager may consider a range of possible performance levels.

The key factors chosen for such an analysis may be either output factors such as crop yield and harvest price or input factors such as input application rates and unit costs.

When conducted manually, the technique of sensitivity analysis usually involves the identification of a limited number of variables considered to be most critical to the performance of the enterprise. Once the key variables are selected, a range of possible outcomes or assumptions for these variables would be incorporated in the enterprise gross margin table, and the resulting enterprise gross margin would be calculated. Other components of the gross margin not affected by the changes are assumed to remain constant. In this way the effect of change (sensitivity) in the key assumptions or performance factors on the resulting enterprise gross margin can be determined.

This analysis is illustrated in Table 9.5, which includes a table below each gross margin. With respect to the wheat gross margin, the table is analysing the sensitivity of the gross margin to changes in the yield per hectare and the on-farm price per tonne. Here the yield ranges from two to three tonnes per hectare and the price received ranges from $100 to $160 per tonne. The body of the table shows the gross margin per hectare resulting from the different combinations of yield per hectare and price per tonne, assuming all the other variables in the gross margin remain constant. Thus at a yield of 3 tonnes per hectare and a price of $130 per tonne, the gross margin is $259 per hectare. If the yield decreases to 2.5 tonnes per hectare and the price decreases to $100 per tonne, the gross margin decreases to $119 per hectare.

It is suggested that these values be substituted into the wheat gross margin in Table 9.5 to verify that the gross margin per hectare is equal to that indicated in the Table.

Because of the practical limitations of manual analysis it is normally only possible to consider a limited number of alternative values for each key factor. In practice this may be restricted to three levels of assumption to cover the range of possible values. These may be chosen in turn as a pessimistic, a best bet and an optimistic level of assumption.

Sensitivity analysis, as described here, is often referred to as '*what if?*' analysis. This arises because the analysis is asking *what* the effect is on the gross margin *if* a particular factor is altered.

Sensitivity analysis is made easier and has a greater practical application through the use of computer spreadsheet software. In this case it would be necessary to format the gross margin such that all calculations in the gross margin table are linked by formula. It is often the practice to identify and list the key factors in a table at the top of the enterprise gross margin, and for each of these factors to again be linked by formula to the gross margin table and calculations required. Once this is done it is then simply a matter of altering one or more variables in the data section and the computer automatically transfers the data change into the gross margin and calculates the new gross margin figures.

Table 9.8 provides an illustration of the screen display for the simple sensitivity analysis of the wheat enterprise previously discussed. It does not show the formula that would be incorporated in the spreadsheet program.

TABLE 9.8   Wheat enterprise gross margin

KEY FACTORS	Assumptions		
Area of crop (ha)	100		
Crop yield tonnes/ha	2.5		
Grain price $/tonne	130		
GROSS MARGIN			
Output		Total	Per ha
Yield (tonnes)		250	2.5
		$	$
Grain sales: 245 tonnes @ $130/tonne net		31 850	318.50
Stocks on hand: 5 tonnes @ $130/tonne net		650	6.50
TOTAL OUTPUT		32 500	325.00
Variable Costs			
Seed: 60 kg/ha @ $0.32/kg		1920	19.20
Fertiliser: 90 kg/ha DAP @ $0.51/kg		4590	45.90
Sprays: Hoegrass 1.5 L/ha @ $20.10/L		3015	30.15
Igran 0.85 L/ha @ $17.50/L		1487	14.87
MCPA 0.3 L/ha @ $6.50/L		195	1.95
Contract casual labour: spraying		250	2.50
Miscellaneous: bags, hessian, insurance		1650	16.50
TOTAL VARIABLE COSTS		13 107	131.07
GROSS MARGIN		19 393	193.93

Spreadsheet programs for analysing and planning farm enterprise performance have become a most valuable aid in agricultural extension activities. They provide the opportunity for analysing farm enterprises based on data most relevant to each individual farm, provide an immediate answer to the 'what if?' questions and, most importantly, enable the benefit of technical advice to be demonstrated in financial terms.

The Use of Gross Margins for Planning

Planning the Reorganisation of the Farm

One of the features of the gross margin system of farm analysis is that the same data can be used both for diagnosing the performance of existing farm enterprises and for comparing enterprise alternatives to determine if changes in husbandry practices or farm organisation will increase income. Having separately considered each enterprise gross margin and determined if changes in production methods would lead to improved financial performance, the farm manager is then in a position to determine if some change in the farm organisation or enterprise structure would result in further increases in farm income.

Thus a farm business may be found to have certain enterprises which make a greater contribution to farm income than other farm enterprises and an expansion in production of the more profitable activities could lead to an increase in whole farm income. Similarly, a manager could be aware of other enterprises, not presently part of the farm program but which, if introduced, could offer returns greater than some existing farm activities.

This comparison between enterprises can be made on the basis of the enterprise gross margin. Formal planning techniques based on the use of gross margins have been described (Rickards and McConnell, 1968; Barnard and Nix, 1980). These farm planning techniques generally assume that the farmer seeks economic optimisation, usually considered to be maximum profitability within the constraints of the resources available to the farm.

It has been found in practice that farmers prefer to seek this position progressively over a period of years by making changes in production methods or enterprise combinations as are dictated by changing circumstances. Thus a farmer is most likely to achieve increases in farm income as a result of a series of management decisions, each requiring economic evaluation, acquisition of new technical knowledge and successful on-farm adoption before further changes are implemented.

This chapter proposes to consider only the simpler planning techniques based on gross margin analysis. These may be used by farmers and those offering technical advice to farmers to evaluate the financial benefit of small changes in production methods or enterprise combination.

In terms of this farm planning problem, the farmer has available three main alternatives for increasing the whole farm income:
1. to make use of spare capacity by expanding existing enterprises or by the introduction of new enterprises;
2. to change production methods, resulting in cost savings and improved enterprise performance; or
3. to substitute new, more profitable enterprises for existing enterprises of lower return.

In planning the reorganisation of the enterprise structure of a farm, it is important to recognise that many technical aspects of production must be considered before one crop is replaced with an alternative crop enterprise.

In particular, different crops may have specific requirements in terms of the place of that crop in the whole farm pasture–crop rotation. For example, canola may be considered as an alternative crop to wheat but is best grown as the first crop in the rotation following the pasture phase because of its high nitrogen requirement. Its susceptibility to disease also dictates that canola should not be grown in the same paddock in consecutive years.

New crop enterprises generally require the farmer to acquire new production skills and experience in the husbandry of the particular crop. This may necessitate the introduction of the new crop on a limited scale with further expansion in the area of production after experience is obtained. The initial yields and financial performance of the crop may not be at the levels achieved by more experienced growers. Such has been the experience in the southern areas of wheat production in New South Wales where farmers have diversified their cropping program to include such alternative crops to wheat as canola and lupins.

It is also important to consider the effect that any change in farm organisation has on the fixed cost structure of the farm. This is most important where enterprise gross margins are being used as the basis of comparison between enterprise returns. Simple substitution of alternative crop enterprises may not result in increased fixed costs. Other changes, however, could be of such a scale that existing labour and machinery become inadequate and additional expenditure will be incurred on these items as a result of the change. Budgets aimed at evaluating the financial benefit of such changes must consider changes in all farm costs, in particular the farm fixed costs.

Decisions to introduce a new crop must be made after consideration of the *markets available* for the particular crop product. It is the market which determines the *price* for the product and ultimately the *value* of the farm output of that product. Many properties may have physical and environmental conditions favouring the production of a certain crop yet may not have access to a suitable market. Production decisions therefore should not be based purely on technical considerations but on gross margins which are based on realistic market expectations.

Partial Budgeting

The technique of analysis used to evaluate changes in farm organisation is known as *partial budgeting*. It is a technique which considers only the changes in income and costs that arise as a direct result of the re-organisation being considered. All other income and cost items that are not affected by the changes are assumed to remain constant. Partial budgeting is thus concerned with both potential increase and loss of income as well as additional costs and costs saved as a result of the changes.

This chapter has previously discussed the accounting distinction between variable and fixed costs, indicating the need to have a consistent classification of these costs for accounting and analysis purposes. It has further been indicated that in a budgeting or planning context there can be no rigid classification of costs as variable and fixed and that the classification depends upon the situation under consideration. With any proposed change in farm organisation, certain costs will change and other cost items will remain unaltered.

Partial budgeting is concerned with change, and the technique of partial budgeting requires identification of all income and cost items which will change. There are four components of any partial budget. These are related in Table 9.9.

TABLE 9.9   General format of a partial budget

REVENUE LOST	$	REVENUE GAINED	$
Income foregone	A	Additional income	C
Additional costs	B	Costs saved	D
TOTAL	(A + B)	TOTAL	(C + D)

A *profit* results in changes where (C + D) exceeds (A + B) and a *loss* where (C + D) is less than (A + B)

If all changes in income and costs have been correctly identified in the partial budget, then the balance item represents the additional profit (or loss) resulting directly from the reorganisation. As all other income and cost items remain constant, this additional profit (or loss) also represents the change in whole farm profit (or loss) resulting from the partial reorganisation of the farm enterprises.

Gross Margins and Partial Budgeting

If the enterprise gross margins are constructed according to the definitions and conventions described in this chapter, the technique of partial budgeting can be greatly assisted by the use of enterprise gross margins.

Where all components of the gross margin, including both output and variable costs, vary in direct proportion to a change in the size of the enterprise, then the gross margin itself must alter in proportion to a change in enterprise size. For example, if the gross margin of a wheat enterprise is estimated to be $194 per hectare (Table 9.5), then a simple increase of one hectare in the area sown to wheat must lead to an increase in income of $194, being the gross margin per hectare of that crop. If the same enterprise is expanded by 100 hectares then the extra income must be $100 \times \$194$ or $19\,400$. Similar calculations apply for any projected increase or decrease in the size of an enterprise.

Thus, given knowledge of the enterprise gross margin, it is not necessary to consider and budget separately the change in output and the change in expenditure for each of the variable inputs of seed, fertiliser, sprays, contract or casual labour and miscellaneous crop costs. As all of these items alter in direct proportion to change in enterprise size, they are automatically accounted for in the use of the gross margin figure as representing the net change in income.

Such a process is only valid if all cost items included in the construction of the enterprise gross margin are true variable costs. It has already been mentioned that many workers construct crop enterprise gross margins including an allocation of permanent labour and machinery costs as variable costs. Many of these items are not true variable costs and do not alter in proportion to change in enterprise size. Such gross margins should *not be used directly* in the partial budgeting procedure. Instead, *each component* of the output and variable cost *needs to be considered separately* to determine if the item will change in the situation under consideration and should therefore be included in the partial budget calculation.

It has been indicated previously that, even if these labour and machinery costs are considered as fixed costs for analysis purposes, they *are not ignored* in partial budgeting. The partial budget technique provides for consideration of the costs saved as well as additional costs resulting from the change. The *costs saved* and *additional costs* items must include any costs, including the fixed costs, which will change but which have not been automatically accounted for in the use of the enterprise gross margin.

Thus if a change in the cropping program results in any change in the costs normally considered as fixed costs (see Table 9.1), then these changes must be identified and included in the partial budget calculation. For example, a wheat farmer may consider expanding the farm's wheat area following a reduction in livestock numbers. The wheat enterprise gross margin would show the *additional income* from wheat, being output less variable costs, but there would be *additional costs* of labour, fuel, oil and machinery costs which would need to be budgeted separately.

While the use of enterprise gross margin figures greatly facilitates the partial budgeting procedure, it is possible to make use of partial budgets even if enterprise gross margins are not known or have not been estimated. This situation requires all changes in income and costs to be budgeted separately.

In selecting gross margin figures to incorporate in partial budgets it is most important that the enterprise gross margins do represent the expected level of physical and financial performance for that crop grown on the farm being planned.

Published enterprise gross margin data are usually presented on an average basis. This single enterprise gross margin may not indicate the year-by-year differences or, importantly, the expected variations in performance of the crop grown at different stages of a rotation in a whole farm system.

While average figures may be a useful guide to budgeting, it is often the case that enterprise gross margins for farm budgeting should be prepared from a zero base and reflect the output and variable costs for that crop on that specific farm.

Some authorities do present enterprise gross margin data acknowledging these factors. This is demonstrated by Walker (1993) in which a range of wheat enterprise gross margins is published for wheat grown under conventional long fallow, conventional short fallow, direct drilled after a break crop, and also chemical fallow.

Examples of Partial Budgets for Cash Crop Production

EXAMPLE 1 — MAKING USE OF SPARE CAPACITY

A farmer may decide, following examination of the farm program, that the land and labour resources of the farm are not being fully utilised by the existing farm enterprises and that the whole farm income could be increased by expanding the scale of the cash cropping program.

The stocking rate on the farm is presently low and the farmer considers that an additional 50 hectares of canola could be grown by removing 50 hectares from the pasture area and concentrating livestock production on the remaining grazing area. No reduction in livestock numbers is envisaged.

A partial budget to evaluate this simple change is given in Table 9.10. In this example the additional costs of fuel, oil, machinery repairs and maintenance need to be budgeted, because these are additional costs that will be incurred as a direct result of the change but have not been accounted for by the enterprise gross margin used in the partial budget.

EXAMPLE 2 — CHANGE IN PRODUCTION METHODS

As a result of diagnosis of enterprise performance, a farmer may recognise that the profitability of the whole farm could be increased by improved performance of some existing farm enterprises. This may be achieved by change in crop production methods aimed at reducing the total labour and machinery costs and improving the timeliness of cultural operations.

For example, a farmer producing 100 hectares of canola may have used contract services for windrowing the crop prior to harvest at a cost of $30 per hectare of crop. The farmer contemplates purchasing a windrower which requires a capital investment of $25 000 but would

TABLE 9.10   Partial budget evaluating the effect of increasing the canola area by 50 ha

REVENUE LOST	$	REVENUE GAINED	$
Income foregone		*Additional income*	
	Nil	Canola–50 ha @ $287 GM/ha	14 350
Additional Cost		*Costs Saved*	
Fuel, oil $8/hr	720	Pasture topdressing 1 yr in	
Machinery repairs		4 @ 125 kg/ha, single	
and maintenace	315	superphosphate @ $194/t	303
$3.50/hr			
		Superphosphate spreading	
		1 yr in 4 @ $30/t	47
TOTAL REVENUE LOST	1035	TOTAL REVENUE GAINED	14 700
		TOTAL REVENUE LOST	1035
		INCREASE IN PROFIT	13 665

TABLE 9.11   Partial budget evaluating the effect of purchasing a windrower for canola production

REVENUE LOST	$	REVENUE GAINED	$
Income foregone	Nil	*Additional income*	
		Contract windrowing	
		200 ha @ $30/ha	6000
Additional Costs		*Costs Saved*	
Depreciation 15% of		Contract windrowing	
$25 000	3750	100 ha @ $30/ha	3000
Fuel, oil			
300 ha @ $2.56/ha	768		
Machinery repairs and			
maintenance 3% of			
$25 000	750		
TOTAL REVENUE LOST	5268	TOTAL REVENUE GAINED	9000
		TOTAL REVENUE LOST	5268
		INCREASE IN PROFIT	3732

Note: The farmer plans to windrow the farm's crop of 100 ha as well as use the machine for contract windrowing of neighbour's crops estimated at 200 ha.

enable the crop to be harvested without the cost of contract labour. It would also be possible to utilise the machinery for contract windrowing of neighbours' crops estimated at 200 hectares per annum.

The farmer also estimates that the timeliness of operation to be achieved by owning a windrower would result in increased yields which would be an additional benefit not included in the budget. The partial budget shown in Table 9.11 considers the effect of these changes.

In Table 9.11 it has been assumed that there is no interest charge on the capital cost of the machinery. In some situations it may be appropriate to include the additional cost of the interest on the full capital cost of the machine.

EXAMPLE 3—SUBSTITUTION WITH MORE PROFITABLE
ENTERPRISE ALTERNATIVES

It may be the case on many properties that spare resources of land and labour do not exist to permit the expansion or introduction of selected enterprises as a means of increasing whole farm income. Opportunity may exist, however, for expanded production of more profitable enterprises in the place of existing enterprises of lower return. This may be a substitution of one crop enterprise for another crop enterprise or a replacement of a livestock enterprise with a more profitable cash crop alternative.

A wheat farmer who has been producing wheat returning an average gross margin of $194 per hectare (Table 9.5) may be interested in developing the cropping program and wishes to consider the effect of substituting 50 hectares of canola with an estimated gross margin of $287 per hectare (Table 9.5) in place of a similar area of wheat.

The partial budget for this proposed change is presented in Table 9.12.

TABLE 9.12   Partial budget evaluating the substitution of 50 ha of wheat with 50 ha of canola

REVENUE LOST	$	REVENUE GAINED	$
Income foregone		*Additional income*	
Wheat—50 ha @		Canola—50 ha @	
$194 GM/ha	9700	$287 GM/ha	14 350
Additional costs	Nil	*Costs Saved*	Nil
TOTAL REVENUE LOST	9700	TOTAL REVENUE GAINED	14 350
		TOTAL REVENUE LOST	9700
		INCREASE IN PROFIT	4650

In this example changes in all Output and Variable Costs associated with each enterprise are automatically accounted for by incorporating the enterprise gross margin figures directly into the partial budget calculations.

It has been assumed that there are no changes in any of the fixed costs as a result of this enterprise substitution. If, however, the farmer considers that there would be changes in any fixed costs, these would be budgeted separately and included in the Additional Costs/Costs Saved part of the partial budget.

Caution in the Use of Enterprise Gross Margins

It is not the intention of this book to develop more formal and sophisticated techniques of farm planning using enterprise gross margins. Simple examples of three common farm planning situations have been used to apply the technique of partial budgeting. These applications involved the consideration of improved enterprise performance and the expansion or substitution of enterprises based on comparison of enterprise gross margin performance levels.

The following three precautions should be recognised in the use of gross margin data for the selection of enterprises.
1. Planning must occur within the limited resources available to the farm and the technical limits that may apply to particular crops.
2. Most farms are mixed farms and produce a range of crop and livestock enterprises based on the available farm resources. One part of the farm cannot be considered in isolation, and changes made to one part of the farm frequently affect other parts of the business.
3. The assumption that land is always the most limiting factor may not always be appropriate.

Makeham and Malcolm (1993) also emphasise that it may not be appropriate in farm planning to assume that land is always the most limiting factor. It is common practice to compare or even rank alternative enterprises on the basis of their level of gross margin per unit of land ($GM/ha). This is based on the assumption that land is the most limiting resource and that farm profits will be maximised by selecting and expanding those enterprises offering the highest return per unit of land.

In many situations in Australian agriculture, land may not be the most limiting factor: other resources such as labour or capital may be more severe limiting factors. If maximising farm profit is the objective, the selection of enterprises should be based on those giving the highest return to the most severe limiting factor.

References

Barnard, C. S. and Nix, J. S. (1973), *Farm Planning and Control,* Cambridge University Press, UK.
Boone, L. E. and Kunz, D. L. (1974), *Contemporary Marketing,* Dryden Press, Hinsdale, Illinois.
Dillon, J. L. (1977), *The Analysis of Response in Crop and Livestock Production,* 2nd edition, Pergamon Press, New York.
Kerr, H. W. T. (1968), *Farm Management Accounting,* Department of Agricultural Economics, University of Nottingham, Farm Report No. 168.

Makeham, J. P. and Malcolm, L. R. (1993), *The Farming Game Now,* Cambridge University Press, UK.

Rickards, P. A. and McConnell, D. J. (1968), *Budgeting, Gross Margins and Programming for Farm Planning,* Professional Farm Management Guidebook No. 3, University of New England, Armidale.

Walker, S. (1993), *Farm Budget Handbook — Winter Crop Budgets 1993 Eastern Riverina and South West Slopes,* New South Wales Agriculture, Wagga Advisory Office.

Wallace, D. B. and Burr, H. (1963), *Planning on the Farm,* School of Agricultural Economics, University of Cambridge, Branch Report No. 60.

Index